Essaïd Sabir · Hicham Medromi
Mohamed Sadik
Editors

Advances in Ubiquitous Networking

Proceedings of UNet'15

Springer

Editors
Essaïd Sabir
Hassan II University of Casablanca
Ecole Nationale Supérieure d'Electricité et de Mécanique
Oasis – Casablanca
Morocco

Hicham Medromi
BP 8118
Hassan II University of Casablanca, Ecole Nationale Supérieure d'Electricité et de Mécanique
Casablanca
Morocco

Mohamed Sadik
BP 8118
Hassan II University of Casablanca, Ecole Nationale Supérieure d'Electricité et de Mécanique
Casablanca
Morocco

ISSN 1876-1100 ISSN 1876-1119 (electronic)
Lecture Notes in Electrical Engineering
ISBN 978-981-13-5740-4 ISBN 978-981-287-990-5 (eBook)
DOI 10.1007/978-981-287-990-5

Printed on acid-free paper

This Springer imprint is published by SpringerNature
The registered company is Springer Science+Business Media Singapore Pte Ltd.

Preface

The International Symposium on Ubiquitous Networking (UNet 2015) was held on September 08–10, 2015 in the fascinating city of Casablanca, Morocco.

The main purpose of this symposium is to serve as a forum that brings together researchers and practitioners from academia and industry to discuss recent developments in pervasive systems and ubiquitous networks. This conference will provide a forum to exchange ideas, discuss solutions, debate on identified challenges and share experiences among researchers and professionals. UNet'15 aims also to promote new methodologies and to provide the participants with advanced and innovative tools able to catch the fundamental dynamics of the underlying complex interactions (e.g., game theory, Mechanism Design theory, Learning theory, SDR platforms, etc.). Accepted papers describe original research on both theoretical and practical aspects of pervasive computing and future mobile computing.

This year, we received 94 submissions from 12 different countries around the globe: Algeria, Brazil, Canada, Finland, France, Ireland, Mali, Mexico, Morocco, Tunisia, Ukraine and United Arab Emirates. Of the submitted papers, we selected 39 full papers and 6 short papers for publication in the proceedings and presentation in the conference.

We are grateful to our Technical Program and Executive Committees for their support. Without their help, expertise and experience we would not have been able to select such an outstanding technical program. The conference also included three technical talks from very prestigious scientists: Marwan Krunz from University of Arizona, Mounir Ghogho from University of Leeds, and Tarik Taleb from University of Aalto. We would like to express our gratitude to them for accepting to give a talk at UNet'15; their presence was a privilege for us all. We finally thank our sponsors and patrons for their effort and good work at making the conference a great success.

September 2015

Essaïd Sabir
Hicham Medromi
Mohamed Sadik

Executive Committee

Honorary Chair

Idriss Mansouri	President of Hassan II University of Casablanca, Morocco

Symposium Chairs

Hicham Medromi	ENSEM, Hassan II University of Casablanca, Morocco
Mohamed Sadik	ENSEM, Hassan II University of Casablanca, Morocco
Essaïd Sabir	ENSEM, Hassan II University of Casablanca, Morocco

Technical Program Chairs

Hamidou Tembine	University of New York, USA
Majed Haddad	LIA-CERI, University of USA Avignon, Avignon, France
Sujit Kumar Samanta	National Institue of Technology Raipur, India
Mohamed El Kamili	FSDM, USMBA, Fez, Morocco

Publication Chairs

Quanyan Zhu	New York University, USA
Daniel Sadoc Menasche	Federal University of Rio de Janeiro, Brazil

Ahmed Errami	ENSEM, Hassan II University of Casablanca, Morocco

Keynotes Chairs

Mohammed Khaldoun	ENSEM, Hassan II University of Casablanca, Morocco
Hicham Behja	ENSEM, Hassan II University of Casablanca, Morocco

Publicity and Patrons Chairs

Mostafa Hefnawi	Royal Military College, Canada
Jean Philippe Leroy	CFA IGS Group, France
Amar Prakash Azad	Samsung Advanced Institute of Technology, Bangalore, India
Abdellatif Kobbane	ENSIAS, Mohammed V University, Rabat, Morocco

Webmaster

Sara Arabi	ENSEM, Hassan II University of Casablanca, Morocco

Technical Program Committee

Abdelilah Karouit	University of Avignon, France
Abdelkarim Haqiq	FST, Hassan First University, Settat, Morocco
Abdelmajid Badri	FST, Hassan II University of Casablanca, Morocco
Abdelouahed Mesbahi	Hassan II University of Casablanca, Morocco
Adel Omar Dahmane	University of Québec à Trois-Rivières, Canada
Adil Sayouti	Royal Naval School, Casablanca, Morocco
Adlen Ksentini	University of Rennes/IRISA, Rennes, France
Alexandre Reiffers	LIA/INRIA-SOAP, University of Avignon, France
Alonso Silva	Alcatel-Lucent Bell Labs, Paris, France
Amar P. Azad	Samsung Advanced Institute of Technology, Bangalore, India
Anas Abou Alkalam	IGS Group, France
Aziz Salah	Quebec University of Montreal, Canada
Corinne Touati	LIG-INRIA, Grenoble, France
Daniel Sadoc Menasche	Federal University of Rio de Janeiro, Brazil

Mohammed Essaaidi	ENSIAS, Mohammed V University of Rabat, Morocco
Mohammed Nabil Saaidi	INSEA, Rabat, Morocco
Mohammed Ouzzif	EST, Hassan II University of Casablanca, Morocco
Mostafa Hefnawi	Royal Military College, Canada
Moulay Lahcen Hasnaoui	FSDM, USMBA, Fez, Morocco
Mounir Ghogho	University of Leeds, UK
Mounir Rifi	EST, Hassan II University of Casablanca, Morocco
Mourad El Yadari	FP, UMI, Errachidia, Morocco
Nabil Benamar	Moulay Ismail University, Meknes, Morocco
Nabil Tabbane	SUP'COM, Tunis, Tunisia
Olivier Brun	LAAS-CNRS, Toulouse, France
Orazio Mirabella	University of Catania, Italy
Othmane Benhmammouch	Mundiapolis, Casablanca, Morocco
Oussama Habachi	University of Limoges, France
Pascale Minet	INRIA, Paris, France
Quanyan Zhu	New York University, USA
Rachid El-Azouzi	LIA, University of Avignon, France
Rachid Saadane	EHTP, Casablanca, Morocco
Sreenath Ramanath	Systems R&D, Lekha Wireless, Bangalore, India
Sujit K. Samanta	National Institue of Technology Raipur, India
Sulan Wong	Faculty of Law Universidad Católica de Temuco, Temuco, Chile
Tarik Taleb	Aalto University, Aalto, Finland
Tayeb Sadiki	International University of Rabat, Morocco
Thierry Peyere	University of Rènnes, France
Tijani Chahed	Telecom SudParis, France
Vijay Kamble	University of California, Berkeley, USA
Yezekael Hayel	Polytechnic School of Engineering, New York University, USA
Yuedong Xu	SIST, Fudan University, China
Zaixin Lu	Marywood University, Scranton, Pennsylvania, USA
Zouheir Rezki	KAUST University, Soudi Arabia

Keynote Speakers

Invited Speaker: Prof. Marwan Krunz
Title: *Exploitation of Full-duplex Communications in Opportunistic Access Systems*
Abstract: Until recently, the ability to achieve simultaneous wireless transmission and reception over the same channel was deemed impossible, due to high self-interference on the receive chain. As a result, bidirectional communications using half-duplex (HD) devices is typically achieved by separating the forward and reverse links in time or frequency. This common practice was recently challenged by several studies, which for the first time demonstrated true full-duplex (FD) operation over the same channel. By employing a variety of conventional and new self-interference suppression (SIS) techniques, a reduction of up to 110 dB in self-interference was demonstrated on single-antenna devices, just enough to enable FD operation on WiFi systems. The impact of these successes on wireless network design and protocols is enormous, as evident by the intensive research activity in this domain.

In this talk, we will discuss a specific application of FD and SIS, in the context of opportunistic dynamic spectrum access (DSA) systems. In such systems, secondary users (SUs) are allowed to access the licensed spectrum opportunistically, provided that they do not cause harmful interference to primary users (PUs). This is often done through a "listen-before-talk" (LBT) strategy, whereby an SU must periodically interrupt its transmission to sense the channel for any PU activity before resuming. During channel sensing, the SU cannot transmit data, which degrades its throughput significantly. If the SU generates real-time or media-streaming traffic, periodic interruption of its transmission results in unacceptable audiovisual artifacts and connection loss. To remedy these issues, our group has been exploring using SIS techniques in one of two possible ways: simultaneous transmission-and-reception (STAR) and simultaneous transmission-and-sensing (STAS). The ability to operate in either mode gives rise to a spectrum awareness/efficiency tradeoff. More specifically, an SU may improve the spectrum utilization by operating in the STAR mode, which will dramatically increase the link throughput. On the other hand, the SU may exploit its SIS capabilities in the STAS mode, enabling it to monitor the PU activity while transmitting and to quickly vacate the channel if such activity is detected. In practice, SIS is imperfect, and residual self-interference may degrade the sensing accuracy of the STAS mode, compared to a classical (HD) sensing-only mode. Our discussion encompasses all these modes of operation, along with a possible channel-switching (CS) mode. We will also address the implications of FD operation on multi-link scenarios (e.g., ad hoc networks), and particularly the potential degradation in network throughput as a result of the additional interference that FD devices inject onto the network. A preliminary channel access protocol that allows nodes to adapt their HD/FD operation, along with its transmission power and rate, will be introduced.

Biography: Matwan Krunz is a Professor of Electrical and Computer Engineering at the University of Arizona. He holds a joint (courtesy) appointment at the same rank in the Department of Computer Science. Since 2008, he has been the UA site director of "Connection One", an NSF Industry/University Cooperative Research Center (I/UCRC) that focuses on RF and wireless communication systems and networks. At present, the center has five participating sites (ASU, UA, OSU, RPI, and the University of Hawaii) and 26+ industrial affiliates. Dr. Krunz is also the UA co-director of BWAC (Broadband Wireless Access and Applications Center), a recently inaugurated NSF I/UCRC that includes UA (lead site), Virginia Tech, University of Virginia, and Auburn University. Dr. Krunz received the Ph.D. degree in electrical engineering from Michigan State University in July 1995. He joined the University of Arizona in January 1997, after a brief postdoctoral stint at the University of Maryland, College Park. In 2010, he was a Visiting Chair of Excellence ("Catedra de Excelencia") at the University of Carlos III de Madrid (Spain), and concurrently a visiting researcher at Institute IMDEA Networks. In summer 2011, he was a Fulbright Senior Specialist, visiting with the University of Jordan, King Abdullah II School of Information Technology. He previously held other visiting research positions at INRIA-Sophia Antipolis, France (2003, 2008, 2011), HP Labs, Palo Alto (2003), University of Paris VI (LIP6 Group, 2006), University of Paris V (LEPADE Group, 2013), and US West Advanced Technologies (1997). Dr. Krunz's research is in the broad area of wireless communications and networking, with particular emphasis on resource management, adaptive protocols, and security issues. In the last 5 years, he has been involved in NSF and DoD funded projects related to cognitive radios and dynamic spectrum access; wireless security (e.g., obfuscation of transmission signatures, insider attacks, selective-reactive jamming/dropping, randomization, game theoretic countermeasures); power-controlled protocols for wireless networks; multi-channel MIMO systems (including virtual/cooperative MIMO); secure satellite communications; energy management in solar-powered WSNs; full-duplex communications with imperfect self-interference suppression; media streaming over wireless links; and fault monitoring/detection in optical networks. Previously, he worked on packet scheduling and buffer management in switches and routers, QoS provisioning, effective-bandwidth theory, traffic characterization, and video-on-demand systems. He has published more than 200 journal articles and peer-reviewed conference papers, and holds three US patents (see Publications for details; see also my Google Scholar Link).

Dr. Krunz is an IEEE Fellow (class of 2010), an Arizona Engineering Faculty Fellow (2011–2014), and an IEEE Communications Society Distinguished Lecturer (2013 and 2014). In 2012, he received the IEEE Communications Society Technical Committee on Communications (TCCC) Outstanding Service Award. He received the National Science Foundation CAREER award (1998–2002). He served and continues to serve on the editorial boards of several journals, including the IEEE/ACM Transactions on Networking (2001–2008), the IEEE Transactions on Mobile Computing (2006–2011), the Computer Communications Journal (2001–2011), the IEEE Communications Interactive Magazine (1998–2003), and the IEEE

Transactions on Network and Service Management (2010–present). He was a guest co-editor for special issues in IEEE Micro and IEEE Communications Magazines. He served as a general chair for the 5th ACM Conference on Wireless Network Security (WiSec'12). He also served as a technical program committee (TPC) chair for INFOCOM 2004 (Hong Kong, March 2004), SECON 2005 (Santa Clara, Sep. 2005), WoWMoM 2006 (Buffalo, New York, June 2006), and Hot Interconnects 9 (San Francisco, August 2001). He has served and continues to serve on the executive and technical program committees of numerous international conferences, and on the panels of several funding agencies. He was the keynote speaker at various conferences, including IEEE CCW (Sedona, Nov. 2012), IFIP Wireless Days 2011 Conference (Niagara Falls, Oct. 2011), and the IEEE Workshop on Wireless Mesh Networks (WiMesh 2009, Rome, June 2009). He was an invited panelist at various international conferences (e.g., INFOCOM 2009, SECON 2009, etc.). He gave tutorials at premier wireless networking conferences (e.g., MobiCom, MobiHoc). He frequently consults for companies in the telecommunications sector.

Invited Speaker: Prof. Mounir Ghogho
Title: *Big Data Empowered 5G networks*
Abstract: A fluid/elastic network architecture which supports "user centric" services will be a central feature for 5G wireless networks. Such a fluid architecture leverages the fact that for ultra-dense networks, the control and the user-planes can be decoupled. Consequently, the network can be treated as a soft entity which can support a variety of user demands by dynamic reconfiguration of the user plane. Furthermore, activities across heterogeneous control planes can be coordinated through the network state awareness and resource virtualization. It is difficult if not impossible to realize such an architecture without the real time network and the user state data. With around 6.2 billion internet-connected devices by the end of 2020, the generated data has enormous volume, high velocity and huge variety. In this talk, the main objective is to highlight how networks can be configured by leveraging big data empowered self-organization. The design framework for the big data empowered self-organized networks under split-plane architecture presents various opportunities and challenges. Such an architecture is an ideal candidate to unify evolving technologies into a unified framework. The key architectural ingredients which will be explored briefly in this talk are:

a) Proactive vs. Reactive Networks;
b) Big Social Data;
c) Information Centric Network Design;
d) Self Organization for 5G Networks;
e) Cloud RAN and HetNets;
f) Energy Efficiency and Harvesting.

Biography: Professor Mounir Ghogho received the MSc degree in 1993 and the PhD degree in 1997 from the National Polytechnic Institute of Toulouse, France. He was an EPSRC Research Fellow with the University of Strathclyde, Glasgow (Scotland), from September 1997 to November 2001. Since December 2001, he has been a faculty member with the school of Electronic and Electrical Engineering (top 5 EEE schools in the UK for Research) at the University of Leeds (UK), where he currently holds a Chair in Signal Processing and Communications. He is also currently the ICT Research Director at the International University of Rabat. He is currently an Associate Editor of the IEEE Signal Processing magazine. He served as an Associate Editor of the IEEE Transactions on Signal Processing from 2005 to 2008, the IEEE Signal Processing Letters from 2001 to 2004, and the Elsevier Digital Signal Processing journal from 2011 to 2012. He is currently a member of the IEEE Signal Processing Society SAM Technical Committee. He served as a member of the IEEE Signal Processing Society SPCOM Technical Committee from 2005 to 2010 and a member of IEEE Signal Processing Society SPTM Technical Committee from 2006 to 2011. He was the General Chair of the eleventh IEEE workshop on Signal Processing for Advanced Wireless Communications (SPAWC'2010), General Chair of the 21st edition of the European Signal Processing Conference (EUSIPCO 2013), the Technical co- Chair of the MIMO symposium of IWCMC 2007 and IWCMC 2008, and a Technical Area co-Chair of EUSIPCO 2008, EUSIPCO 2009 and ISCCSP'05. His research interests are in wireless communication and information/signal processing. His research work has partly but continuously been funded by the US Army since 1998. He held invited scientist/professor positions at many institutions including the US Army Research Lab (USA), Télécom Paris-Tech (France), National Institute of Informatics (Japan), the University Carlos Third of Madrid (Spain), ENSICA (France), Technical University of Darmstadt (Germany), the University of Minnesota (USA), and Beijing University of Posts and Telecommunication (China). He is the EURASIP Liaison in Morocco. He was awarded the five-year UK Royal Academy of Engineering Research Fellowship in September 2000. He is also one of the recipients of the internationally competitive 2013 IBM Faculty award.

Invited Speaker: Prof. Tarik Taleb
Title: *Towards Cloud-Native Mobile Networks*
Abstract: The telecom industry keeps reinventing itself. Soon, the world will be experiencing the 5th generation mobile networks (5G), also referred to as beyond 2020 mobile communication systems. Major obstacles to overcome in 5G systems are principally the highly centralized architecture of mobile networks along with the static provisioning and configuration of network nodes built on dedicated hardware components. This has resulted in lack of elasticity and flexibility in deployment of mobile networks; rendering their run-time management costly, cumbersome and time-consuming. Software Defined Networking, Network Function Virtualization, and Cloud Computing, along with the principles of the latter in terms of service elasticity, on-demand features, and pay-per-use, could be important enablers for various mobile network enhancements, to specifically virtualize and decentralize mobile networks using general-purpose COTS (commercial of the shelf) hardware. For this purpose, different requirements have to be met and numerous associated challenges have to be subsequently tackled. This talk will touch upon the recent trends the mobile telecommunications market is experiencing and discuss the

challenges these trends are representing to mobile network operators. To cope with these trends, the talk will then showcase the feasibility of on-demand creation of cloud-based elastic mobile networks, along with their lifecycle management. The talk will introduce a set of technologies and key architectural elements to realize such vision, turning end-to-end mobile networking into software engineering.

Biography: Professor Tarik Taleb is an IEEE Communications Society (ComSoc) Distinguished Lecturer and a senior member of IEEE. He is currently a Professor at the School of Electrical Engineering, Aalto University, Finland. Prior to his current position, he was working as Senior Researcher and 3GPP Standards Expert at NEC Europe Ltd, Heidelberg, Germany. He was then leading the NEC Europe Labs Team working on R&D projects on carrier cloud platforms. He was also serving as technical leader of the main work package, Mobile Core Network Cloud, in EU FP7 Mobile Cloud Networking project, coordinating among 9 partners including NEC, France Telecom, British Telecom, Telecom Italia, Portugal Telecom Innovation, SAP, & Intel. Before joining NEC Europe and till Mar. 2009, Prof. Taleb worked as assistant professor at the Graduate School of Information Sciences, Tohoku University, Japan, in a lab fully funded by KDDI, the second largest mobile network operator in Japan. From Oct. 2005 till Mar. 2006, he was working as research fellow with the Intelligent Cosmos Research Institute, Sendai, Japan. Prof. Taleb received his B.E. degree in Information Engineering with distinction, MSc and Ph. D degrees in Information Sciences from GSIS, Tohoku Univ., in 2001, 2003, and 2005, respectively.

Prof. Taleb's research interests lie in the field of architectural enhancements to mobile core networks (particularly 3GPP's), mobile cloud networking, network function virtualization, software defined networking, mobile multimedia streaming, inter-vehicular communications, and social media networking. Prof. Taleb has been also directly engaged in the development and standardization of the Evolved Packet System as a member of 3GPP's System Architecture working group. Prof. Taleb is a board member of the IEEE Communications Society Standardization Program Development Board. As an attempt to bridge the gap between academia and industry, Prof. Taleb has founded and has been the general chair of the "IEEE Workshop on Telecommunications Standards: from Research to Standards", a successful event that got awarded "best workshop award" by IEEE Communication Society (ComSoC). Based on the success of this workshop, Prof. Taleb has also founded and has been the steering committee chair of the IEEE Conference on Standards for Communications and Networking (IEEE CSCN).

Prof. Taleb is/was on the editorial board of the IEEE Transactions on Wireless Communications, IEEE Wireless Communications Magazine, IEEE Transactions on Vehicular Technology, IEEE Communications Surveys & Tutorials, and a number of Wiley journals. He is serving as chair of the Wireless Communications Technical Committee, the largest in IEEE ComSoC. He also served as Secretary and then as Vice Chair of the Satellite and Space Communications Technical Committee of IEEE ComSoc (2006 – 2010). He has been on the technical program committee of different IEEE conferences, including Globecom, ICC, and WCNC, and chaired some of their symposia.

Prof. Taleb is the recipient of the 2009 IEEE ComSoc Asia-Pacific Best Young Researcher award (Jun. 2009), the 2008 TELECOM System Technology Award from the Telecommunications Advancement Foundation (Mar. 2008), the 2007 Funai Foundation Science Promotion Award (Apr. 2007), the 2006 IEEE Computer Society Japan Chapter Young Author Award (Dec. 2006), the Niwa Yasujirou Memorial Award (Feb. 2005), and the Young Researcher's Encouragement Award from the Japan chapter of the IEEE Vehicular Technology Society (VTS) (Oct. 2003). Some of Prof. Taleb's research work have been also awarded best paper awards at prestigious conferences.

Contents

Part I
Advances in Wireless/Mobile Networking

Generalized Slotted Aloha Revisited: A Throughput and Stability Analysis

Essaid Sabir, Sidi Ahmed Ezzahidi and El Houssine Bouyakhf

Abstract We consider the uplink case of a cellular system where m bufferless mobiles transmit over a common channel to a base station, using the slotted aloha medium access protocol. We study the performance of this system under a new transmission scheduling scheme. In contrast to slotted aloha, each new arrival packet is first transmitted at a probability p_a and retransmitted later, if needed, with probability q_r. Under this proposal, we analyze the cooperative team in which a common goal is jointly optimized. Later, we derive the average throughput and the expected delay and use them as the objectives to optimize. We further study the impact of parameters on the average throughput as well as the expected delay. Furthermore, we address the stability behavior of our new scheme. We finally carry out extensive numerical examples to illustrate the obtained gain which concerns, in particular, the expected delay minimization.

1 Introduction

Aloha [1] and slotted aloha [11] are without doubt the most studied protocols in the literature of telecommunication systems. In fact, they have always been of great interest and are used for understanding and analyzing wireless transmission systems. These access methods are widely implemented in the satellite networks and mobile phone technology for the sporadic transfer of data packets received or as mechanisms for resource reservation. In aloha, new packets are immediately transmitted, and if several packets are sent simultaneously by more than one mobile, they imperatively

E. Sabir
UBICOM Research Group, ENSEM, Hassan II University, Casablanca, Morocco
e-mail: e.sabir@ensem.ac.ma

S.A. Ezzahidi(✉) · E.H. Bouyakhf
LIMIARF Laboratory, FSR, Mohammed V University of Rabat, Rabat, Morocco
e-mail: ezzahidi@gmail.com, bouyakhf@fsr.ac.ma

E. Sabir et al. (eds.), *The International Symposium on Ubiquitous Networking*,
Lecture Notes in Electrical Engineering 366,
DOI: 10.1007/978-981-287-990-5_1

collide. After a packet is sent, the sender receives the acknowledgment (ACK) about the state of the transmission (success or collision). All unreceived packets are corrupt and will be retransmitted after some random time. In this paper, we will focus on slotted aloha, an improved version invented by Lawrence G. Roberts [11], in which time is divided into equal units called slots. The time of vulnerability has therefore been halved. The transmission always takes place in the early slot, ie the transmissions are synchronized with the system clock, and the ACK will be received at the end of the current slot. Collided packets are put on hold (we call then throughout this article the backlogged) until their retransmissions are scheduled after a random number of slots.

When several mobiles share the same channel (frequency,code$\cdots$), a conflict arises and mobiles compete to transmit their packets. Recently, the selfish behavior of mobile in MAC protocols has been widely analyzed using game theory with all its mathematical formalism and the power of its solution concepts. It has been shown in [2, 3, 5] and [10] that the mobile individualism often leads to a tragic network inefficiency. This leads the mobiles to fall into the similar dilemma as the prisoners widely known by game theorists and economists.This shows in fact that the Nash[1] equilibrium is not effective in some situations. Thus, a better use of the wireless system would require coordination between the mobiles using explicit signaling messages or the deployment of an arbitration mechanism. These solutions may be unrealistic given the distributed and arbitrary nature of the topology (due to mobility, end / new calls, environment ,...) of the wireless networks on one side, and the eventual overload due to the exchange of signaling messages. To achieve a good performance without using coordination mechanisms, mobiles need to cooperate or comply with certain priority / hierarchy [12, 13].It is therefore advantageous to put in the system a set of mobiles, whose mission is to provide incentives for other mobiles to cooperate and / or comply with the defined priority. The latter mechanism may limit the level of aggressiveness (Channel Access) of mobiles and solve the problem of competition. Another way to reduce the concurrency is to define a transmission cost (transmission pricing). Indeed, it has been shown in [2] that the costs have a stabilizing effect, the mobiles would tend to delay the transmission of packets when the cost of the transmission is high. Unfortunately, doing so could increase the waiting time and the blocking probability. The authors of [12], showed that the introduction of the hierarchy between mobiles improves the performance of the entire system depending on the offered traffic. At low traffic, this improvement is due to the fact that mobiles retransmit with a higher probability compared to slotted aloha, hence the backoff time is reduced. At medium and high loads, the mobiles, or more specifically the leaders (masters) are generally less aggressive than the case of the Nash equilibrium. In that way, the probability of collision is reduced hence the positive impact of the hierarchy on the performance of the leaders and the followers (slaves). An interesting phenomenon is that when the number of followers is higher than the number of leaders, the latter become more pacific and reduce their probability of retransmission, while the others become too aggressive with a probability close to 1. Many authors have presented algorithms of channel access where multiple

[1] : A strategy profile is a Nash equilibrium if no player has incentive to deviate unilaterally.

power levels can be used. So, a mobile transmission will succeed when it transmits at a highest power compared to those chosen by the other mobiles in the cell [2]. But, in [5], a mobile transmission is successful if it chooses a higher level of power than those chosen by the other mobiles while having a signal to noise ratio superior than the given threshold. In this paper, we are developing a protocol with a delayed first transmission similar to CSMA (Carrier Sense Multiple Access), it has the advantage of not having to listen to the channel or follow the backoff algorithm for future transmissions. This allows us not to consume an extra bandwidth.

The following article is organized as follows: Section 2 presents the problem's formulation and develops the associated Markov model. We calculate the metrics of performance; investigate the existence of a solution and the stability of this new variant of slotted aloha in Section 3. Next, we discuss some numerical examples in Section 4. The article ends with a conclusion and some perspectives.

2 Formulation of the Problem

We consider a collision channel shared by m without buffer and transmitting to a central receiver, ie, the mobiles do not generate new packets until the current one is successfully transmitted. This assumption is obviously realistic in the signaling context; Indeed, it is natural to assume that the source does not begin to generate new signaling packets (for example, a new reservation) as long as the current packet is not successfully transmitted. In what comes next, we treat slotted aloha as an optimization problem (team's problem), where all mobiles seek to optimize the same objective function (maximize the total throughput or minimize the average delay).We define the following mechanism of priority: Each mobile i transmits its new packet with the probability p_a^i when it retransmits its backlogged packets with the probability q_r^i (it is a Bernoulli process of parameter q_r^i); This new mechanism allows 1) to better manage the scheduling of transmissions, 2) prioritize packets based on the requested application and the system traffic and 3) relieve the instantaneous system load. To study this new channel access method, we extend the Markov model introduced for the first time in [4].The probability of packet arrival at the source i follows a Bernoulli process of parameter q_a (ie at the beginning of each slot, a new packet arrives at a free source with probability q_a). As one source fails the transmission of the packet in progress, the new arrivals will be blocked and lost (because we consider sources without buffer). As a result, the arrivals are independent.

We consider the stochastic process that represents the number of backlogged packets noted n as a global state of the system.$\bar{\mathbf{q}}_r$ (respectively $\bar{\mathbf{p}}_a$)is the vector of retransmission probabilities (respectively first transmission) of all the mobiles, the jth component is q_r^j (respectively p_a^i).$\pi(\bar{\mathbf{q}}_r, \bar{\mathbf{p}}_a)$ is the stationary distribution of the Markov chain where the n^{me} component $\pi_n(\bar{\mathbf{q}}_r, \bar{\mathbf{p}}_a)$ represents the probability that n packets are backlogged. The solution of our problem is to find *the pair of vectors* $(\bar{\mathbf{q}}_r^*, \bar{\mathbf{p}}_a^*)$ *which maximizes the objective function*.This is clearly a problem of vector optimization. Without losing the generality and only for the purpose of the calculation feasibility, we limit our problem to the symmetric case where all mobiles

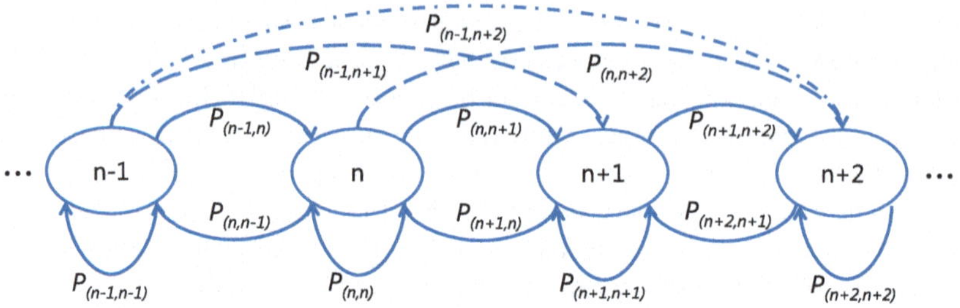

Fig. 1 Global Markov Chain of the system. The state of the system is the number of backlogged packets (backlog state), it can reduce up to one (transition to the left), but may increase by less than or equal to $m - n$ (Transition to the right).

use the same transmission policy $(q_r^i,\ p_a^i) := (q_r,\ p_a)$. So we must find an optimal symmetric solution, ie , a couple of probabilities of (re)transmission $(q_r,\ p_a)$ that does not depend of i. When all the components of $\bar{\mathbf{q}}_r$ are the same, noted q, we'll write (with some notation abuse) $\pi(q_r,\ p_a)$ instead of $\pi(\bar{\mathbf{q}}_r,\ \bar{\mathbf{p}}_a)$.

Arrival Rate and Retransmission Rate: $Q_r(i, n)$ is the probability that i mobiles among $n$waiting packets retransmit in the current slot. $Q_a(i, n)$ is the probability that i free sources (with $i \leq m - n$) generate packets in the current slot (ie the probability that i new arrivals occurred to sources that have no waiting packet). In the same way, $A_p(j, i))$ is the probability that t new arrivals are directly transmitted in the beginning of the current slot. We have :

$$Q_r(i, n) = \binom{n}{i}(1 - q_r)^{n-i}(q_r)^i. \tag{1}$$

$$Q_a(i, n) = \binom{m-n}{i}(1 - q_a)^{m-n-i}(q_a)^i. \tag{2}$$

$$A_p(j, i) = \binom{i}{j}(1 - p_a)^{i-j}(p_a)^j. \tag{3}$$

It follows that $Q_r(i, 0) = 0$, $Q_a(i, m) = 0$ and $A_p(j, 0) = 0$.

The probability that the system transits from state n to state $n + i$ is given by

$$P_{n,n+i} = \begin{cases} Q_a(m-n, n) \sum\limits_{\substack{j=0,t=0 \\ j+t\geq 2}}^{n,i} Q_r(j, n)A_p(t, i), i = m - n, i \geq 2 \\ Q_a(i, n) \sum\limits_{\substack{j=0,t=0 \\ j+t\geq 2}}^{n,i} Q_r(j, n)A_p(t, i) + \\ \quad Q_a(i+1, n)\Big[Q_r(1, n)A_p(0, i+1)+ \\ \quad Q_r(0, n)A_p(1, i+1)\Big], \qquad 0 \leq i < m - n \\ Q_a(0, n)Q_r(1, n), \qquad i = -1 \end{cases} \tag{4}$$

3 Social Welfare in Terms of Throughput and Delay

3.1 Performance Metrics

Let's explain now some performance indicators (average throughput and waiting time) as functions of the stationary state of the Markov chain. $\pi_n(q_r, p_a)$ is the probability at equilibrium that the system is in state n (waiting state at the beginning of a slot). The optimal solution is obtained by solving the following optimization problem.

$$\max_{q, p_a} objective(q_r, p_a) \text{ s.t.} \begin{cases} \pi(q_r, p_a) = \pi(q_r, p_a) \cdot P(q_r, p_a), \\ \sum_{n=0}^{m} \pi_n(q_r, p_a) = 1, \\ \pi_n(q_r, p_a) \geq 0, \qquad n = 0, 1, ..., m. \end{cases} \tag{5}$$

Where $objectif(q_r, p_a)$ is replaced by the average throughput or the opposite of the average waiting time. We find that the solution can be obtained by calculating recursively the steady state of probabilities, as in Problem 4.1 in [4].

Singularity at $q_r = 0$: The only point where the Markov chain P has no unique stationary distribution is in $q_r = 0$, where it has an absorbing state: $n = m$ (we recall that $q_r = 0$, slotted aloha has two absorbing states $n = m$ et $n = m - 1$ [2, 3, 5]). All other states are transitory (for all $q_a > 0$), and the probability of completing at one of the absorbing states depend on the initial distribution of the Markov chain. When the state m is reached, the throughput is 0, which means that it is a state of deadlock. For all $q_a > 0$ and $q_r = 0$, this state is reached with a positive probability from any other initial state, that is why we will exclude the case of $q_r = 0$ and optimize only on the interval $\epsilon < q_r \leq 1$.

Existence of a Solution: The stationary distribution $\pi(q_r, p_a)$ is continuous on $0 < q_r \leq 1$ that is not a closed interval, so a solution might not exist. As we restrict ourselves to the closed interval $[\epsilon, 1]$, where $\epsilon > 0$, an optimal solution always exist. Therefore, for every $\gamma > 0$, it exists a certain $q_r^* > 0$, that is γ-optimal. $q_r^* > 0$ is said γ-optimalfor the maximization of the throughput, if it satisfied $T(q_r^*) \geq thp(q_r, p_a) - \gamma$ for any value of $q_r \in [\epsilon, 1]$. The same definition remains valid for any objective function (the minimization of the average delay in particular). Solving the system (5) provides the stationary distribution that will allow us to derive other metric measurements.

The average number of backlogged packets is simply calculated by

$$S(q_r, p_a) = \sum_{n=0}^{m} \pi_n(q_r, p_a) \cdot n. \tag{6}$$

Similarly, the average total throughput (defined as the average number of packets that are successfully transmitted) is almost certainly given by the constant.

$$\begin{cases} \sum_{n=0}^{m}\sum_{i=0}^{m-n}\sum_{j=0}^{min(1,n)}\sum_{t=0}^{min(1,i)} \pi_n(q_r, p_a)Q_a(i,n)Q_r(j,n) \\ \quad A_p(t, i+1)\cdot\delta(1-t-j), & \text{Generalized aloha} \\ \sum_{n=0}^{m}\pi_n(q_r, p_a)\Big[Q_a(0,n)Q_r(1,n)+ \\ \quad Q_a(1,n)Q_r(0,n)\Big]. & \text{Slotted aloha} \end{cases}$$

$\delta(\cdot)$ being the Dirac distribution. Using the balance equation (i.e. input = output) at equilibrium, the average total throughput satisfied (and therefore can be more easily calculated by :

$$T(q_r, p_a) = q_a\sum_{n=0}^{m}\pi_n(q_r, p_a)(m-n) = q_a(m - S(q_r, p_a)). \tag{7}$$

The average number of packets in the system is equal to $S(q_r, p_a) + T(q_r, p_a)$ (sum of backlogged packets and new arrivals). The rate of departure is exactly equal to the average throughput thp (q_r, p_a). We define the average time to transmit a packet D as the average time, in slots, that takes a packet to go from its source to its destination. The application of the Little formula gives :

$$\begin{aligned} D(q_r, p_a) &= \frac{T(q_r, p_a) + S(q_r, p_a)}{T(q_r, p_a)}, \\ &= 1 + \frac{S(q_r, p_a)}{T(q_r, p_a)}, \\ &= 1 + \frac{S(q_r, p_a)}{q_a(m - S(q_r, p_a))}. \end{aligned} \tag{8}$$

By analyzing the equations (7) and (8), we find that maximizing throughput is equivalent to minimizing the average delay packet transmission. So, we will restrict ourselves to maximize the average total throughput. However, we will consider the backlogged packets delay as another objective functions to optimize.

3.2 Performance Metrics of Backlogged Packets

The analysis of the system based on the ability to serve packets awaiting retransmission is indeed an interesting alternative for measuring system performance. It measures the system's ability to serve the packets already in the system, hence its interest especially for real-time applications. $\Delta(q_r, p_a)$ is the average throughput of new arrivals, ie, the average number of arrivals whose first attempt is successful. So, the average throughput of backlogged packets is given by:

$\widetilde{T}(q_r, p_a) = T(q_r, p_a) - \Delta(q_r, p_a)$ o $\Delta(q_r, p_a)$ is calculated from the Markov chain.

$$\Delta(q_r, p_a) = \begin{cases} \sum_{n=0}^{m} \sum_{i=1}^{m-n} \pi_n(q_r, p_a) Q_a(i, n) Q_r(0, n) A_p(1, i), \\ \qquad \text{Generalized Aloha} \\ \sum_{n=0}^{m} \pi_n(q_r, p_a) Q_a(1, n) Q_r(0, n). \\ \qquad \text{Slotted aloha} \end{cases} \tag{9}$$

The delay of backlogged packets $\widetilde{D}$, which is defined as the average time in slots that a waiting packet takes to go from the source to the destination, can be calculated using the formula of Little.

$$\widetilde{D}(q_r, p_a) = 1 + \frac{S(q_r, p_a)}{\widetilde{T}(q_r, p_a)}. \tag{10}$$

3.3 Stability

A qualitative approach to tackle the performances of our new protocol is to study its stability. Slotted aloha is known by its bi-stable behavior, we therefore ask the following questions: Does our variant of slotted aloha suffers from the same bi-stability problem? And if so, is there any difference between the two methods of access? If not, under which conditions our protocol would be better than slotted aloha?

We are defining p_n^{succ} as the probability of a successful transmission during a time slot when the system remain in the state n, Based on the markov chain, p_n^{succ} is given by:

$p_n^{succ}(q_r, p_a) =$

$$\begin{cases} \sum_{i=0}^{m-n} \sum_{j=0}^{min(1,n)} \sum_{t=0}^{min(1,i)} Q_a(i, n) Q_r(j, n) A_p(t, i) \cdot \delta(1 - t - j) \\ , \qquad \text{Generalized aloha} \\ Q_a(0, n) Q_r(1, n) + Q_a(1, n) Q_r(0, n). \quad \text{Slotted aloha} \end{cases}$$

We now define D_n in the State n, as the change of a backlog from one slot to the next (drifting), which is the number of arrived, i.e.,$q_a(m - n)$, minus the numbers of successful departures p_n^{succ}, which is:

$$D_n = q_a(m - n) - p_n^{succ}. \tag{11}$$

In slotted aloha standard, It was shown that three equilibrium exist. The first one is effective because it is a number of small backlogged packets. The second is unstable and allows the system to switch from one equilibrium to another. on the other hand the third equilibrium is characterized by a very large number of waiting packets n (potentiellement $n = m$), and then corresponds to a situation of congestion (low flow and large delay) which reduces the efficiency of the system. The equilibrium points of the system occur when the drift is zero, i.e. the curve p_n^{succ} and the right $q_a(m-n)$ intersect. When the drift is positive the system state tends to increase, because the rate of input to the system is higher than the rate of output, and decreases when the drift is negative. This explains why the equilibrium of the environment is unstable and stable for the others. A bi-stable situation as in standard aloha is not desirable because it means, in practice, the system is likely to be absorbed by the inefficient stable equilibrium.

4 Numerical Investigations

We present some numerical examples evaluating the performances of the new access channel solution slotted aloha taking as reference. We solve the system (5) for $m = 3$ and $m = 10$ mobiles. The average time of backlogged packets being the objective function, We find that the flow, Fig. 2 (a), and the average delay, Fig. 2 (b) improved significantly especially in the medium and heavy traffic. This result can be explained by analyzing the optimal transmission policy $(q_r, \ p_a)$. Indeed, at low-traffic, mobiles retransmit with high probability q_r while promoting the transfer of new packages.

At heavy traffic, the base station recommends mobiles to reduce their probability of retransmission $(q_r, \ p_a)$, Fig. 3, which reduces the number of collisions and thereafter continues to increase the flow rate when the traffic becomes high. When total throughput is considered as a function to be optimized, notice that we rate that the flow of our model coincides with that of slotted aloha. On the other hand, the generalized aloha average delay was found to be far lower than that guaranteed by slotted aloha. Indeed, the corresponding generalized aloha average delay remains almost constant regardless of system traffic q_a, slotted aloha tends to increase exponentially when the traffic becomes medium or high. This (aloha generalized behavior) is mainly due to the fact that when q_a increases, q_r decreases (i.e. which gives more priority to new packages, where a small number of backlogged packets.) Which leads us to say that our model is more suited to services and real-time applications that require constraints on the delay.

We turn now to the illustration of stability of generalized aloha. La figure 4 (a) shows the behavior of our model compared to standard slotted aloha $q_a = 0.01$ and $q_r = 0.1$. for static p_a, we note that at low load (saturation), aloha is better than aloha generalized. Indeed, when the average number of backlogged packets is small, i.e. only some of mobiles have packets to be transmitted, the fact of the new transmission delay reduces the success rate. On the other hand when the number of saturated sources becomes large, aloha generalized becomes more powerful as the mechanism

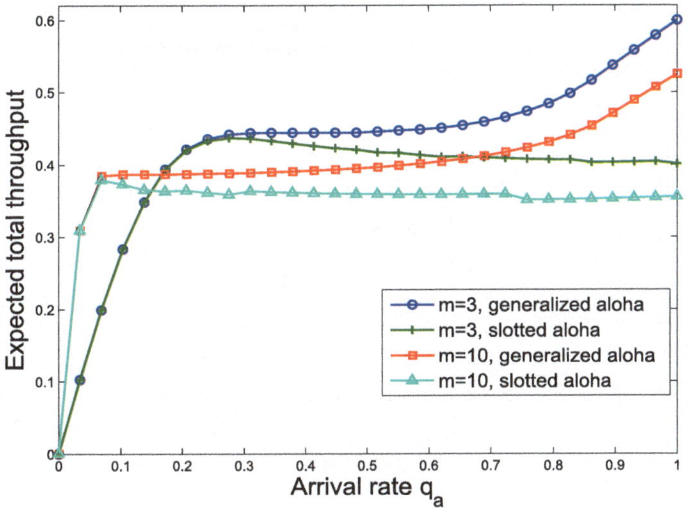

(a)

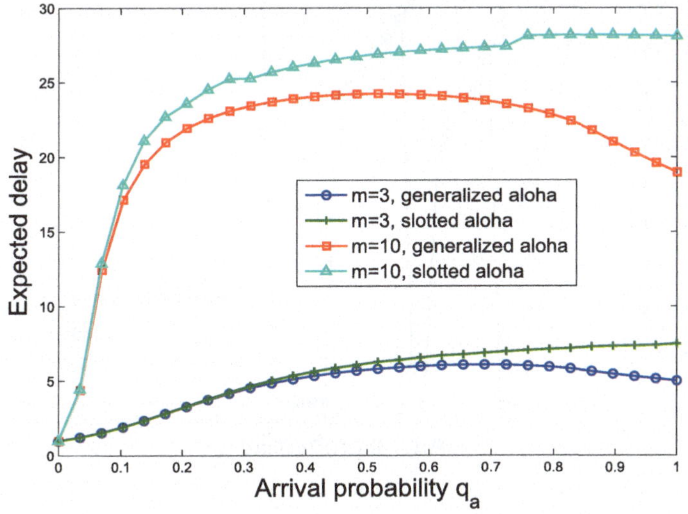

(b)

Fig. 2 Figure 2 - total flow (a) and wait time of backlogged packets when delay is minimized (b).

to delay the new transmissions avoid / reduce collisions. Figure 4 (b) shows the same pattern with dynamic p_a. We note that performances improves when $p_a \to 1$ at low load and when $p_a \to 0$medium and heavy load. This finding inspired us to offer a dynamic selection of p_a parameter such as sigmoidal functions (S-shaped) type

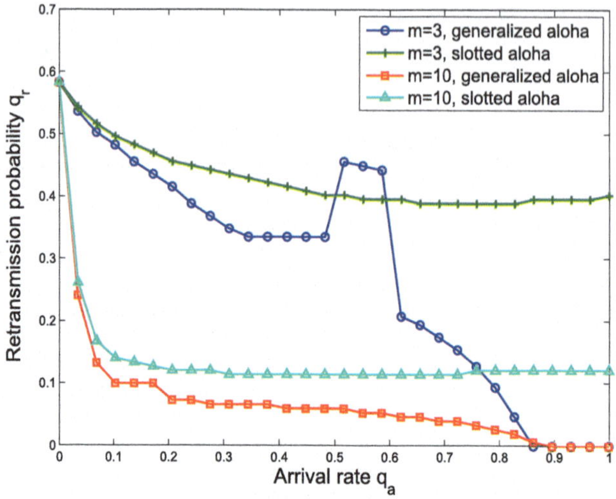

(a)

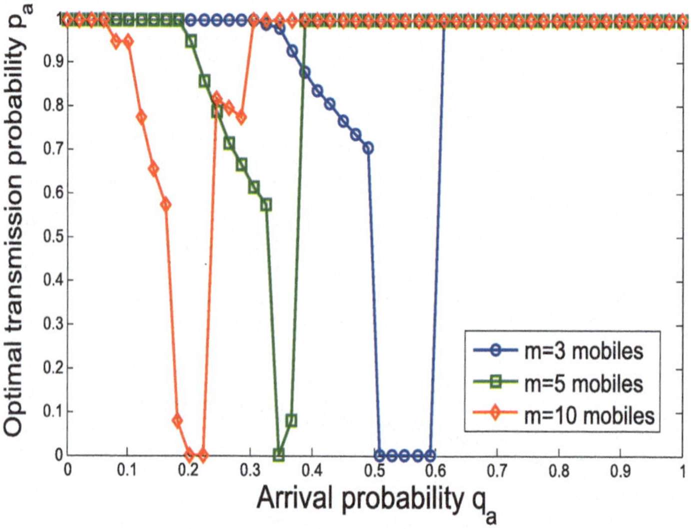

(b)

Fig. 3 Optimal probability of transmission when the delay is minimized (a). A similar result is obtained, total flow is taken as an objective function (b).It should be underlined that mobiles are somewhat less aggressive.

$1-(1-e^{-n})^M$ where M is a given constant controlling the function's speed $p_a(n)$. The major problem with this adaptive mechanism is that mobiles must estimate the actual value of the number of backlogged packets n. This can be easily solved using a learning algorithm based on the observation of events "success", "collision" and

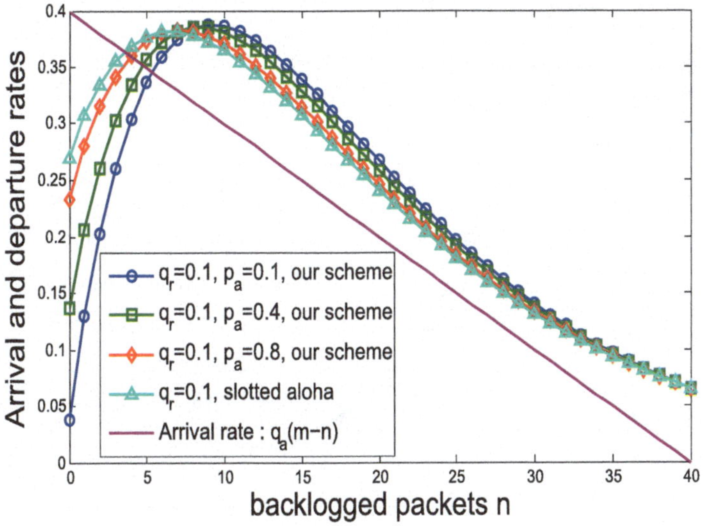

(a)

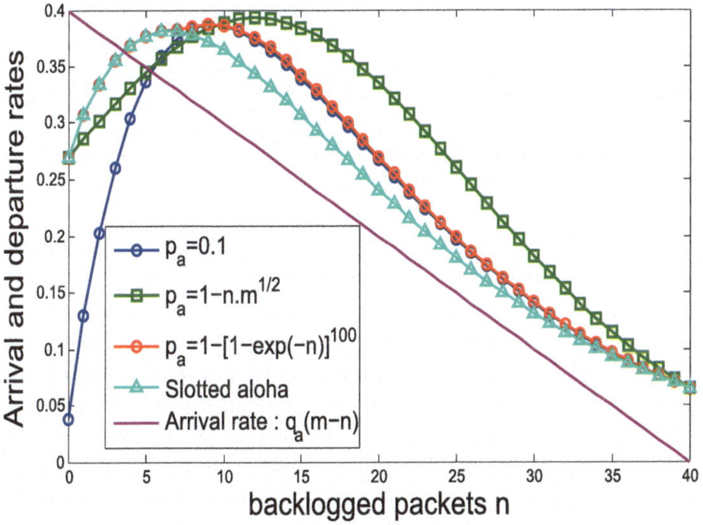

(b)

Fig. 4 Stability of our model and that of slotted aloha for static (a) and dynamic $p_a(n)$ (b).

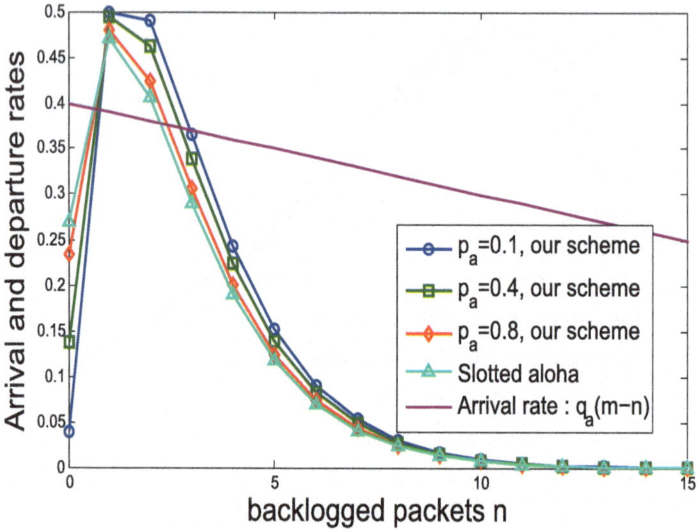

Fig. 5 Case when all of the schemas are instable.

"free" Problem 4.1 [4], then

$$\hat{n}_{k+1} = \begin{cases} max(\lambda, \hat{n}_k + \lambda - 1), & \text{Free or success.} \\ \hat{n}_k + \lambda + (e-2)^{-1}, & \text{collision.} \end{cases} \quad (12)$$

where $\lambda = q_a/m$. And the dynamic case of generalized aloha is almost always above the slotted aloha for any state n of system. At low q_r, it is clear that none of the alternatives of slotted aloha are suffering from the problem of bi-stability. So the two schemes can only be absorbed by the efficient equilibrium point, i.e. by the non-congested balance. Figure 5 shows the impact of retransmission probability q_r on protocol stability (here q_r=0.5). Overall, we affirm that our protocol is also bi-stable, except that it is more efficient in terms of probability of success. One way to address the problem of bi-stability would be to define a dynamic control on q_r depending on the instantaneous number of backlogged packets. Introduce a transmission cost (energy consumption) is also a solution elegant for mobiles reducing their transmission rates to better manage their energy budget.

5 Conclusion

We introduced an improved version of slotted aloha protocol based on transmission of new packet with probability p_a. Instead of immediate transmission. This mechanism allows better management of transmission by reducing the number of collision. We built a Markovian Model, We have calculated the Stationary Distribution of associated process. This has allowed us to calculate different metric performance as of total flow, Average delivery time, and probability of success. This performance indicators gives the possibility of to study the stability of protocol in a qualitative approach and flow increase in a quantitative approach. Our solution is far more being promising and more effective than slotted aloha as it allows a larger flow and especially wait time has been shortened. It turned out to be a robust, effective and easier implementation solution. And could make this protocols families capable of carrying interactive applications and Real-time.

References

1. Abramson, N.: The aloha system - another alternative for computer communications. In: AFIPS Conference Proceedings, vol. 36, pp. 295–298 (1970)
2. Altman, E., Barman, D., Benslimane, A., El-Azouzi, R.: Slotted aloha with priorities and random power. In: Proceedings of IFIP Networking, Ontario, Canada, pp. 610–622 (2005)
3. Altman, E., El-Azouzi, R., Jimenez, T.: Slotted Aloha as a stochastic game with partial information. In: Proceedings of WiOpt 2003, Sophia Antipolis, France, March 3–5, 2003
4. Bertsekas, D., Gallager, R.: Data Networks. Prentice Hall, Englewood Cliffs (1987)
5. El-Azouzi, R., Sabir, E., Benarfa, S., Jimenez, T., Bouyakhf, E.H.: Cooperative and noncooperative control for slotted aloha with random power level selection algorithms. In: Proceedings of Valuetools 2007, Nantes, France, October 23–25, 2007
6. LaMaire, R.O., Krishna, A., Zorzi, M.: On the randomization of transmitter power levels to increase throughput in multiple access radio systems. Wireless Networks **4**, 263–277 (1998)
7. MacKenzie, A.B., Wicker, S.B.: Selfish users in aloha: a game theoretic approach. In: Proceedings of 54th Vehicular Technology Conference (VTC 2001), Atlantic City, NJ, vol. 3, pp. 1354–1357 (2001)
8. MacKenzie, A.B., Wicker, S.B.: Stability of slotted aloha with Multi-packet Reception and Selfish Users. Proceedings of IEEE INFOCOM **3**, 1583–1590 (2003)
9. Metzner, J.: On improving utilization in ALOHA networks. IEEE Transaction on Communications **24**(4), 447–448 (1976)
10. Park, J., van der Schaar, M.: Stackelberg contention games in multi-user networks. EURASIP Journal on Advances in Signal Processing (2009, to appear)
11. Roberts, L.G.: Aloha packet system with and without slots and capture. Tech. Rep. Ass Note 8, Stanford Research Institute, Advance Research Projects Agency, Network Information Center (1972)
12. Sabir, E., El-Azouzi, R., Hayel, Y.: A hierarchical slotted Aloha game. In: Proceedings of IEEE/ICST International conference on game theory for networks (GameNets 2009), Istanbul, Turkey, May 13–15, 2009
13. Sabir, E., El-Azouzi, R., Hayel, Y.: Hierarchy Sustains Partial Cooperation and induces a Braess-like Paradox in Slotted Aloha-based Networks. Elsevier International Journal of Computer Communications **35**(3), 273–286 (2012)

Enhanced Hybrid Uplink Time Difference of Arrival and Assisted Global Positioning System for UMTS Users

**Ilham El Mourabit, Aïcha Sahel,
Abdelmajid Badri and Abdennaceur Baghdad**

Abstract Due to the growing needs of positioning solutions serving for both emergency cases and daily life uses, many researches are carried out to enhance the existing techniques in a way to find a robust method that meets the expectations of both areas. In this paper, we propose an enhanced hybrid approach between the up-link time difference of arrival and Assisted GPS positioning methods for UMTS users by improving the accuracy of UTDOA using the NLMS adaptive filter instead of usual filter before performing the cross-correlation.

Keywords UTDOA · NLMS · A-GPS · Position · UMTS · Adaptive filters

1 Introduction

Location Based services (LBS) refers to services that utilizes the position estimate of a mobile station. As specified in [1], there are four categories of location services: commercial LBS, internal LBS, emergency LBS, and lawful intercept LBS. The demand to locate mobile phones in the case of emergency calls is commonly accepted as the main driving force for LBS regarding the great benefit of such services in rescuing operations. There are different methods to obtain the position of a Mobile Station, Based on the entity which measures and/or calculates the position, we define the following categories [2]:

- **Handset-based:** In this category the handset measures the data needed for the approximate location and calculates its own position out of this data. the methods belonging to his type do not need any access to the network, so no changes are imported to this one, only an upgraded phone is needed to

I. El Mourabit(✉) · A. Sahel · A. Badri · A. Baghdad
EEA and TI Laboratory, Faculty of Sciences and Techniques Mohammedia,
Hassan II University Casablanca, Casablanca, Morocco
e-mail: elmourabit.ilham@gmail.com

E. Sabir et al. (eds.), *The International Symposium on Ubiquitous Networking*,
Lecture Notes in Electrical Engineering 366,
DOI: 10.1007/978-981-287-990-5_2

calculate the position estimation. As an example of these techniques we find the Global Positioning System GPS, the Time of Arrival TOA and the fingerprint method.

- **Network-based:** The network measures all data needed for the handset location and calculates it. The handset is passive in the whole progress, this type is the more suitable for emergency positioning services. The big advantage of this category is that no changes or upgrades are applied to the mobile phone, however, expensive location measurement units are needed to be added (if not exist) to the network structure. Among the commonly known approaches of this type we have Cell-Id, Angle Of Arrival AOA, and Up-link Time Difference Of Arrival UTDOA.
- **Hybrid positioning:** In hybrid positioning, the network and the handset work together to first measure and then calculate the device position. In this category we find the Assisted GPS and enhanced Cell-Id + RTT (Round Trip Time).

Among these location estimation techniques we are interested in UTDOA and A-GPS positioning methods to build a hybrid solution. In the next section we will explain the main idea of each approach then present the principle and the simulation of the proposed method.

2 Uplink Time Difference of Arrival Technique

It is a real time locating technique for mobile phone that uses multi-lateration (hyperbolic positioning) based on timing of received signals. Location Measurement Units (LMUs) are co-located at the Base Stations (BSs) to calculate the time difference measurements used to determine the location of a mobile phone. The technique is a network-based, so it can locate even legacy phones [3].

For hyperbolic lateration, the mobile terminal's position is determined by the distance differences (di -dj) instead of the absolute distance d. A hyperbola is defined to be the set of all points for which the difference from two fixed points is constant [4]. Then two BSs with known locations determine a hyperbolic curve. The intersection of two hyperbolic curves can determine a unique point. Therefore, three BSs are required to locate a mobile terminal in 2D coordinates. The handset position (x, y) can be derived from equation (1) :

$$\begin{cases} d_2 - d_1 = \sqrt{(x_2 - x)^2 + (y_2 - y)^2} - \sqrt{(x_1 - x)^2 + (y_1 - y)^2} \\ d_3 - d_1 = \sqrt{(x_3 - x)^2 + (y_3 - y)^2} - \sqrt{(x_1 - x)^2 + (y_1 - y)^2} \end{cases} . \tag{1}$$

Knowing that the speed of propagation is estimated to be c = 3.10^8 ms-1, we can replace the distances in (1) with their corresponding times according to the following formula $c = \frac{d_i}{t_i}$. After some simplification we can write the previous equations in the form:

$$\frac{x^2}{a} - \frac{y^2}{b} = 1 . \tag{2}$$

Which describes an hyperbolic curves. The figure 1 shows the results of a hyperbolic lateration with three base stations to determine the location of an MS.

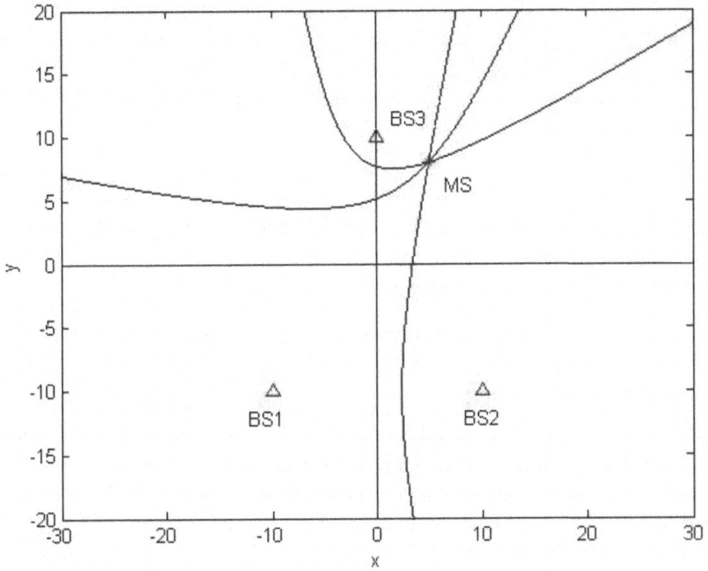

Fig. 1 Time Difference Of Arrival hyperbolic solution

UTDOA is the best network based method in term of accuracy compared to other solutions as shown in the table 1 [5].

Table 1 Comparing Positioning techniques accuracy.

Technique	network	handset	accuracy
Cell-ID	Both	Both	100m - 3 km
AOA	Both	Both	100 - 200 m
UTDOA	Both	Both	<50m
EOTD	GSM	Upgrade	50 - 200 m
A-GPS	Both	Upgrade	5 - 30 m

2.1 Enhanced UTDOA

A cross correlation is performed between the same signal received at two different BSs, one of them should be currently serving the mobile user. A peak detection correspond to the searched time difference value.

The received signal at the ith base station can be written as:

$$r_i[n] = A_i\, s[n - \tau_i] + n_i[n] \tag{3}$$

Where

A_i : signal amplitude.

$s[n - \tau_i]$: delayed version of the sent signal $s[n]$.

$n_i[n]$: propagation noise (additive white Gaussian noise).

Considering that BS_1 is the one actually the MS (Mobile Station) is connected to, so it has the shortest time of arrival then the received signals equations can be written as:

$$r_1[n] = s[n] + n_1[n] \tag{4}$$

$$r_2[n] = A\, s[n - \tau_d] + n_2[n] \tag{5}$$

Where $\tau_d = \tau_2 - \tau_1$ is the searched time difference of arrival between the two base stations and A is the amplitude ratio.

The formula of cross correlation equation is given as:

$$R_{r_1,r_2}[k] = \sum_{n=-\infty}^{+\infty} r_1[n]\, r_2[n-k] \tag{6}$$

Where k is the estimation of TDOA which correspond to a peak detection.

As shown in the modeling equations of the received signal an additional term referring to the propagation noise is attached to every version of this signal, this term will affect the precision of time difference measurements if we perform the cross correlation directly. From here came the idea of pre-filtering the received signal with adaptive filters before the cross correlation so we can minimize the noise effect to have more accurate position [6] [7].

Previous work has been carried out to enhance the UTDOA technique using adaptive filters as a noise cancellation system before the time estimation by cross correlation [8]. This kind of filters is controlled by an algorithm to update the filter's coefficients in function of the received signal and the error signal. As presented in [8] the Normalized Least Mean Square algorithm has shown better performances and less complexity comparing to other type of controlling algorithms (LMS and RLS).

2.2 *Normalized Least Mean Square Algorithm*

In the standard LMS algorithm the filter's coefficients are updated following the formula

$$w(n+1) = w(n) + \mu\, e(n)\, r(n) \tag{7}$$

And

μ: the convergence factor.

r (n): received signal.

e(n) : error signal defined as e(n) = r(n) - y(n) where y(n) = w(n) r(n-Δ).

When the convergence factor μ is large, the algorithm experiences a gradient noise amplification problem. In order to solve this difficulty, we can use the NLMS algorithm. The correction applied to the weight vector w(n) at iteration n+1 is "normalized" with respect to the squared Euclidian norm of the input vector r(n) at iteration n. We may view the NLMS algorithm as a time-varying step size algorithm [9], defining the convergence factor μ as

$$\mu = \frac{\alpha}{c + \|r(n)\|^2} \tag{8}$$

Where α is the NLMS adaption constant, which optimize the convergence rate of the algorithm and should satisfy the condition $0< \alpha<2$, and c is the constant term for normalization and is always less than 1. In NLMS algorithm, the filter weights are updated by the equation

$$w(n+1) = w(n) + \frac{\alpha}{c+\|r(n)\|^2} e(n)\, r(n) \qquad (9)$$

3 Assisted Global Positioning System (AGPS)

Global Positioning System was invented to localize a mobile device outdoor where a clear LOS (Line Of Sight) to the satellites exist. The Time To First Fix is related to the usage behavior of the GPS device and the actuality of the almanac that is stored on the device. To get a fix on its position an MS needs to obtain certain data from the satellites and this can takes some time. In case of an emergency, the user wants the position as fast as possible and without limitations. To overcome these drawbacks of GPS, A-GPS was developed. In A-GPS not only the satellites but also the terrestrial cellular network (GSM, UMTS or the internet) are used to estimate the position [10]. If the MS is not able to receive the needed data from the satellites the request is directed toward the cellular network to provide the data needed [11].

A-GPS technology follows the principle of D-GPS with additional assistance data from cellular network which is used to reduce acquisition time. Assistance data is done using variety of UMTS mobile network positioning types such as Cell-ID, TDOA, O-TDOA and UTDOA explained previously. The accuracy of this technique is estimated to be around 5-30m in clear areas with a direct LOS. As shown in fig. 2, the structure of AGPS network requires a reference station to provide assistance data through cellular network to the Mobile station [12].

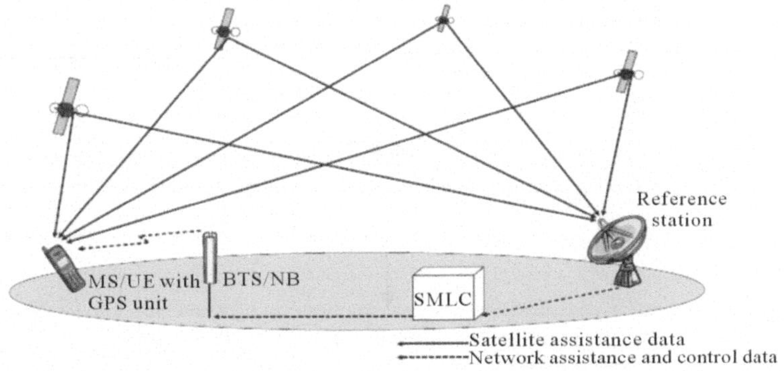

Fig. 2 Structure of AGPS network.

4 Proposed Hybrid Solution and Simulation Results

The GPS is a very accurate positioning system but it suffers from the weakness of satellites signals which makes it difficult to receive it at dense urban environment with Non Line Of Sight, unlike UTDOA, the most accurate technique based on cellular network which signals can be received even in indoor areas.

The Flow chart of the proposed technique as shown in fig. 3 has the possibility to locate a MS either with hybrid process or standalone UTDOA in case where the mobile phone cannot receive GPS signal (ex. legacy phones).

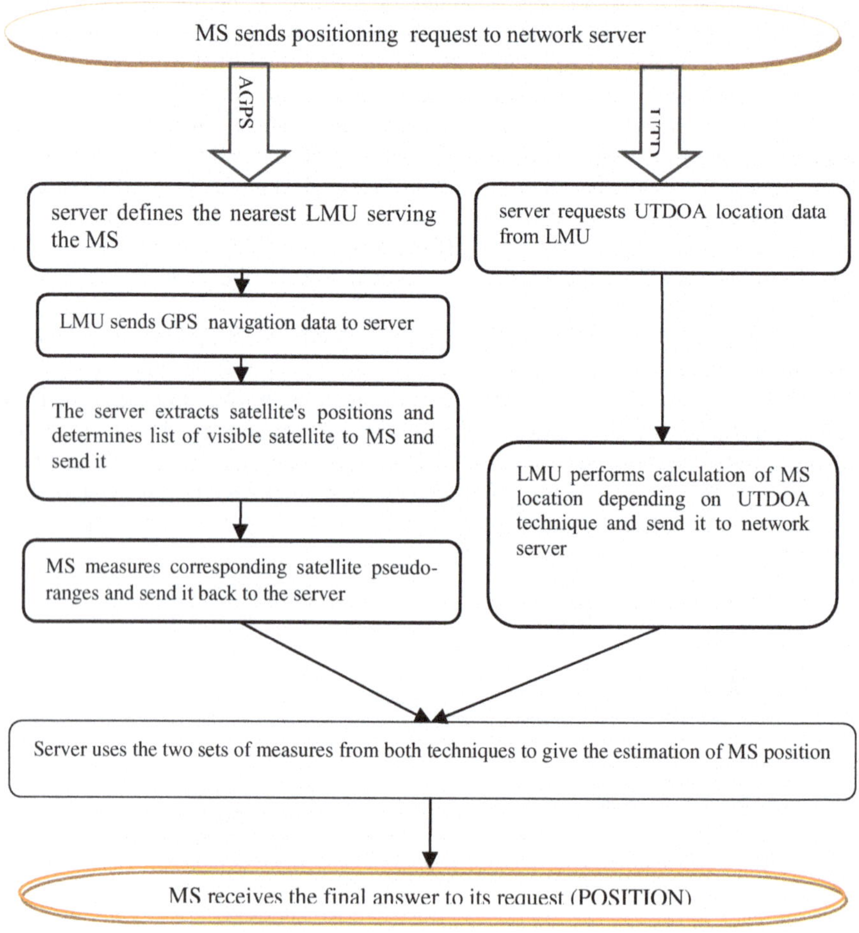

Fig. 3 Proposed scenario of the hybrid positioning

Computer simulation was done in MATLAB software, using the DSP and signal processing toolboxes, to build conventional UTDOA block and those of

NLMS filter and propagation channel according to the COST-231 model. To perform this simulation we consider the sent signal to be a sin wave, the attached noise is an additive white Gaussian noise and the received signal is modeled by the following formula:

$$r_i[n] = \sum_{j=1}^{P_N} A_{ij}\, s[n - \tau_i] + n_i[n] \tag{10}$$

where P_N is the number of paths.

The accuracy of this method is evaluated using the Root Mean Square Error, which can be defined as the difference between the estimated and the true position of the mobile user

$$\mathrm{RMSE} = \sqrt{\frac{1}{N}\sum_{k=1}^{N}[x_{measured}(k) - x_{true}]^2} \tag{11}$$

In figure 4, we investigated the cumulative probability of different RMSEs of the standard UTDOA, Enhanced UTDOA and the hybrid solution. It can be clearly seen that the positioning error has been reduced by applying the adaptive filters to the localization method and more when associating it with AGPS. According to the FCC's requirements for E911 location services the RMSE of the proposed approach should be less than 100m in 67% of cases, this criteria is fulfilled at respectively 58%, and 70% of cases for Enhanced UTDOA and Hybrid techniques filters while the same condition can be achieved at only 15% for the conventional UTDOA.

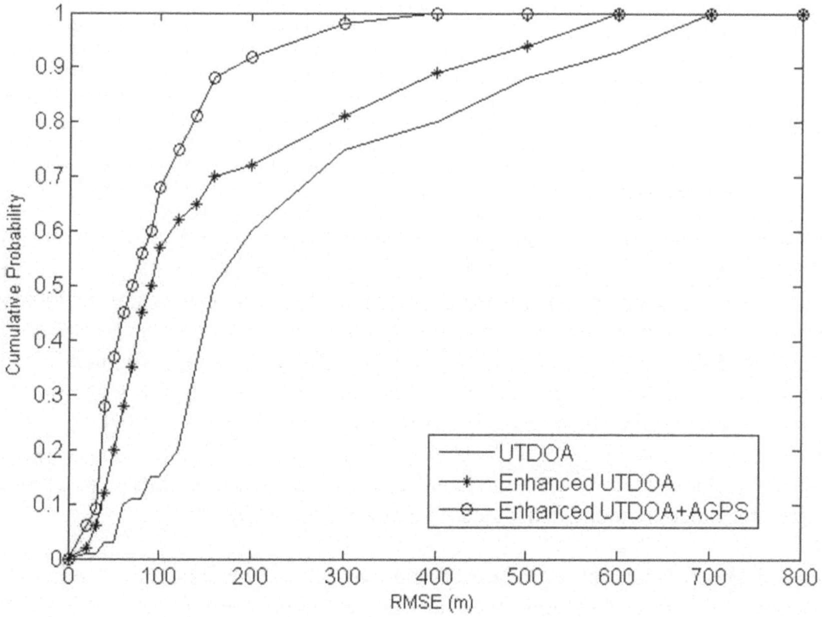

Fig. 4 RMSE cumulative probability

Figure 5 look into the effect of the paths number in the multipath propagation channel on the accuracy of the proposed positioning methods. Despite having the same number of signal's paths, the conventional UTDOA is more influenced than the Enhanced UTDOA and hybrid techniques. it is noticed that the adaptive filtering and AGPS compensates the noise and multipath effect on the measurements precision.

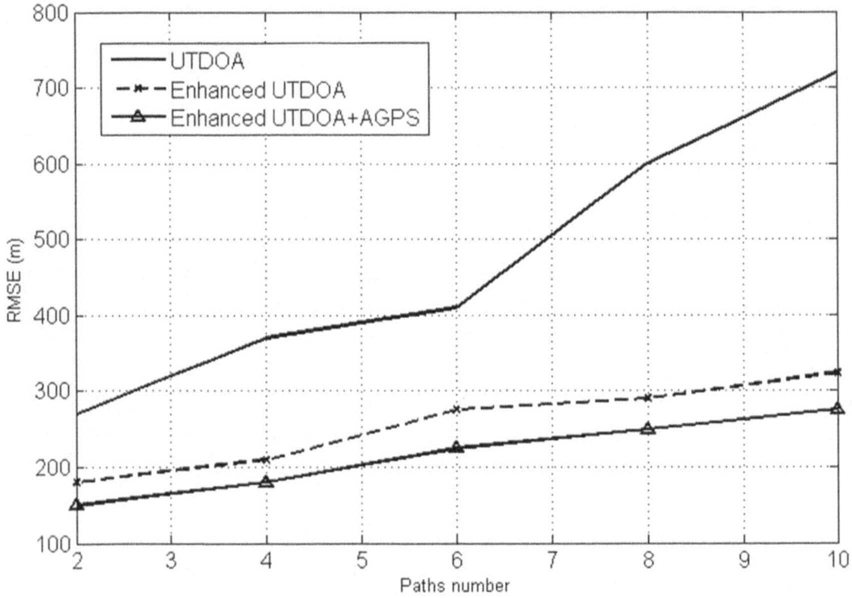

Fig. 5 Multipath effect

5 Conclusion

As shown in the study carried out through this paper, hybrid solution based on enhanced UTDOA using adaptive filtering and Assisted GPS can increase enormously the accuracy of mobile positioning, and compensate the degradation caused by the propagation noise and multipath. In our case the enhancement was investigated in a fixed and known mobile position in outdoor environment using a sine wave to model the sent signal, for future work, the cellular signal will be defined following the standards and this technique would be generalized to take in charge the mobility of users and extended to indoor areas by defining a moving path for the handset, managing the handover between cells and the use of different propagation models with Distributed Antenna Systems (DAS).

Related Work. The concept of using adaptive filters to enhance the Uplink time difference of arrival technique was the subject of previous papers [13][14], but this work did not limit the study on the enhancement of UTDOA by the adaptive filtering and introduced a hybrid solution with the Assisted-GPS.

Acknowledgment This work falls within the scope of telecommunication projects. Our sincere thanks to the Faculty of Sciences and Technology, Hassan II University, Mohammedia, Morocco, for providing us an opportunity to carry out our work in a well-equipped laboratory (EEA&TI). We are also thankful to all our colleagues who helped us while working on this project.

References

1. Deligiannis, N., Louvros, S.: Hybrid TOA-AOA location positioning techniques in GSM networks. Wireless Personal Communications **54**, 321–348 (2010). doi:10.1007/s11277-009-9728-x
2. Willaredt, J.: Wi-Fi and Cell-ID based positioning - Protocols, Standards and Solutions
3. Baloch, Y.A.: UMTS positioning methods and accuracy in urban environments
4. Sayed, A.H., Tarighat, A., Khajehnouri, N.: Network-Based Wireless Location: Challenges faced in developing techniques for accurate wireless location information. Signal Processing Magazine, 24–40. IEEE (2005)
5. Schmidt-Dannert, A.: Positioning Technologies and Mechanisms for mobile Devices. Seminar Master Module SNET2 TU-Berlin
6. Mardeni, R., Anuar, K., Shahabi, P., Manesh, M.R.: Efficient uplink time difference of arrival mobile device localization in cellular networks. In: 2013 Asia-Pacific on Microwave Conference Proceedings (APMC), pp. 1124–1126. IEEE (2013)
7. Mardeni, R., Shahabi, P., Riahimanesh, M.: Mobile station localization in wireless cellular systems using UTDOA. In: 2012 International Conference on Microwave and Millimeter Wave Technology (ICMMT), vol. 5, pp. 1–4. IEEE (2012)
8. El Mourabit, I., Sahel, A., Badri, A., Baghdad, A.: Performance of multiples adaptive algorithms for Uplink Time Difference of arrival positioning technique. International Journal of Emerging Trends in Engineering and Development **1**(5), 143–155 (2015). ISSN: 2249-6149
9. Tandon, A., Ahmad, M.O., Swamy, M.N.S.: An efficient, low-complexity, normalized LMS algorithm for echo cancellation. In: The 2nd Annual IEEE Northeast Workshop on Circuits and Systems, NEWCAS 2004, pp. 161–164. IEEE (2004)
10. Ratti, C., Williams, S., Frenchman, D., Pulselli, R.M.: Mobile landscapes: using location data from cell phones for urban analysis. Environment and Planning B Planning and Design **33**, 727–748 (2006)
11. Australian Communication Authority: Location, location, location - the future use of location information to enhance the handling of emergency mobile phone calls, January 2004. www.acma.gov.au/webwr/consumer_info/location.pdf
12. Abo-Zahhad, M., Ahmed, S.M., Mourad, M.: New Technique for Mobile User's Location Detection, Future Prediction and their Applications. International Journal of Engineering and Innovative Technology (IJEIT) **3**(4), 307–321 (2013)
13. Mardeni, R., Anuar, K., Shahabi, P., Manesh, M.R.: Efficient uplink time difference of arrival mobile device localization in cellular networks. In: 2013 Asia-Pacific on Microwave Conference Proceedings (APMC), pp. 1124–1126 (2013)
14. Mardeni, R., Shahabi, P., Riahimanesh, M.: Mobile station localization in wireless cellular systems using UTDOA. In: 2012 International Conference on Microwave and Millimeter Wave Technology (ICMMT), vol. 5, pp. 1–4 (2012)

Efficient Incentive Scheme for Wireless Random Channel Access with Selfish Users

Abdelillah Karouit

Abstract In the decentralized networks, interactions among selfish users sharing a common transmission channel can be modeled as a non cooperative game using game theory. When selfish users choose their transmission probabilities independently without any coordination mechanism, Nash equilibria leads to a suboptimal use of the channel resource and to the degradation of the performance of MAC protocols. In this paper we transform the non cooperative game in a general Stackelebrg game when multiple leaders and followers coexist and competitively maximize their own function utilities. Under a simple multiple power levels scheme, the new Stackleberg equilibrium can overcome the deficiency of the Nash equilibrium as well as the limited efficiency of the Sackelberg contention game when the users transmit using a single power level. The power control is defined in such way that the leaders choose the lowest power to transmit their packets among N available levels whereas the followers retransmit at the random power levels picked from $N - 1$ higher distinct. Furthermore, we investigate the impact of the implementation of our scheme on the equilibrium and discuss the optimal partition of leaders/followers to achieve better performance. We further show that the hierarchical system is performing better as the number of leaders is small.

Keywords Wireless random access protocol · Power diversity · Capture effect · Markov chain · Nash equilibrium · Stackelberg equilibrium · Performance evaluation

When multiple users share a common channel and contend for access, a typical conflict problem arises. To resolve the contention problem many different medium access control (MAC) protocols have been designed assuming that the users comply with the protocol. Recently, the selfish behavior of users in MAC protocols has

A. Karouit(✉)
LIA/CERI University of Avignon, Agroparc, BP 1228, 84911 Avignon, France
e-mail: akarouit@gmail.com

E. Sabir et al. (eds.), *The International Symposium on Ubiquitous Networking*,
Lecture Notes in Electrical Engineering 366,
DOI: 10.1007/978-981-287-990-5_3

been widely analyzed using game theory with all its powerful solution concepts. There have been attempts to understand the existing MAC protocols as the local utility maximizing behavior of selfish users by reverse-engineering the current protocols. It has also been investigated whether existing protocols are vulnerable to the existence of selfish users who pursue their self-interest in a non-cooperative manner. Non-cooperative behavior often leads to inefficient outcomes. For example in [1], the 802.11 distributed MAC protocol, DCF, and its enhanced version, EDCF, competition among selfish users can lead to an inefficient use of the shared channel in Nash equilibria [1]. Similarly, a prisoner's dilemma phenomenon arises in a non-cooperative game for a generalized version of slotted-Aloha protocols [2, 3, 4]. This illustrates, in fact, that Nash equilibrium is not efficient in some situations and more appropriate solution concepts need to be considered. A full system utilization requires coordination among users using explicit message exchanges or presence of an arbitration mechanism, which may be impractical given the distributed nature and arbitrary topology changes of wireless collision channels. To achieve a better performance without coordination schemes, users need to sustain cooperation. It is promising to introduce a set of special users whose mission is to provide incentives for other users to behave cooperatively, this mechanism may limit the aggressiveness level (access to the channel) and leads to design more robust protocols in the future.

In addition, the work [7] studied slotted Aloha as Stackleberg game, in which the feasible outcome can be achieved as Stackelberg equilibrium with independent strategy. Under the proposed game, the leaders are considered to be totally selfish, i.e., competitively maximize their individual throughput. This behavior resulted on a network collapse, in particular for average and relatively high loads.

To improve the system performance when multiple leaders and followers coexist and competitively maximize their own function utilities, we introduce a prioritization between the users in terms of transmit power. In particular we assume that the leaders first chooses the lowest power among N available levels whereas the followers retransmit at random power levels picked from $N - 1$ distinct power levels. The use of multiple power levels is studied with a general capture model based on the experienced signal to noise plus interference ratio in cooperative and non cooperative framework. In [2, 8, 9], power diversity is studied with a general capture model based on the experienced signal to noise plus interference ratio. In addition to the power diversity, the authors in [9] have associated a cost to each transmission attempt. The results show that such pricing could be used to enforce an equilibrium whose throughput corresponds to the team optimal solution [8].

Without loss of generality, the system is split in two non cooperative and coupled sub-games. The first game concerns the set of leaders and the second implies the followers group. But, the two sub-games are implicitly interdependent. In this case, the followers can observe the strategies used by the leaders and optimize their strategies accordingly, i.e., to maximize the payoff. This class of bi-level games is called a Stackelberg game [5, 6]. In our setting, the base station has an important role to play. Indeed, the first role of the base station is to decide the set of users that should be leaders, this decision may be taken based on the QoS of users (e.g., users requiring high rate or having hard constraints on throughput/delay would be selected as leaders). Then, the base station computes the optimal decision vector of the leaders and communicates it to the followers group.

The rest of the paper is organized as follows. We first describe the design principles of multiple power levels as related to its use in conjunction with the Wireless Random Access mechanism (WRA) in Section 1. In this way, we set the basis to understand the existing interaction between the physical and the MAC layers of the protocol stack following a cross-layer approach. Next we provide the Stackelebrg formulation in Section 2. In particular we introduce the equilibrium analysis principles and the performance metrics of interest used in the evaluation of the mechanisms under study. Section 3 provides a numerical evaluation of the proposed mechanism. Finally, our concluding remarks are drawn in Section 4.

1 System Model

We consider a wireless multiple access system composed of one central receiver base station (BS) and $m = m^L + m^F$ geographically dispersed mobile stations communicating with the BS, where m^L and m^F represent the number of leaders and followers respectively. The users use a common wireless access to send data to the BS. Time is divided into multiple equal and synchronized slots. Transmission feedback (success or collision) is received at the end of the current slot. As mentioned before, the use of multiple power levels, also referred as power diversity, is to be considered in conjunction with the Wireless Random Access mechanism. Throughout this paper, we will refer to the resulting mechanism as the Hierarchical Multiple Power Level Wireless Random Access mechanism or simply HMPL-WRA. The use of such facility will give place to a capture effect. Due to this effect, a receiver may be able to decode a message even in the presence of a collision. In fact, the unsuccessful concurrent messages are lost and treated as interference.

1.1 Multiple Power Levels

In the MAC/Physical cross-layer design, the leaders contending for a message transmission, choose the lowest power level T_1 among N available levels $T = \{T_1, T_2, \ldots, T_N\}$. Whereas the followers retransmit on the random power levels picked from $N - 1$ higher distinct power levels. The power levels selection follows the probability vector $X = [x_1, x_2, \ldots, x_N]$, where the $j\text{-}th$ entry x_j is the probability of choosing the power level T_j. We consider a general capture model where a message transmitted by a user i is received successfully when and only when its SINR exceeds some given threshold Θ_{th}. Let σ^2 be the variance of the thermal noise and denote by $S = [S_1, S_2, \ldots, S_m]$ the vector of selected power level at the beginning of the current slot. Note that the components of S are selected from the vector T. The received power on the BS can be related to the transmitted power by the propagation relation $h_i \cdot S_i$, where h_i is the channel gain experienced by the base station when receiving a message transmitted by user i. Note that h_i does not depend on the value of using power level S_i. Thus, the instantaneous SINR of user i transmitting at power level S_i experienced by the receiver is

$$\Theta_i(S) = \frac{h_i \cdot S_i}{\sum\limits_{k=1,k\neq i}^{m} h_k \cdot S_k \cdot \mathbf{1}_k + \sigma^2}, \tag{1}$$

We denote by $A_{s',s}$, the probability of a successful transmission when $s + s' \geq 2$ attempt transmission. $A_{s',s}$ is calculated using the following expression:

$$A_{s',s} = \sum_{l=0}^{N-2} \sum_{s_1=0}^{s-1} \cdots \sum_{s_{N-l-1}=0}^{s-1} \prod_{i=1}^{N-l-1} x_i^{s_i} x_{N-l}^{1} \\ .u\left(\frac{T_{N-l}}{\sum\limits_{r=1}^{N-l-1} T_r s_r + s' T_1 + \frac{\sigma^2}{h}} - \Theta_{th} \right) \cdot \delta\left(s - 1 - \sum_{r=1}^{N-l-1} s_r \right), \tag{2}$$

where $A_{0,0} = 0$, $A_{0,1} = 1$ and $A_{1,0} = 1$. $\delta(t)$ is the Kronecker's delta function and $u(t)$ is the Heaviside function (unit step function) and are given by the following expressions:

$$\delta(t) = \begin{cases} 1, & \text{if } \ t = 0, \\ 0, & \text{otherwise} \end{cases}$$

and (3)

$$u(t) = \begin{cases} 1, & \text{if } \ t \geq 0, \\ 0, & \text{otheriwse} \end{cases}$$

2 Stackelberg Game Formulation

In the Stackleberg Game, the system is split in two non cooperative and coupled sub-games. The first game concerns the set of leaders and the second implies the followers group with the two sub-games being implicitly interdependent. The followers game is played after the leaders game, and its outcome depends on the strategies profile of the leaders. The leaders fix their optimal strategy and lets the followers optimize their own utility according to the leaders strategy.

2.1 Stackelberg Equilibrium

Define $\mathbf{q}^F = (\mathbf{q}^F_{-i}, q^F_i)$ to be a retransmission policy where each follower j with $j \in \mathcal{N}^{\mathcal{F}} \setminus \{i\}$ retransmits at any slot with probability q^F_j for all $j \neq i$ and where tagged follower i retransmits with probability q^F_i and when the maximization is taken respect to q^F_i. Let $\mathcal{NE}^*(\mathbf{q}^L)$ be the set of Nash equilibria of the followers sub-game given the leaders strategy $\mathbf{q}^L = (q^L_1, q^L_2, \ldots, q^L_{m_L})$ where q^L_i is the transmission probability of the leader i. Given that the leaders maximize their utility function which depends on the followers strategy, i.e., the Nash equilibrium. The definition of a Stackelberg equilibrium can be formally specified as follows:

Definition 1. A policies vector $\mathbf{q} = (\mathbf{q}^L, \mathbf{q}^F)$ is called s Stackelberg equilibrium if $\mathbf{q}^F \in \mathcal{NE}^*(\overline{\mathbf{q}}^L)$ and $\mathbf{q}^L$ is the Nash equilibrium for leaders group, i.e., q^L_i maximizes the utility function U_i of leader i: $q^{L*}_i \in \underset{q^L_i}{\text{argmax}}\, U_i \left((q^L_{-i}, q^L_i), \mathbf{q}^F(q^L_{-i}, q^L_i)\right)$,

We restrict to a symmetric policy $(\mathbf{q}^L, \mathbf{q}^F)$ and we identify $\mathbf{q}^L$, $\mathbf{q}^F$ with the actual retransmission probabilities which is the same for all leaders and followers. We assume that the first $m^F - 1$ followers adopt for transmission the vector of probabilities $(\mathbf{q}^F)^{-m^F} = (q^F, q^F, \ldots, q^F)$ whereas tagged follower m^F retransmits with probability p^F. Similarly, we assume that the first $m^L - 1$ followers adopt for transmission the vector of probabilities $(\mathbf{q}^L)^{-m^L} = (q^L, q^L, \ldots, q^L)$ whereas tagged leader m^L retransmits with probability p^L. Under rationality, each user seeks to maximize its own function utility.

Define the set $\mathcal{BR}^{m^F}_F((\mathbf{q}^F)^{-m^F}, (\mathbf{q}^L)^{-m^L}, p^L)$ and $\mathcal{BR}^{m^L}_L((\mathbf{q}^L)^{-m^L}, (\mathbf{q}^F)^{-m^F}, p^F)$ as the set of best response strategies of tagged follower m^F and tagged leader m^L respectively, it can be written as

$$\mathcal{BR}_F\left((\mathbf{q}^F)^{-m^F}, (\mathbf{q}^L)^{-m^L}, p^L)\right) = \underset{p^F}{\text{argmax}}\, U_{m^F}\left((\mathbf{q}^F)^{-m^F}, p^F, (\mathbf{q}^L)^{-m^L}, p^L\right) \quad (4)$$

$$\mathcal{BR}^{m^L}_L\left((\mathbf{q}^L)^{-m^L}, (\mathbf{q}^F)^{-m^F}, p^F\right) = \underset{p^L}{\text{argmax}}\, U_{m^L}\left((\mathbf{q}^F)^{-m^F}, p^F, (\mathbf{q}^L)^{-m^L}, p^L\right) \quad (5)$$

Definition 2. The policy $(\overline{\mathbf{q}}^L, \overline{\mathbf{q}}^F)$ is a symmetric Stackelberg equilibrium iff $\overline{\mathbf{q}}^F \in \mathcal{BR}^{m^F}_F\left((\mathbf{q}^F)^{-m^F}, (\mathbf{q}^L)^{-m^L}, p^L\right)$ and $\overline{\mathbf{q}}^L \in \mathcal{BR}^{m^L}_L\left((\mathbf{q}^L)^{-m^L}, (\mathbf{q}^F)^{-m^F}, p^F\right)$.

Remark 1. Where multiple Stackelberg equilibria may exist, we assume along this paper that the equilibrium with the lowest congestion level would be selected. Indeed, in this game, there are natural reasons that motivate the users to focus on a Stackelberg equilibrium that yields lowest congestion level.

To compute the performance metrics of leaders and followers, we introduce a 4-D Markov chain **P**. The first state component corresponds to the number of backlogged packets among the $m^L - 1$ leaders, and the second component is the number of backlogged packets among the $m^F - 1$ followers, the third component is the number of backlogged packets (either 1 or 0) of the tagged follower m^F and the last component is the number of backlogged packets (either 1 or 0) of the tagged leader m^L.

Let $Q_r(j, n)$ be the probability that j out of the n backlogged messages are retransmitted in the current slot. Then

$$Q_r(j, n) = \binom{n}{j} (1 - q)^{n-j} (q)^j . \tag{6}$$

Similarly, let $Q_a(j, n)$ denote the probability that j unbacklogged users generate new messages in the current slot. Thus

$$Q_a(j, n) = \binom{m - n - 1}{j} (1 - \lambda)^{m-n-j-1} (\lambda)^j . \tag{7}$$

Where m is the number of users in the system.

Replacing λ and q in expressions 6 and 7 respectively by (λ^L, q^L) and (λ^F, q^F) we get the arrival and retransmission flows for both, leaders and followers, groups. The transition probabilities $P_{(n^L,n^F,i,a),(n^L+k_1,n^F+k_2,j,b)}$ of our schemes is given in appendix 5. We note that the transition probabilities of the hierarchical scheme using a single power level is obtained from the transition probabilities defined in appendix 5 when $N = 1$.

2.2 Performance Metrics

We are particularly interested in deriving the average throughput and the expected delay of transmitted packets. We first discuss the procedure to obtain the steady state probabilities. Then we derive the expressions of the performance metrics of interest functions of the steady state equations. Denote by $\pi_{(n^L,n^F,i,a)}((\mathbf{q}^F)^{-m^F}, p^F, (\mathbf{q}^L)^{-m^L}, p^L)$ the steady state of the Markov chain. In the rest of the paper we refer to the steady state of the Markovian process by $\pi_{(n^L,n^F,i,a)}$.

Then the steady state is given by the following system:

$$\begin{cases} \bar{\pi} = \bar{\pi}\mathbf{P}, \\ \sum_{n^L=0}^{m^L-1} \sum_{n^F=0}^{m^F-1} \sum_{i=0}^{1} \sum_{a=0}^{1} \pi_{(n^L,n^F,i,a)} = 1, \\ \pi_{(n^L,n^F,i,a)} \geq 0, \forall (n^L, n^F, i, a), \end{cases} \tag{8}$$

Next we give the mean number of backlogged packet for each group as follows:

$$\begin{cases} B_L\left((\mathbf{q}^F)^{-m^F}, p^F, (\mathbf{q}^L)^{-m^L}, p^L\right) = \sum_{n^L=0}^{m^L-1} \sum_{n^F=0}^{m^F-1} \sum_{a=0}^{1} \pi_{(n^L,n^F,1,a)} \\ B_F\left((\mathbf{q}^F)^{-m^F}, p^F, (\mathbf{q}^L)^{-m^L}, p^L\right) = \sum_{n^L=0}^{m^L-1} \sum_{n^F=0}^{m^F-1} \sum_{i=0}^{1} \pi_{(n^L,n^F,i,1)} \end{cases} \tag{9}$$

Average Throughput: Based on the balance rate equation, i.e., at equilibrium the arrival and departure rates of each user are balanced, then the individual throughput of the tagged leader m^L and tagged follower m^F are given by:

$$\begin{cases} \Gamma_L\left((\mathbf{q}^F)^{-m^F}, p^F, (\mathbf{q}^L)^{-m^L}, p^L\right) = \lambda^L \sum_{n^L=0}^{m^L-1} \sum_{n^F=0}^{m^F-1} \sum_{a=0}^{1} \pi_{(n^L,n^F,0,a)} \\ \Gamma_F\left((\mathbf{q}^F)^{-m^F}, p^F, (\mathbf{q}^L)^{-m^L}, p^L\right) = \lambda^F \sum_{n^L=0}^{m^L-1} \sum_{n^F=0}^{m^F-1} \sum_{i=0}^{1} \pi_{(n^L,n^F,i,0)} \end{cases} \tag{10}$$

Expected Delay: The expected delay experienced by the packets at tagged Follower/Leader can be easily obtained using Little's result, namely

$$D_F\left((\mathbf{q}^F)^{-m^F}, p^F, (\mathbf{q}^L)^{-m^L}, p^L\right) = 1 + \frac{B_F\left((\mathbf{q}^F)^{-m^F}, p^F, (\mathbf{q}^L)^{-m^L}, p^L\right)}{\Gamma_{m^F}\left((\mathbf{q}^F)^{-m^F}, p^F, (\mathbf{q}^L)^{-m^L}, p^L\right)}. \tag{11}$$

$$D_L\left((\mathbf{q}^F)^{-m^F}, p^F, (\mathbf{q}^L)^{-m^L}, p^L\right) = 1 + \frac{B_L\left((\mathbf{q}^F)^{-m^F}, p^F, (\mathbf{q}^L)^{-m^L}, p^L\right)}{\Gamma_{m^L}\left((\mathbf{q}^F)^{-m^F}, p^F, (\mathbf{q}^L)^{-m^L}, p^L\right)}. \tag{12}$$

We note that the existence of a solution has been argued in [7]

3 Performance Evaluation

We turn now to compare the performance of the two schemes HMPL-WRA and H-WRA, taking the protocol WRA as reference. The symmetric Stackelberg equilibrium is computed in three steps as follows:

1 The set of leaders should select a retransmission profile.
2 The base station broadcasts the transmission policy of the leaders to the followers group. The best response, which results naturally in a symmetric Nash equilibrium, of the followers is then computed.
3 Once the base station gets the followers decision, the leaders group checks their selected profile; if it is not a symmetric Nash equilibrium, they should update their strategies profile by unilateral deviation till getting absorbed by a symmetric equilibrium.

3.1 Numerical Results

We perform here the numerical results with $m = 10$ users ($m^L = 3$ and m^F=7). Each user acts selfishly and maximizes its own throughput. We consider the throughput as the utility function. Similar trends are obtained when minimizing expected delay. We set $\Theta_{th} = 4.8dB$, $\epsilon = 10^{-4}$ and consider four selectable power levels T = [1, 5, 25, 125] mWatts according to the distribution [0.25, 0.25, 0.25, 0.25].

Figure 1 depicts the global throughput and the expected delay at Stackelbreg equilibrium (SE) and Nash equilibrium (NE) as a function of the packet arrival rate. We remark that the equilibrium throughput of the hierarchical scheme without power control outperforms the Nash-WRA for low and average loads. We remark also that the expected delay of the hierarchical scheme is efficiently reduced for low loads. However in both cases the expected delay increases exponentially as the network load increases due to an increase on the number of channel collisions. We note that either in terms of aggregate throughput and the expected delay, the hierarchical scheme HMPL-WRA outperforms the H-WRA and Nash-WRA schemes for all loads. Indeed the equilibrium throughput of HMPL-WRA is unimodal function of arrival probability, at low load values it increases until achieving a maximum throughput of $\Gamma_{max} \simeq 0.6$ at $\lambda \simeq 0.27$, for HMPL-WRA and $\Gamma_{max} \simeq 0.32$ at $\lambda \simeq 0.12$ for H-WRA and Nash-WRA. An important result is that the expected delay of the hierarchical scheme HMPL-WRA is efficiently reduced whereas the expected delay of the H-WRA and Nash-WRA increases exponentially as the network load increases.

Figure 2 shows the retransmission probability at SE and NE as function of the packets arrival rate. From Figure 2(a) it is clear that under the HMPL-WRA, the followers are less aggressive which results on the efficiency gain of the game formulation with hierarchical scheme with multiple power levels. Under the H-WRA

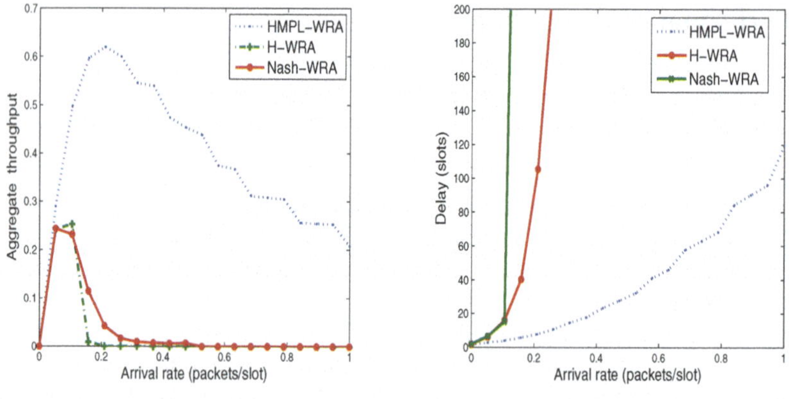

(a) (b)

Fig. 1 The global throughput and the mean expected delay as function of the arrival rate

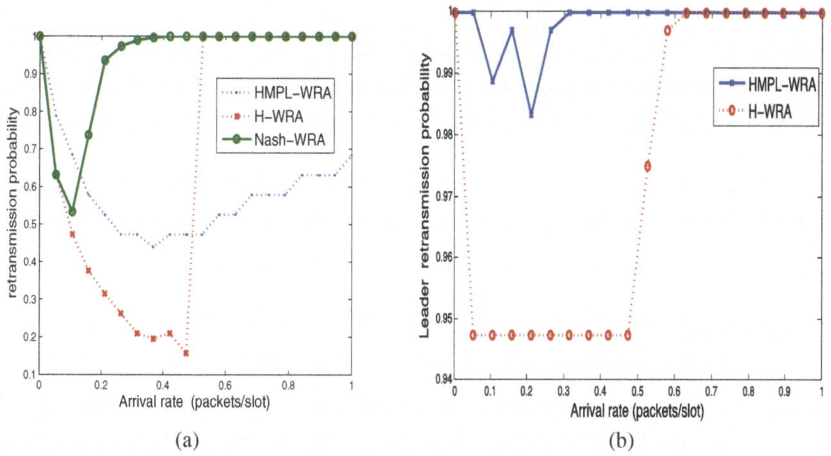

Fig. 2 The equilibrium retransmission probability as function of the arrival rate

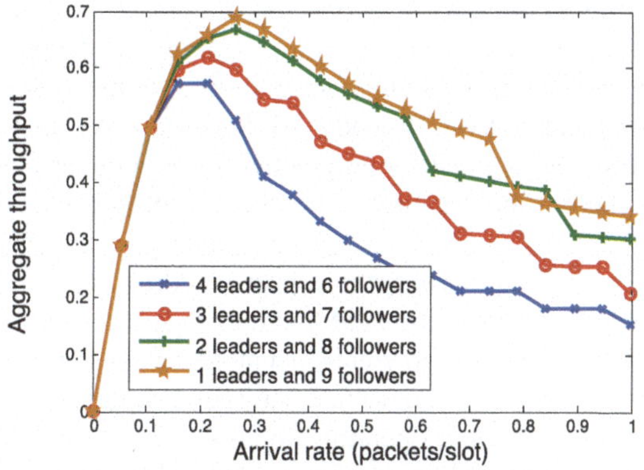

Fig. 3 The global throughput under different partitions of users

we see that the users become more and more aggressive as the arrival probability increases (at $\lambda = 0.5$), which leads to a dramatic decrease in the network throughput and the respective huge delay. Moreover, the equilibrium retransmission drastically increases to 1 at heavy load.

Figure 2(b) depicts the equilibrium retransmission probability of leaders. We note that under our new scheme the leaders choose to transmit at a probability close to one., i.e., to take over the channel, thus reducing the success rates of all the other users. Such result has been argued in [7]. But thanks to the capture effect, the power diversity solution helps to reduce the congestion level over the channel and also it breaks the impact due to the selfish behaviors of followers as they become less aggressive. This is due to the fact that the follower transmits using the highest power level, play the role of "dominant station" during that slot. Since the power level is chosen randomly, this role is fairly shared by all the followers in the subsequent slots.

Next we plot in Figure 3 the aggregate throughput for different partitions of users (followers and leaders). From the figure we remark that the aggregate throughput is improved with a decreasing number of leaders. Hence the scheme with only one leader outperforms all other schemes with different partitions. Which means that the case of a single leader is the best setting in our game. This result can be explained by the fact that in our game the leader become a maximiser of the contention game, when the followers competitively maximize their individual throughput [10].

4 Conclusion

We have proposed a general hierarchical game where multiple leaders coexist and competitively maximize their individual throughput. We have shown that the Stackelberg game contention with single power level leads to a networks collapse in particular at heavy load. In order to keep the performance system reasonable and take advantage from hierarchy we have developed a power diversity solution to enhance the performance metrics. Under the proposed scheme the performance, either in terms of throughput and delay is improved compared to the system with hierarchy using single power level and for all loads. We have observed that the hierarchical system with power diversity is performing better as the number of leaders is small, which means that there is an existence of optimal partition between all users in the networks.

As perspective, we propose to study the impact of adding a cost on the performance of new scheme. Indeed we have observed that the leaders act selfishly and transmit with probability close to one, which yields to a more energy consumption. To do so, we study the effect of adding cost for each transmission attempts, which is alternative mechanism to obtaining efficient operating solutions. The cost could be expressed in terms of the energy consumption which is a scare resource.

References

1. Tan, G., Guttag, J.: The 802.11 mac protocol leads to inefficient equilibruim. In: Proceeding of the 24th Annual Joint Conference of IEEE Computer and Communications Societies (INFOCOM 2005), vol. 1, pp. 1–11, Miami, FL, USA, March 2005
2. El-Azouzi, R., Sabir, E., Jimènez, T., Bouyakhf, E.H.: Modeling slotted aloha as a stochastic game with random discrete power selection algorithms. International Journal of Computer Systems, Networks, and Communications (2009)
3. Altman, E., Barman, D., El-Azouzi, R., Jimenèz, T.: A game theoretic approach for delay minimization in slotted aloha. In: Proceedings of the IEEE International Conference on Communications, Paris, France, June 2004
4. MacKenzie, A.B., Wicker, S.B.: Selfish users in aloha: a game theoretic approach. In: Proceedings of the Fall 2001 IEEE Vehicular Technology Conference (2001)
5. Stackelberg, V.: Marketform und gleichgewicht. Oxford University Press (1934)

6. Bloem, M., Alpcan, T., Basar, T.: A stackelberg game for power control and channel allocation in cognitive radio networks. In: Proceedings of the ACM/ICST GameComm Workshop, Nantes, France, October 2007
7. Sabir, E., El-Azouzi, R., Hayel, Y.: Hierarchy sustains partial cooperation and induces a braess-like paradox in slotted Aloha-based networks. Computer Communications
8. Karouit, A., Sabir, E., Ramirez-Mireles, F., Barbosa, L.O., Haqiq, A.: A Team study of a Multiple power Wireless Random Channel Acces Mechanism with Capture effect. International Mathematical Problems in Engineering Journal Hindawi **2013**, Article ID 187630, 16 pages (2013)
9. Karouit, A., Sabir, E., Ramirez-Mireles, F., Barbosa, L.O., Haqiq, A.: A Stochastic Game Analysis of Binary Exponential Backoff Algorithm with Multi-Power Diversity and Transmission cost. International Journal of Mathematical Modelling and Algorithms. Springer, May 10, 2012
10. Karouit, A., Ramirez-Mireles, F., Barbosa, L.O., Haqiq, A.: Hierarchical Slotted Wireless Random Channel Access with Power Control. International Mathematical Problems in Engineering Journal Hindawi (in press, 2015). Englewood Cliffs, New Jersey (1987)

5 Transition Probabilities of General Hierarchical Game with Power Diversity

– $P_{(n^L,n'^F,i,a),(n^L-1,n'^F,j,b)}=$

$$\begin{cases} (1-\lambda^L)(1-\lambda^F)\psi\,(0,0,1,0) & a=b=0, i=j=0 \\ (1-\lambda^L)\lambda^F psi\,(0,0,1,0) & a=b=0, i=0, j=1 \\ (1-\lambda^L)(1-p^F)\psi\,(0,0,1,0) & a=b=0, i=j=1 \\ \lambda^L(1-\lambda^F)\psi\,(0,0,,1,0) & a=0, b=1, i=j=0 \\ \lambda^L\lambda^F\psi\,(0,0,1,0) & a=0, b=1, i=0, j=1 \\ \lambda^L(1-p^F)\psi\,(0,0,1,0) & a=0, b=1, i=j=1 \\ (1-q^L)(1-\lambda^F)\psi\,(0,0,1,0) & a=b=1, i=0=j=0 \\ (1-q^L)\lambda^F\psi\,(0,0,1,0) & a=b=1, i=0, j=1 \\ (1-p^L)(1-p^F)\psi\,(0,0,1,0) & a=b=1, i=j=1 \end{cases}$$

– $P_{(n^L,n'^F,i,a),(n^L,n'^F-1,j,b)} =$

$$\begin{cases} (1-\lambda^L)(1-\lambda^F)\psi\,(0,0,0,1) & a=b=0, i=j=0 \\ (1-\lambda^L)\lambda^F\psi\,(0,0,0,1) & a=b=0, i=0, j=1 \\ (1-\lambda^L)(1-p^F)\psi\,(0,0,0,1) & a=b=0, i=j=1 \\ \lambda^L(1-\lambda^F)\psi\,(0,0,0,1) & a=0, b=1, i=j=0 \\ \lambda^L\lambda^F\psi\,(0,0,0,1) & a=0, b=1, i=0, j=1 \\ \lambda^L(1-p^F)\psi\,(0,0,0,1) & a=0, b=1, i=j=1 \\ (1-q^L)(1-\lambda^F)\psi\,(0,0,0,1) & a=b=1, i=0=j=0 \\ (1-q^L)\lambda^F\psi\,(0,0,0,1) & a=b=1, i=0, j=1 \\ (1-p^L)(1-p^F)\psi\,(0,0,0,1) & a=b=1, i=j=1 \end{cases}$$

- If $0 \le k_1 \le m^L - n^L - 1$ and $0 \le k_2 \le m^F - n^F - 1$,

$$P_{(n^L,n'^F,i,a),(n^L+k_1,n'^F+k_2,j,b)} =$$

$$\begin{cases}
(1-\lambda^L)(1-\lambda^F)[\phi(k_1,k_2,0,0,0)+\psi(k_1+1,k_2,1,0)+ & \\
\psi(k_1,k_2+1,0,1)], & i=j=0, a=0, b=0 \\
(1-\lambda^L)\lambda^F[\phi(k_1,k_2,0,0)+ & \\
\psi(k_1+1,k_2,1,0)+\psi(k_1,k_2+1,0,1)], & i=j=0, a=0, b=1 \\
(1-\lambda^L)p^F\xi(k_1+1,k_2,0,1), & i=j=0, a=1, b=0 \\
(1-\lambda^L)[p^F\phi(k_1,k_2,0,1,1)+(1-p^F)(\phi(k_1,k_2,0,0,0)+ & \\
\psi(k_1+1,k_2,1,0)+\psi(k_1,k_2+1,0,1))], & i=j=0, a=b=1 \\
\lambda^L(1-\lambda^F)[\phi(k_1,k_2,0,0,0)+\psi(k_1+1,k_2,1,0)+ & \\
\psi(k_1,k_2+1,0,1)] & i=0, j=1, a=b=0 \\
\lambda^L\lambda^F[\phi(k_1,k_2,0,0,0)+\psi(k_1+1,k_2,1,0)+ & \\
\psi(k_1,k_2+1,0,1)], & i=0, j=1, a=0, b=1 \\
\lambda^L p^F\xi(k_1,k_2,0,1), & i=0, j=1, a=1, b=0 \\
\lambda^L[p^F\phi(k_1,k_2,0,1)+(1-p^F)(\phi(k_1,k_2,2,0,1)+ & \\
\psi(k_1+1,k_2,1,0)+\psi(k_1,k_2+1,0,1))], & i=0, j=1, a=b=1 \\
p^L(1-p^F)\xi(k_1,k_2,1,0), & i=1, j=0, a=b=1 \\
p^L(1-\lambda^F)\xi(k_1,k_2,1,0), & i=1, j=0, a=b=0 \\
p^L\lambda^F\xi(k_1+1,k_2,1,0), & i=1, j=0, a=0, b=1 \\
(1-\lambda^F)[p^L\phi(k_1,k_2,1,0,1)+(1-p^L)(\phi(k_1,k_2,2,0,0)+ & \\
\psi(k_1+1,k_2,1,0)+\psi(k_1,k_2+1,0,1))], & i=j=1, a=b=0
\end{cases}$$

$$\begin{cases}
\lambda^F[p^L\phi(k_1,k_2,1,0,1)+(1-p^L)(\phi(k_1,k_2,0,0,0)+ & \\
\psi(k_1+1,k_2,1,0)+\psi(k_1,k_2+1,0,1))], & i=j=1, a=0, b=1 \\
p^F(1-p^L)\xi(k_1,k_2,0,1), & i=j=1, a=1, b=0 \\
(1-p^L)(1-p^F)[\phi(k_1,k_2,2,0,0)+\psi(k_1+1,k_2,1,0)+\psi(k_1,k_2+1,0,1)]+ & \\
p^L(1-p^F)\phi(k_1,k_2,0,0,1)+p^F(1-p^L)\phi(k_1,k_2,0,1,0)+ & \\
p^F p^L Q_a(k_2,n^F)Q_a(k_2,n^L), & i=j=1, a=b=1
\end{cases}$$

The probabilities ψ, ϕ and ξ are given as follow:

$$\psi(k_1,k_2,i,j) = Q_a(k_2,n^F)Q_a(k_1,n^L)\sum_{s=i}^{n^L}\sum_{t=j}^{n^F}\frac{s.i+t.j}{s+t}Q_r(t,n^F)Q_r(s,n^L)A_{s+i,t+j} \quad (13)$$

$$\phi(k_1,k_2,x,i,j) = Q_a(k_2,n^F)Q_a(k_1,n^L)\sum_{s=i}^{n^L}\sum_{t=j}^{n^F}Q_r(t,n^F)Q_r(s,n^L)\left(1-A_{s+i,t+j}\right)\delta(i+j\geq x) \quad (14)$$

$$\xi(k_1,k_2,i,j) = Q_a(k_2,n^F)Q_a(k_1,n^L)\sum_{s=i}^{n^L}\sum_{t=j}^{n^F}Q_r(t,n^F)Q_r(s,n^L)A_{s+i,t+j} \quad (15)$$

A Pricing-Based Spectrum Leasing Framework with Adaptive Distributed Learning for Cognitive Radio Networks

Sara Handouf, Essaid Sabir and Mohammed Sadik

Abstract In this paper, we consider the decentralized scenario of spectrum leasing, whereby a primary user (PU) who owns the spectrum resource, may lease a part of her licensed spectrum to a secondary (SU). We propose a pricing-based spectrum leasing framework between the PU and the SU. The spectrum leasing problem can be depicted by a non-cooperative game where: on one hand, the PU plays the seller and attempts to maximize its own utility by setting the price of spectrum. On the other hand, the SU (i.e., the buyer) has to decide whether to accept the leasing offer or to decline it, while seeking to maximize her own utility. Next, we characterize the Nash equilibria of the induced game for both pure strategies and mixed mixed strategies. We also propose numerous learning algorithms that allow cognitive users to learn their optimal strategies and payoffs for both continuous and discontinuous actions. Simulation results evaluate our model and show the behaviour (accuracy and speed of convergence) of the proposed learning algorithms.

Keywords Cognitive radio · Spectrum leasing · Nash equilibrium · Automatic learning

1 Introduction

The concept of spectrum leasing in cognitive radio networks is currently under investigation as a promising paradigm to achieve efficient spectrum utilization. That prescribes the coexistence on the same spectral resource of primary user (licensed user) and secondary user. Unlike previous works, our paper tackles a new spectrum leasing framework. For this, a game-theoretic approach to spectrum leasing is proposed as an effective tool to model interactions between PU and SU. In the same context, distributed learning algorithms are introduced leading users to converge to their Nash Equilibrium status.

S. Handouf(✉) · E. Sabir
UBICOM Research Group, ENSEM, Hassan II University, Casablanca, Morocco
e-mail: {sara.handouf,e.sabir}@ensem.ac.ma

M. Sadik
COMSYS Research Group, ENSEM, Hassan II University, Casablanca, Morocco
e-mail: m.sadik@ensem.ac.ma

E. Sabir et al. (eds.), *The International Symposium on Ubiquitous Networking*, Lecture Notes in Electrical Engineering 366,
DOI: 10.1007/978-981-287-990-5_4

1.1 Literature Review

In this section we review some notable works on spectrum leasing and automatic learning,

Authors in [1] look for optimizing opportunistic sensing in cognitive radio when SU has the choice between accessing licensed or unlicensed channels. [6] proposes a pricing based spectrum leasing framework. In [7], [11] and [9] cooperative relaying for cognitive radio networks is investigated with a leasing strategy in exchange of SUs' power. [8] draw a cooperative communication-aware spectrum model. Cooperative and non-cooperative sharing methods are analyzed and compared in [10]. [12] presents a leasing model that enhances the profit of secondary operator and spectrum owner. the paper [2] is about developing cognitive sensing and access strategies that learn from observations. Learning algorithms are proposed in [3], [4] when authors of [5] study the channel selection issue of SUs and propose some algorithms to perform the accessibility check based on measurements of primary signals.

1.2 Purpose of this Paper and Contribution

Unlike previous works, our paper tackles a new spectrum leasing framework. The main contributions are summarized as follows.

First we model the spectrum leasing problem as a seller-buyer game.

Second we propose adaptive learning schemes that enable users to learn theirs strategies and payoffs.The convergence property of the algorithms to a pure strategy NE point as well as mixed strategy NE is verified theoretically as well as numerically.

1.3 Organization of the Paper

The paper is organized as follows. In Section 2 we present the system model. In Section 3 we present and study pure NASH EQUILIBRIUM status. Mixed NASH EQUILIBRIUM is investigated in section 4. Section 5 studies the equilibrium efficiency. Section 6 proposes learning algorithms for both pure and mixed Nash equilibrium. Simulations are illustrated in section 7. Finally, a conclusion is drawn in section 8.

2 System Model and Problem Formulation

In the following, we detail the proposed game-theoretic model of spectrum leasing and the main system parameters. Consider a cognitive radio network consisting of one PU and one SU as shown in fig.1. We assume that there is only one channel with capacity B however it is under-utilized, i.e., there exist idle periods when it is not being used by the PU. The case for multiple channels with variable capacities and multiple SUs is in the list of our future work. The primary user occupies the channel just in a proportion of time denoted by the probability of activity π and in order to enhance his revenues, PU being the licensed owner of the channel may lease the idle periods to SU and propose a pricing-based spectrum leasing framework with some fee f, we assume existence of a ceiling price f_{max}. The PU for its part, has to pay an indemnity α_p which is the cost of spectrum leasing, the purpose of this cost is to make the PUs' spectrum leasing proposition efficient. Note that the SUs can chose either to accept the leasing and buy the service or not. If it decides to buy, it uses the channel directly with the same throughput as the PU. whereas if it does not accept the PU's leasing proposition it senses the channel and opportunistically utilize it when it is not being used by the PU and in this case the channel is used for free, the cost is only the energy spent for sensing. We denote by P_a the probability that the SU accepts to buy the licensed channel in idle period. This probability may be considered as the proportion of time when SU chooses to accept to pay the fee. In case if it decides to sense, we assume that α_s is the cost for sensing the licensed channel(here SU has to be energy efficient).We define here a parameter P_{fa} which represent the false alarm probability when sensing and finding the PU active. The system model is depicted in Figure 1 and is composed of parts. The first one, namely Part I, represents the primary part of the system which is licensed for PU and open for SU' opportunistic access, and the secondary one, namely part II represents the secondary part of the system. By differentiating the payoff functions, we obtain discontinuous utility function for each user. let us begin by defining the utility function for the primary user. Two cases occur, the first one when SU accepts the leasing proposition; hence the PU can make a benefit thanks to the price of service. whereas the second case is when rejects the leasing propositions and choose to sense himself the channel to opportunistically use it thus the PU will not make benefit but lose the energy of making proposition.

The payoff function obtained whether the secondary accepts the service proposition or not, is represented in the following discontinuous utility function as follow.

$$U_p = \begin{cases} (1-\pi).(f-\alpha_p) \text{ if } SU \ \ accepts \ \ leasing \\ -\alpha_p.(1-\pi).f \ \ \text{if } SU \ \ rejects \ \ leasing \end{cases} \tag{1}$$

Where α_p is a positive number representing the indemnity the PU pays for proposing. Thus, regardless of whether the proposition is accepted or not α_p is taxed.

π is the probability that the primary player is active(is transmitting over the given channel), this Probability can be computed using standard tools of Markov theory.

The primary user switches between two actions, from active to inactive and the inverse with the probability p (resp, q). Applying the Markov Model Methodology we obtain the following expression of π

$$\pi = \frac{q}{p+q}$$

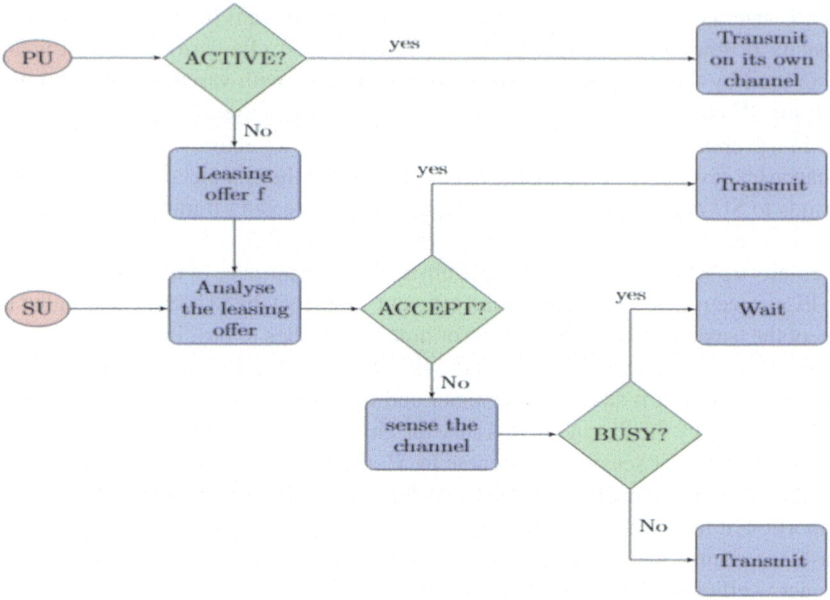

Fig. 1 System model.

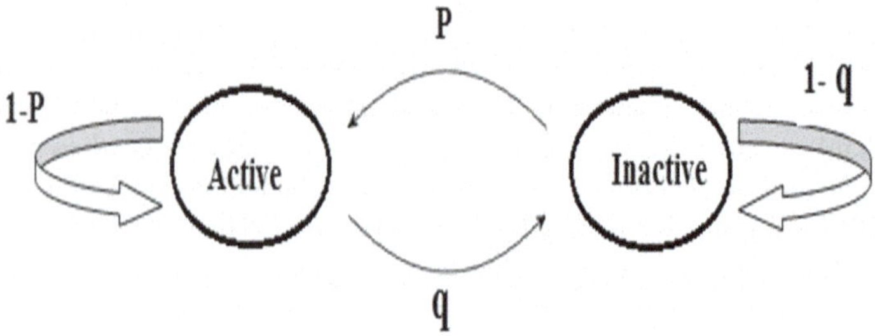

Fig. 2 Markov chain of primary activity

Now, let us define the utility function for the secondary user. If the Secondary user chooses to pay for the proposed solution, thus it will gain direct access to the licensed channel without sensing, whereas if it chooses not to pay for the service, there will not be any loss of money, nonetheless the SU will lose the energy α_s dissipated on sensing the channel and when this channel is idle it can transmit trough it. The payoff function is represented by the following discontinuous utility function

$$U_s = \begin{cases} (1-\pi).(B-f) & \text{if } SU \ \ accepts \ \ leasing \\ (1-\pi).(1-P_{fa(\tau)})B - \alpha_s & \text{if } SU \ \ rejects \ \ leasing \end{cases} \quad (2)$$

B here is the throughput gain obtained when paying the fee f. (we assume for simplification that B is in a currency unit).and $P_f a(\tau)$ is defined as probability of false alarm when sensing and finding the channel busy, τ is the period of sensing.

3 Pure Nash Equilibrium Analysis

In this section, we consider a non-cooperative game involving two players, primary user referred to, by PU and secondary user referred to, by SU, and we study their pure strategy Nash equilibrium. In game theory, The Nash Equilibrium is defined as the set of strategies according to which no player can benefit by unilaterally changing his strategy.PU and SU decide on how to win more and lose less. Their payoff functions are the profit they make (we assume for simplicity that the payoff functions are in a currency unit).

With the assumption of f is between 0 and f_{max}, we can define the best response of PU in both cases when SU chooses to lease channel and when it prefer sensing.

Lemma 1. *The strategy profile (Accept, f_{max}) is a pure Nash Equilibrium if the following condition holds*

$$P_{fa(\tau)} \geq \frac{\alpha_s + (1-\pi) f_{max}}{(1-\pi)B} \quad (3)$$

Proof. the best response of PU while the SU strategy is accepts the leasing, is obtained with

$$B_U = argmax(Up = (1-\pi).(f - \alpha_p), withf = f_{max}) \quad (4)$$

clearly this function is linear with f then $f^E = f_{max}$ and the result follows

$$B_U(accept) = f_{max} \quad (5)$$

Lemma 2. *The strategy profile (Rejects, $f = 0$) is a pure Nash Equilibrium if the following condition holds*

$$P_{fa(\tau)} \leq \frac{\alpha_s}{(1-\pi)B} \quad (6)$$

Proof. When the SUs' action is reject leasing, the best response is the maximum value of U_p, thus

$$B_U = argmaxUp = -\alpha_p.(1-\pi).f, \quad (7)$$

this function is linear and decreasing with f, then the maximum is obtained with $f = 0$, and the result follows

$$B_U(reject) = 0 \tag{8}$$

4 Mixed Nash Equilibrium Analysis

In this section we formulate the expected utility functions for either PU and SU as continuous functions where the actions are the fee f and the probability of acceptance P_a.In the mixed strategy, each player has a probability distribution over the two possible actions, i.e., player i can choose to pay f with probability P_a, and choose to sense and not to pay with probability $1 - P_a$. At first consider the SU: At mixed strategy Nash equilibrium in both cases SU accepts and SU rejects the utility function should have the same expected payoff.

- If SU chooses to accept proposition:

It will receive a payoff with probability P_a.Therefore its expected payoff $U_p(a)$ from accepting leasing proposition is

$$U_s(a) = P_a(1 - \pi)(B - f) \tag{9}$$

.

- If SU chooses not to accept:

it will receive a payoff with probability $(1 - P_a)$. Therefore its expected payoff $U_s(r)$ from rejecting leasing proposition is

$$U_s(r) = (1 - P_a)[(1 - P_{fa(\tau)})B - \alpha_s] \tag{10}$$

Using the fact that the fee f of mixed strategy must yield the same payoff to SU at the equilibrium , $U_s(a) = U_s(r)$ Therefore SU will mix between the two strategies only if these two expected payoffs are the same, and:

$$f^E = B - \frac{(1 - \pi)(1 - P_{fa(\tau)})B + \alpha_s}{1 - \pi} \tag{11}$$

Next lets consider PU, the mixed strategy Nash equilibrium is f^E and P_a^E where f^E takes the value predetermined above. The mixed Nash equilibrium payoff is obtained with:

$$U_p = P_a(1 - \pi)(f - \alpha_p) - (1 - P_a)\alpha_p(1 - \pi)f \tag{12}$$

In the equilibrium state, all players (here PU and SU) choose strategies that maximize their expected payoff.

Proposition: There exists a unique fully mixed Nash Equilibrium, that is the solution to:

$$\frac{\partial U_p}{\partial f} = 0 \tag{13}$$

We have found the equilibrium probability P_a^E as follows

$$P_a^E = \frac{\alpha_p}{1+\alpha_p} \tag{14}$$

5 Equilibrium Efficiency: Price of Anarchy

In order to evaluate the efficiency of Nash equilibria of the non-cooperative leasing game, we are consider as a metric the well-known Price of Anarchy (PoA). Yet, PoA measures the gap of performance between the centralized and the decentralized setting. It is explicitly defined as the ratio between the optimal utility or the maximum achievable social utility U_T (obtained when considering a centralized setting) versus the worst social utility achieved when PU and SU selfishly maximize their respective profits and reach a Nash equilibrium (decentralized setting). In other words, PoA help to compare the performance of users at the worst NE, taking as benchmark the best performance of the global system. In our context, we define the PoA as follows:

$$PoA = \frac{U_T(p_{opt}, f_{opt})}{U_T(p^E, f^E)} \geq 1 \tag{15}$$

The optimal values of the fee and the probability of acceptance when considering a centralized system are obtained through the maximum achievable social utility which are simply given by:

$$f_{opt} = \frac{\alpha_p(1-\pi) - P_{fa(\tau)}(1-\pi)B - \alpha_s}{\alpha_p(1-\pi)} \tag{16}$$

and

$$P_{a.opt} = 1 \tag{17}$$

by finding the optimal values that maximise the social utility U_T, the maximun can be calculated as follows

$$U_{T.opt} = (1-\pi)(B-\alpha_p) \tag{18}$$

We characterize the worst-case PoA in terms of the fee payed and the probability of acceptance P_a. After some algebra (while taking $B = 1$ and $\alpha_s \approx \alpha_p$), the expression obtained for the PoA is the following:

$$PoA = \frac{X}{X - P_{fa(\tau)}(1 + \alpha_p - \pi) - \alpha_p \alpha_s} \tag{19}$$

where

$$X = (B - \alpha_p)(1 - \pi) + \alpha_p(B - \alpha_p)(1 - \pi).$$

Analyzing the expression above, we easily show that the difference between the numerator and the denominator of the is negative. Therefore, the Price of Anarchy is greater than 1, and as expected we have some loss in efficiency due to due to selfish behavior of users. Now, this quantity can be efficiently controlled by fine-tuning parameters such B and α_p.

6 Distributed Learning Algorithms

In this section we introduce learning schemes depending on equilibrium and strategy natures. The goal is that users learn their own payoffs and determine the fee f(for PU) and the probability of acceptance Pa(for SU)that lead to the equilibrium state. To accomplish this, we propose adaptive learning schemes.

Learning is a fundamental component of intelligence and cognition, it is defined as the ability of synthesizing the acquired knowledge through practice, training, or experience in order to improve the future behavior of the learning agent.

6.1 Learning Scheme for Primary User (Continuous Action)

In our leasing game, the proposed fee represents a continuous action, one of the suitable class of learning schemes is the Combined Fully Distributed Payoff and Strategy learning (CODIPAS) for continuous actions.

Let f_t be the leasing fee, proposed at time $t.f_t^\star$ is an intermediary variable. a ω ϕ represent the amplitude, frequency and phase of the sinus perturbation signal respectively [5]. $U_{p,t}$ represents the reward or the utility of primary user at time $t+1$. At time t primary user updates the intermediary variable $f_t^\star$ and makes the leasing proposition with f_t.the fee for the next round f_{t+1} is then updated. The convenient algorithm is provided below.

With $t^\star = \sum_{t'=1}^{t} \lambda_t$ and $\lambda_t = \frac{1}{1+t}$.

Algorithm 1. PU:Nash equilibrium seeking algorithm

1.initialisation:

Primary user initializes $f_0^\star$ and makes a leasing proposition;

2.Learning pattern:

Observes the payoff $U_{P,t}$;
$f_{t+1}^\star = f_t^\star + \lambda_t a sin(\omega t^\star + \phi) U_{p,t}$
$f_{t+1} = [f_{t+1}^\star + a sin(\omega t^\star + \phi)]_0^{f_{max}}$;

6.2 *Learning Scheme for Secondary User (Discrete Actions)*

For the secondary user we differentiate equilibrium states (Pure and mixed equilibrium) using the convenient classes of learning schemes.

Learning Scheme Leading to Pure NE. In this context a convenient learning scheme is provided that in the class of linear reward inaction LRI algorithm, in which the learner (SU) considers its action and payoff from the environment, and ignores any other information [1]. That can guarantee the convergence to the Nash equilibrium with pure strategies. The linear reward inaction for SU is defined in algorithm 3:

Algorithm 2. SU:Linear Reward Inaction(LRI)

Learning pattern:
for each action $r \in$ ("$Accept$", "$Reject$") do;
$Pa_{t+1}^r = Pa_t^r + \epsilon_t U_{s,t}(Ind_{a_t=r} - Pa_t^r)$

Where $a_t = r$ is the action chosen by SU at round t and $Ind_{a_t=r}$ denotes the indicator function and $\epsilon_t = \frac{1}{1+t}$.

Learning Scheme Leading to Mixed NE. Based on learning by imitation, the imitative combined fully distributed payoff and learning strategy (imitative CODIPAS)is the convenient class of learning scheme, where SU learns his probability of acceptance and payoff by observing his previous actions. Su estimates randomly his payoff at time iteration t to make a strategy for the next iteration. Let Pa_{t+1}^r be the probability of SU to choose action r at round $t+1$. λ_t is learning rate for strategy-dynamics and μ_t represents the learning rate for payoff-dynamics. $\hat{U}_{s,t}$ is the estimated payoff at time t [1]. the algorithm is given as follows:

Algorithm 3. SU:Imitative CODIPAS

1.initialisation:
for Secondary user SU do:

$\hat{U}_{s,0} = (\hat{U}^1_{s,0}, \hat{U}^2_{s,0})$;
$Pa_0 = (Pa^1_0, Pa^2_0)$;
Define the sequence up to T:$\lambda, \mu, t \in 1, 2, ..., T$;
2.Learning pattern:for secondary user SU do
for each action $r \in 1, 2$ do;
$Pa^r_{t+1} = \frac{Pa^r_t (1+\lambda_t)^{\hat{U}^r_{s,t}}}{\sum_{r' \in 1,2} Pa^{r'}_t (1+\lambda_t)^{\hat{U}^r_{s,t}}}$
$U^r_{s,t+1} = U^r_{s,t} + \mu_t Ind_{a_t=r}(U^r_{s,t} - \hat{U}^r_{s,t})$
Where $a_t = r$ is the action chosen by SU at round t and $Ind_{a_t=r}$ denotes the indicator function.

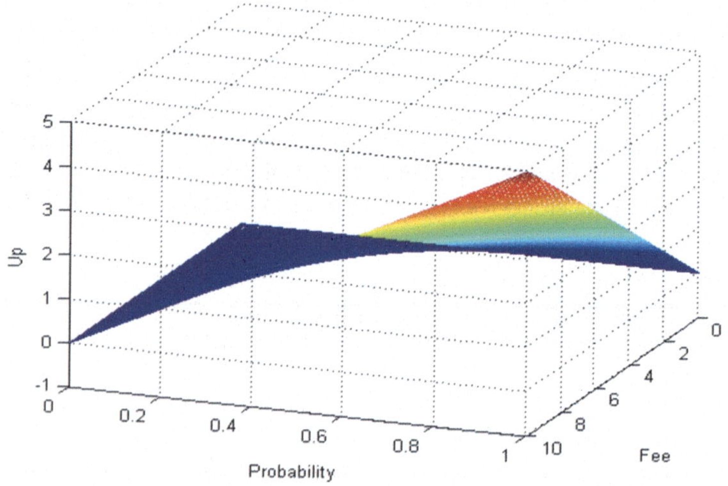

Fig. 3 Payoff function of PU

7 Performance Evaluation

We now present some simulation results to check the efficacy of the proposed schemes. The false alarm probability i.e., the probability that the channel is sensed to be busy while it is actually idle is set for example to 0, 5. The parameters of simulations are set to be: $\pi = 0, 5$, $\alpha_p = \alpha_s = 0, 001$, $\phi = 0$, $a = 1$.

Figures 3 and 4 depict the payoff functions for PU (resp. SU) with respect to variables Pa and $fee \in (0, f_{max})$.

We turn now to our learning algorithms, we run the simulations for 10000 iterations(T=10000) for Pure NE and 5000 iterations for mixed NE. Figure 5. illustrates evolutions of CODIPAS scheme for primary user and LRI for secondary, with time

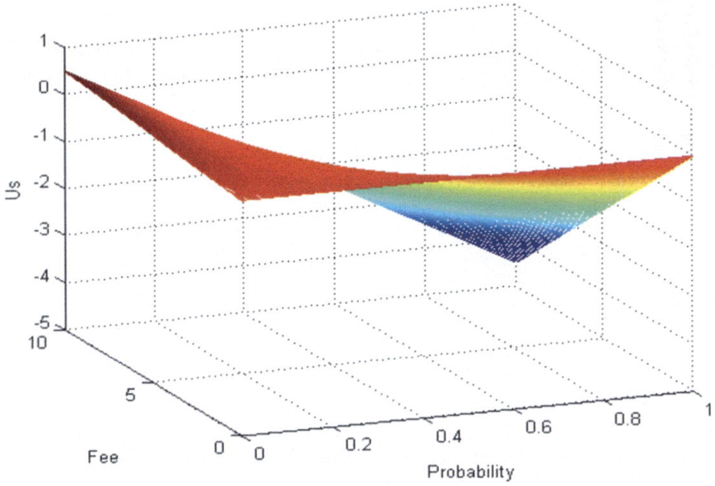

Fig. 4 Payoff function of SU

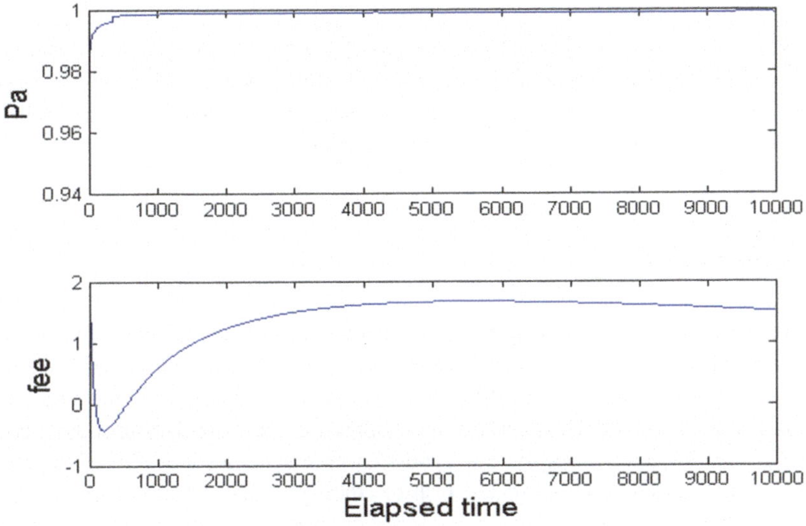

Fig. 5 convergence of CODIPAS and LRI to pure NE

iterations. After some rank, either the probability Pa and the proposed fee f converge to constant values, that are exactly the pure Nash equilibrium status when the pure strategy is (accept, f_{max}). The simulations is repeated several times, and the convergence is always guaranteed proving robustness of the proposed algorithms.

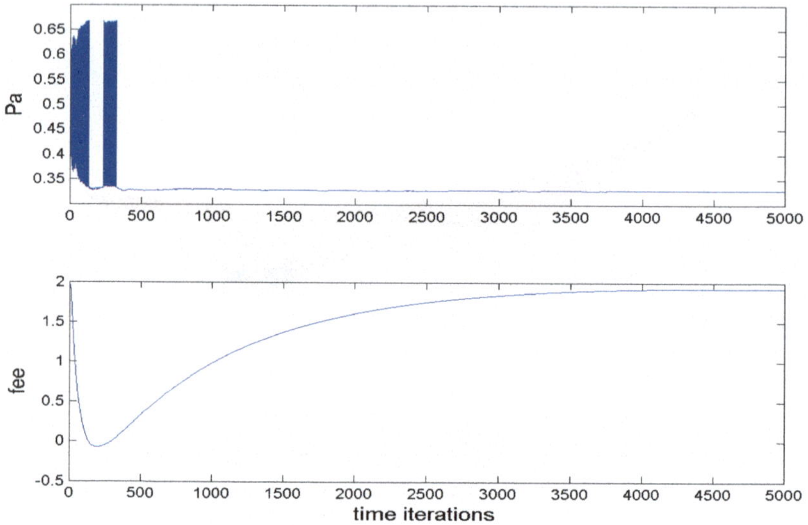

Fig. 6 convergence of CODIPAS and imitative CODIPAS to mixed NE

Figure 6. provides an illustrative example of Pa and f curves convergence, under CODIPAS scheme and Imitative CODIPAS. It discovers the mixed equilibrium status, assisting our finding of unique mixed equilibrium. Then the proposed scheme is an appropriate one to reach the mixed Nash equilibrium.

8 Conclusion

In this paper we have studied a pricing-based spectrum leasing problem. We considered a network with one spectrum owner, with licensed band to lease and one secondary user willing to lease spectrum to enhance its throughput. We have modeled the interaction between the spectrum owner(PU) and the secondary user as a seller-buyer game. We have proposed numerous learning schemes that allow both users to learn, in a completely distributed fashion, their payoffs and strategies. We have shown with both analysis and extensive simulations that the equilibrium state can be efficiently reached, our distributed algorithms are accurate and fast.

References

1. Habachi, O., Hayel, Y.: Optimal opportunistic sensing in cognitive radio networks. IET Communications **6**(8), 797–804 (2012)
2. Sabir, E., Haddad, M., Tembine, H.: Joint strategic spectrum sensing and opportunistic access for cognitive radio networks. In: 2012 IEEE Global Communications Conference (GLOBECOM), pp. 1368–1373, December 3–7, 2012
3. Lu, X., Schwartz, H.M.: Decentralized learning in two-player zero-sum games: a LR-I lagging anchor algorithm. In: American Control Conference (ACC), 2011, San Francisco, CA, pp. 107–112 (2011)
4. Bouferda, S., Sabir, E., Hayar, A., Rifi, M.: Equilibrium sensing time for distributed opportunistic access incognitive radio networks. In: Proceedings of the 16th ACM International Conference on Modeling, Analysis and Simulation of Wireless and Mobile Systems(MSWiM 2013). ACM, New York, pp. 229–236 (2013)
5. Liu, X., Shankar, N.S.: Sensing-based opportunistic channel access. Mobile Networks and Applications **11**(4), 577–591 (2006)
6. Wang, X., Guan, X., Han, Q., Liu, Z., Ma, K.: A Stackelberg Game for Spectrum Leasing in Cooperative Cognitive Radio Networks. International Journal of Automation and Computing, 125–133 (2014)
7. Vassaki, S., Poulakis, M.I., Panagopoulos, A.D., Constantinou, P.: An auction-based mechanism for spectrum leasing in overlay cognitive radio networks. In: 2013 IEEE 24th International Symposium on Personal Indoor and Mobile Radio Communications (PIMRC), pp. 2733–2737, September 8–11, 2013
8. Yi, Y., Zhang, J., Zhang, Q., Jiang, T., Zhang, J.: Cooperative communication-aware spectrum leasing in cognitive radio networks. In: 2010 IEEE Symposium on New Frontiers in Dynamic Spectrum, pp. 1–11, April 6–9, 2010
9. Ma, K., Yang, J., Hu, G., Guan, X.: Cooperative Relay-Aware Spectrum leasing based on Nash bargaining solution in cognitive radio networks. Int. J. Commun. Syst. **28**, 1250–1264 (2015)
10. Salim, S., Baek, C., Moh, S., Chung, I.: Cooperative and Non-Cooperative Games for Spectrum Sharing in Cognitive Radio Networks: A Comparative Stud. HIKARI Ltd Contemporary Engineering Sciences **7**(29), 1633–1639 (2014)
11. Stanojev, I., Simeone, O., Bar-Ness, Y., Yu, T.: Spectrum leasing via distributed cooperation in cognitive radio. In: IEEE International Conference on Communications, ICC 2008, pp. 3427–3431, May 19–23, 2008
12. Feng, X., Zhang, Q., Zhang, J.: Dynamic spectrum leasing with user-determined traffic segmentation. In: 2013 IEEE International Conference on Communications (ICC), pp. 6096–6100, June 9–13, 2013
13. Taleb, T., Anastasopoulos, M.P., Nasser, N.: An Auction-based Pareto-optimal Strategy for Dynamic and Fair Allotment of Resources in Wireless Mobile Networks. IEEE Trans. on Vehicular Technology **60**, 4587–4597 (2011)
14. Elmachkour, M., Sabir, E., Kobbane, A., Ben-Othmane, J.: Greening the Spectrum Sensing: A Minority Game-based Mechanism Design. IEEE Communications Magazine, 150–156, December 2014

Theoretical Analysis of BER Performance for Asynchronous FBMC Based Multi-cellular Networks with Non Linear Distortions

Brahim Elmaroud, Ahmed Faqihi, Mohammed Abbad and Driss Aboutajdine

Abstract In this paper, we present a theoretical analysis of bit error rate (BER) for asynchronous filter bank multicarrier (FBMC) based multi-cellular networks in the presence of high power amplifiers (HPA) nonlinear distortion (NLD). A promising class of FBMC modulation called Cosine Modulated Multitone (CMT) is considered and the analytical BER is derived based on the signal to interference plus noise ratio (SINR) of a cellular network consisting of one reference mobile user (MU), one reference base station (BS) and K interfering BSs. The proposed model is evaluated and it is found in very good agreement with simulation results.

Keywords Multi-cellular networks · Filter bank multicarrier · Timing offset · HPA NLD · Signal to interference plus noise ratio · Bit error rate

1 Introduction

Filter Bank Multicarrier is a multi-carrier modulation that consists of transmitting simultaneously N elementary symbols via N parallel orthogonal subchannels. The first developments in FBMC systems were introduced by Chang and Saltzberg. In particular, Chang proposed a special way to transmit a set of parallel (real valued) pulse amplitude modulated (PAM) symbols using Vestigial side-band modulation (VSB) [1]. This method was called Cosine Modulated Multitone (CMT) [2]. Saltzberg, on the other hand, has followed up the original work of Chang and proposed another method called Staggered Modulated Multitone (SMT) [1]. SMT consists on transmitting a set of complex-valued data symbols (in general QAM symbols) whose

B. Elmaroud(✉) · A. Faqihi · M. Abbad · D. Aboutajdine
LRIT, Associated Unit to CNRST (URAC29), FSR, Mohammed V University, Rabat, Morocco
e-mail: b.elmaroud@gmail.com

A. Faqihi
ENSIAS, Mohammed V University, Rabat, Morocco
e-mail: faqihi@ensias.ma

E. Sabir et al. (eds.), *The International Symposium on Ubiquitous Networking*,
Lecture Notes in Electrical Engineering 366,
DOI: 10.1007/978-981-287-990-5_5

real and imaginary parts are separated and time staggered by one half of the symbol duration.

Despite their appealing features, FBMC systems are very sensitive to timing errors and carrier frequency offsets (CFO). This sensitivity has been illustrated in terms of signal to interference ratio (SIR) in [3],[4] and in terms of signal to noise ratio (SNR) degradation in [5] by deriving accurate and approximate expressions of the inter-carrier interference (ICI) and intersymbol interference (ISI). In addition the authors of [6] have developped a theoretical derivation of the average error expressions for an asynchronous FBMC based multi-cellular network and it was concluded that timing errors between BSs cause a severe degradation in system performance.

Another real issue for FBMC systems comes from the fact that the transmitted signal is the sum of a large number of independently modulated subcarriers. Thus, it suffers from high Peak-to-Average Power Ratio (PAPR) which makes the system very sensitive to nonlinear distortion (NLD) caused by nonlinear devices such as high power amplifiers (HPA). This problem has been largely studied for OFDM systems and the impact of NLD on OFDM signals was presented [7] as well as some solutions to estimate and compensate this NLD [8],[9]. For FBMC systems, the authors in [10] have carried out a theoretical analysis of bit error rate (BER) performance for nonlinearly amplified FBMC/OQAM signals under additive white Gaussian noise (AWGN) and Rayleigh fading channels. In this paper, we aim to study jointly the effects of timing errors and HPA nonlinear distortions on CMT based multi-cellular networks. A theoretical expression of the BER will be derived for an unsynchronized nonlinearly distorted CMT signal. The proposed BER expression will be compared to the one obtained for an unsynchronized CMT signal without considering NLD distortions [6].

The remainder of this paper is organized as follows: In section 2, the considered multi-cellular network is introduced and the transmission conditions are presented. Section 3 describes the CMT transceiver and derives the theoretical expression of the BER. Section 4 includes the evaluation of the obtained BER expression through simulation results. Finally, section 5 concludes this paper.

2 Considered System Model

The considered system is shown in Figure 1, where MU_0 is the reference mobile user who communicates with the reference base station BS_0. The other base stations $BS_i, i = 1 \dots 6$ represent an interfering sources for MU_0.

We assume a perfect synchronization between MU_0 and BS_0. However MU_0 is not necessarily synchronized with the other base stations. In addition we consider the presence of HPA non linear distortions. These distortions can be modeled using the Bussgang theorem which states that for a gaussian signal $x(t)$ which undergoes a non-linear distortion and produce an output signal $y(t)$, the cross-correlation function R_{xy} is related to the auto-correlation function R_{xx} by the following expression

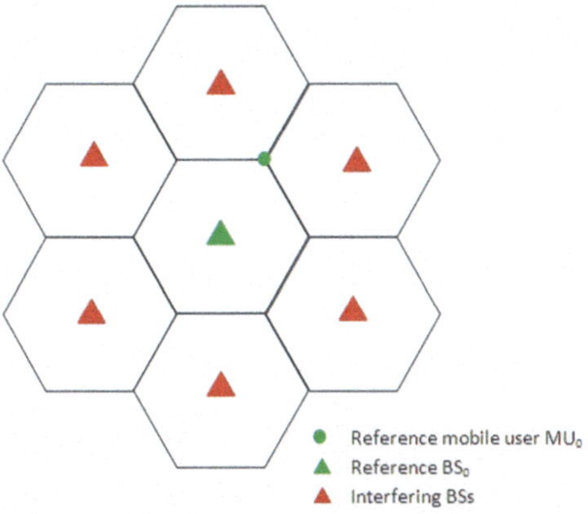

Fig. 1 Considered system.

$$R_{xy} = \alpha R_{xx} \tag{1}$$

Where α is a complex factor.

An interesting result from the Bussgang theorem is that the output y(t) can be written as [7]

$$y(t) = \alpha x(t) + d(t) \tag{2}$$

where $d(t)$ is a zero mean additive noise, which is uncorrelated to $x(t)$.

It is worth noting that, in the case of a multicarrier system and especially for CMT signals, if the number of subcarriers N is sufficiently large, it follows then from the central limit theorem that x(t) can be considered as a complex Gaussian random process, thus, the Bussgang theorem can be applied in our case.

On the other hand, the received signal $z(t)$ by MU_0 from an interfering base station BS_i is affected by a timing offset τ_i. Therefore, $z(t)$ can be written as follows

$$z(t) = \alpha s_i(t - \tau_i) + d(t) \tag{3}$$

Where s_i is the interference signal transmitted by the base station BS_i.

3 BER Analysis

In this section we will investigate the BER of the CMT based cellular network presented in the last section. The block diagram of a CMT transceiver is depicted in

Figure 2. According to this figure, the transmitted signal is given by the following expression

$$x(t) = \sum_{n=-\infty}^{\infty} \sum_{k=0}^{N-1} s_n^k \gamma_{k,n}(t) \tag{4}$$

where $\gamma_{k,n}(t) = h(t - nT)e^{j\frac{\pi}{2T}(t-nT)}e^{j\Phi^k(t)}$ and $\Phi^k(t) = k((\pi/2T)t + \pi/2)$.

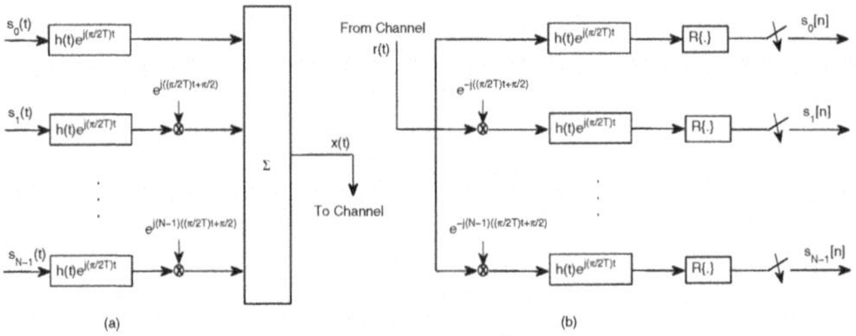

Fig. 2 Block diagram of a CMT transceiver: (a) transmitter; (b) receiver.

In the presence of HPA nonlinear distortions and a timing offset τ, the interference signal coming from the interferer on his m^{th} subcarrier can be written as follows

$$\begin{aligned} s_m(t-\tau) &= \sum_{n=-\infty}^{\infty} \alpha s_n^m \gamma_{k,n}(t-\tau) + d(t) \\ &= \sum_{n=-\infty}^{\infty} \alpha s_n^m h(t-nT-\tau)e^{j\frac{\pi}{2T}(t-nT-\tau)}e^{j\Phi^m(t-\tau)} + d(t) \end{aligned} \tag{5}$$

At the receiver side, the demodulated signal at the m_0^{th} subcarrier on the 0th time index (for simplicity's sake) is given by

$$\begin{aligned} z &= \left\langle s_m(t-\tau), \gamma_{m_0,0}^*(t) \right\rangle \\ &= \sum_{n=-\infty}^{\infty} \alpha s_n^m e^{-j\frac{\pi}{2T}(nT+\tau)} e^{-jm\frac{\pi}{2T}\tau} \int_{-\infty}^{\infty} h(t-nT-\tau)h(t)e^{j\Phi^l(t)}dt \\ &+ \int_{-\infty}^{\infty} d(t)h(t)e^{-j\frac{\pi}{2T}t}e^{-j\Phi^{m_0}(t)}dt + n \\ &= Z(\tau, l) + D + n \end{aligned} \tag{6}$$

with $l = m - m_0$ is the subcarrier offset and n is the AWGN noise.

The interference power can thus be written as follows

$$I'(\tau, l) = I(\tau, l) + \sigma_d^2 \quad (7)$$

Where

$$I(\tau, l) = E\left[|\mathcal{R}\{Z(\tau, l)\}|^2\right] \quad (8)$$

and we assumed that

$$E\left[|D|^2\right] = E\left[\left|\int_{-\infty}^{\infty} d(t)h(t)e^{-j\frac{\pi}{2T}t}e^{-j\Phi^{m_0(t)}}dt\right|^2\right] = \sigma_d^2 \quad (9)$$

The approximation made in equation 9 will be considered in this paper. An accurate study on the impact of the prototype filter impulse response on the additive noise $d(t)$ will be done in our future work.

According to [6] and considering the presented system in section 2, the asynchronous and non linearly distorted interference power coming from the kth base station can be calculated using the following expression

$$P_i^k(\tau, l) = d_k^{-\beta}\sigma_s^2 I'(\tau_k, l) = d_k^{-\beta}\sigma_s^2(I(\tau_k, l) + \sigma_d^2) \quad (10)$$

Where d_k is the distance between the reference mobile user and the kth interfering base station, σ_s^2 is the transmitted symbol power and β is the path loss exponent.

The total interference power coming from all the K base stations is then given by

$$P_i(\tau, l) = \sigma_s^2 \sum_{k=1}^{K} d_k^{-\beta}(I(\tau_k, l) + \sigma_d^2) \quad (11)$$

Using the following expression of the signal to interference plus noise ratio

$$SINR = \frac{\sigma_s^2 d_0^{-\beta}}{P_i(\tau, l) + \sigma_n^2} \quad (12)$$

And by substituting equation 11 in 12, the final expression of the SINR can be written as follows

$$SINR(\tau, l) = \frac{d_0^{-\beta}}{\sum_{k=1}^{K} A_{k,l} + \sigma_d^2 \sum_{k=1}^{K} d_k^{-\beta} + b} \quad (13)$$

Where $A_{k,l} = d_k^{-\beta} I(\tau_k, l)$ and $b = \frac{\sigma_n^2}{\sigma_s^2}$.

It is known in the literature that the BER of 4-PAM modulation in AWGN channel is given by the following expression

$$BER(SNR) = \frac{3}{4} erfc\left(\sqrt{\frac{1}{5} SNR}\right) \quad (14)$$

By substituting 13 in 14, we obtain the final expression of the BER for the proposed model

$$BER(SNR) = \frac{3}{4} erfc\left(\frac{1}{5} \frac{d_0^{-\beta}}{\sum_{k=1}^{K} A_{k,l} + \sigma_d^2 \sum_{k=1}^{K} d_k^{-\beta} + b}\right)^{1/2} \quad (15)$$

4 Simulation Results

In this section we will show the effect of non linearity on the BER of a nonsynchronized CMT based multi-cellular network. The validity of the proposed theoretical model will also be tested through simulation results. We have considered the 6 neighboring cells surrounding the reference mobile user as interfering base stations. The cell radius in our simulation is $R = 1km$ and the path loss exponent β is equal to 3.76.

On the other hand, we consider a CMT system with N=64 subcarriers transmitting 4-PAM modulated symbols. PHYDYAS prototype filter (PF) is used with overlapping factor of 4. The PHYDYAS PF is obtained using the frequency sampling technique described in [11]. Regarding the simulation curves, they are obtained using a traveling-wave tube amplifier (TWTA) model with AM/AM and AM/PM distortion given respectively by the following expressions

$$f_{am}(s_n^k) = \frac{|s_n^k|}{1 + \left(\frac{|s_n^k|}{A}\right)^2} \quad (16)$$

$$f_{pm}(s_n^k) = \phi_0 \frac{|s_n^k|^2}{|s_n^k|^2 + A^2} \quad (17)$$

where A is the amplitude saturation of the high power amplifier and ϕ is a parameter which controls the maximum phase distortion introduced by this latter. A numerical computing environment (MATLAB) is used to simulate the interference CMT signals coming from the 6 interfering base stations. A 4-PAM CMT receiver is also simulated and the BER is directly computed by comparing the decoded symbols with the real transmitted symbols from the reference base station. The timing offset τ is taken equal to T/8.

Figure 3 shows the BER of the studied CMT system as a function of the signal to noise ratio using expression 15 (proposed model) and the expression obtained for CMT using the derivations of Medjahdi et al. [6] (which ignores HPA effects). The perfect synchronized scenario without HPA nonlinear distortions is also considered for comparison purposes.

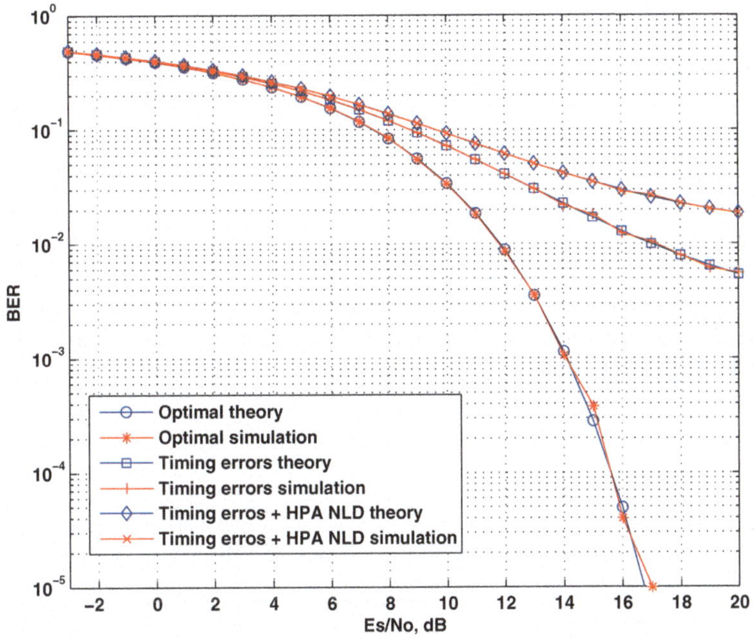

Fig. 3 BER vs. SNR for CMT with and without timing errors and/or HPA NLD.

As shown in Figure 3, timing synchronization errors cause a severe degradation in the BER. This result is found in agreement with the results of [6] where no HPA NLD effect is considered. Moreover, when the signals coming from the different interfering base stations are affected by a HPA nonlinearity as illustrated by the proposed model (equation 15), the BER degradation is more significant. As an example, to achieve a BER of 2×10^{-2}, the unsynchronized nonlinear system requires a value of SNR equal to $19dB$ but the unsynchronized system without HPA nonlinearities requires only $14dB$ of SNR, which means a loss of $5dB$ in the transmitted signal power for the former. Furthermore, we can see an error floor at high SNR values in both cases.

Figure 3 also shows that, for relatively low SNR (i.e., $SNR < 5dB$), the BER performance of the unsynchronized and nonlinearly distorted signal is very close to the BER performance of the perfectly synchronized system. This is due to the fact that the residual degradation of the BER, which is caused by the nonlinear HPA and timing offset, is negligible compared to the AWGN interference. On the other hand, when the SNR level increases the BER degradation becomes important for both models (Equation 15 and Medjahdi's model [6]). This can be explained by the fact that the interference caused by timing errors and/or HPA nonlinear distortions are prominent compared to the noise level at high SNR values. Finally, we can see from Figure 3 a very good match between theoretical expression of the bit error rate given by equation 15 and simulation results.

5 Conclusion

In this paper, we have studied jointly the impact of timing errors and HPA nonlinear distortions on asynchronous CMT based multi-cellular networks. A theoretical analysis of BER was derived for a scenario which consists of a reference mobile user, a reference base station and K interfering base stations. The proposed model was validated through simulation results and compared with existing models considering only timing misalignments. We have concluded that the HPA nonlinear distortions could not be ignored for unsynchronized multi-cellular networks since it affects severely the BER performance of the system.

References

1. Farhang-Boroujeny, B., (George) Yuen, C.: Cosine modulated and offset qam filter bank multicarrier techniques: A continuous-time prospect. EURASIP Journal on Applied Signal Processing, Article ID 165654 (2010) 16 pages
2. Farhang-Boroujeny, B.: Ofdm versus filter bank multicarrier. IEEE Signal Processing Magazine **28**(3), 92–112 (2011)
3. Saeedi-Sourck, H., Wu, Y., Bergmans, J.W., Sadri, S., Farhang-Boroujeny, B.: Sensitivity analysis of offset QAM multicarrier systems to residual carrier frequency and timing offsets. Signal Processing **91**(7), 1604–1612 (2011)
4. Sourck, H., Wu, Y., Bergmans, J., Sadri, S., Farhang-Boroujeny, B.: Effect of carrier frequency offset on offset qam multicarrier filter bank systems over frequency-selective channels. In: 2010 IEEE Wireless Communications and Networking Conference (WCNC), pp. 1–6, April 2010
5. Fusco, T., Petrella, A., Tanda, M.: Sensitivity of multi-user filter-bank multicarrier systems to synchronization errors. In: 3rd International Symposium on Communications, Control and Signal Processing, pp. 393–398, March 2008
6. Medjahdi, Y., Terre, M., Le Ruyet, D., Roviras, D., Dziri, A.: Performance analysis in the downlink of asynchronous ofdm/fbmc based multi-cellular networks. IEEE Transactions on Wireless Communications **10**(8), 2630–2639 (2011)
7. Dardari, D., Tralli, V., Vaccari, A.: A theoretical characterization of nonlinear distortion effects in ofdm systems. IEEE Transactions on Communications **48**(10), 1755–1764 (2000)
8. Tellado, J., Hoo, L., Cioffi, J.: Maximum-likelihood detection of nonlinearly distorted multicarrier symbols by iterative decoding. IEEE Transactions on Communications **51**(2), 218–228 (2003)

9. Dohl, J., Fettweis, G.: Iterative blind estimation of nonlinear channels. In: 2014 IEEE International Conference on Acoustics, Speech and Signal Processing (ICASSP), pp. 3923–3927, May 2014
10. Bouhadda, H., et al.: Theoretical analysis of ber performance of nonlinearly amplified fbmc/oqam and ofdm signals. EURASIP Journal on Advances in Signal Processing **2014**(1) (2014)
11. Bellanger, M.: Specification and design of a prototype filter for filter bank based multicarrier transmission. In: 2001 IEEE International Conference on Acoustics, Speech, and Signal Processing, Proceedings, (ICASSP 2001), vol. 4, pp. 2417–2420 (2001)

Towards a Zero-Failure Distributed Access for Wireless Collision Channels

Sara Arabi, Ahmed Errami, Mohamed Khaldoun, Essaid Sabir and Jelloul El Mesbahi

Abstract Currently in the literature, there is a significant number of access methods to the shared communication channel. For this purpose, two major classes of protocols are defined: controlled access protocols and competing protocols, in which our work is registered. In our paper, we define an access protocol where each station possesses a different and unique contention sequence CS, with which they enter in a battle with other stations. In the end of the competition, only one station is identified as a winner, and will be allowed to access to the channel and transmit its information successfully without any risk of failure. This technique is named zero failure access ZFA.

The object of this paper is to compare the present technique ZFA with the other existing techniques and access methods, for instance slotted ALOHA and DCF protocols. For this issue, we derive the throughput for the three different access methods and finally illustrate the results by simulation.

Keywords Zero Failure Access ZFA · Competition · Battle · DCF · Slotted ALOHA · Contention Sequence CS · Fairly · Uniqueness

1 Introduction

Recently, wireless networks have been playing an important role in our daily life. Telecom operators and service providers interest in implementing wireless network takes more and more importance. This is due to the added value that can provide this

---

S. Arabi(✉) · A. Errami · M. Khaldoun · J. El Mesbahi
COMSYS Research Group, ENSEM, Hassan II University, Casablanca, Morocco
e-mail: {sara.arabi,a.errami,m.khaldoun,j.elmesbahi}@ensem.ac.ma

E. Sabir
UBICOM Research Group, ENSEM, Hassan II University, Casablanca, Morocco
e-mail: e.sabir@ensem.ac.ma

E. Sabir et al. (eds.), *The International Symposium on Ubiquitous Networking*,
Lecture Notes in Electrical Engineering 366,
DOI: 10.1007/978-981-287-990-5_6

type of network, by the diversification of interconnection means calculators (Wireless LAN), domotics network construction (Bluetooth), internet access through the local radio relay. The concept of wireless is closely associated with the transmission medium, the air. To implement a network you have to define first a relatively large coverage area, then install in it a base station, which serves as an access point to users operating in the area. To cope with the growing number of users of wireless networks, increasing the system capacity was necessary, and so having a larger number of relays. In order to exploit effeciently communication support for various flows (digital data, sound, images), an establishment of a protocol that manages the channel access is necessary. Medium access control protocol has a large impact on the achievable network throughput and stability of wireless networks.

1.1 Related Work

Many wireless network protocols have been developed over the last decade, and there have been a lot of researches intersted in the efficiency of the IEEE 802.11 protocol, investigating the maximum throughput that can be achieved under various network configurations. For instance, the interest was growing since the publication of the famous article of Bianchi [2], which presents a simple analytical model to compute the saturation throughput performance assuming a finite number of stations and ideal channel conditions. Many researches propose an extension of Bianchi's model such [4] that characterize the channel activities governed by IEEE 802.11 DCF in multi-hop wireless networks of an individual sender. Authors in [3] propose an interesting framework model for analyzing the throughput, the delay and fairness characteristics of IEEE 802.11 DCF multi-hop networks. There have been quite a number of studies on the performance of wireless networks [5][7][8] with ZigZag decoding to enhance the system throughput, by combating the problem of hidden nodes in order to minimize the rate of packet loss. Also several studies analyze the throughput bounds in three different idealized slotted multi-access system models and provide results for the Aloha and DCF models, and other analyse the performance of carrier sense multiple access with enhanced collision avoidance [9]. All these studies for the purpose of improving network performance, precisely to have a higher throughput.

In this paper, our major aim is to build a model and present an extended work of [1] that presents a new algorithm to manage the channel access between stations in a fair manner. The proposed solution tries to resolve problems related to access to a same shared communication channel. The idea is to put stations trying to access to the channel into a combat, where only the winner can use the channel to send its data.

The Article Structure: We formulate the problem in section 2, which we present the proposed solution. After that we explain the procedure of our proposal in section 3 and section 4. Next the throughput for different access method to the channel is derived in section 5. Before concluding the paper, we validate our approach by simulation and discuss our results in section 6.

2 Problem Formulation

Bandwidth in wireless local networks is shared between users. This requires an access management protocol that manages the access to the channel, in order to resolve conflicts that may occur if all users are allowed to transmit without any rules. The principle function of a medium access protocol is to allow only one station to transmit its packets over a shared physical support. Two major protocol families that manage access to the communication channel are distinguished: controlled access protocols and competing protocols. Controlled access protocols operate on time multiplexing, where each station has a part of the available communication, to use the bandwidth numerous drawbacks are inherent in this technology: management difficulties, inefficient and limited number of stations. On the contrary, competing access methods do not require any prior reservation of bandwidth, each station can transmit at anytime. This implies a risk of access contention with other stations and failure to obtain the channel. For this purposewe must have resolution's policy for these conflicts. The main role of access policy is to ensure the distribution of the network load between all stations depending on the relative importance of their data and the requested access type.

The most known solution is the CSMA/CA mechanism, that is an access method in which stations listen to the channel before starting any transmission. The number of collisions is reduced but not completely avoided because for example if two stations listen and transmit at the same time collision cannot be avoided. Therefore a considerable degradation of the overall transmission rate can be observed in the channel. Indeed, the mechanisms used in CSMA/CA overloads the network, what makes the network performance worse than a wired network. Also, as and when the number of devices connected to the network increases, the probability of collisions become more important.

2.1 Proposed Solution

Our present work focuses on resolving conflicts in an environment where several stations try to access to the same shared communication channel. It involves a method using a decentralized comparison technique which goal is to pick out a single communicating device from several that are participating in the contest to obtain the channel.

In this proposed solution, each communication on the channel must pass through three successive phases:

- The first phase is the phase of an unoccupied channel: when stations initiate the checking process before entering in the combat.
- The second phase corresponds to the combat phase between stations, where each station transmits its CS (Contention Sequence).

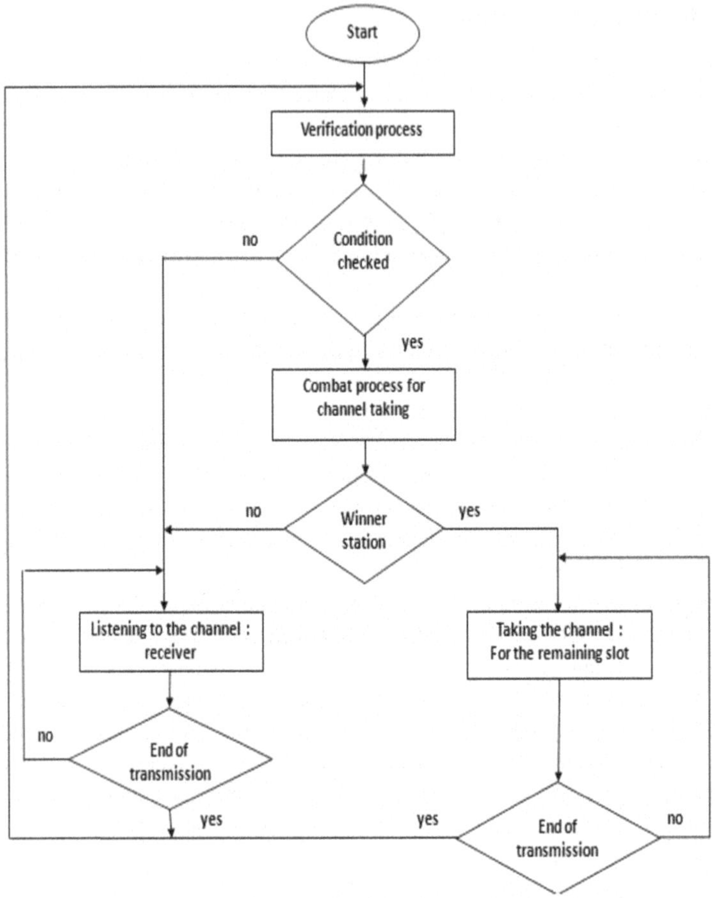

Fig. 1 Procedure followed by stations

- The last phase is the phase of obtaining channel: in which the winner station can carry on transmitting on the channel safely without any risk of collision with other transmitters, but stations that lost in the competition become receivers and wait for the next combat.

2.2 *Combat Readiness*

The solution to determine who can access to the channel is to put stations in competition. Normally, when a source station uses the channel as soon as it is ready to transmit, it ignores if there are other transmitters tying to access to the channel or not, which risks to cause collisions between sent packets. In order to limit the number of collisions, the presented technique uses the principle of listening to the channel, by using a detector of the channel state. More precisely, this technique present a condition to participate in the combat, based on time of canal inoccupation. In other words, stations desiring to transmit must listen first to the channel. If it remains unoccupied for a time T, the station can try to access to the channel. However, let suppose that there are other stations that sensed the channel to be unoccupied for a time T and try at the same time to access to the medium. Then, all these channels will enter in combat to determine which of them can transmit.

To summarize, if the combat condition is checked, the station can initiate the combat process by using the CS code.

3 Combat Process

We consider an infinite number of M nodes forming a wireless network, on which we are applying the competition technique. The main advantage of the technique is that it makes possible to always pick out a unique winner that obtain the channel. In order to achieve this, each station possesses a CS code which consists of a series of m element X_i (size of CS is m) where $1 < i < m$. Each element X_i may have one of two logics states: A or B corresponds to 0 and 1. The principle of combat is based on a decentralized method of comparison between the values of CSs of different stations with the unique purpose of extricating a single winner. This property is always satisfied regardless of the number of contestants but on one condition. Each code CS of the different stations must be different from all other codes. The win of the combat, and consequently the channel, is specified according to one of two logics. If the maximum logic is followed, the winner is the station possessing the greatest CS. On the other hand, if the minimal logic is followed, the winner is the station possessing the smallest CS. In both cases, at the end of the contest, only one station is declared as the winner and obtains the channel, where other stations, losers, became receivers and wait the next combat to try to obtain the channel.

For example, as shown in the fig.2, we consider five stations trying to access to the channel, after sensing the channel unused for a time T, and verifying accordingly their access condition to combat. The CS assigned to these stations is mentioned on fig.2. Then, considering the maximal logic, the competition taking part between the five stations will follow the following evolution:

- **During the period T1**, the eliminated combatants are station 1 and station 5 because they emit each the state B in the first element X1, in the presence of state A on the channel. The remaining combatants are stations 2, 3 and 4 because each of they send in their element X1 the state A.
- **During the period T2**, the eliminated combatant is station 4 because it emits the element X2 set to the state B in the presence of state A on the channel. Then, the stations 2 and 3 will continue the combat for the same reason mentioned during the period T1.

Fig. 2 Combat proceeding

- **During the period T3**, station 2 and 3 will continue in the combat because each of them emits the state B in the element X3 in absence of state A.
- **During the period T4**, with the same reason the stations 2 and 3 continue again in the combat and sent their elements.
- **During the period T5**, the stations 2 and 3 will continue the combat because each of them emit the element X5 which contains state A.
- **During the period T6**, the stations 2 and 3 will still again on the combat for the same reason mentioned during the period T4.
- **During the period T7**, station 4 is eliminated because it emit the element X7 which contains state B in the presence of state A on the channel. Then, the station 2 will continue the combat because it sent its element X7 which is set to state A.
- **During the period T8**, the station 2 sent its last element X8 which is to state A. So it is the only station that was not eliminated, and consequently, it is the winner of the combat.

4 Selection of CS Code

4.1 Static Code Allocation

Our present proposal uses a decentralized comparison technique and it does not require the presence of an arbitrator to manage communications through the channel. It based on a contest technique in order to pick out a single device from several that are participating in the contest to obtain the channel as explained before.

The contention sequence CS code must be unique as said before, and for this, the following uniqueness condition must be satisfied:

$$n = log_2(M)$$

with n: size of CS

The present technique ensures that the overall throughput is constant and independent from the number of communicating devices sharing the channel. To model the present technique and competition between stations, we note p as the probability that a station is active. Wireless networks must always satisfy a trade-off between the average rate and equitable access to the channel. Subsequently our aim is to provide a solution to improve this trade-off, in order to ensure that only one station accesses the channel, and on the same time the equitable access to the channel. This can be ensured as explained later by using the cyclic code.

4.2 Cyclic Code Allocation

According to our method, any acquisition of the channel by a station has to pass through three successive phases: 1- Phase of an unoccupied channel, 2- The combat phase between stations to obtain the channel. 3- Phase of transmitting data on the channel by the winner station.

This strategy ensures that only one station accesses the channel, but that does not ensure definitely that the access to the channel is equitable for all stations. That is the aim of the cyclic code allocation. To ensure the fairness among stations, the bits of the CS code chosen by each station will rotate in every new battle in a cyclic way. However, the CS's bits cannot be all zeros or ones, otherwise the equity will not be ensured, because station with a code with all bits equal to 1 will always win the contest, while a station with a code with all bits equal to 0 will always lose. This condition implies that if we have codes of n bits can be allocated $2^n - 2$ stations (not taking into account the codes 000 and 111). This new strategy is illustrated with the

Stations	CS Codes
Station 1	0 0 1
Station 2	0 1 0
Station 3	0 1 1
Station 4	1 0 0
Station 5	1 0 1
Station 6	1 1 0 Winner

First combat

Stations	CS Codes
Station 1	1 0 0
Station 2	0 0 1
Station 3	1 0 1
Station 4	0 1 0
Station 5	1 1 0 Winner
Station 6	0 1 1

Second combat

Stations	CS Codes
Station 1	0 1 0
Station 2	1 0 0
Station 3	1 1 0 Winner
Station 4	0 0 1
Station 5	0 1 1
Station 6	1 0 1

Third combat

Fig. 3 Battle with cyclic CS

example on the fig.3, in which we consider codes of 3 bits, and in each combat the bits of each rotate, and thus all stations obtain the channel fairly.

Remark: As the stations' energy is limited, having a permutation of the bits of CS code on each slots for the non-participant stations affects their battery life is an inconvenient. However, if the non-participant stations do not permute their codes until their back to participate in the combat may cause a problem, as their code may become identical to another belonging to an already active station that have permuted its code. That is why the following solution is proposed. Each of the non-participant stations may not permute the bits code until it is back to be active. Then, it has to listen to the channel and exploits the current frame sequence to figure out the number of permutation that other active stations have perform and apply it on its own CS code. This will assure that the new obtained code is different from other stations' codes.

5 Throughput Analysis

5.1 *Derivation of ZFA Throughput*

The purpose of this study is to compare the random access protocols Slotted ALOHA, DCF and the present algorithm. The objective is to try to determine which protocol appears to be the most robust to the scalability of simultaneously active stations to transmit their packets. To compare the different protocols we consider the case of saturated network, which we assume a fixed number of stations, each always having a packet available for transmission.

We aim in this section to calculate the system throughput. We consider a network formed by M stations, each having packets of m, including n bits reserved to CS code as shown in the fig.4. Let θ be the normalized throughput defined as the probability that a station wins the contest multiplied by the payload of packet size (DATA), without counting the bits of CS code. As explained above, once the battle condition is checked the station will be active and participate in the contest. Then we note A the

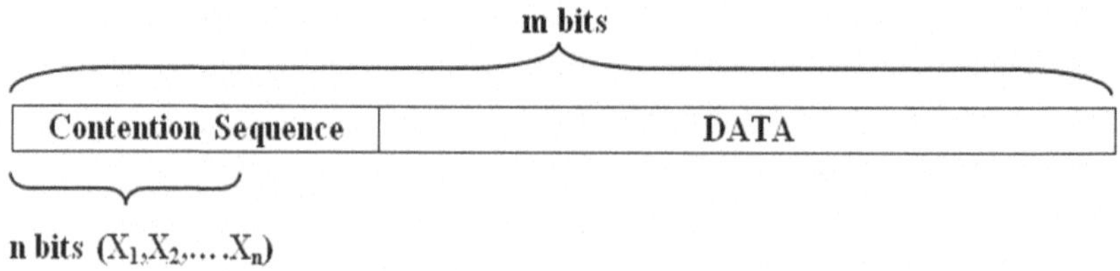

Fig. 4 Size of packet

set of active stations. Therefore, in the case of unsaturated network, the throughput can be written as follow:

$$\theta_{ZFA} = \sum_{A \in M} \prod_{i \in A} p_i \prod_{j \in M|A} (1 - p_j) \frac{(m - n)}{m} \tag{1}$$

And in the opposite case, when all stations have a packet to transmit, the achievable optimal throughput in saturation is given by:

$$\theta_{ZFA} = \frac{(m - n)}{m} \tag{2}$$

5.2 Derivation of S.ALLOHA Throughput

In this paper we have compared our result with that obtained when using the access methods slotted ALOHA and DCF. Considering that the throughput achieved in the Slotted ALOHA case is the fraction of slots that contain a successful transmission, then:

$$\theta_{SA} = Mp(1 - p)^{(M-1)} \tag{3}$$

To maximize the throughput, we need to solve $\frac{\partial D}{\partial p} = 0$:

$$\frac{\partial \theta_{SA}}{\partial p} = M(1 - p)^{(M-1)} - Mp(M - 1)(1 - p)^{(M-2)}$$

$$= M(1 - p)^{(M-2)}[1 - p - (M - 1)p]$$

$$= M(1 - p)^{(M-2)}(1 - Mp) = 0$$

It can be shown that the optimal transmission probability is $p = 1/M$, then the throughput is obtained as follow:

$$\theta_{SA} = (1 - \frac{1}{M})^{(M-1)} \tag{4}$$

5.3 Derivation of DCF Throughput

After that we examine the saturation throughput of DCF system. The normalized throughput is defined as the fraction of time when the channel used to transmit successfully payload bits. As mentioned in Bianchi's model [2], we can obtain the expression of throughput by using the following definition:

$$\theta_{DCF} = \frac{E[payload\ information\ transmitted\ in\ a\ slot\ time]}{E[length\ of\ a\ slot\ time]}$$

Hence, the throughput can be stated as follow:

$$\theta_{DCF} = \frac{P_{tr} P_s E[P_s]}{P_I \delta + P_s P_{tr} T_s + P_{tr}(1 - P_s) T_c} \tag{5}$$

Where E[P] is the average packet payload size, the average payload information successfully transmitted in a slot time is $P_s P_{tr} E[P]$. And P_{tr} the probability that there is at least one transmission in the considered slot time. Then $1 - P_{tr}$ is the probability that a slot time is empty, $P_{tr} P_s$ the probability of a successful transmission and $P_{tr}(1 - P_s)$ the probability that a slot time contains a collision. Also, δ is the duration of an idle slot time, T_s is the average time the channel is sensed busy by a successful transmission, and T_c is the average time the channel is sensed busy resulting in a failure.

As shown in the fig.5, the times T_c and T_s is written as follow:

$$\begin{cases} T_s = H + dataP_s + SIFS + \tau_p + ACK_S + DIFS + \tau_p \\ T_c = H + data + DIFS + \tau_p \end{cases} \tag{6}$$

According to Bianchi's model, the size of packet in our proposal is given by:

$$m = P_I \delta + P_s T_s + T_c(1 - P_s - P_I)$$

6 Evaluation

We turn now to study a typical example of wireless network in which our present algorithm is implemented. We consider a simple network formed by M stations. The goal

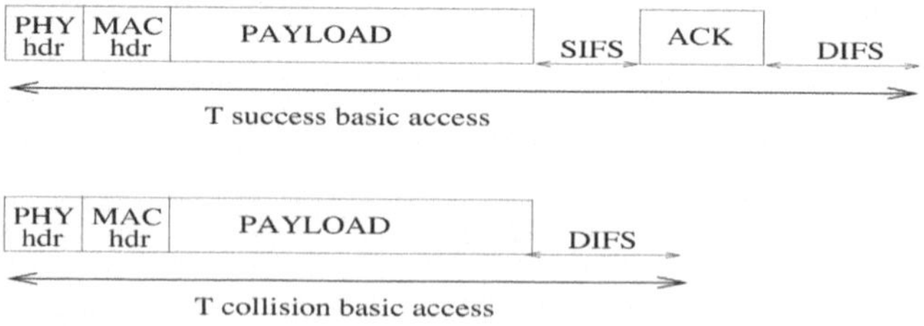

Fig. 5 T_s and T_c for DCF basic access mechanism

Table 1 FHSS system parameters

Packet payload	$8184 bits$
MAC header	$272 bits$
PHY header	$128 bits$
ACK	$112 bits + PHY header$
Propagation delay	$1 \mu s$
Slot time	$50 \mu s$
SIFS	$28 \mu s$
DIFS	$128 \mu s$

of this section is to investigate the throughput improvement that our decentralized comparison technique can achieve, specifically, in the case of saturated network.

So in order to simplify the numerical results, we will evaluate the performance of FHSS system parameters, which are summarized in table 1.

Other parameters are fixed as follow: Contention window W=32, back-off stage s=3 and the CS length values depend on the network size [0,50].

The normalized throughput for all zero failure access schemes (ZFA), slotted aloha and DCF is depicted in fig.6. Our final comparison results show that the present proposal ZFA, in which the principle is based on competition between users, seems to be better than both methods DCF and S.ALLOHA. Also as it is known, as the number of stations increases on the network, the probability of collisions increases too for both schemes DCF and Slotted ALLOHA, and it negatively affect the MAC throughput performances. On the contrary, even when the size of network increases, and as

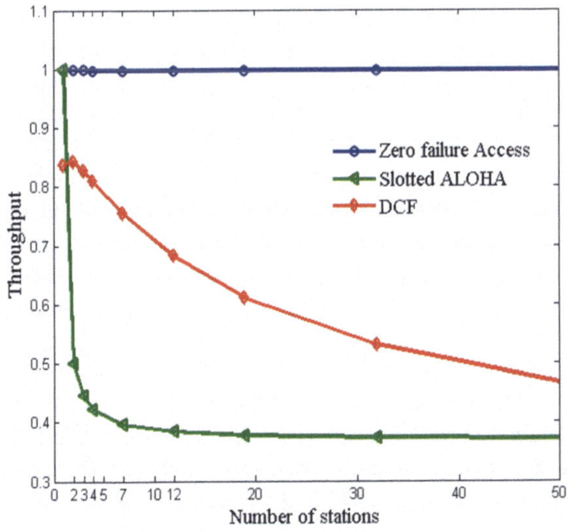

Fig. 6 Throughput versus number of stations with a high size of packet

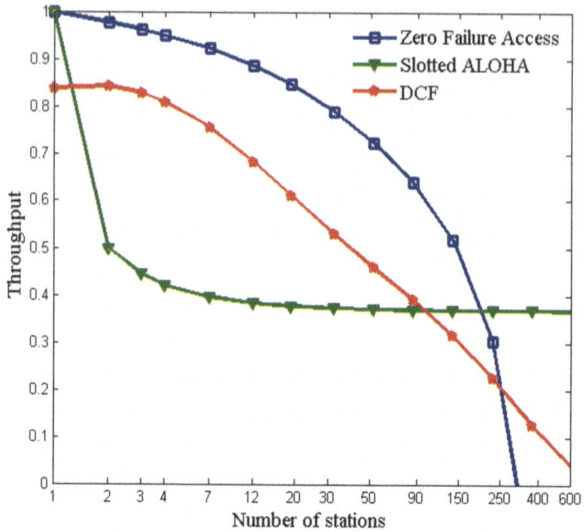

Fig. 7 Throughput versus number of stations with a small size of packet

long as both fairness and uniqueness conditions are checked in order to participate in the battle, the our zero failure access technique will always pick out a single station among all stations' participating in the contest. So the winner station takes the channel for the remaining time of slot without any risk of collision, and transmits its data successfully. However, as the contention sequence CS depends on the number of station as mentioned above $n = log_2(M)$, the throughput is affected when a great network size is considered. In fact, the last relation proves that the contention sequence size n is more consistent when the station number M is high. Thus the ratio of useful information transmitted in the packet will decrease in proportion to the overall size of transmitted data, i.e. useful data plus contention sequence. However, this does not affect the throughput that much when great size of packets is considered. As shown in figure 6, the throughput of ZFA is not affected by great number of stations as slotted aloha and DCF are, because of increasing number of collisions. But, if we consider a small size of packet as in figure 7, the throughput is clearly affected by great size of network, because the ratio of useful data will be significantly reduced.

Also depending on the Bianchi's model [2], the slot time can take one of three cases: Idle, collision and success. So, the size of the packet changes according to the case. However, for our comparison to be fair enough, and as the throughput decreases when the packets size decreases, we take the minimal case $P_I\delta$. Then the size of packet will be included in the interval [1,50] bits. However, in common networkd, the size of packet is normally much higher. In instance, the Ethernet network has a size of 1500 octets, and ever for ATM which packets are smaller, their size is 50 octets. That means the size of packet considered is not realistic, but it was taken just to prove the fact that the ZFA throughput decreases if the CS length exceeds the size

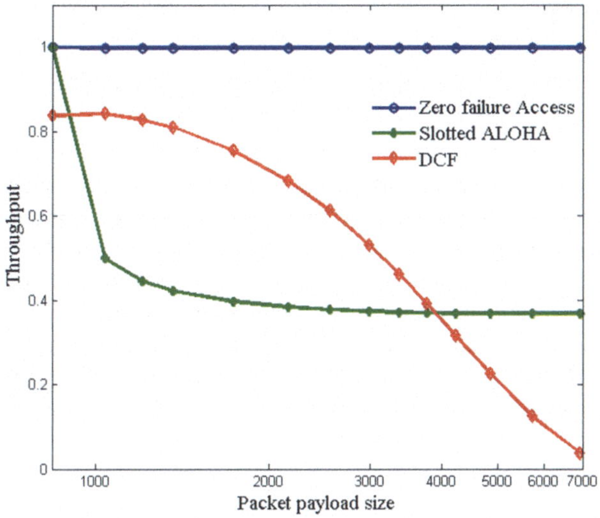

Fig. 8 Throughput versus payload packet size

frame. Hence, as shown in the fig.7, when the network size exceeds M=300 stations, the CS code size exceeds the frame size, and this definitely affect the throughput. That is because the throughput depends on the length of CS as proved in equation (2). And this is could be explained more simply by the fact that when the CS code size exceeds the packets size, we will have more CS code data than users data, and so the network will be overloaded with control information. So to rectify this issue, we define a maximal value that the code size cannot exceed. With this approach, we must limit the network's size too, because in the reverse case, some stations may have the same code, and this does not agree with the uniqueness condition of our solution. Similar issues related to the network size are very common in reality as it becomes difficult to maintain the same performances with millions users connected to the same access point for example.

Next, we plot in fig.8 the normalized throughput for all schemes as function of the packet payload size. Here, the ZFA throughput is perfectly improved in comparison with both schemes, slotted aloha and DCF. This is due first and as we have said before to the fact that ZFA does not suffer from collisions issue, as one user only is assured of sending it packet. So every sent packet is successfully transmitted. Second, as we can observe in the fig.5, slotted aloha and DCF needs an acknowledgment to be sent after a successful transmission of packet. These acknowledgement packets can overload the network, in comparison with the CS code in ZFA schemes, not counting the bandwidth loss that can be caused by losses of acknowledgement messages. However, and as we said before, in dense networks, the CS codes size become higher and overloads the network more than the acknowledgements do.

7 Conclusion

In wireless networks, several studies was carried to accomplish successful packet transfer and to improve the throughput. In this context, we have developed an analytical model using a decentralized comparison technique. Our framework has been derived in term of throughput and included into a comparative study with other methods of shared communication channel access. Relying on different values of stations number and packets size in the wireless network, the analytical result has demonstrated that the present proposal improves perfectly the throughput, without any packet losses as long as the uniqueness condition is respected.

As a future work, we can have as a perspective for next works to try to overcome the issue of limited network size, and add to our approach the case of many packets having the same code.

References

1. El Mesbahi, J., Errami, A., Khaldoun, M.: Method and System for Resolving the Problem of Conflict in a Shared Communication Channel, WO 2007/123383 A1, Novembre 1, 2007
2. Bianchi, G.: Performance analysis of the IEEE 802.11 distributed coordination function. IEEE Journal of Selected Areas in Communications **18**, March 2000
3. Medepalli, K., Tobagi, F.A.: Towards performance modeling of IEEE 802.11 based wireless networks: a unified framework and its applications. In: Proceedings of IEEE INFOCOM (2006)
4. Yang, Y., Hou, V., Kung, L.C.: Modeling the effect of transmit power and physical carrier sense in multi-hop wireless networks. In: Proceedings of IEEE INFOCOM (2007)
5. Gollakota, S., Katabi, D.: ZigZag decoding: combating hidden terminals in wireless networks. In: SIGCOMM 2008 Proceedings of the ACM SIGCOMM 2008 Conference on Data Communication, pp. 159–170 (2008)
6. Paek, J., Neely, M.J.: Mathematical analysis of throughput bounds in random access with ZIGZAG decoding. In: 7th International Symposium on Modeling and Optimization in Mobile, Ad Hoc, and Wireless Networks, WiOPT 2009, pp. 1–7, June 2009
7. Arabi, S., Sabir, E., Sadik, M.: ZigZag decoding-enabled multi-hop Ad hoc networks: taking benefit from collisions. In: 2014 Fifth International Conference on Next Generation Networks and Services NGNS, May 28–30, 2014
8. Arabi, S., Sabir, E., Sadik, M.: Towards a ZigZag-decoding-enabled distributed coordination function for WLANs. In: Fifth International Workshop on Codes, Cryptography and Communication Systems, IWCCCS, November 27–28, 2014
9. Barcelo, J., Bellalta, B., Cano, C., Sfairopoulou, A., Oliver, M.: Carrier sense multiple access with enhanced collision avoidance: a performance analysis. In: Proceedings ACM IWCMC 2009 (2009)

Part II
Advances in Cloud Computing and Could Networking

Toward a New Extension of the Access Control Model ABAC for Cloud Computing

Maryam Ed-Daibouni, Adil Lebbat, Saida Tallal and Hicham Medromi

Abstract Cloud computing is the new technology in open and distributed systems, recently adopted by many IT companies and business organizations. In such systems, there are usually a large number of users in different domains having their own policies, which make many problems arise as degradation or loss of information access control, theft or unauthorized use of information, as well as quality of services problems, traceability and accountability. Data access using various resources requires a user authentication, access control model for integrated management and control in cloud computing environments. The traditional access control models that are identity based are closed and inflexible. In this paper we present an overview of access control models with comparative and analysis to retrieve a suitable model for such environment. An extended Attribute based Access Control (ABAC) model is proposed by introducing the concept of privacy aware to retrieve a new efficient advanced model that can provide a more flexible, privacy and integrity model.

Keywords Cloud computing · Security · Access control model · Privacy aware

1 Introduction

Today, modern societies are highly dependent on information of technology and communications, especially on Cloud Computing which becomes an emerging IT technology being used increasingly in industry, government and academia.

The concept of ''Cloud Computing'' brings together under a single name a set of concepts in which we find the concept of open environment. Cloud computing is a new technology in distributed systems based on the principle of services on request, and the enormous groups of linked computers, those computers are serv-

M. Ed-Daibouni(✉) · A. Lebbat · S. Tallal · H. Medromi
System Architecture Team, ENSEM, Hassan II University, Casablanca, Morocco
e-mail: {eddaibounim,s.tallal}@hotmail.com, adil.lebbat@gmail.com, hmedromi@yahoo.fr

E. Sabir et al. (eds.), *The International Symposium on Ubiquitous Networking*,
Lecture Notes in Electrical Engineering 366,
DOI: 10.1007/978-981-287-990-5_7

ers or personal computer, and they can be structured private or public. Cloud computing services are differentiated according to the security policy component [1, 2] due to the differences in the permitted access right between service providers and users.

This technologie provides large number of benefits, some major issues exist in such as security, data confidentiality, access management, interoperability and continuous evolution. Security and privacy come as key concerns in this domain, as a trusted environment is required for complete control by the owner of data residing in another platform. In particular, data access using different resources needs a user authentication and access control model for integrated management and control in cloud computing environments [1, 3, 4].

Cloud Computing provide three kinds of services: [5] Software as a Service (SaaS); Platform as a Service (PaaS) and Infrastructure as a Service (IaaS).

According to the National Institute of Standards and Technology (NIST) [6], Cloud Computing is composed of four deployment models: (a) public clouds in which the physical infrastructure is owned and managed by the service provider; (b) community clouds in which the physical infrastructure is owned and managed by a consortium of organizations; (c) private clouds in which the infrastructure is owned and managed by a specific organization and (d) hybrid clouds which include combinations of the previous three models.

The Cloud network characteristics vary from one provider to another. The NIST define five essential characteristics of Cloud Computing system: [6] on-demand Self-service; Broad Network Access; Resource Pooling; Rapid Elasticity and Measured service.

Different security issues have been identified at each service layer of cloud like data availability, data access, data segregation, data locality etc. Access control mechanism is considered as an obligatory part of security in order to prevent sensitive data from unauthorized access of malicious users. So there is a need to specify some factors according to which access control systems are evaluated for cloud environment.

The remainder of the paper is organized as follows. In Section 2, we introduce a study about the access control and an overview of the access control models. In Section 3 we have a comparative analysis of access control models on Cloud Computing. Section 4 presents a proposal for improvement of ABAC model by introducing the element of privacy-aware and Section 5 closes this paper with conclusion.

2 Access Control

In security parlance, Access Control is the enforcement of a predefined access policy on resources in order to confine the actions of an entity to only those requests and resources to which it is entitled. A fine-grained Access Control system provides control over methods, resources, and collections of resources. The Access Control system for the cloud must enforce fine-grained access control at both the function and object level to ensure that users with permission to call a function also have appropriate access to target objects required to successfully perform a particular task [7].

In addition, access control is the mechanism by which services decides whether to accept or deny requests according to the permissions affected to users. There are four pieces to the problem:

Identification: Assigning a responsible party for actions. A responsible party may be a person or a non-person entity (NPE), such as a computer or a router. We'll use the term user to cover both cases.

Authentication: The means used to prove the right to use an identity, take on a role, or prove possession of one or more attributes.

Authorization: The means of expressing the access policy by explicitly granting a right.

Access Decision: Using some combination of the other three to decide whether or not a request should be honored. [8]

2.1 Access Control Models

Access control mechanism is considered as an essential part of security in order to prevent sensitive data from unauthorized access of malicious users. Various access control models are in use, including Mandatory Access Control (MAC), Discretionary Access Control (DAC), Identity Based Access Control (IBAC), Role Based Access Control (RBAC) and Attribute Based Access Control (ABAC).

a. Discretionary Access Control (DAC)

DAC is method of restricting access to an object based on the entity identity (e.g. user, process, and group). This type of Access Control is called discretionary because the entity with a certain privilege is capable of passing its privileges or security attributes to another entity. This results, however the problem of loss confidentiality of information. DAC is widely used in many networks and operating systems [7].

b. Mandatory Access Control (MAC)

MAC is a method of constraining an entity from accessing or performing operations on a resource, based on predefined security attributes or labels assigned to the entity and the target object [7].

c. Identity Based Access Control (IBAC)

Access control based on the identity (IBAC - Identity Based Access Control) is among the first access control models. This model introduces the basic concepts of subject, action and object. The objective of this model IBAC is to control direct access of subjects to objects through the use of actions. This control is based on the identity of the subject and the object identifier.

In IBAC, permission has the following format: a subject has permission to perform an action on an object. The authorization policy for specifying the permissions defined through the use of an access control matrix in which the rows and columns of the matrix correspond respectively to the set of the subjects and objects of the system information. The table 1 below describes the use of access control matrix.

Table 1 Access control matrix

	Object 1	Object 1	Object 1
Subject 1	*rw*	*r*	*w*
Subject 2	*r*	-	*rw*
Subject 3	*w*	-	***R***

Where ***r*** mean read action and ***w*** mean write action.

However, the limitation of this model is its scaling. Indeed, the policy becomes complex to maintain when the number of entities is important. Furthermore, While IBAC could manage centralized monolithic systems, distributed systems proved to be problematic in such model.

d. Role Based Access Control (RBAC)

The RBAC model [9–12] is versatile and conforms closely to the organizational model used in firms. RBAC meets this requirement by separating users from roles. Access rights are given to roles, and roles are assigned to users. Here the role combines users and privileges [13]. Roles are created for various job functions, and users are assigned to roles based on their qualifications and responsibilities. RBAC is thus more scalable than user based security specifications and greatly reduces the cost and administrative overhead.

The following terms are used to describe RBAC reference model:

- Component – component refers to one of the major blocks of RBAC features, core RBAC, hierarchical RBAC, Static Separation of Duty (SSD) relations, and Dynamic Separation of Duty (DSD) relations.
- Objects (OBS) – an object can be any system resource subject to access control, such as a file, printer, terminal, database record, etc.
- Operations (OPS) - An operation is an executable image of a program, which upon invocation executes some function for the user.
- Permissions (PRMS) - Permission is an approval to perform an operation on one or more RBAC protected objects.
- Role - A role is a job function within the context of an organization with some associated semantics regarding the authority and responsibility conferred on the user assigned to the role.
- User - A user is defined as a human being. Although the concept of a user can be extended to include machines, networks, or intelligent autonomous agents.[14]

The Figure 1 shows the functionality of a Role Based Access Control Model (RBAC).

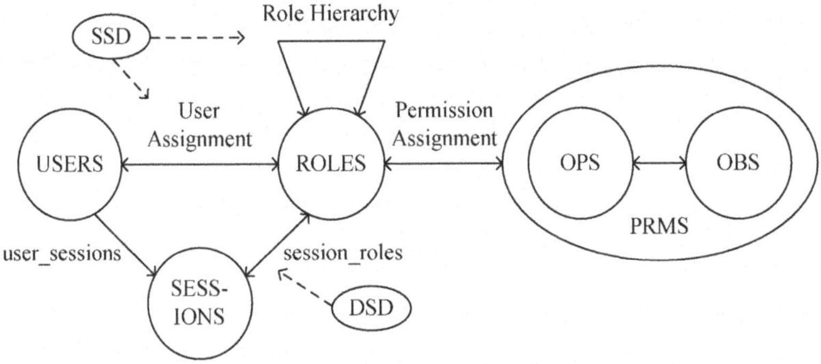

Fig. 1 Role based access control model

This model is very simple and easy to use and it considered the best access control model for the local domain. The roles are assigned to user staticlly by the security administrator that is not preferable in dynamique environment. In addition it is very difficult to change the privilege of the user without changing the role of the user. Furthermore, RBAC becomes problematic in distributed and dynamic environment and has no delegation models which are required in such environment. Also there is a problem of "Role Explosion" that can arise when used to support dynamic attributes in large organizations, in thousands of separate roles for different collections of permissions [9].

e. Attribute Based Access Control (ABAC)

The authorization process of RBAC makes use of only one user attribute namely role. This model is not designed to express flexible and dynamic access control, and shows its advantages in administration because of the assumption that role-permission assignment is static.

ABAC is a logical access control model that is distinguishable because it controls access to objects by evaluating rules against the attributes of the entities (subject and object) actions and the environment relevant to a request. In addition, Attribute based access control (ABAC) is regarded as the candidate to achieve flexible and dynamic access control. Attributes are associated with users (e.g., organization, department, role), objects (i.e., resources) (e.g. owner, size) and environment (e.g. time, location). Access requests are evaluated based on the attributes of involved entities. For instance, a university temporarily allows all full time student users who are currently registered in course A to login the university cloud (with their school id which has been stored in the cloud) to create virtual machines with only Ubuntu images. With ABAC, the IT administrator only needs to insert one policy, compared with the work of creating role, assigning permissions to role, and assigning users to roles explicitly in RBAC model [15, 16].

The functionality of ABAC model is shown in figure 2. Permissions contain the combination of an object descriptor and operations, where Object Descriptor is a combination of a set of attributes and conditions. Operation describes the instructions denoted by the descriptor which is executed on the objects. Access rights can be defined between a subject descriptor and permission. Using descriptors we can dynamically assign permissions to subjects and objects. ABAC uses subject, object, and their environment attributes. Before using these attributes for making access control decision the attribute document is checked for the integrity and validated [17][18].

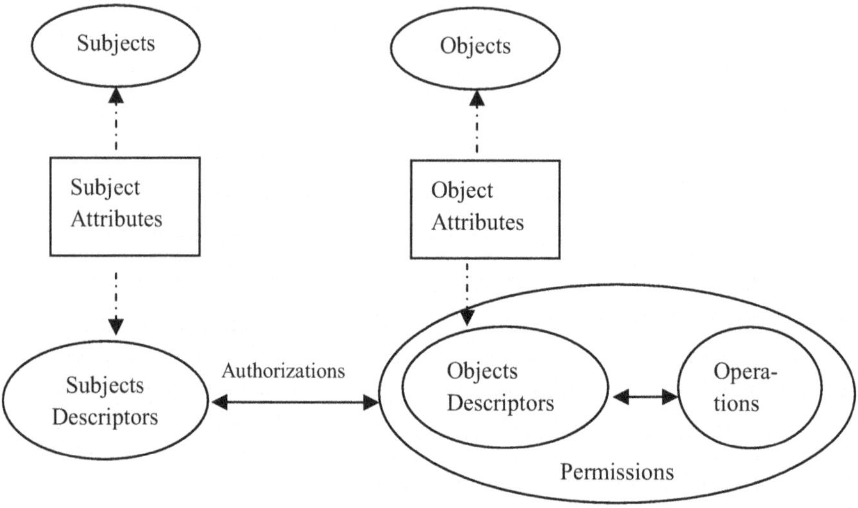

Fig. 2 Attribute Based Access Control

ABAC is very flexible model that is considered easy to administer than RBAC. It works well in an open, distributed and dynamic environment as it has a higher flexibility.

This model is very flexible and supportive in an environment which is large, open, distributed, sharable and collaborative and where the numbers of users are very high and almost of the users are unknow before and the roles of the users are statically or not defined in advance. Further more, it support the global agreement feature such as the user attributes that are provided in one domain can forwarded to the other domain at the point of domain to domain interaction.

ABAC has the high complexity due to the specification and maintenance of the policies. In addition, there is a problem occur of mismatching and confusing attributes especially when those attributes are provided by the user don't necessarily match to those used by the service provider of a web based system or service. Furthermore, it increases the privacy, flexibility, sharing, and global agreemement, and provides the interoperability among several services provides which can use these attributes data dynamically and can decide upon user rights.

3 Comparative Analysis of Access Control Model on Cloud Computing

In this section, we present a comparative analysis of MAC, DAC, IBAC, RBAC and ABAC with some point of issues.

This comparison is based on different criteria, which will help us to choose the most efficient access control model suitable for Cloud Computing environment. Those criteria are: dynamicity, distributed systems, global agreement, flexibility, simplicity, authorization decision, granularity, manageability, trust, changing priviliges, policies specification and maintenance, role explosion, contextual information, scalability.

The following are comparative discussion of MAC, DAC, IBAC, RBAC and ABAC:

Dynamicity: The classical models (MAC, DAC, IBAC, RBAC) has not this feature because allow to model only security policies that are limited to static permissions whereas ABAC is supportive in dynamic environment where user attributes are assigned at the time of request and used to make an access decision.

Distributed Systems: The classical models proved to be problematic for distributed system, whereas ABAC supported for distributed environment where the number of users is very high.

Global Agreement: The models MAC, DAC, IBAC and RBAC not support the global agreement because the permission is assigned in the local domain whether ABAC model supports global agreement due to domain to domain interaction and sharable user attribute database, so that attributes which are provided in one domain can be forward to the other domain at the point of domain to domain interaction.

Flexibility: MAC, DAC, IBAC and RBAC models are not flexible in the open and distributed environment because of its static nature characteristic. But ABAC model is more flexible in an open, distributed, sharable and dynamic environment.

Simplicity: MAC, DAC, IBAC and RBAC are simple in administration and easy to use access control model in which permissions are assigned statically according to the static established policies. Whereas ABAC model is very complex because it is very flexible, sharable, support global agreement and user attributes are heterogeneous.

Authorization Decision: In MAC, DAC and IBAC authorisation decision are in local domain. In RBAC authorization decision are established in advance at the time of permission role assignment in local domain. But in ABAC model, the authorization is made globally due to the user credential provided dynamically.

Granularity: The classical access control models (MAC, DAC, IBAC and RBAC) has low granularity due to its local domain and user has least privileges in RBAC model. Whereas ABAC has the high granularity due to centralize user attributes database, in addition it provide more privileges.

Manageability: Manageability in MAC and DAC is simple. RBAC is too simple to manage because the subject has the permission to execute an action according to his role and the policies are made for the roles actions and permissions, but in case of ABAC model the management of the user's attributes is more complex due to the heterogeneity of the user attributes. Furhtermore there is a problem of protection the user attributes in centralized database in an open and distributed environment. Whereas in IBAC model Managing access to the system and resources became hard and vulnerable to errors.

Trust: Trust is one of the most important issues in access control mechanism. In MAC, DAC, IBAC and RBAC trust is highly obtained in local domain. Although in ABAC trust constitution is more complex due to the global agreement of the attributes in sharable environment.

Changing Privileges: changing privileges in MAC, DAC and IBAC is simple. But it is very diffecult to change the privilege of a user in RBAC system without changing the role of the user. To solve this problem, there is a need to change the policy specification. On the other hand in ABAC system, changing the privilege of a user not need to change the user identity or the policy specification because the user has the permission according to his/her attributes values.

Policy Specification and Maintenance: In a MAC, DAC, IBAC and RBAC system policy specification is not much complex but ABAC system is very complex due to the flexibility, dynamicity, sharing in open and distributed environment, heterogeneity of user attributes, different formats of the attributes causes inadequate and confusing of attributes. Sharing the attributes in open system is dangerous it can create the problem of integrity or privacy of the user information and policy defined for this purpose are much more complex.

Role Explosion Problem: In RBAC model there is a role explosion problem that is caused by the situation where each role requires different sets of permissions and large numbers of roles have to be defined in advance.

Contextual Information: Context is one of the most important characteristics of any collaboration especialy contextual information that is utilized by the access control model in order to secure the system. The classical models (MAC, DAC, IBAC and RBAC) do not allow the opportunity to express the contextual rules. On other hand ABAC model take into account the environmental information.

Scalability: Scalability in MAC is lower and also it won't be adapt to all type of applications. RBAC provides scalability but it fails when having large number of users in the system [19]. But IBAC is problematic when implementing it in large distributed system. Whereas ABAC is more scalable in Cloud Computing environment.

The table 2 presents a summary of what we have said before:

Table 2. Comparative analysis of access control models

	MAC	DAC	RBAC	IBAC	ABAC
Dynamicity	no	no	no	no	yes
Distributed Systems	proved to be problematic for distributed systems	proved to be problematic for distributed systems	proved to be problematic for distributed systems	proved to be problematic for distributed systems	Supported for distributed environment
Global Agreement	no	no	no	no	yes
Flexibility	no	no	no	no	yes
Simplicity	yes	yes	yes	yes	no
Authorization Decision	locally	locally	locally	locally	globally
Granularity	low	low	low	low	high
Manageability	simple	simple	simple	complex	complex
Trust	locally	locally	locally	locally	Globally
Changing Privileges	Simple individual user cannot change access rules	Simple	complex	Simple	simple
Policies specification & Maintenance	low	low	simple	low	complex
Role Explosion Problem	no	no	yes	no	no
Contextual Information	low	low	low	low	high
Scalability	low	low	low	low	yes

Traditional models mentionned in the table above (MAC, DAC, IBAC and RBAC) are generally intended for static operation, they hardly allow dynamic evolution. Whereas, the ABAC model is the most suitable for architectures operating in an open environment such Cloud Computing where different organizations can ensure both access to information and the protection of their resources.

4 Extended Attribute Based Access Control Model

From the above discussion and the comparative table it is clear that Attribute-Based Access Control (ABAC) is the most suitable for dynamic environment like cloud computing. This model granted access to users based on the attributes of the requesting user. It uses multiple attributes for authorization decision, which enables the system to be highly flexible, scalable, interoperable, and multifunctional access control that may deal with diverse security requirements in a Cloud Computing environment.

However, ABAC has drawbacks with reference to privacy-aware concerns. Indeed, because of the descriptive nature of subject attributes, implementation of attribute by sharing capabilities causes a problem of increasing the risk of privacy violation of personally identifiable information by dint of involuntary exposure of attribute data to entrusted third parties or aggregation of sensitive information in environments less protected. A second consideration is that releasing attributes to the policy evaluating engine is a sensitive activity as the third party may not be trusted.

Although diverse techniques have been proposed to regulate the release of information, still, least privilege has to be enforced in ABAC. It means the least set of user attributes are released for the purpose of request evaluating. Figure 3 shows the scenario of ABAC model by introducing privacy-aware element.

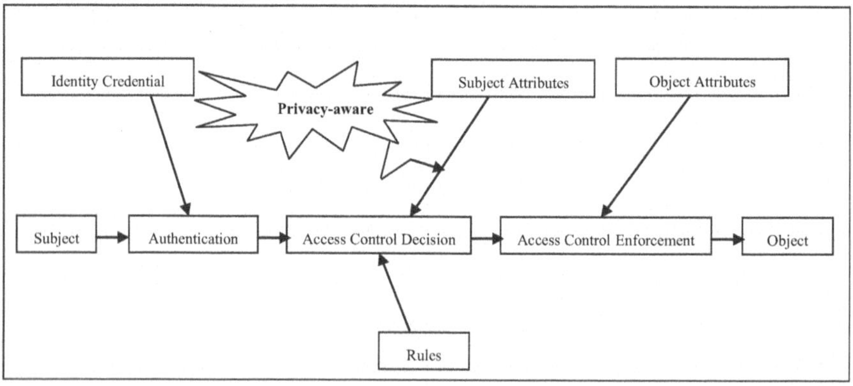

Fig. 3 Basic ABAC Scenario with privacy-aware element

5 Conclusion

This paper presented a comparative study of various access control models that are popularly used in cloud computing environment such as DAC, MAC, IBAC, RBAC and ABAC. This study led us to choose the ABAC model as the most suitable model for a cloud computing environment, although this model present some drawbacks such us privacy aware violation.

In this respect and as a future work, we propose a privacy aware element to control attribute release with optimal set of user attributes and exposing the minimum of sensitive user's informations.

References

1. Bowen, B.M., Ben Salem, M., Hershkop, S.: Designing host and network sensors to mitigate the insider threat. IEEE Security Privacy Mag. **7**(6), 22–29 (2009)
2. Jung, J.J.: Service chain-based business alliance formation in service-oriented architecture. Expert Syst. Appl. **38**(3), 2206–2211 (2011)
3. Kiyomoto, S., Fukushima, K., Miyake, Y.: Towards secure cloud computing architecture—a solution based on software protection mechanism. J. Internet Serv. Inf. Secur. **1**(1), 4–17 (2011)
4. Pieters, W.: Representing humans in system security models: an actor-network approach. J. Wirel. Mobile Netw. Ubiquitous Comput. Depend. Appl. **2**(1), 75–92 (2011)
5. Bhaskar, P., Admela, J., Dimitrios, K., Yves, G.: Architectural Requirements for Cloud Computing Systems: An Enterprise Cloud Approach. J. Grid Computing **9**(1), 3–26 (2011)
6. Li, X., He, J.: A user-centric method for data privacy protection in cloud computing. In: 2011 International Conference on Computer, Electrical, and Systems Sciences and Engineering, pp. 355–358 (2011)
7. Faraji, M.S.: Identity and Access Management in Multi-tier Cloud Infrastructure, thesis (2013)
8. Karp, A.H., Haury, H., Davis, M.H.: From ABAC to ZBAC: The Evolution of Access Control Models. HP Laboratories-2009-30
9. Ahn, G.-J., Sandhu, R.: Role-based authorization constraints specification. ACM Trans. Inf. Syst. Secur. **3**(4), 207–226 (2000)
10. Bertino, E., Bonatti, P.A., Ferrari, E.: Trbac: a temporal role-based access control model. ACM Trans. Inf. Syst. Secur. **4**(3), 191–233 (2001)
11. Joshi, J.B.D., Bertino, E., Latif, U., Ghafoor, A.: A generalized temporal role-based access control model. IEEE Trans. Knowl. Data Eng. **17**(1), 4–23 (2005)
12. Li, N., Tripunitara, M.V.: Security analysis in role-based access control. ACM Trans. Inf. Syst. Secur. **9**(4), 391–420 (2006)
13. Kalajainen, T.: An access control model in a semantic data structure: case process modelling of a bleaching line. Department of Computer Science and Engineering (2007)
14. Nabeel Tahir, M.: C-RBAC: Contextual role-based access control model. Ubiquitous Comput. Commun. J. **2**(3), 67–74 (2007)
15. http://csrc.nist.gov/projects/abac/
16. https://blueprints.launchpad.net/keystone/+spec/attribute-based-access-control
17. Stermsek, G., Stremmbeck, M., Neumann, G.: Using Subject- and Object-specific Attrubutes for Access Control in Web-based Knowledge Management System
18. Ubale, S.A., Apte, S.S.: Comparison of ACL Based Security Models for securing resources for Windows operating system. IJSHRE **2**(6), 63
19. Punithasurya, K., Jeba Priya, S.: Analysis of Different Access Control Mechanism in Cloud. International Journal of Applied Information Systems (IJAIS) **4**(2) September 2012. ISSN: 2249-0868 Foundation of Computer Science FCS, New York, USA

A Bargaining Nash Game Based Adaptive Negotiation of Context Level Agreements for Pervasive Systems

Hayat Routaib, Elarbi Badidi, Essaid Sabir and Mohammed ElKoutbi

Abstract With the growing popularity of Internet-enabled devices, the impressive progress in sensing technology, and the adoption of cloud computing for provisioning services, users increasingly demand services that can adapt to their recent context. In this paper, we propose a multi-attributes and adaptive approach for Context Level Agreements (CLAs) negotiation between a context provider and a context consumer using a context broker. The approach employs a Nash bargaining model and evaluates the global utility of each party as a linear function of normalized Quality of Context (QoC) attributes during the rounds of negotiation. The ultimate goal is to improve context-based adaptation of context-aware applications and services. One of the advantages of this approach is that it permits to resolve conflicts of interests between the context provider and the context consumer when the global utility of each party reaches a Pareto optimum.

Keywords CLA · QoC · Nash bargaining · Pareto optimum

1 Introduction

Over the last two decades, the continuous increase in the number of worldwide deployed wireless networks and the impressive progress in sensing technology have

---

H. Routaib(✉) · M. ElKoutbi
MIS Team, ENSIAS, Mohammed V University, BP 713, Rabat, Morocco
e-mail: hayat.routaib@um5s.net.ma, elkoutbi@ensias.ma

E. Badidi
College of Information Technology, UAE University, PO Box. 15551, Al Ain, UAE
e-mail: ebadidi@uaeu.ac.ae

E. Sabir
UBICOM Research Group, ENSEM, Hassan II University, Casablanca, Morocco
e-mail: e.sabir@ensem.ac.ma

E. Sabir et al. (eds.), *The International Symposium on Ubiquitous Networking*,
Lecture Notes in Electrical Engineering 366,
DOI: 10.1007/978-981-287-990-5_8

led to an extremely growing number of users who are using a variety of mobile internet-enabled devices to consume online services. Also, the demand for pervasive applications and cloud services, which can offer anytime and anywhere access to information and services according to recent context of users and their environment, is growing rapidly. Context is commonly defined as: *"any information that can be used to characterize the situation of an entity. An entity is a person, place, or object that is considered as relevant to the interaction between the user and the application, including the user and the applications themselves"* [5]. *Geo-location, time, temperature, humidity, pressure,* and *mobile user activity* are the most widely used context indicators by applications [1]. Context information might be incomplete and inconsistent due to the massive amount of sensed context information and to variations in the quality and reliability of used sensors. Furthermore, enforcing privacy requirements of users regarding their context information could reduce its quality significantly [12]. Buchholz et al. [3] defined Quality of Context (QoC) as: *"any information describing the quality of information that is used as context."* The most important QoC attributes are: *precision, trust-worthiness, freshness* and *probability of correctness.* Using context information with bad quality might result in incorrect decisions and inappropriate behavior of context-aware applications and services. One of the challenging issues in context-aware systems is evaluating QoC in the different layers of context management frameworks. As stated in [5], QoC might be used for: (a) selecting sensing devices and context providers; (b) rejecting raw data that do not meet the minimum quality requirements fixed by some global QoC thresholds; (c) improving context-aware reasoning and decision making; (d) reducing the probability of incorrect context-aware adaptation. With regards to context information provisioning to context consumers, two main activities have to be considered: context negotiation and the monitoring of delivered QoC.

In this paper, we focus on the context negotiation process and we propose a Nash bargaining based model for the negotiation of multiple QoC attributes between context consumers, which require context information to adapt their services, and context providers. Our motivation is to design an automated CLA negotiation mechanism between these two adaptive and rational entities using concepts from bargaining game theory.

The remainder of the paper is organized as follows. Section 2 describes related work on the issues of automated negotiation and QoC modeling. Section 3 defines the concept of Context Level Agreement (CLA) and describes the CLA negotiation scenario. Section 4 describes our proposed multi-attributes negotiation model, which is based on the bargaining model, and the Nash-based solution. Section 5 presents the fair solution to the multi-attributes negotiation problem and illustrates the findings with numerical results. Finally, Section 6 concludes the paper and highlights future work.

2 Related Work

Several research efforts investigated the issue of automated negotiations in several application domains such as game theory, Service-Oriented Architecture (SOA), and cloud-based applications. Service Level Agreement (SLA) negotiation has been the focus of several efforts. Copil et al. [4] proposed an automated SLA negotiation protocol, which is based on particle swarm optimization techniques, for obtaining a balance between consumed energy and offered performance in the cloud. The two parties of the defined negotiation protocol are the performance-oriented cloud customer and the energy-oriented cloud provider. The agreement resulting from the negotiation process is close to a Pareto optimal solution. Maarouf et al. [2] determined a Quality of Service (QoS) which meet consumer's Service Level Agreements (SLA) specifications. They proposed the concept of monitoring and respect of the QoS in order to manage of the relationships, between, cloud providers and their services customers. Bouterse et al. [10] analyzed the tradeoff of service availability versus utilization using a two-dimensional Pareto analysis, such as utilization versus system availability or the impact of service level agreement timescale on utilization, in order to provide a cloud resources with a available cost. Muller et al. [5] provided a conflict classification for SLAs that includes new conflicts derived from the use of conditional and optional term sets; and a novel language-agnostic technique based on constraint satisfaction problems to automatically detect and explain these conflicts.

With the growing interest in adaptive context-aware services, context information representation and automated context information negotiation have received growing consideration over the last few years. Manzoor et al. [8] suggested automated negotiation of context information and presented a context aggregation system that detects and eliminates replicates and conflicts from the system using QoC-based policies. Badidi et al. [1] proposed a framework for context information provisioning, which relies on deploying context services on the cloud and uses context brokers to negotiate context information provisioning between context consumers and context services. Furthermore, they described a multi-attributes decision algorithm for the selection of potential context services that can fulfill context consumers' requirements. Kim et al. [7] proposed quality measurement methods and QoC indicators from the perspective of data quality dimensions: reliability, completeness, representation consistency, access security, and up-to-dateness. Furthermore, they descrided an aggregation method for evaluating the QoC completeness of a set of raw context information. The work of Xianrong et al. [11] is the closest to our work as they represent the one-to-one Web services negotiation as a bargaining game and determine a Nash equilibrium that can be regarded as the fair solution to a two players bargaining game. Katsuhide et al. [6] focused on negotiation with several interdependent issues, in which agent utility functions are nonlinear. They suggest a Distributed Mediator Protocol (DMP) to securely find the agreements that satisfy the Pareto optimality. They proposed approximated fairness approach for each agent and employ deviation for measuring the difference between agents' utilities. They also determined the rate of the Nash bargaining solution, which represents the closeness of the contract to the Nash bargaining solution.

The Nash bargaining problem is the most essential problem in Economics which, involves the possibility of bargaining and side payments. Two approaches are used to deal with this classical economic problem. The first is the strategic approach, while the second is the axiomatic method. Both Nash [9] follows the strategic approach in his bargaining models.

To the best of our knowledge, none of the existing approaches for context information negotiation relies on the Nash bargaining model to reach a compromise between context consumer and context provider. In this work, we propose an automated context negotiation process that uses a Nash bargaining model and takes into account the interests of the negotiating parties, context broker (on behalf of context consumer) and context provider. The model considers an adaptive and automated context negotiation process between a Context Broker consumer and one or several context providers through a context broker can improve the marketing of context-aware services. Determining a Pareto optimum can prove the feasibility of this adaptive and automated negotiation process. For this reason, we consider in this paper a bargaining model that represents and evaluates utility functions of both the context broker and the context provider's offer as functions of the normalized QoC. The aim is to reach a Pareto optimum.

3 Context Level Agreement

A CLA is an arrangement between a context provider and a context consumer concerning the guarantees of delivered context information. It describes common understandings and expectations between the two parties. The guarantees concern the context information and the QoC levels to be delivered. The negotiation of context information provisioning involves the following components:

- ***Context Consumer***: the consumer of context information offered by context providers. It may be implemented as a Web service that uses context information to adjust its behavior according to the changing situation conditions.
- ***Context Provider***: the provider of one or several types of context information (location, temperature, user-activity, etc.) to context consumers.
- ***Context Broker***: the mediator party that decouples context consumers from context providers. It is in charge of handling subscriptions of context consumers in which they express their requirements concerning context information and their preferences in terms of QoC levels. It is also in charge of the registration of context-providers that are willing to provide some types of context information. Since context consumers do not have the capabilities to select appropriate context providers and negotiate the terms of the contracts, they delegate this task to the context broker.

The typical sections of a CLA are:

- *Parties*: the parties involved in the CLA and their respective roles (context consumer and context provider).

- *Activation time*: the period of time at which the CLA will be valid.
- *Scope*: the types of context information covered in the agreement.
- *Context-level objectives (CLOs)*: the levels of QoC that both parties agree on, and they habitually include a number of quality indicators such as accuracy and freshness.
- *Penalties*: the penalties for not meeting the stated context level objectives, such as getting discount or having the right to terminate the contract in light of unsatisfactory context levels.
- *Exclusions*: specifies what is not covered in the CLA.
- *Administration*: defines the processes to assess the CLA objectives, and describes the responsibility of the context provider regarding the control of each of these processes.

4 Context Broker

Figure 1 depicts the architecture of the Context Broker, which includes several components that cooperate in order to deliver personalized services to context providers and context consumers. These components are the *Context Request Dispatcher*, the *CLA Manager*, the *QoC Information Manager*, the *Profile Manager*, and the *Policy Manager*. These components allow carrying out various management operations such as QoC-based context-provider selection, CLA negotiation, user profile management, and policies management. The back-end databases maintain information about context-providers' policies, customers' profiles and preferences, CLAs, and dynamic QoC information. the *Context Request Dispatcher* is in charge of implementing different policies for the selection of context providers, based on the context-consumer's requirements in terms of context information and QoC, and the context-providers' QoC offerings. The *CLA Manager* is in charge of carrying out the CLA negotiation process between a context-consumer and a context provider. The *Profile Manager* is responsible for managing context consumers' profiles, including their preferences in terms of context information and required QoC. The *Policy Manager* is in charge of managing rules and policies such as authorization policies.

The CLA negotiation process is triggered when, the context consumer sends a CLA Request to the context broker. Then, the context broker attempts to negotiate with an appropriate context provider the CLA terms and conditions in a series of rounds of proposals and counter-proposals until an agreement or a timeout is reached. The parties sign a contract when an agreement has been reached.

5 The Multi-attributes Negotiation Model

Context-consumers often have different preferences concerning QoC levels for each category of context information they need. For example, a context consumer may

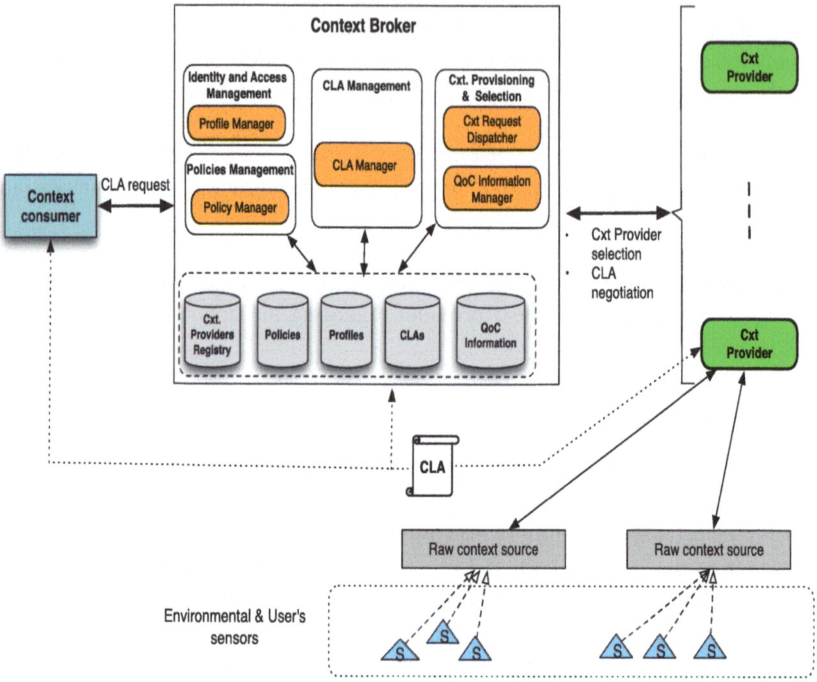

Fig. 1 Architecture of the Context Broker.

require temperature information with a freshness smaller than 10 minutes and a probability of correctness greater that or equal to 98%. Likewise, context providers may have several QoC offerings for each category of context information they provide. As the objectives of both parties with regards to context information provisioning may conflict, it is necessary to find a compromise. To deal with this issue, we propose in this section a multi-attributes negotiation model that relies on the Nash Bargaining Problem to solve the case of just one kind of context information. The model can be extended to several types of context information that the context consumer requires.

5.1 *The Nash Bargaining Problem*

Nash bargaining is a cooperative game, where the two players have a cooperative behaviour. This does not mean that they fully cooperate. They can, however, discuss the different situations and agree on some rational strategies by exchanging relevant information. It can be used to illustrate a negotiation about divisible payoffs. We define the Nash Bargaining Problem for CLA negotiation as follows:

Definition 1. *A two-agents bargaining problem is the quadruple* (P, U, D, Σ), *where:*

- $P = \{C_B, C_P\}$ *is the set of two players or agents, the Context Broker and the Context Provider, who bargain over context information provisioning.*
- $U = \{u_B, u_P\}$ *is the set of two utilities and payoff functions* u_B *and* u_P *of* C_B *and* C_P.
- $D = \{d_B, d_P\}$ *is the set of the disagreement payoff points or a status quo. When no agreement is reached, the context broker gets* d_B *and the context provider gets* d_P.
- $\Sigma = \{\sigma_B, \sigma_P\}$ *is the set of two reservation values* σ_B *and* σ_P *of the players* C_B *and* C_P. *A reservation value is a threshold for negotiation termination.*

Research on multi-attributes negotiation tries to find a Nash solution when many outcomes are possible. In the next sections of the paper, we describe the concept of Nash solution and we formulate the modeling of the problem to find out a Pareto-optimal frontier for CLA negotiation.

5.2 Nash Solution

The Nash solution to a bargaining problem focuses on some fundamental axioms [9]. The key ones among them are:

- *Pareto-efficient* (EFF) (Pareto-optimal): the negotiation solution should not be weakly dominated by any point in the solution space except itself, i.e. the Pareto optimal solution requires that no party increases its utility without decreasing the utility of the other party.
- *Symmetry (SYM)*: the negotiation solution does not depend on the party that triggers the negotiation. The only significant differences between parties are their strategy spaces and utility functions. Symmetry means that the two parties get the same utility.
- *Invariance (INV)*: the utility functions do not change the shape of the solution space as well as the relative position of the solution although numerical values might change.
- *Independence of irrelevant alternatives* (IIA): Let $G = (U, d)$ be the bargaining game where U is the utility space, d is the disagreement payoff, and $f(U, d)$ is the solution of G. Let $G' = (U', d)$ be another bargaining game where $U' \subseteq U$. If $f(U, d) \in U'$ then the two bargaining games G and G' have the same solution $f(U, d) = f(U', d)$

According to these axioms, a Nash bargaining solution corresponds to the payoff pair $u = (u_B, u_P)$ that maximizes the Nash product $(u_B - d_B)(u_P - d_P)$. d_B and d_P are the disagreement points. They represent Nash equilibrium values for a non-cooperative game. Nash proves in [9] that this payoff u is the only solution that satisfies the above axioms and represents a fair solution to a bargaining game.

6 The Fair Solution to Multi-attributes Negotiation

Context consumers typically have different preferences with regards to QoC levels for each category of Context information they require. Likewise, Context providers may have several QoC offerings for each category of context information they can offer. A context consumer may then require several types of context information from one or multiple context providers. For example, a context provider may offer current *location* information of a subject while another one may offer current *temperature* information. Therefore, to satisfy all the requirements of a context consumer in terms of context information and QoC levels, different CLAs may be signed with different context providers. In our framework, a context provider negotiates with context broker, on behalf of a context consumer, to reach an agreement, which is a Pareto optimum solution to the bargaining problem.

6.1 System Model

In the following, we describe the negotiation model of multiple QoC attributes in the case of just one kind of context information to find a fair solution for both parties.

Let $C = \{c_1, c_2, ..., c_m\}$ be the list of context information that a context consumer wants to obtain from one or several context providers.

We assume that for c_j with $j \in [1, 2, .., m]$, the context consumer requires a list of QoC attributes $X = \{X_1, X_2, ..., X_n\}$ that the Context Broker C_B negotiates with an appropriate context provider that can offer c_j. For instance, the context consumer requires the temperature information with the following quality levels: 83% for the precision and 50% for the freshness. In our previous work [12], we proposed an auction-based algorithm for the selection of suitable context services.

Let $Q_B = \{q_B^1, q_B^2, ..., q_B^n\}$ be the quality levels that the context consumer demands for c_j.

Let $Q_P = \{q_P^1, q_P^2, ..., q_P^n\}$ be the quality levels that the context provider suggests to offer for c_j.

We consider that q^i, $i \in [1, 2, .., n]$, is in the range [0, 1]. A higher value corresponds to better quality.

We suppose that the utility payoff u^i for X_i is a linear function, which takes as input the attribute X_i and returns a utility value for each player.

The utility function used by the player C_B is expressed by the following expression:

$$u_B^i(X_i) = \omega_B^i q_B^i. \tag{1}$$

With $i \in [1, 2, .., n]$ and ω_B^i is the importance weight given by C_B to the attribute X_i.

Similarly, the utility payoff of the attribute X_i offered by the player C_P to the other player C_B is expressed by the following expression:

$$u^i_P(X_i) = \omega^i_P q^i_P \quad (2)$$

With $i \in [1, 2, .., n]$ and ω^i_P is the weight given by C_P to the attribute X_i.

If the player C_B does not agree on the utility offered by C_P while negotiating the attribute X_i, then C_B will get value d^i_B which is expressed by:

$$d^i_B = \beta^i \sigma^i_B \quad (3)$$

With $i \in [1, 2, .., n]$ and σ^i_B is the maximum offer that the context consumer can tolerate. In other words, it is similar to a buyer's reservation price or the maximum price that the buyer is willing to pay, and β^i is the penality of the disagreement of the player C_B.

Therefore, we can express σ^i_B as:

$$\sigma^i_B = max_{1 \leq k \leq l}(q^{i,k}_P - q^{i,k}_B) \quad (4)$$

k is the number of negotiation iterations.

$$d^i_P = \alpha^i \sigma^i_P. \quad (5)$$

With $i \in [1, 2, .., n]$ and σ^i_P is the minimum offer that the context provider can tolerate. It is similar to a seller's reservation price, which is the minimum price that the seller is ready to accept. α^i is the penality of the disagreement of the player C_P

We can express σ^i_P as:

$$\sigma^i_P = min_{1 \leq k \leq l}(q^{i,k}_P - q^{i,k}_B) \quad (6)$$

with k is the number of negotiation iterations. α^i and β^i are negotiated too.

In this case, we assume that if $\sigma^i_B < 50\%$ (respectively. σ^i_P), the player incurs no loss but he may not prefer having an agreement with the opponent player. Therefore, the penality β^i (resp. α^i) will be between 20% and 30%. If $\sigma^i_B \geq 50\%$ (respectively. σ^i_P), the player incurs a loss. Therefore, the penality β^i (resp. α^i) will be between 10% and 20%.

The above negotiation is a cooperative game between the context broker C_B and the context provider C_P. So, it is a *nonzero-sum two player game*. That is:

$$u^i_B + u^i_P \leq 1 \quad (7)$$

Where $u^i_B \in [0, 1]$ and $u^i_P \in [0, 1]$.

Therefore, the Nash Product for the attribute X_i is determined by:

$$\begin{cases} max(u^i_B - d^i_B)(u^i_P - d^i_P) \\ s.t \\ u^i_B + u^i_P \leq 1 \end{cases} \quad (8)$$

To find the maximum of Nash Product, we use the Lagrangian method, which is expressed by the following function:

$$
\begin{aligned}
L(q_B^i, q_P^i, \lambda) &= (u_B^i - d_B^i)(u_P^i - d_P^i) + \lambda(u_B^i + u_P^i - 1) \\
&= (\omega_B^i q_B^i - \beta^i \sigma_B^i)(\omega_p^i q_p^i - \alpha^i \sigma_P^i) + \lambda(\omega_B^i q_B^i + \omega_p^i q_p^i - 1)
\end{aligned} \tag{9}
$$

$$
\begin{cases}
\frac{\delta L}{\delta q_B^i} = \omega_B^i((\omega_P^i q_P^i - \alpha^i \sigma_P^i) + \lambda) = 0 \\
\frac{\delta L}{\delta q_P^i} = \omega_P^i((\omega_B^i q_B^i - \beta^i \sigma_B^i) + \lambda) = 0 \\
\frac{\delta L}{\delta \lambda} = (\omega_B^i q_B^i + \omega_p^i q_p^i - 1) = 0
\end{cases} \tag{10}
$$

Then, we conclude that the only values (q_B^{*i}, q_P^{*i}) to satisfy the maximum of the Nash Product are presented as follow:

$$
\begin{cases}
q_B^{*i} = \frac{1+\beta^i \sigma_B^i - \alpha^i \sigma_P^i}{2\omega_B^i} \\
q_P^{*i} = \frac{1+\alpha^i \sigma_P^i - \beta^i \sigma_B^i}{2\omega_P^i}
\end{cases} \tag{11}
$$

Finally, to find the Pareto Optimum, we resolve the following Nash Product for the global utility. Let $u = (u_B, u_P)$ be the global utility for all attributes where $u_B = \sum_{i=1}^{n} \omega_B^i q_B^i$ and $u_P = \sum_{i=1}^{n} \omega_P^i q_P^i$

The Nash Product for the global attribut is:

$$
\begin{cases}
max(u_B - d_B)(u_P - d_P) \\
s.t \\
u_B + u_P \leq n
\end{cases} \tag{12}
$$

Where $d_B = \sum_{i=1}^{n} \beta_i \sigma_B^i$, $d_P = \sum_{i=1}^{n} \alpha_i \sigma_P^i$ are the global disagreement payoff for all attributes and n is the total number of attributes.

Therefore, the global pareto optimum is:

$$
\begin{cases}
u_B^* = \sum_{i=1}^{n} u_B^{*i} = \sum_{i=1}^{n} \omega_B^i q_B^{*i} = \frac{n+\sum_{i=1}^{n} \beta^i \sigma_B^i - \sum_{i=1}^{n} \alpha^i \sigma_P^i}{2} \\
u_P^* = \sum_{i=1}^{n} u_P^{*i} = \sum_{i=1}^{n} \omega_P^i q_P^{*i} = \frac{n+\sum_{i=1}^{n} \alpha^i \sigma_P^i - \sum_{i=1}^{n} \beta^i \sigma_B^i}{2}
\end{cases} \tag{13}
$$

Hence, if the global utility of each player reaches a Pareto optimum, it means that the context broker and the context provider have reached an agreement and will then sign a contract. Otherwise, the negotiation will be canceled and the context broker will try to find another suitable context provider.

6.2 Numerical Results

This section presents a simulations of the CLA negociation, which are collected using Matlab Simulator. We use Formula (12) with 100 QoC attributes and each importance weight ω_B^i (respectively ω_P^i) takes a random value between 10% and 90% depending on the preference of player C_B (respectively on the offering ability of player C_P) for the attribute X_i.

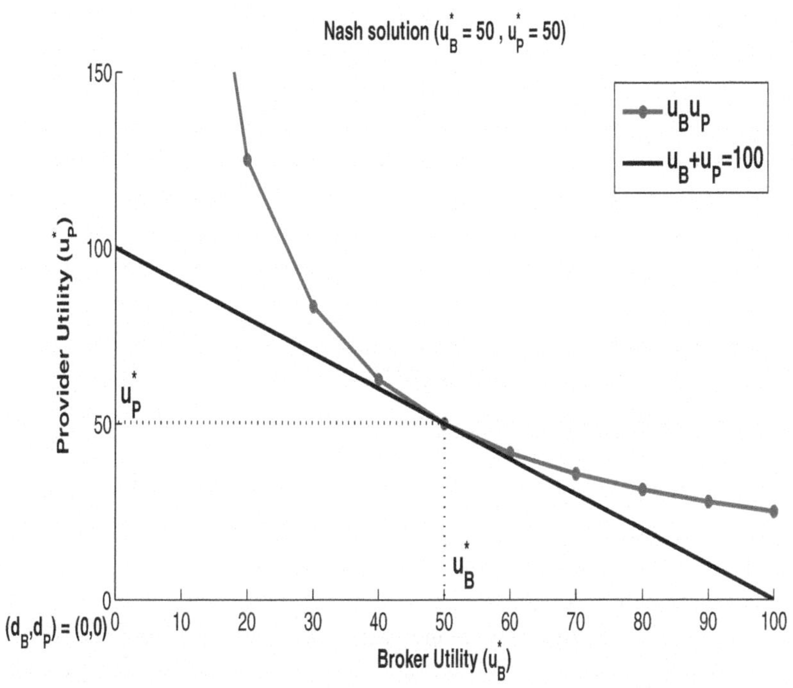

Fig. 2 Nash Bargaining Solution in the absence of disagreement points.

Fig. 2 depicts the Nash bargaining solution in the case of absence of disagreement points between the two players. From the figure, we can see that the negotiation process is able to converge to a close optimum result when $d_B = d_P = 0$, which means that there is no disagreement points between the two players. The objective function in this case is: $f(u_B, u_P) = (u_B - d_B)(u_P - d_P) = u_B u_P$, which is an hyperbolic function. We can also notice that the Pareto optimum ($u_B^* = 50$, $u_P^* = 50$) is a fair solution to the problem. It means that at this point: (1) 50% of QoC attributes demanded by the context consumer, through a context broker, are satisfied; and (2) 50% of delivered QoC attributes by the context provider are offered with high quality. This point corresponds to the intersection of the hyperbolic curve and the constraint line for the following expression: $u_B + u_P \leq 100$.

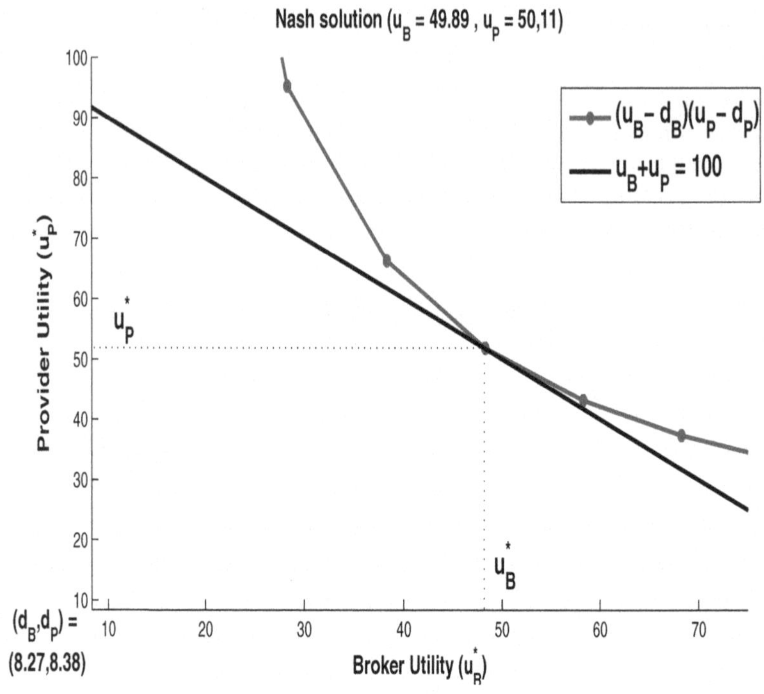

Fig. 3 Nash Bargaining Solution with status quo.

When $d_B > 0$ and $d_P > 0$ the reservation value will change randomly during the negotiation rounds. That's why we test values simultaneously. If $\sigma_B^i < 50\%$ (resp. σ_P^i), the penality β^i (resp. α^i) is between 20% and 30%. Otherwise, the penality β^i (resp. α^i) is between 10% and 20%.

Fig. 3 depicts the Nash Bargaining solution with status quo. We deduce from Fig. 3 that the objective function $f(u_B, u_P) = (u_B - d_B)(u_P - d_P)$ is also an hyperbolic function for which (d_B, d_P) is the origin point and the Nash solution is the common point between the constraint line and the function curve. This solution corresponds to the point ($u_B^* = 49.89$, $u_P^* = 50.11$). It means that if the context provider tries to maximize its global utility u_P to more than 50%, then the global utility of the context consumer will not exceed the average utility (i.e. no player can increase its utility without decreasing the utility of the other player). This is one of the fundamental characteristics of a Pareto optimum. In other words, the global utility of the context provider grows as long as the context provider refuses the context brokers proposals while this latter does not reject the context providers proposals, this result shows that the context broker and the context provider coorporate each other dynamically.

Fig. 4 depicts the Nash bargaining solution with equal disagreement points between the two players. When both players express the same disagreement during multiple rounds of negotiation of different QoC attributes, they get similar values for their global utility functions (e.g $u_B^* = 50$, $u_P^* = 50$). In this situation, the Pareto optimum ($u_B^* = 50$, $u_P^* = 50$) is a fair solution between the two parties.

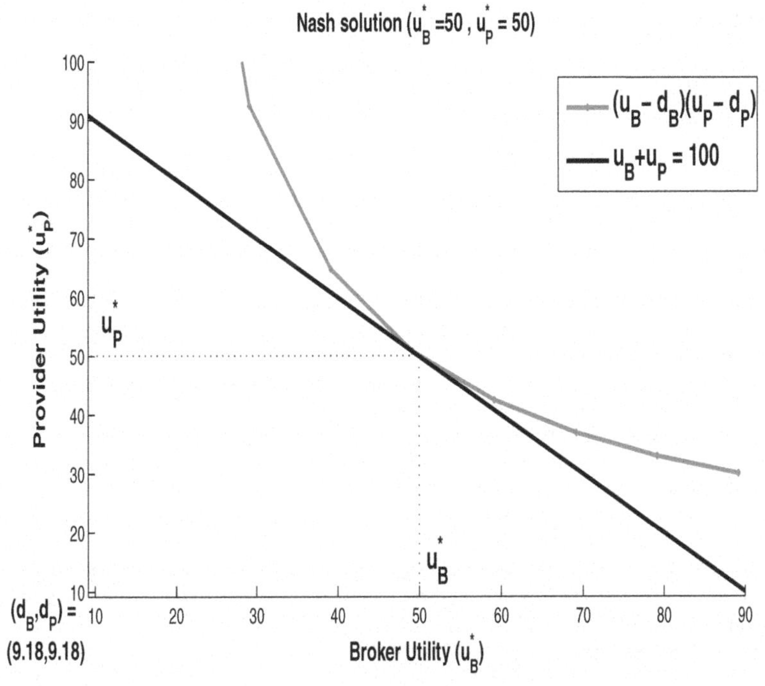

Fig. 4 Nash Bargaining Solution with equale disagreement points.

7 Conclusion

With the proliferation of mobile and Internet-enabled handheld devices, users are increasingly demanding context-aware services, which can adapt to their current context and circumstances. These services may obtain context information from several context providers. However, it is essential to find appropriate context providers that are capable to supply high quality context information and negotiate with them the terms and quality levels of context information delivery. In this paper, We have used a bargaining model for evaluating the global utility of each party as a linear function of Quality of Context (QoC) attributes during the negotiation phase using Nash bargaining game. The results of this model show that the two parties of the negotiation approve the proposal on QoC levels when the global utility of each party converges to a Pareto optimum.

As a future work, we intend to investigate how the CLA negotiation involves when the importance weights, given by the negotiation parties to the QoC attributes, vary in time. We are also interested in studying the effect of the interdependence between QoC attributes on the CLA negotiation.

References

1. Badidi, L., Esmahi, L.: A Cloud-based Approach for Context Information Provisioning. World of Computer Science and Information Technology Journal (WCSIT) **1**, 63–70 (2011)
2. Maarouf, A., Marzouk, A., Haqiq, A., El Hamlaoui, M.: Towards a MDE approach for the establishment of a contract service level monitoring by third party in the cloud computing. In: Signal-Image Technology and Internet-Based Systems (SITIS), pp. 715–720 (2014)
3. Buchholz, T., Kpper, A., Schiffers, M.: Quality of context: what it is and why we need it? In: Proceedings of the 10th International Workshop of the HP OpenView University association (HPOVUA) (2003)
4. Copil, G., Moldovan, D., Salomie, I., et al.: Cloud SLA negotiation for energy saving - a particle swarm optimization approach. In: IEEE International Conference on Intelligent Computer Communication and Processing (ICCP), pp. 289–296 (2012)
5. Muller, C., Resinas, M., Ruiz-Cortes, A.: Automated Analysis of Conflicts in WS-Agreement. IEEE Transactions Services Computing **7**, 530–544 (2014)
6. Katsuhide, F., Takayuki, I., Klein, M.: Finding nash bargaining solutions for multi-issue negotiations: a preliminary result. In: The 1st International Working Conference on Human Factors and Computational Models in Negotiation (HuCom 2008), pp. 63–70 (2008)
7. Kim, Y., Lee, K.: A quality measurement method of context information in ubiquitous environments. In: ICHIT 2006, pp. 576–581. IEEE Computer Society, Washington, DC (2006)
8. Manzoor, A., Truong, H.L., Dustdar, S.: Quality aware context information aggregation system for pervasive environments. In: International Conference on Advanced Information Networking and Applications Workshops (WAINA 2009), pp. 266–271 (2009)
9. Nash, J.F.: The bargaining problem. Econometrica **18**, 155–162 (1950)
10. Bouterse, B., Perros, H., Thuente, D.: Multiobjective cloud capacity planning for time-varying customer demand. In: High-capacity Optical Networks and Emerging/Enabling Technologies (HONET), pp. 84–88 (2014)
11. Xianrong, Z., Martin, P., Powley, W., et al.: Applying bargaining game theory to web services negotiation. In: IEEE International Conference on Services Computing (SCC), pp. 218–225 (2010)
12. Badidi, E., Esmahi, L.: A Scalable Framework for Provisioning Context-aware Application Services with High-quality Context Information. The International Journal of ACM Jordan **1**, 86–97 (2011)

A New Secure Network Architecture to Increase Security Among Virtual Machines in Cloud Computing

Zakaria Elmrabet, Hamid Elghazi, Tayeb Sadiki and Hassan Elghazi

Abstract Cloud computing is a new model of computing which provides scalability, flexibility and on-demand service. Virtualization is one of the main components of the cloud, but unfortunately this technology suffers from many security vulnerabilities. The main purpose of this paper is to present a new secure architecture of Virtual Network machines in order to increase security among virtual machines in a virtualized environment (Xen as a case study). First, we expose the different network modes based on Xen Hypervisor, and then we analyses vulnerabilities and security issues within these kind of environment. Finally, we present in details new secure architecture and demonstrate how it can face the main security network attacks.

Keywords Cloud computing · Virtualization · Virtual network security · Xen hypervisor · Spoofing · Sniffing · Mac flooding

1 Introduction

The Cloud Computing is a new model and computing paradigm which offer scalable on-demand services to consumer with greater flexibility and lesser infrastructure investment. Economic benefits of the cloud computing are the main factors that encourage consumers to adopt it, according to [1] 91% of organizations in US and Europe agreed that reduction in cost is a major reason for them to migrate to this new environment.

Z. Elmrabet(✉) · H. Elghazi
National Institute of Posts and Telecommunications, Rabat, Morocco
e-mail: zakaria.elmrabet2@gmail.com, elghazi@inpt.ac.ma

H. Elghazi · T. Sadiki
International University of Rabat, Rabat, Morocco
e-mail: {hamid.elghazi,tayeb.sadiki}@uir.ac.ma

E. Sabir et al. (eds.), *The International Symposium on Ubiquitous Networking*,
Lecture Notes in Electrical Engineering 366,
DOI: 10.1007/978-981-287-990-5_9

Virtualization is one of the main components of the cloud computing, this component is important because it enables multi-tenancy and on-demand use of scalable shared resources [2]. Virtualization means create a virtual version of a device or resources such as Server, Storage or Network [3] and implements it using a "Hypervisor".

However, vulnerabilities in a virtual environment expose the security of the information stored in the cloud computing to many big challenges [4]. This vulnerability is classified on three main issues: availability, integrity and confidentiality of the information (or the CIA triad).

Fig. 1 The CIA Triad of information security

The CIA triad, are the three fundamental pillars of information security [5]. They are defined as:

- **Availability:** The data is available when needed.
- **Integrity:** The data is not modified without being detected.
- **Confidentiality:** The data remains undisclosed to unauthorized parties

Thus, any global solution for security problems within Cloud Computing shall take into account these three parameters. However, not every attack compromises every attribute of the triad, there are some attacks affect the three attributes, whereas some others only one or two attributes.

The table 1 shows the impact of the most known dangerous attacks in Cloud Computing on the availability, integrity and confidentiality of the stored information: spoofing attack, sniffing attack and the Mac flooding attack.

Table 1 Security Attributs affected by network attacks

	Spoofing/ Poisoning	Sniffing	Mac Flooding
Availability	X		X
Integrity	X		
Confidentiality	X	X	X

We believe that in the context of Cloud Computing and in order to guarantee security among virtual machines, hypervisor should ensure isolation of each virtual machine (VM) by using dedicated physical channel. However, and as presented previously hypervisor (In our case Xen Hypervisor) uses software channel to create bridge and in order to link virtual machines. In fact, software channels isolation is easy to be broken [6].

The main purpose of this paper is to propose a new architecture to increase security in virtual environment and especially in virtual network; the hypervisor used in our study is the Xen Hypervisor. The proposed architecture is composed of two main layers: Switch Layer and Firewall Layer. The purpose of the switch layer is to prevent from sniffing attack, whereas the firewall layer will prevent from spoofing and Mac flooding attack.

In this paper, we start first by introducing the virtual network modes in Cloud Computing. Then, in the next sections we expose a critical analysis of security issues within these modes. Finally, we present in more details our new architecture model.

2 Related Work

In order to secure the communication among virtual machines the Xen Hypervisor environment, [12] proposes a model composed of three layers: Routing layer, Firewall and shared Network.

The routing layer connects the physical network and creates a logical dedicated channel in order to establish the communication between the virtual network and the physical network. The administrator assigns a set of unique static IDs to each shared network, these ID are stored in a configuration file and they could be used to monitor the source of packets sent from each shared network [12].

The second layer is the Firewall layer; the purpose of this layer is to prevent spoofing attacks between virtual machines belongs to a different shared network, by identifying the network ID specified in the configuration file [12].

The third and the last layer in the proposed model is the shared network layer, in this layer the authors assume that the VMs belong to a same virtual shared network are trustful to each other and they are working for the same company or organization [12].

The [12] model is a good approach to increase security among virtual machines within a virtual network. However, it can prevent only spoofing attack. Other attacks, for instance: Sniffing, and Mac flooding still exist and a malicious person could use a virtual machine to launch these kinds of attacks. In fact, it could affect the availability and the confidentiality of the information hosted in the cloud.

3 Critical Analysis of Vulnerabilities in Virtual Network

3.1 Virtual Network

Virtual network is a technique used to create independent or isolated logical network within a shared physical network. Existing hypervisors (such as Xen and Vmware) offer this mechanism to share the access of physical network [12]. Xen is open source software and we will use it as an example of application of our approach.

Xen is a project developed in 2003 at the University of Cambridge Computer Laboratory. There are both commercial and free versions of Xen.it is a Hypervisor that provides a platform for running multiple instance of the operating system on one physical hardware. It supports several operating systems such as: Linux, FreeBSD and also Microsoft Windows.in this paper the free version of Xen will be used [11].

Xen Hypervisor refers to each virtual machine as a domain, and there are two types of domains: the first one called Domain0 (dom0) is a privileged domain that can access the hardware resources and also contains some tools for managing other domains in other word other virtual machines, while the second type is called domU, and it is an unprivileged virtual machines that created and managed by dom0 [11].

All requests from domU instances for hardware are passed via dom0, and then they will be forwarded to the actual hardware. The figure below illustrates an example of 3 virtual machines: dom0, doml and dom2 running on Xen hypervisor.

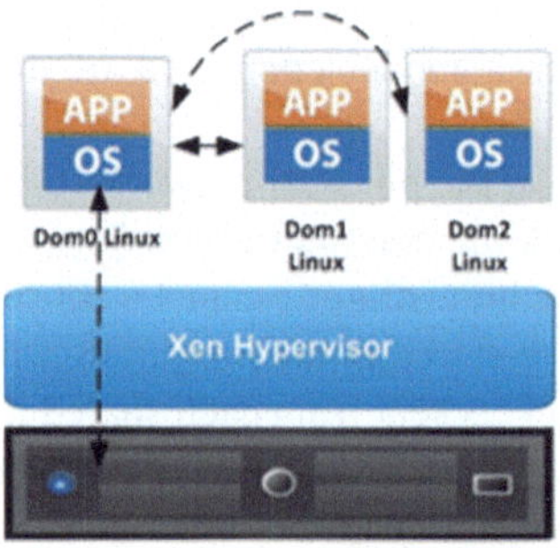

Fig. 2 Three virtual machines: dom0, doml and dom2 running on Xen Hypervisor

Xen Hypervisor provides different networking modes to configure virtual network:

a. Bridged Networking

Bridge is a technique used to connect two LANs (Local area network) together and forwards frames using their MAC (media access control) address. It is the default option for Xen networking.

Before forwarding frames, the bridge checks its bridge table where MAC addresses are stored. In order to fill up this table with the MAC addresses the bridge uses broadcasting.

The following figure shows an example of two virtual machines (dom0, domU:1 and domU:2) connecting to the bridge.

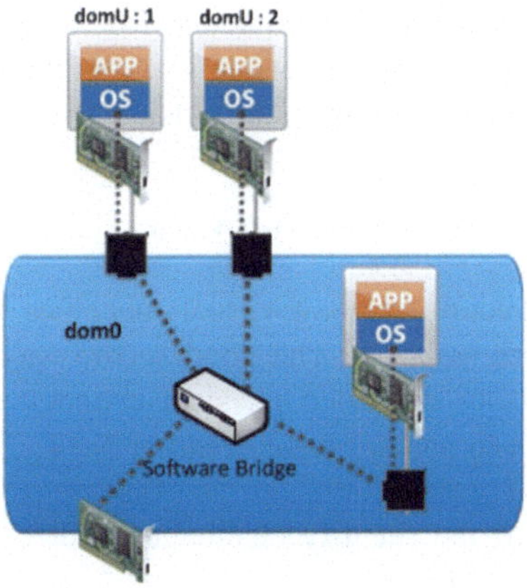

Fig. 3 Xen Bridged Networking

b. Routed Networking

Routed networking is the second mode offered by Xen Hypervisor, is a technique that uses IP addresses to send and receive network traffic from one segment to another. As shown in the figure below, dom0 in Xen acts as a default gateway for the other domains to communicate with the outside Word.

In addition, the domU machines are visible from the outside and can be directly accessed via the default gateway in this case dom0.

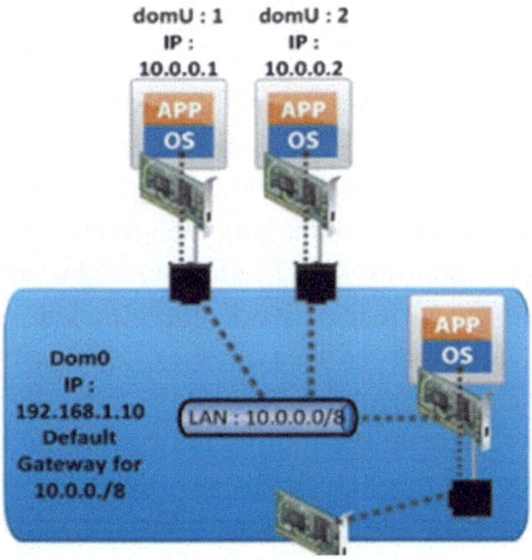

Fig. 4 Xen Routed Networking

c. Virtual Local Area Network (VLAN) with Network Address Translation (NAT)

The VLAN is the third mode for configuring virtual network in Xen Hypervisor. In this mode the domU machines are created on a Private LAN, and they use the dom0 as a default gateway to reach the outside.

The mean difference between this mode and Routed mode, in the first one the domU machines are visible from the outside while in the second mode the domU machines are hidden and protected from the outside.

3.2 Vulnerabilities in Virtual Network

In order to guarantee security among virtual machines within virtual network, hypervisor should ensure isolation. The good secure way to isolate each virtual machine (VM) is by using dedicated physical channel as presented previously hypervisor uses software to create bridge, route or NAT. Thus, isolation is easy to be broken [12].

Bellows are presented the security challenges related to virtual network and especially in Xen Hypervisor:

- **Spoofing and Poisoning**

Arp (Address Resolution protocol) is a protocol used in the local Network to resolve an IP Address into MAC (Medium Access Controllers) address. A Virtual Machine (VM) invokes an Arp resolution when it needs a MAC address of a new IP, and then the result is saved in its cache [7].

Arp poisoning is a malicious technique used to modify the association between an IP address and its corresponding MAC address [7]. Arp poisoning utilizes Arp Spoofing, a VM sends a spoofed ARP message with a forged IP address to the "Virtual Switch", once the Virtual Switch receives the request it will dynamically update its cache. As a result, frames intended for the legitimate VM can be mistakenly sent to the Attacker VM.

To illustrate the processes of spoofing within a virtual network, assuming that a VM-A would communicate with VM-B and the attacker VM-C will launch a spoofing attack in order to capture the communication. The figure 5 (a) illustrates the state of the routing table before launching the attack.

Once the routing table records the IP address, MAC address and the port of each virtual machine in the virtual network, the attacker virtual machine VM-C sends an ARP request with a forged IP addresses, for the first time with the same IP address as VM-A and for a second time with the same IP address as VM-B, and then the virtual route will update the routing table with the new information coming from the attacker-virtual machine VM-C as illustrated in figure 5 (b). In fact, any traffic coming to or from the VM-A or VM-B would be mistakenly sent to VM-C instead.

The spoofing attack can be used to conduct other attacks resultant a serious damage, such as:

- **Denial-of-service (Dos)** : it is an attack where the malicious person launches a spoofing attack, and then he/she chooses to block the frames passing through he/she machine instead of just sniffing or modifying its contents.as a result the regulars virtual machines could not communicate with each other.

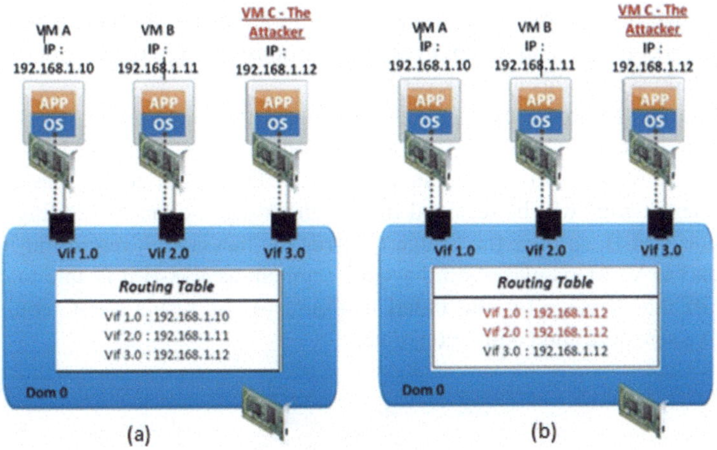

Fig. 5 Spoofing Attack in Xen Hypervisor (a) Content of the routing table before launching the attack, (b) Content of the routing table before launching the attack

- **Man-in-the –middle:** it is an attack where the malicious person launches a spoofing attack, and then he/she chooses to listen or to modify the communication between two virtual machines, and unfortunately these ones believe that they communicate directly with each other.
- **Session Hijacking:** it is an attack where the malicious person stole a cookie after launching a spoofing attack. Recall that cookies are used by the client to maintain a session on web servers. Therefore, if the attacker stole the cookie, he/she could connect easily to the web server with the legitimate user's privileges, and he/she could access to sensitive data.

In Xen Hypervisor, the route mode plays role as a "Virtual switch"[12], therefore a malicious VM can launch an ARP spoofing attack against the "virtual switch" in order to redirect the traffic to its machine and then conduct other malicious attacks.

- **Sniffing**

Sniffing or listen to the traffic, it is an attack where the attacker insert itself between two communicating hosts in order to capture the frames and then retransmit it [8][9][10].

In Xen Hypervisor, Bridge mode is a technique used to connect two LAN (Local Area Network) together and it forwards frames using their MAC (media access control) address [11]. Therefore the bridge plays a role as a "Virtual Hub" for each segment [12], in which a malicious VM is able to sniff easily the virtual network by first configuring the "promiscuous mode " of its network card and then using a free sniff tool such as "tcpDump" and "Wireshark"[13].

Recall that the "promiscuous mode" refers to an operation mode where a network card is configured to accept every packet transmitted in the network whether they are addressed to this network card or not.

To illustrate the processes of sniffing within a virtual network in this case: Xen Hypervisor, assuming that VM-A would communicate with VM-B and the Attacker VM-C will sniff the communication.so the VM-A sent a message to a VM-B, the message will be received first by the bridge before forwarding it to its destination VM-B, so once the bridge receives the message, it retransmit it to the VM-B and as VM-C is placed on the same segment as VM-B it is able to capture or to sniff the message (as illustrated in figure 6). As result VM-C could have access to a sensitive and confidential data.

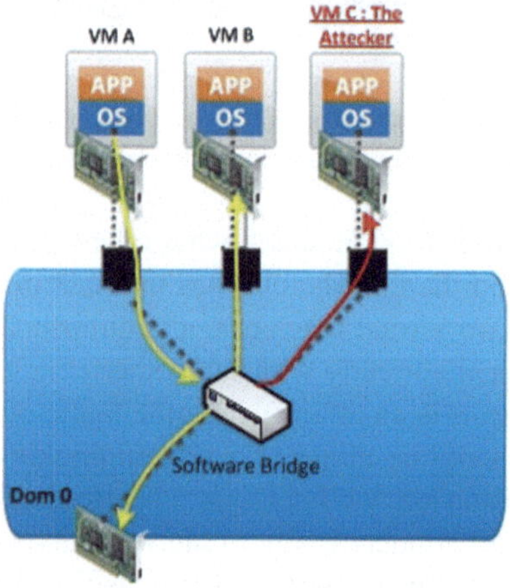

Fig. 6 Sniffing attack in Xen Hypervisor

- **Mac Flooding**

The Content-Addressable Memory (CAM) table stores MAC addresses in the switch, and it is fixed size. This attack consists of flooding a switch with MAC addresses using forged ARP packets until the CAM table is full. Then, the switch operates as a hub and starts broadcasting the traffic without a CAM entry [14].

In Xen Hypervisor, the route mode plays role as a "Virtual switch"[12], therefore a malicious VM can launch a MAC flooding attack against the "virtual switch" in order to fill up it CAM table and then sniff the traffic coming from and going to other VMs.

3.3 *Security Vulnerabilities on Virtual Network Mode*

The three modes available on Xen Hypervisor to configure network are: Bridge Mode, Route Mode and NAT, and as described above these modes are vulnerable. The table below resumes the security vulnerabilities related to each virtual network mode of Xen Hypervisor. The sign (-) means that we can launch an attack in that mode, and the sign (+) means we cannot lunch an attack in that mode.

Table 2 Vulnerabilities in virtual network

Attack \ Network Mode	Routed Mode	NAT Mode	Bridged Mode
Spoofing	-	-	-
Sniffing	+	+	-
Mac Flooding	-	-	-

From the table 2 we think that a new architecture shall be defined to protect virtual network from attacks. The main purpose of this paper is to provide architecture inspired from classical network security. This kind of architecture proved their efficiency and confidence in classical network, thus they can be reproduced in the context of virtual network.

4 A New Architecture Model

In order to increase security among virtual machines, we propose a new architecture model to prevent users from network attacks. Especially, from spoofing, sniffing and mac flooding attack. This architecture is composed of three layers: VLans Layer, Virtual Switch layer and firewall layer as shown in figure 7.

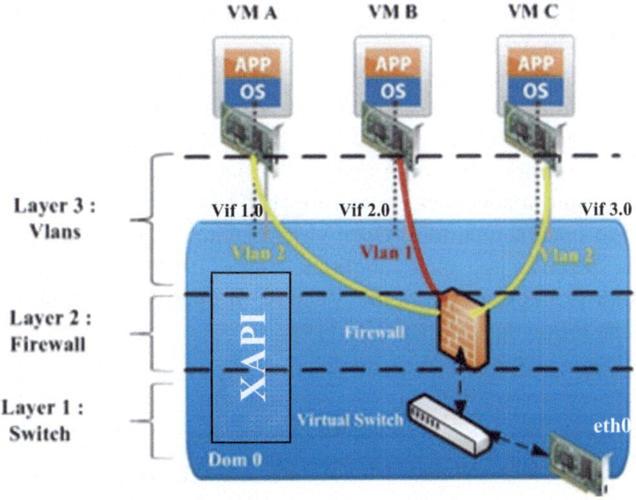

Fig. 7 A secure virtual network model

- **Layer 1: Virtual Switch,** the main purpose of this layer is to prevent sniffing attack. First, it allows virtual machines to communicate with the outside network and also permits virtual machines belong to the same VLAN to communicate securely with each other. The virtual machines belong to the same organization are assigned by the administrator to a specific VLAN, in fact the communication among virtual machines within this VLAN could not be sniffed form other VLANs.

Open virtual Switch [16] is an open source tool and it could be used with Xen Hypervisor to implement a virtual switch.

The integration of Open vSwitch with the Hypervisor is similar to a bridged configuration; however, instead of connecting each virtual network card (vif) to a bridge the open virtual switch will be used. As an example to set up a virtual switch (in this case the Open Virtual switch) we could use the following steps :

- After the installation, we create a new bridge named vswith0:

```
ovs-vsctl add-br vswith0
```

- And we attach the physical network interface card (eth0) to the new bridge (vswith0) :

```
ovs-vsctl add-port vswith0 eth0
```

- **Layer 2: Virtual Firewall,** will be used in order to prevent spoofing, poisoning and MAC flooding attack a firewall is crucial. At this level, we define a set of security policies mainly including: any packet send from one VLAN and attempts to reach another VLAN will be dropped. Once the virtual switch records information about all virtual machines on the virtual network in its CAM table, any packet tries to modify or to fill up this table will be dropped.

  Iptable [15] an open source tool could be used with Xen Hypervisor to implements these policies.
- **Layer 3 : Vlans**, provides a logical segmentation of switch ports, allowing communication as if all ports were on the same physical LAN segment. Limiting the broadcast traffic to a subset of the switch ports saves significant amounts of network bandwidth and also to increase security among users by preventing some attacks such as: sniffing[17][18].

  As an open source tool, Open virtual switch could be used to either implement a virtual switch and to configure Vlans [19].

  As an example to configure VLANs on open Virtual Switch, we could perform the following configuration [19]:

- Create an Open Virtual Switch

```
ovs-vsctl add-br br0
```

- Add the physical network interface (eth0) to the virtual switch (br0)

```
ovs-vsctl add-port br0 eth0
```

- Add VM A (which has the virtual interface vif1.0) and VM C (which has the virtual interface vif3.0) on VLAN 2. This means that traffic coming into virtual switch from VM A or VM C will take the tag number 2 and will be considered part of VLAN 2:

```
ovs-vsctl add-port br0 vif1.0 tag=2
ovs-vsctl add-port br0 vif3.0 tag=2
```

- Add VM B (which has the virtual interface vif 2.0) on VLAN 1.this means that traffic coming to the virtual switch from VM B will belong to the VLAN 1.

```
ovs-vsctl add-port br0 vif2.0 tag=1
```

As a result, the virtual machines belong to the same Vlan, for instance VM A and VM C witch belong to the VLAN 2 should succeed, however virtual machines belong to a different VLANs, for example VM A and VM B should not succeed. Therefore the isolation of traffic is assured by this method.

In order to manage all aspect of Xen, including VMs, Storage, and especially networking, we have used XAPI (Xen Server API), it provides an external interface for configuring the system and also enables the hypervisor to establish a communication with other plug-in such as Open Virtual Switch[20].

In order to evaluate the efficiency of this new model regarding attack within a virtual network and specially: sniffing, spoofing, poisoning and MAC flooding attacks. The table below presents a comparison between the proposed model in this paper and the existing modes.

Table 3 Comparaison between the proposed mode and the existing virtual network modes in Xen Hypervisor in term of security

Network Mode / Attack	Routed Mode	NAT Mode	Bridged Mode	Proposed Mode
Spoofing	-	-	-	+
Sniffing	+	+	-	+
Mac Flooding	-	-	-	+

5 Conclusion

The main purpose of this paper is to improve security in the cloud and especially in the virtual network, which constitute the main component of the cloud. Through our work we presented a critical security analysis of the different mode of virtual network (based on Xen Hypervisor). Then we expose the main vulnerability related to those modes. Finally, we described in more details a new architecture witch composed of three layers, Vlans, Firewall and Switch. This new architecture can prevent from some attacks such as: spoofing, sniffing and mac flooding.

As future work, we will implement the new architecture in Xen hyervisor, with the Firewall Iptable and the Open virtual Switch, in order to evaluate its efficiency in term of security and performance.

References

1. Chirag, M., Patel, D., et al.: A survey on security issues and solutions at different layers of Cloud computing. The Journal of Supercomputing **63**(2), 561–592 (2013). Springer
2. Salah, K., et al.: Using Cloud Computing to Implement a Security Overlay Network. Security & Privacy **11**(1), 44–53 (2013). IEEE
3. Vangie, B.: (website visited April 2015). http://www.webopedia.com/TERM/V/virtualization.html
4. National vulnerability database version 2.2.NIST (website visited mars 2015). http://web.nvd.nist.gov/view/vuln/searchresults?query=virtual&search%_type=all&cves=on
5. Mariam, K.: A Methodology for Cloud Security Risks Management. Cloud Computing: Challenges, Limitations and R&D Solutions, 75–104 (2014). Springer
6. Grover, J., Shikha, Sharma, M.: Cloud computing and its security issues — a review. In: 2014 International Conference on Computing, Communication and Networking Technologies (ICCCNT), pp. 1–5. IEEE (2014)
7. Bruschi, D., Ornaghi, A., Rosti, E.: S-ARP: a secure address resolution protocol. In: Proceedings of the 19th Annual Computer Security Applications Conference, 2003, pp. 66–74. IEEE (2003)
8. Chin, T.S., Singh, Y.P.: Single-hop wavelength assignment using an ant algorithm in WDM MESH network. WSEAS Transactions on Computers **5**(7), 294–300 (2006)
9. Zheng, W., Li, C.: An Algorithm Against Attacks Based on ARP Spoofing. Journal of Southern Yangtze University (Natural Science Edition) **2**(6), 167–1696 (2003)
10. Tian, Z.H., Fang, B.X., Li, B., et al.: Avulnerability-driven approach to active alert verification for accurate and efficient intrusion detection. WSEAS Transactions on Communications **4**(10), 1002–1009 (2005)
11. Chaganti, P.: Xen Virtualization, p. 74 (2007)
12. Wu, H., et al.: Network security for virtual machine in cloud computing. In: 2010 5th International Conference on Computer Sciences and Convergence Information Technology (ICCIT), pp. 18–21. IEEE (2010)
13. CERT Training and Education, Carnegie Mellon University (2009). http://science.hamptonu.edu/compsci/docs/iac/packet_sniffing.pdf
14. Hayriye, A., Sven, K., Henry, L., et al.: Securing layer 2 in local area networks. In: Networking - ICN 2005, 4th International Conference on Networking, Reunion Island, France, Proceedings, Part II, pp. 699–706. Springer (2005)
15. http://www.netsecurity.org/software.php?id=4 (website visited March 2015)
16. Open vswitch. http://openvswitch.org/ (website visited April 2015)
17. http://www.omnisecu.com/cisco-certified-network-associate-ccna/advantages-of-vlan.php (website visited April 2015)
18. Vmweare Virtual Networking Concept (2007). http://www.vmware.com/files/pdf/virtual_networking_concepts.pdf
19. Vlan configuration. http://openvswitch.org/support/config-cookbooks/vlan-configuration-cookbook/ (website visited April 2015)
20. Justin, P., et al.: Virtual Switching in an Era of Advanced Edges. http://benpfaff.org/papers/adv-edge.pdf (website visited July 2015)

Text Mining for Suspicious Contents in Mobile Cloud Computing Environment

Salim Alami and Omar Elbeqqali

Abstract Mobile devices, such as tablets and smartphones, have become the main computing platform for many people. Ubiquitous computing, as a concept, has developed with the emerging of cloud computing technology that has forced the mobile devices industry to prerequisite the bringing of cloud computing to mobile domain. Mobile Cloud Computing MCC is a service that allows users of mobile devices high availability of their personal applications as well as their own content, everywhere and anytime. Unfortunately, malicious people take advantage of this technological achievement in the sense that they store all illegal information on cloud in order to hide all digital illegal records justifying their illicit acts served by their mobile devices. What is more, mobile forensic expertise on those mobile devices cannot be accomplished by digital investigators of law enforcement, simply because all the storage is done in the cloud. In this vein, Mobile Cloud Computing MCC technology is a double-edged weapon; it has made life easier on one hand. And, it has complicated the work of law enforcement authorities to find truth, on the other hand. In cloud environment, malicious users can be stored several and various formats of suspicious content (text, image, video…), so in this work we will focus only on textual content. Text mining is an effective way to add semantics aspect to this communication's form presenting a significant research challenge. Similarity approach is used in text analysis to detect suspicious text contents in cloud storage. So, in this paper we will present a state-of-the-art and research challenges of mobile cloud computing. We will also discuss the problem of data management and data analysis on a cloud environment. Ultimately, we will suggest an approach to come up with the aforementioned problems.

Keywords Cloud computing · Mobile Cloud Computing · Digital forensics · Mobile forensics · Cloud forensics · Text analysis · Text mining · Profiling

---

S. Alami(✉) · O. Elbeqqali
LIIAN Laboratory, Sidi Mohammed Ben Abdellah University, Fez, Morocco
e-mail: {salim.alami,omar.elbeqqali}@usmba.ac.ma

E. Sabir et al. (eds.), *The International Symposium on Ubiquitous Networking*,
Lecture Notes in Electrical Engineering 366,
DOI: 10.1007/978-981-287-990-5_10

1 Introduction

The development of cloud computing is one of the major advances in computing history and the most transformative computing technologies [1]. The number of consumer cloud-based service users in worldwide increases day after day. According to the report published by STATISTA[1] website, the presented statistics show the number of consumer cloud-based online service users worldwide. In 2018, approximately 3.6 billion internet users are estimated to use cloud computing services, up from 2.4 billion users in 2013.
In this vein, mobile devices play a role of paramount importance and have gained speed by the emerging cloud computing technologies, as these devices take an important role in the human behavior and habits as communication and entertainment with neglecting time and place. Nowadays, all mobile devices use cloud that empowers users to:

- Synchronize photographs, music and files across multiple devices;
- Continue working where they left work on one device, upon opening a different one;
- Synchronize user settings such as passwords and browser settings across all devices;
- Apply online storage;
- Leverage pushed updates to applications across all devices…etc.

The use of mobile devices and the emergence of cloud computing technology gave birth of Mobile Cloud Computing. There are several definitions of mobile cloud computing, and diverse research refers to different concepts of the 'Mobile Cloud Computing'.

The Mobile Cloud Computing Forum [2] defines MCC as "Mobile Cloud computing at its simplest refers to an infrastructure where both the data storage and the data processing happen outside of the mobile device. Mobile cloud applications move the computing power and data storage away from mobile phones and into the cloud, bringing applications and mobile computing to not just smartphone users but a much broader range of mobile subscribers".

In this definition we are going to focus on "data storage and the data processing happen outside of the mobile device", in particular the kind of data treated and stored on cloud, because there is malicious people that take advantage of this hi-tech achievement in the sense that they store all illicit information on cloud in order to hide all digital illegal records justifying their illegal acts served by their mobile equipments. Mobile Cloud Computing offers several and different kind of services allowing users to store different formats of data, including image, video, text…etc.

[1] http://www.statista.com/statistics/321215/global-consumer-cloud-computing-users/

As example, malicious people can store document containing a terrorist attack plan, this document that can be recover by another person from the same terrorist cell. Mobile Cloud Computing can help criminals to coordinate and manage online suspect actions.

The paper is organized as following: in section 2, we present background and challenges including problematics, then the proposal approach in section 3. In section 4 an evaluation of results; finally we provide a conclusion and perspectives in section 5.

2 Background and Challenges

2.1 Digital Forensics

Digital forensics is a new science, branch of forensic science that has implemented to include the forensics of all digital technology. Digital forensics is the application of computer science principles to recover electronic evidence for presentation in a court of law [3]. To achieve its goals, digital forensics require the respecting of chain of custody that is defined as a verifiable provenance or log of the location and possession history of evidence from the point of collection at the crime scene to the point of presentation in a court of law.

Digital forensics has been defined as "the use of scientifically derived and proven methods toward the preservation, collection, validation, identification, analysis, interpretation, documentation, and presentation of digital evidence derived from digital sources for the purpose of facilitation or furthering the reconstruction of events found to be criminal, or helping to anticipate unauthorized actions shown to be disruptive to planned operations" [4].

Mobile device forensics is a branch of digital forensics relating to recovery of digital evidence or data from mobile devices under forensically sound conditions [5].

2.2 Cloud Forensics

Cloud forensics is a cross discipline of cloud computing and digital forensics. Cloud computing is a shared collection of configurable networked resources (e.g., networks, servers, storage, applications and services) that can be reconfigured quickly with minimal effort [6].

ISO 27037 is an international standard that seeks to create a baseline for the practice of digital forensics. It is relatively new standard and only addresses the initial stages of a digital investigation, but it represents an international public and private sector consensus of how potential digital evidence should be handled in the critical initial steps of an investigation. There are many complex challenges of digital forensics in a cloud environment and how CSA[2] (Cloud Security Alliance)

---

[2] https://cloudsecurityalliance.org/CSA is a not-for-profit organization with a mission to "promote the use of best practices for providing security assurance within Cloud Computing.

mapped and reinterpreted the ISO 27037 guidance for a cloud context is explained. For some parts of the standard, no changes are necessary for cloud environments. For others, including identification and acquisition of evidence, cloud requires special considerations [7].

2.3 Cloud Computing Services

Cloud computing typically revolves around the concepts of Infrastructure as a Service (IaaS), Platform as a Service (PaaS) and Software as a Service (SaaS), the table below (refer to table 1) explain the characteristics of each cloud service models:

Table 1 Cloud services models characteristics

Cloud Service Models	Characteristics
Cloud Software as a Service (SaaS)	Use provider's applications over a network
Cloud Platform as a Service (PaaS)	Deploy customer-created applications to a cloud
Cloud Infrastructure as a Service (IaaS)	Rent processing, storage, network capacity, and other fundamental computing resources

2.3.1 SaaS: Software as a Service

Cloud application services, or Software as a Service (SaaS), represent the largest cloud market and are still growing quickly. SaaS uses the web to deliver applications that are managed by a third-party vendor and whose interface is accessed on the clients' side. Most SaaS applications can be run directly from a web browser without any downloads or installations required, although some require small plugins [8].

2.3.2 PaaS: Platform as a Service

Cloud platform services — or platform-as-a-service (PaaS) — are used for applications, and other development, while providing cloud components to software. PaaS makes the development, testing, and deployment of applications quick, simple, and cost-effective. With this technology, enterprise operations, or third party providers, can manage operating systems, virtualization, servers, storage, networking, and the PaaS software itself. Developers, however, manage the applications [9].

2.3.3 IaaS: Infrastructure as a Service

Cloud infrastructure services — known as infrastructure-as-a-service (IaaS) — are self-service models for accessing, monitoring, and managing remote data center infrastructures, such as compute (virtualized or bare metal), storage, networking, and networking services (e.g. firewalls). Instead of having to buy hardware outright, users can purchase IaaS based on consumption, similar to payment for electricity or other utilities [10].

All the presented models propose different services in cloud environment, the table below (refer to table 2) shows the comparison within some relevant criteria:

Table 2 All methods comparison of text's analysis

	Saas	PaaS	IaaS
Application	X		
Runtime	X	X	
Database	X	X	
OS	X	X	
Virtualization	X	X	X
Server	X	X	X
Storage	**X**	**X**	**X**
Networking	X	X	X

The analysis of this table shows that presented cloud models offer typical characterization services in cloud environment. All these presented models take into account the storage service capacities. Compared to SaaS and PaaS, IaaS users are responsible of managing applications, data, runtime, middleware and OS. Providers still manage virtualization, servers, hard drives, storage, and networking. Go back to "the storage service" and we can wonder if the CSP know the nature of data stored on their environment, especially the illegal data (child pornography, terrorist documents...etc). CSP must involve security aspect and the content analysis aspect should be done, surely with respecting the privacy.

-Cloud services providers have yet to found adequate forensic capabilities that could support criminal investigations activities online in the cloud platforms since their use are responding to real needs.

2.4 Problematic and Cloud Forensic Challenges

Digital forensic faces various challenges in the cloud computing environment. Malicious people take advantage of this technological achievement, especially storage service, in the sense that they store all illegal information on cloud in order to hide all digital illegal records justifying their illicit acts served by their mobile devices. What is more, mobile forensic expertise on those mobile devices cannot be accomplished by digital investigators of law enforcement, simply because all the storage is done in the cloud.

Ruan et al. identified three dimensions in cloud forensics technical, organizational, and legal. Cloud forensics procedures will vary according to the service and deployment model of cloud computing. The authors mention that SaaS and PaaS services, the digital investigators have very limited control over process or network monitoring. But, they have gain more control in IaaS and can deploy some forensic friendly logging mechanism. They also talked about the collection procedure of SaaS and IaaS. For SaaS, the investigations depend on the CSP to get the application log, but in IaaS, the acquisition of the virtual machine instance from the customer is required before examination and analysis phase. On the other hand, in the private deployment model, the digital investigators have physical access to the digital evidence [11].

Cloud services providers have yet to found adequate forensic capabilities that could support criminal investigations activities online in the cloud.

A criminal can also keep her suspicious files (e.g., child pornography, terrorist documents) in cloud storage and can destroy all his local evidence to remain clean. As example, malicious people can store document containing a terrorist attack plan, this document may be recover by another person from the same terrorist cell. Mobile Cloud Computing can help criminals to coordinate and manage online suspect actions.

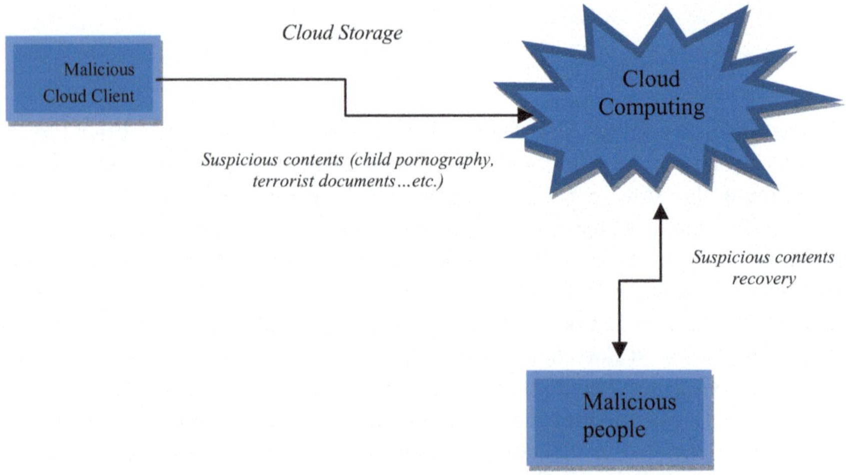

Fig. 1 Example of bad use of cloud storage by malicious people

In this case, when law enforcement investigates such a suspect, the suspect can deny having the illegal content in her cloud storage, and mobile forensic expertise on this mobile equipment cannot be accomplished by digital investigators, simply because all the storage is done in the cloud. At present, there is no way to claim that an adversary owns certain data at a given time.

Researchers are working to protect the cloud environment from different types of attacks. But in case of an attack, we also need to investigate the case, i.e., we need to carry out a digital forensic investigation in the cloud. Besides protecting the cloud, we need to focus on this issue. Unfortunately, there have been little research on adapting digital forensics for use in cloud environments. In this paper, we address this problem, which has significant real-life implications in law enforcement investigating cybercrime and terrorism.

3 Proposal Approach

The main objective behind our research project is to develop a framework (Suspicious Cloud Storage –C2S-), shown in Fig. 2, allowing an efficient investigation of suspicious contents in the cloud environment, through which we may uncover suspicious behavior and interests of cloud users as well.

In this work we will focus only on text contents. Text mining is an effective way to add semantics aspect to this communication's form presenting a significant research challenge [12].

This presented framework can detect secret files (e.g., child pornography, terrorist documents) before their definitely stored on cloud. This framework is destined to CSP to analyze all data coming from their customers to their Data Centers.

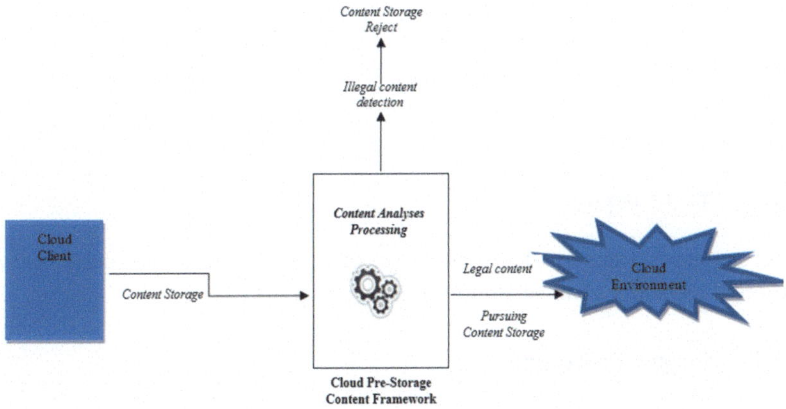

Fig. 2 Overview of our research project

As known, Mobile Cloud Computing offers several and different kind of services allowing users to store different formats of data, including image, video, text …etc. In this paper we are going to focus only on text content storage.

Our proposed framework is mainly based on similarity approach using the text analysis to detect suspicious contents stored on the cloud.

Text mining is based on the calculation of a similarity distance, between the data that will be stored in the cloud and predefined database containing suspicious words, to distinguish suspicious contents. The figure below (Fig. 3) shows three stages of our proposition:

-Text corpus.
-Corpus processing.
-Classification process using similarity approach.

We note that the similarity between words is calculating with a predefined suspicious words database.

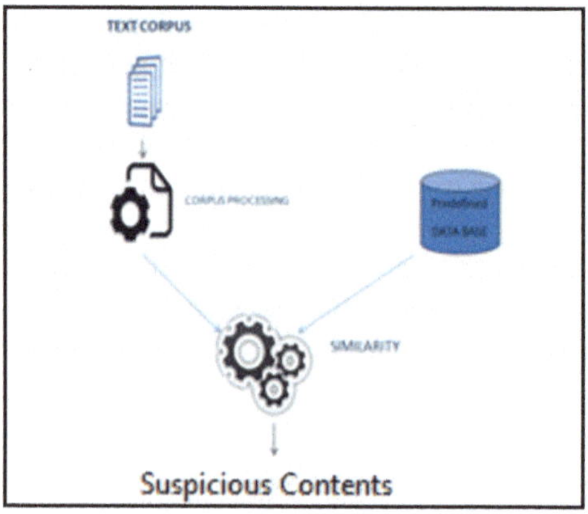

Fig. 3 Our proposed approach

3.1 Processes

3.1.1 Text Corpus

Text corpus is a huge and structured set of texts posted in the social media, and different techniques can be employed in this step.

3.1.2 Corpus Processing

This stage consists to remove stop words and stemming. In computing, stop words are words which are filtered out prior to, or after, processing of natural language data (text). To simplify the study we have to eliminate stop words[3] that contains no useful information, as stop word remove stemming can simplify the processing and reduce errors.

3.1.3 Classification Process Using Similarity

The classification stage aims to well organize a set of texts in two classes:

- Automatic classification method is based generally on the following idea of similarity;
- Two close elements are in the same class and two distant elements are into different classes.

To measure the similarity or dissimilarity between objects, the notion of distance is mandatory and depending on the nature of the data [13].

To measure the similarity or dissimilarity between objects, a notion of distance, depending on the nature of the data is requisite. In this paper, we are going to focus only on text contents.

3.2 Mathematical Formulation

In this paper we use the Normalized Compression Distance to detect the similarity between terms that a post contains and suspicious terms collected in a data base.

C (xy) will have the same number of bytes as C(x) when x = y.

The more y looks like x the more redundancy will be met by the compressor, resulting in C (xy) bytes coming closer to the number of bytes of C(x) [14].

The obtained distance of similarity is expressed by:

$$\boldsymbol{NCD(x,y)} = \frac{\boldsymbol{C(xy) - \ min\ \{C(x), C(y)\}}}{\boldsymbol{max\{C(x), C(y)\}}}$$

Where **0 <= NCD(x, y) <= 1.**

If NCD(x, y) = 0, then x and y are similar, if NCD(x,y) = 1, they are dissimilar. The distance is used to cluster objects.

The idea of our approach is to analyze text contents that will stored on cloud environment. We decompose each post in terms and compare them automatically to suspicious terms.

We defined a threshold that we call "a" determining the maximum values of the distance comparison allowing us to conclude that the two terms are similar. If a sentence contains two terms (suspicious words) which presents similarity with the terms of our database we classify as suspicious post. The figure below (refer to Fig. 4) shows an example of detecting of suspicious content using similarity processing.

[3] http://www.lextek.com/manuals/onix/

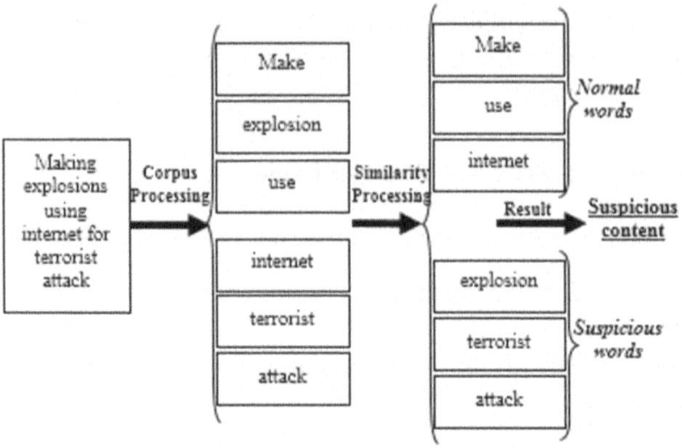

Fig. 4 Example of detecting suspicious content using similarity processing

4 Evaluation of Results

We consider this example “Making explosions using internet for terrorist attack”. After Corpus Processing step, we tested our system using this sentence and we detected three suspicious words which are: explosion, terrorist and attack.

This table (refer to table 3) shows the obtained result of NCD calculations, between words posted and predefined words in our database.

Table 3 Results of NCD calculating between similar words

Term 1	Term 2 (Database)	NCD
explosion	explosion	0
terrorist	terrorist	0
attack	attack	0

If we compare these three words with other words that aren’t similar we find that NCD tends to 1, like shown in table 4.

Table 4 Results of NCD calculating between different words

Term 1	Term 2 (Database)	NCD
make	explosion	0,3
internet	terrorist	0,34
use	attack	0,8

In evaluation of results, presented in table 3 and 4, are conducted based on textual description of each terms, we note that the similarity distance is important when the two terms are not similar and tends to 0 if the two terms are equals. The purpose of our approach is to decompose each post in terms and compare them automatically to predefined suspicious terms database by using similarity distance calculation.

5 Conclusion and Perspectives

Mobile Cloud Computing MCC is a service that allows users of mobile devices high availability of their personal applications as well as their own content, everywhere and anytime. Malicious people take advantage of this technological achievement in the sense that they store all illegal information on cloud in order to hide all digital illegal records justifying their illicit acts served by their mobile devices.

The main idea of our global research project is to develop an automatic system for detecting suspicious contents stored in the Cloud Computing Environment, through which we can uncover suspicious behavior and interests of users as well. The presented framework can detect secret files (e.g., child pornography, terrorist documents) before their definitely stored on cloud.

Our proposed approach is based on the calculation of a similarity distance to distinguish suspicious contents in the cloud. The purpose of our approach is to decompose each content in terms and compare them automatically to predefined suspicious terms database by using similarity distance calculation.

For future work, we plan to improve the system in term of execution time, developing automated classification and using other knowledge resources in order to improve the precision rates, the semantic of exchanged information will be used to identify more significant suspicious profiles.

References

1. Perry, R., Hatcher, E., Mahowald, R., Hendrick, S.: Force.com cloud platform drives huge time to market and cost savings, IDC White Paper, International Data Corporation, Framingham, Massachusetts (2009). (thecloud.appirio.com/rs/appirio/images/IDC_Force.com_ROI_Study.pdf)
2. http://www.mobilecloudcomputingforum.com (accessed on July 15, 2012)
3. Kent, K., Chevalier, S., Grance, T., Dang, H.: Guide to Integrating Forensic Techniques into Incident Response, Special Publication 800-86, National Institute of Standards and Technology, Gaithersburg, Maryland (2006)
4. Digital Forensics Research Workshop. A Road Map for Digital Forensics Research (2001). www.dfrws.org
5. http://en.wikipedia.org/wiki/Mobile_device_forensics (accessed on January 5, 2015)

6. Mell, P., Grance, T.: The NIST Definition of Cloud Computing (Draft), Special Publication 800-145 (Draft), National Institute of Standards and Technology, Gaithersburg, Maryland (2011)
7. Dykstra, J.: Mapping the forensic standard ISO/IEC 27037 to cloud computing. In: Proceedings of the AAFS 66th Annual Scientific Meeting, Seattle, WA, February 2014
8. http://apprenda.com/library/paas/iaas-paas-saas-explained-compared/ (accessed on January 05, 2015)
9. http://cloudcomputingnews.in/making-sense-of-the-cloud-saas-paas-and-iaas-explained/ (accessed on January 5, 2015)
10. http://www.clouderpc.com/making-sense-of-the-cloud-saas-paas-and-iaas-explained/#.VOciQubz3fI (accessed on January 5, 2015)
11. Ruan, K., Carthy, J., Kechadi, T., Crosbie, M.: Cloud forensics: an overview. In: Proceedings of the 7th IFIP International Conference on Digital Forensics (2011)
12. Alami, S., Elbeqqali, O.: Detecting Suspicious Profiles Using Text Analysis Within Social Media. Journal of Theoretical and Applied Information Technology **73**(3), March 31, 2015
13. Porter, M.F.: An algorithm for suffix stripping. Program **14**(3), 130–137 (1980)
14. Kaufman, L., Rousseeuw, P.J.: Finding groups in data: An introduction to cluster analysis. WILEY-Interscience (1990)

Evaluation of a Security Policy Based on OrBAC Model Using MotOrBAC: Application E-learning

Asmaa Kassid and Najib El Kamoun

Abstract E-learning is a new method of learning which depends on the Internet in its execution. Nowday's e-learning's popularity is increasing as more and more people are taking online courses. It becomes the need of the hour, since it is being used as a big platform for enhancing and upgrading knowledge by increasing flexibility in various fields and providing ways to make learning easier. However E-learning has several challenges, one of these major challenges is Information Security. The security aspect is even more important for controlling access to information resources intended to specific users depending on several predefined contexts. Defining such access control is directly related to the appliance of a control access policy, responsible of securing learning sessions in an e-learning platform.

The purpose of this paper is to propose how to adapt ORBAC (organization role based access control) model which is considered as one of the most developed access control security models, to improve the highest degree of security in a concrete e-learning scenario for educational purpose, and to prove how the expressive power and flexibility of this model work. The proposed approach is implemented and evaluated by simulation using "MotOrbac" tool in order to define its validity context and limitations for a large and extended deployment.

Keywords Access control · OrBAC model · Security policies · E-learning platform · Spatial metaphor

1 Introduction

Technology has changed the way people communicate and revolutionized education and training in the 21st century, E-learning is one of the fastest growing markets in virtual world. E-learning is a flexible term used to describe the newest

A. Kassid(✉) · N. El Kamoun
STIC Laboratory, Chouaib Doukkali University, B.P: 20, El Jadida, Morocco
e-mail: {kassid.asmaa87,elkamoun}@gmail.com

E. Sabir et al. (eds.), *The International Symposium on Ubiquitous Networking*,
Lecture Notes in Electrical Engineering 366,
DOI: 10.1007/978-981-287-990-5_11

method of teaching throughout the online internet technology; this method incorporates self-motivation, communication, efficiency. Although it has a relative short history, eLearning become an important part of learning system to achieve its goals, it has gained most of its popularity in the recent years. It is a convenient and inexpensive way to gain knowledge and learn while living everyday's ordinary life. Almost all e-learning researchers are focusing on e-learning system design, course development and delivery, system interoperability and scalability, the security concerns did not have enough attention from these e-learning research initiatives; knowing that every technology brings a new set of threat to the user and the application security the same is applied to the e-learning system [1]. The e-learning security concerns become inevitable.

A secure learning platform should incorporate all the aspects of security without affecting too much of the system performance [2].One of the important aspects that take into account this constraint is Access control, it is manifested by the selective restriction of access to resources, during the authentication time when the user is granted with all the necessary rights, In this way the user have permissions, of performing certain actions assigned to some specific roles in the system.

The traditional access control policies like DAC[3], MAC[4] are limited according to permissions of performing certain actions assigned to some specific roles in the system, R-BAC [5]: role-based access model is it has been proposed to overcome this limitations, in that it takes advantage of the concept of role relations. In such models, the user's rights to access computer resources (objects) are determined by the user's membership to roles and by these roles' permissions to perform operations on objects, this model allows a flexible and granular control of the operations of each user. Several models were inspired by RBAC to organize policies with additional concepts like TMAC(team)[6], TBAC(task)[7] , However, the cited models become limited to the current network development, in terms of technology and communication aspects, and the expansion of the information system in the organizations and enterprises, as we know to achieve a satisfying level of security, it is necessary to define a security policy that meets the needs of the organizations, Orbac simplifies the specification of security policy and provides a richer and more modular Model.

The organization of this paper is as follow. Section 2 is dedicated to present the basic concept of e-learning platform oriented spatial metaphor. Section 3 recalls main concepts of OrBAC model. Section 4 presents a use case of OrBAC model in a concrete e-learning scenario for educational purpose. Related works are presented in section 5. Concluding remarks are presented in section 6.

2 The E-learning Platform

In learning platforms, the learner's isolation is the main cause of abandonment. This abandonment is usually important and throws on this type of training an unbearable shadow. Ensure that the user process is not to "connect to a platform", but to go "a place to invest Virtual Training" to move there, to act or interact to simulate the fact to go to a place of work, is involved in the breaking of learner isolation and make him forget for a moment the distances is the final objective of an e-learning is oriented to a spatial metaphorical platform.

2.1 The Virtual Space

The metaphor of a virtual campus (consisting of meeting space, group space, team space, private space, Public space ....), the graphic interface of this kind of platform allows, by hyperlinks, access to different spaces according to the needs and the rights of actors and activities specific to each virtual space. Fig. 1 shows an example of the different virtual spaces.

The spatial context that is based on the spatial metaphor in the e-Learning platform is defined in each space (Sx) with specific security requirements that are applied to it, for example, it is possible to deposit or consult certain documents on Team space, but not in the public space [8].

If a given URL is assigned to this space, the IP-URL who makes a request is enough to locate it. Therefore Sx corresponds to the name of this space, considering that the entity has assigned IP address on a URL providing space on which a user's IP address is connected (URLs are allocated to specific spaces).

2.2 Roles and Rights of Actors

Each actor is assigned a space and rights determined by declaring a customary name, a password and selecting Profile user that corresponds to this identification in the virtual campus, The teacher-designer enters into relationship with students only through the mediation of its course, while the teacher-tutor offers a methodological and psychological support to students, knowing, that he has the rights of access in all platforms, all public spaces of each group and each team except the private spaces.

Students are grouped into groups of ten people and teams of 2 to 4 people and accompanied by a tutor who organizes tasks and facilitate their implementation, Inside virtual spaces granted to them, each group / each team will timely meet and exchange the instructions of the task as each team meets to plan, control, regulate collaborative tasks in support of the tools available (file repository space, chat, messaging, ...).

The ***coordinator*** is responsible for the administrative side of the platform.

The ***technician*** is responsible for the maintenance as well as purely technical aspects.

2.3 The Pedagogical Scenario

As we mentioned in the previous paragraph that any e-learning platform is manipulated by different actors: tutors, learners, and teachers where each actor play a specific role in the learning process.

- The teacher sets through the learning platform its pedagogical scenario, and uses a set of tools and resources offered by the platform.
- The tutor supports the students in their learning activity, to answer questions, raise discussions, and evaluate progress.
- The student uses on his side the pedagogical scenario defined by the teacher, for example at the solution of an exercise and trying to understand the course content.

3 The OrBAC Model

Organization Based Access Control [9] is an access model that took over the previous models by adding the idea of abstract entities. Originally OrBAC was proposed in [10] to meet security policy requirements in the health care fields (clinics, hospitals, etc.).

Using OrBAC in different domains confirms the expressive power of this model, contrary to other traditional access control models that are limited to the expression of permissions in a static way, OrBAC includes contextual rules not only based on permissions but also prohibitions, obligations and recommendation are making the security policy rich, modular and dynamic.

3.1 Basic Concepts of OrBAC Model

The central entity in Or-BAC is the Organization. An organization can be seen as an well sorted group of subjects playing a role or another,worth knowing that a group of subjects does not necessarily correspond to an organization. More specifically, the fact that each object plays a role in the organization is an agreement between the materials to form an organization.

One of the main objectives of ORBAC model is to allow the designer to define an independent security policy from its implementation. The chosen method to achieve this objective is the introduction of an abstract level where the role, activity and view concepts abstract the subject, action and object concepts. A view is a set of objects that satisfies a common property. An activity includes actions which are involved in the same principles and privileges applied only in a specific context. Each organization org specifies its own security rules, some role may have the permission, prohibition, and obligation or recommendation to do some activity on some view given an associated context is true: *Permission (org, role, activity, view, context).*

OrBAC also propose a formal representation linking all these concepts and to derive the access decisions (rules at the concrete level) from those of the organizational level and context: ***Is_permitted*** *(subject,action,object).*

Similar rules are defined to infer the is_prohibited, is_obliged and is_recommended predicates which represent concrete prohibitions, obligations and concrete recommendations, Fig. 1 shows the relationship between different entities of the model OrBAC.

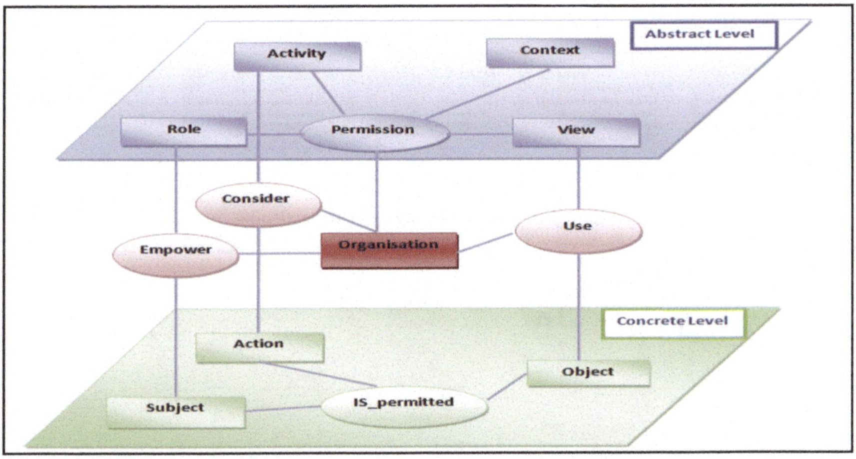

Fig. 1 The ORBAC model

3.2 Notion of Hierarchy

Managing a large number of entities (abstract or concrete) In Orbac model is not an easy task, although their sorting is in a coherent sub-entities form (as defined in the organization), it means that an organization can be divided in sub-organization, a role in sub-role, activity in sub-activity, and a view in sub-view. These decompositions generate hierarchical relationships between the parent entity (generalization) and the Child entity (specialization).

In order to express this hierarchy OrBAC introduces the predicates ***sub_role*** (org, R1, R2) which means that in organization org, role R1 is a sub-role of the role R2, same thing with ***sub-activity***(org, A1, A2) and ***sub-view***(org, V1, V2).

3.3 Notion of Context

Contexts is used to express different types of extra conditions or constraints that control activation of rules expressed in the access control policy [11] , OrBAC model represents the contextual constraints allocation rights, brings together the different contexts by type, the acceptance of a request needs the evaluation of the context, this requires having at our disposal a number of information to test the activation of the context, the following figure 3 resume and presents the taxonomy of contexts and describes all the information necessary that the system should be able to provide for their evaluation.

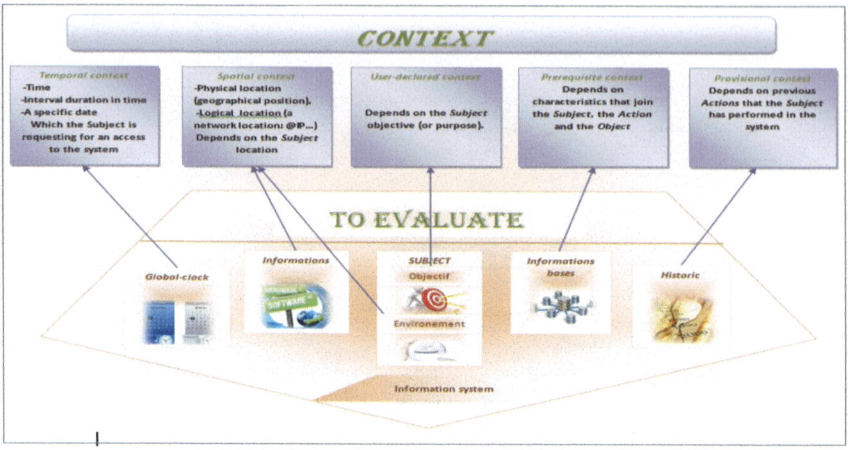

Fig. 2 Context taxonomy and required data

3.4 Notion of Conflict

Since the Or-BAC model provides means to specify both permissions and prohibitions, it is possible that some conflicts arise. This occurs when a given user is both permitted and prohibited to perform a given action on a given object. To model such a situation, we introduce a predicate called conflict (). The following rule specifies a situation of conflict:

$$Is\text{-}permitted(s, \alpha, o) \wedge Is\text{-}prohibited(s, \alpha, o) \rightarrow \boldsymbol{conflict}\ ()$$

4 Application

As the first step in our work, we choose to model an example of a pedagogical scenario, so as to bring out how Or-BAC provides a natural statement of various entities and concepts used in the security of this proposed scenario. In this section we show how to express a simple example of security policy in an e-learning platform through the OrBAC model.

4.1 Example of Pedagogical Scenario

The pedagogical activity fits into the context of database training for the student of the 2nd year specialty networks and telecommunications at the University , it includes 22 learners to find a model Entity / Association and to implement it on a computer software . This training takes 3 sessions of 2 days for each. At the end of each session the learner must take a test and validate it so he can access the resources of the 2nd. Figure 4 summarizes the scenario in accordance with the constraint of time. The distribution of groups and teams is done by the ticket of teacher-tutor.

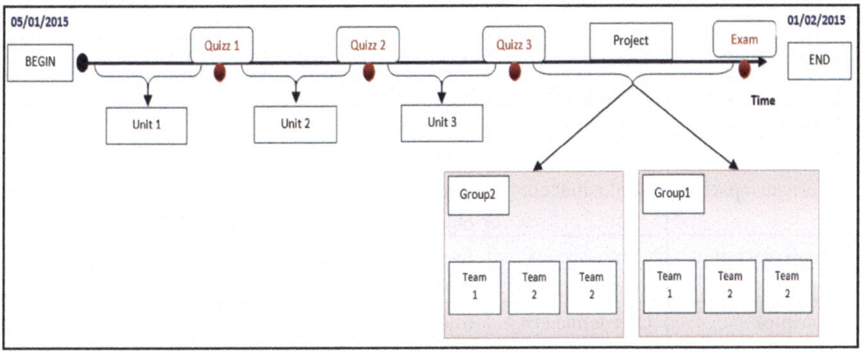

Fig. 3 An example of a pedagogical scenario

4.2 Managing Organisation, Subjects / Roles, Objects / Views and Actions / Activities

We define some abstract/concrete entities of our pedagogical scenario, considering that the following table resumes some of needed entities to model our approach:

Notation	Role-scenario1	Subject
$R_{T\text{-}d}$	Teacher-designer	{Mr Najib}
$R_{T\text{-}c}$	Teacher-tutor	{Mlle Fatima , Mr Khalid , Mr Yassine}
R_S	Student	{Asmaa,Chaimaa,zahira,hind...}
G_1	Group1/2	9 students per group
T_1	Team1/2/3	3students per team
Notation	**Activity-scenario1**	**Actions**
A_{11}	Creation	- writing courses / exercises -filed courses / exercises
A_{12}	Update	-Modification -Suppression- courses / exercises
A_{13}	Follow	-Explain-course/project – Answer questions
A_{14}	download	-download-course/TD/Quiz/Exam
A_{15}	Resigter	-Registration-formation
Notation	**View-scénario1**	**Objects**
V_{11}	course	Course-X.doc/html/pdf/ppt
V_{12}	Resource-sup	- video/-shema.jpg / …
V_{13}	Resource-Test	-Quiz.doc/-Exam-module-x.doc
V_{14}	Doc-team	- Doc-team1/2/3 ….
V_{16}	training	-Training-BD
V_{17}	URLs	$URL_{1,2,3,4}$ <::::> private ,public, group ,team space

Notation	Type-context	Explanation
Inscrip-delay	Temporal.ctx	mo=01 & d<=5 & d>=1 & y=2015
weekend	Temporal.ctx	dow=6 & dow= 7
Not-weekend	Composed.ctx	! weekend
Unit1-end	Temporel.ctx	d>=10 & y=2015 & mo=01
In-private-space	GeoSpatial.ctx	result = object.GetAttributeAsString("url1").equals("SP");
GroupSpace-time	Composed.ctx	In-group-space & unit1-end
Depo-doc	Temporal.ctx	(h >= 8 && h<= 12) || (h >= 14 && h= 18)
Inscrip-ok	Userdefine.ctx	true

5 Specifying the Security Rules and Evaluating the Security Policy

5.1 *The Security Rules*

We can now specify different rules (permissions/obligations/prohibitions) the different actors involved in the advancement of the scenario.

Student	* The student are authorized to download the course resources for their study ***Perm-Rs-do**: **Permission (master-RT, Student, download, course, in-private-space)*** * Students cannot have the right to change the teacher's course **Prohi-Rs-update:** ***Prohibition (master-RT, Student, update, course, default)***
Teacher-T	The teacher tutor does not have the right to download document Concerns the two groups ***Prohib-Tt-dow: Prohibition (Master-RT, Tt, download, doc-T1, default)***
Teacher-C	* Teacher designer must deposit the course once the training starts ***obli-Tc-depo: Obligation (master-RT, T_C, deposit, course, in-public-space)*** * The teacher designer can change the course on weekends **Perm-Tc-update:** ***Permission (master-RT, T_C, update, course, weekend)***
Team	Teams can deposit their documents or consult some of the other teams in the same group according to time and space constraints **** Prohib-Tea-depo: Prohibition (Master-RT, Team1, deposit, doc-T1, Groupsapce-time)*** **** Prohib-Tea-consul: Prohibition (Master-RT, Team2, consult, doc-T3, Teamsapce-time)*** **** Perm-Tea-consul: Permission (Master-RT , Team3, consult, doc-T2, Groupsapce-time)***
Group	Groups can deposit their documents or consult some of the other Group according to time and space constraints **** prohib-Grp-depo :Prohibition (Master-RT, Group1, deposit, doc-G1, in-Group-sapce)*** **** Perm-Tea-consul: Permission (Master-RT, Group2, consult, doc-G1, Groupsapce-time)***

5.2 The Evaluation of Security Policy with MotOrBAC

Designers of the Or-BAC model have developed MotOrBAC [12][13] a security policy tool which can be used to specify, simulate, evaluate and administrate the security policies, the following figures shows the different steps:

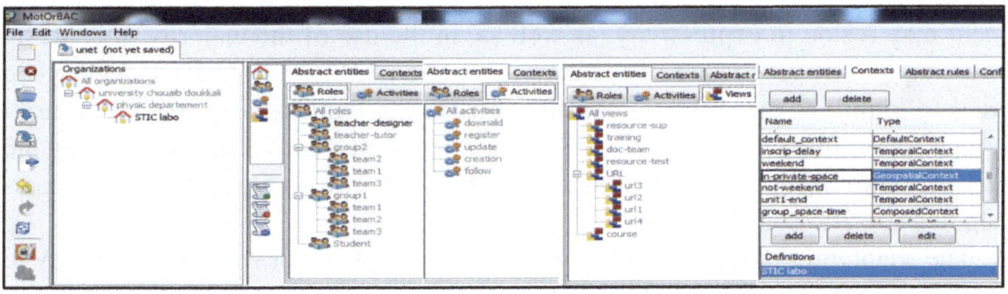

Fig. 4 The creation of the abstract entities

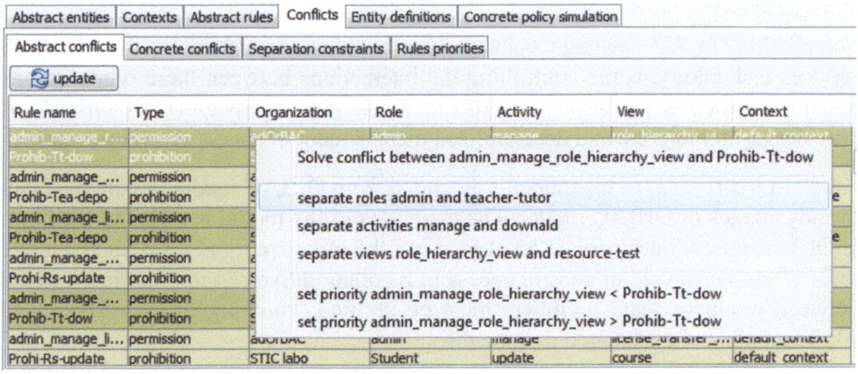

Fig. 5 Abstract conflicts/Separation constraints added in the policy

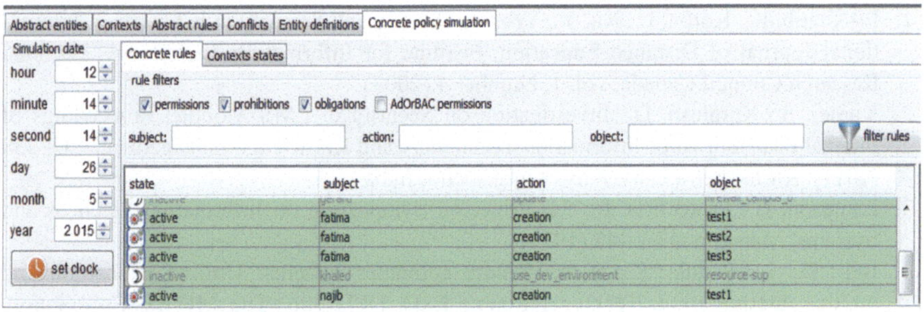

Fig. 6 Simulation of the policy

6 Conclusion and Perspectives

Our main goal in this paper, is to improve the level of security in e-learning platform, taking into account the different interactions of these actors and the notion of context, because it is one of the most interesting contributions of OrBAC , it allows to set certain conditions for the application of safety rules.This improvement may reside in expressivity, as well as, in the use of the majority of contexts and adaptability as needed of the activity in the platform. In OrBAC model, we tried to highlight the different contributions in the expressivity , then a security policy was expressed in an e-learning platform oriented spatial metaphor using the OrBAC formalism, which allowed us to unfold the various aspects of this formalism in a specific concrete case.

The adaptation of this approach to an e-learning platform will allow not only to finalize this study, but also to extend it to a set of e-learning platforms as a part of a global security policy. The OrBAC model promotes this through several extensions answering different scenario in the platform as PolyOrBAC and session-OrBAC:

Poly-OrBAC[14,15]: manage collaboration between organizations through Web services technology, while controlling the interactions between these organizations conform to their expectations and their internal policies specified by OrBAC. It provides an architecture and decentralized management of access control policies.

Session-OrBAC [15]: This model is extended from OrBAC model, it benefits from the advantages of OrBAC model. The main idea of the model is added to the model OrBAC a new intermediate level between the organizational level and concrete level : "session-level" to control access in a collaborative session, the interactions between group or team members must be secured, knowing the sensitivity and confidentiality of exchanged documents.

References

1. El-Khatib, K., Korba, L., Xu, Y., Yee, G.: Privacy and Security in E-Learning. International Journal of Distance Education, Institute for Information Technology, National Research Council Canada, vol. 1, Number 4 (2003)
2. Kumar, S., Kamlesh, D.: Investigation on Security in LMS Moodle. Proceedings of International Journal of Information Technology and Knowledge Management, 233–238 (2011). Kurukshetra University, Kurukshetra, India
3. Lampson, B.: Protection. In: 5th Princeton Symposium on Information Sciences and Systems, Mars, pp. 437–443 (1971)
4. Bell, D., La Padula, L., et al.: Secure computer systems: Unified exposition and multics interpretation. Technical Report ESD TR73-306, The MITRE Corporation, Mars (1976)
5. Sandhu, R., Coyne, E., Feinstein, H., Youman, C.E.: Role-based access control models. IEEE Computer **29**(2), 38–47 (1996)

6. Thomas, R.: TMAC: a primitive for applying RBAC in collaborative environment. In: 2nd ACM, Workshop on RBAC, Fairfax, Virginia, USA, pp. 13–19, November 1997
7. Thomas, R., Sandhu, R., et al.: Task-based authorization controls (TBAC): a family of models for active and enterprise-oriented authorization management. In: 11th IFIP Working Conference on Database Security, Lake Tahoe, California, USA, pp. 166–181 (1997)
8. Bousmah, M., Elkamoun, N., Berraissoul, A.: Conception et réalisation d'un environnement virtuel d'apprentissage collaboratif, orienté métaphore spatiale, couplé avec un système observateur d'usage. In: Proceedings of the 6th IEEE International Conference on Advanced Learning Technologies, July 2006. IEEE Computer Society, Kerkrade (2006)
9. Abou El Kalam, A., El Baida, R., Balbiani, P., Benferhat, S., Cuppens, F., Deswarte, Y., Miege, A., Saurel, C., Trouessin, G.: Organization based access control. In: Proceedings of IEEE 4th International Workshop on Policies for Distributed Systems and Networks (POLICY 2003), Lake Como, Italy, June 2003
10. www.orbac.org
11. Cuppens, F., Cuppens-Boulahia, N.: Modeling Contextual Security Policies. International Journal of Information Security **7**(4), 285–305 (2007)
12. Autrel, F., Cuppens, F., Cuppens-Boulahia, N., Coma, C.: MotOrBAC 2: a security policytool. In: 3rd Conference on Security in Network Architectures and Information Systems (SAR-SSI 2008), Loctudy, France, pp. 273–288 (2008)
13. http://motorbac.sourceforge.net/index.php?page=home&lang=en
14. El Kalam, A., Deswarte, Y., Baina, A., Kaaniche, M., et al.: Access control for collaborative systems: a web services based approach. In: IEEE International Conference on Web Services (ICWS 2007), pp. 1064–1071 (2007)
15. Baina, A.: Contrôle d'Accès pour les Grandes Infrastructures Critiques: Application au réseau d'énergie électrique. Thèse de doctorat, Université de Toulouse, Septembre 29, 2009
16. Madani, M.A., Erradi, M.: Network Security and Systems (JNS2), 2012 Networking and Distributed Systems Research Group, SIME Lab University Mohammed V-Souissi, ENSIA

Applying Encryption Algorithm for Data Security in Cloud Storage

Zaid Kartit, Ali Azougaghe, H. Kamal Idrissi, M. El Marraki, M. Hedabou, M. Belkasmi and A. Kartit

Abstract This paper proposes a simple, secure, and privacy-preserving architecture for inter-Cloud data sharing based on an encryption/decryption algorithm which aims to protect the data stored in the cloud from the unauthorized access.

Keywords Cloud storage · Data security · Cryptography · RSA · AES

1 Introduction

Cloud computing is the concept implemented to remedy the Daily Computing Problems. Cloud computing is basically virtual pool of resources and it provides these resources to users via internet. It provides IT services as on-demand services, accessible from anywhere, anytime and by authorized user. It offers a range of services for end users; among which there's Storage as a service. Storage as a service (STaaS) is a Cloud business model in which a service provider rents space in its storage infrastructure to individuals or companies. The data stored in the cloud can be sensitive to the business. The problematic is that these data are likely to be exploited by the provider or other unauthorized persons. Currently, most of cloud storage users protect their data with SLAs contracts and are based on the

Z. Kartit(✉) · H. Kamal Idrissi · M. El Marraki
LRIT, Faculty of Sciences, Mohammed V University, Rabat, Morocco
e-mail: z_kartit@yahoo.fr

A. Azougaghe · M. Belkasmi
SIME, Mohammed V University, ENSIAS, Rabat, Morocco
e-mail: azaling@gmail.com

M. Hedabou
ENSA de Safi, University of Cadi Ayyad, Marrakech, Morroco

A. Kartit
LTI, University Chouaib Doukkali, ENSA, El Jadida, Morocco

E. Sabir et al. (eds.), *The International Symposium on Ubiquitous Networking*,
Lecture Notes in Electrical Engineering 366,
DOI: 10.1007/978-981-287-990-5_12

trust and reputation of the provider. This weakness has motivated us to think about solutions that enable users to secure their data to prevent malicious use.

In recent years, STaaS in Cloud gained popularity among both companies and private users [1]. It allows the end-user to take advantage of the maximum computing capability with minimum hardware requirement. However, data privacy, security, reliability and interoperability issues still have to be adequately solved. But the most important problem is security and how cloud provider assures it. Data security in cloud storage is a major obstacle limiting its spread. There are various opinions on the security of cloud computing with pros and cons [2].

Our contribution aims to provide a solution that ensures the storage of data securely in the cloud. The data must be encrypted before sending them to the cloud. We used the symmetric encryption algorithm AES in order to benefit from its advantages in terms of robustness and speed. The AES key will be encrypted by the asymmetric encryption algorithm RSA and will be stored in a private server away of the cloud. The integrity and confidentiality of the data is ensured by providing access to the data only on successful authentications (authentication in cloud and authentication in private server). The authorized user can also download the file and read it on the system. This hybrid model that we have proposed allows to secure the data and to enhance the resistance to attacks.

This document is organized as follows: The first section gives a comprehensive definition and the characteristics of cloud computing. The second section describes layers and their technologies related to this concept. The third section describes the different types of cloud computing and their characteristics. The fourth section describes our model proposed of securing data in cloud storage algorithm for encryption/decryption for outsourcing data in cloud storage and then the general conclusion.

2 About Cloud Computing

2.1 Definition

Cloud Computing is an important concept in computer development in recent years. This concept refers to the use of computing capacity and storage of computers and servers in the world over the Internet. Cloud services allow individuals and businesses to use software and hardware that are managed by third parties at remote locations. Examples of cloud services include online file storage, social networking sites, webmail, and online business applications. Cloud computing provides a shared pool of resources, including data storage space, networks, computer processing power, and specialized corporate and user applications.

2.2 Essential Characteristics

Cloud model promotes availability and is composed of five essential characteristics [4]:

- **On-demand self-service:** A consumer can unilaterally provision computing capabilities, such as email, applications, and network or server service, as needed automatically without requiring human interaction with each service provider.
- **Broad network access:** Capabilities are available over the network and accessed through standard mechanisms that promote use by heterogeneous thin or thick client platforms (e.g., mobile phones, tablets, laptops, and workstations).
- **Resource pooling:** The provider's computing resources are pooled to serve multiple consumers using a multi-tenant model, with different physical and virtual resources dynamically assigned and reassigned according to consumer demand. Examples of resources include storage, processing, memory, and network bandwidth.
- **Elasticity:** Capabilities can be elastically provisioned and released, in some cases automatically, to scale rapidly outward and inward commensurate with demand.
- **Measured service:** Cloud systems automatically control and optimize resource use by leveraging a metering capability at some level of abstraction appropriate to the type of service (e.g., storage, processing, bandwidth, and active user accounts). Resource usage can be monitored, controlled, and reported, providing transparency for both the provider and consumer of the utilized service.

3 Layers of Cloud Computing

There are different layers of cloud services that refer to different types of *service model* (fig. 1), each offering discrete capabilities. Apart from management and administration, the major layers are [2]:

3.1 Infrastructure as a Service (IaaS)

Infrastructure as a service delivers computing resources as a service, servers, network devices, and storage disks are made available to organizations as services on a need-to basis.

Virtualization, allows IaaS providers to offer almost unlimited instances of servers to clients, while making cost-effective use of the hosting hardware. Companies can use IaaS to build new versions of applications or environments without having to invest in physical IT assets. Some cloud solutions also rely solely on this layer like the Amazon's product EC2, Amazon S3.

3.2 Platform as a Service (PaaS)

This layer provides a platform for creating applications. PaaS solutions are essentially development platforms for which the development tool itself is hosted in the

Cloud and accessed through internet. With PaaS, developers can build Web applications without installing any tools on their computers and then deploy those applications without any specialized systems administration skills.

Examples include Google App Engine, Force.com and Microsoft Azure.

3.3 Software as a Service (SaaS)

This layer includes applications that run off the Cloud and are available on demand to Web and paid for on a per-use basis, anytime-anywhere basis. There is no need to install and run the special software on your computer if you use the SaaS. A more efficient form is fine grained multi-tenancy [5].

The concept of SaaS is attractive and some software runs well as cloud computing, but the delay of network is fatal to real time or half real time applications such as 3D online game [6].

Examples include online word processing and spreadsheet tools, customer relationship management (CRM) services and web content delivery services (Salesforce CRM, Google Docs, etc.)

These three are the main layers, although there can also be other forms of service provided, such as business process as a service, data as a service, security as a service, storage as a service (object of our paper), etc.

4 Cloud Deployments Models

4.1 Private Cloud

Private cloud is a new term that some vendors have recently used to describe offerings that emulate cloud computing on private networks. It is set up within an organization's internal enterprise datacenter. In the private cloud, scalable resources and virtual applications provided by the cloud vendor are pooled together and available for cloud users to share and use. Only the organization and designated stakeholders may have access to operate on a specific Private cloud [7].

4.2 Public Cloud

A public cloud is a model which allows users access to the services and infrastructure and are provided off-site over the Internet [8]. It's typically based on a pay-per-use model, similar to a prepaid electricity metering system which is flexible enough to cater for spikes in demand for cloud optimization. Public clouds are managed by third parties or vendors over the Internet. Public clouds are less secure than the other cloud models because it places an additional burden of ensuring all applications and data accessed on the public cloud are not subjected to malicious attacks. However, security and governance issues must be well planned and ample security controls was put in place.

4.3 *Hybrid Cloud*

A new concept combining resources from both internal and external providers will become the most popular choice for enterprises. A hybrid cloud is a combination of public and private cloud models that tries to address the limitations of each approach. In a hybrid cloud, part of the service infrastructure runs in private clouds while the remaining part runs in public clouds. Hybrid clouds offer more flexibility than both public and private clouds. Specifically, they provide tighter control and security over application data compared to public clouds, while still facilitating on-demand service expansion and contraction. On the down side, designing a hybrid cloud requires carefully determining the best split between public and private cloud components [9].

4.4 *Community Cloud*

This model is rarely offered; the infrastructure is shared by several organizations for a shared cause and may be managed internally or a third party service provider. It brings together, in general, the structures with same interest (mostly security) and may even be in the same field of activity.

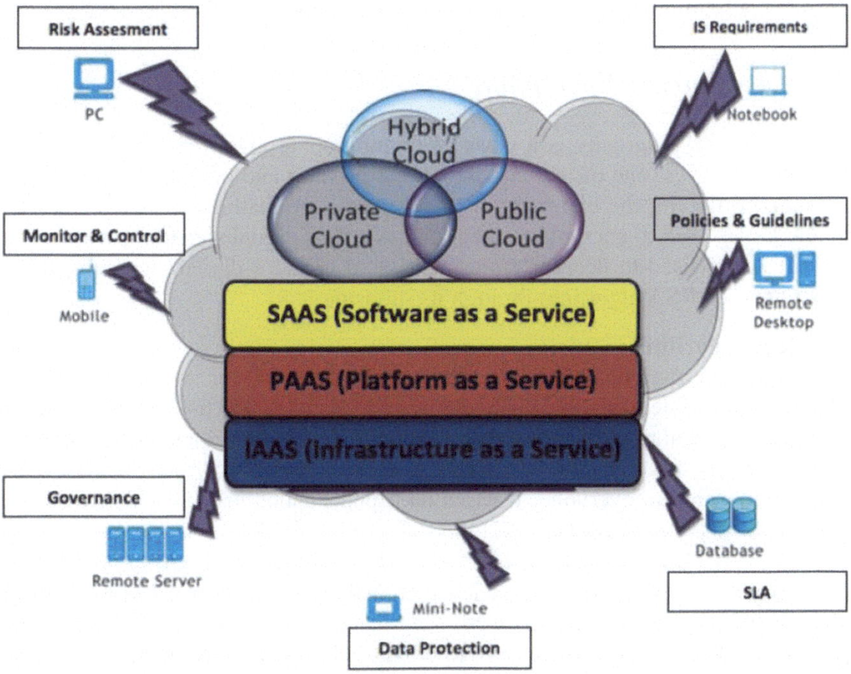

Fig. 1 Cloud computing map [10]

5 Security

Security in cloud computing involves concepts such as network security, equipment and control strategies deployed to protect data, applications and infrastructure associated with cloud computing. An important aspect of cloud is the notion of interconnection with various materials which makes it difficult and necessary securing these environments. Security issues in a cloud platform can lead to economic loss, also a bad reputation if the platform is oriented large public and are the cause behind the massive adoption of this new solution. The data stored in the cloud for customers represents vital information. This is why the infringement of such data by an unauthorized third party is unacceptable. There are two ways to attack data in Cloud. One is outsider attack and the other is insider attack. The insider is an administrator who can have the possibility to hack the user's data. The insider attack is very difficult to be identified. So the users should be very careful while storing their data in cloud storage. Hence, the need to think of methods that impede the use of data even though the data is accessed by the third party, he shouldn't get the actual data. So, all the data must be encrypted before it is transmitted to the cloud storage [11].

Security allows the confidentiality, integrity, authenticity and availability of information. The development of technologies and their standardization makes available a set of algorithms and protocols for responding to these issues.

5.1 Asymmetric Encryption

Asymmetric cryptography is a class of cryptographic algorithms which requires two separate keys, one of which is secret (or private) and one of which is public. Although different, the two parts of this key pair are mathematically linked. The public key is used to encrypt plaintext or to verify a digital signature; whereas the private key is used to decrypt cipher text or to create a digital signature. In our paper we used RSA algorithm through its robustness.

- **RSA Algorithm**

The most common Public Key algorithm is RSA, named for its inventors Rivest, Shamir, and Adelman of MIT. RSA is basically an asymmetric encryption/decryption algorithm. Public key distributed to all through which one can encrypt the message and private key which is used for decryption is kept secret and is not shared to everyone. It based on exponentiation in a finite field over integers modulo a prime numbers.

RSA uses Euler's Theorem: $a^{\emptyset(n)} mod(n) = 1$ where $gcd(a,n)=1$ in RSA we have to initially calculate $n=p.q$ such that $\emptyset(n)=(p-1)(q-1)$ one has to carefully chose e and d to be inverses $mod\ ø(n)$.

To encrypt a message M we have to obtain public key of recipient $Pu=\{n,e\}$ to calculate the cipher: $C=M^{e} mod(n)$, where $0 \leq M < n$. It is important that the message M must be smaller than the modulus n. Similarly for decryption the recipient uses their private key $Pr=\{n,d\}$ and computes: $M=C^{d} mod(n)$.

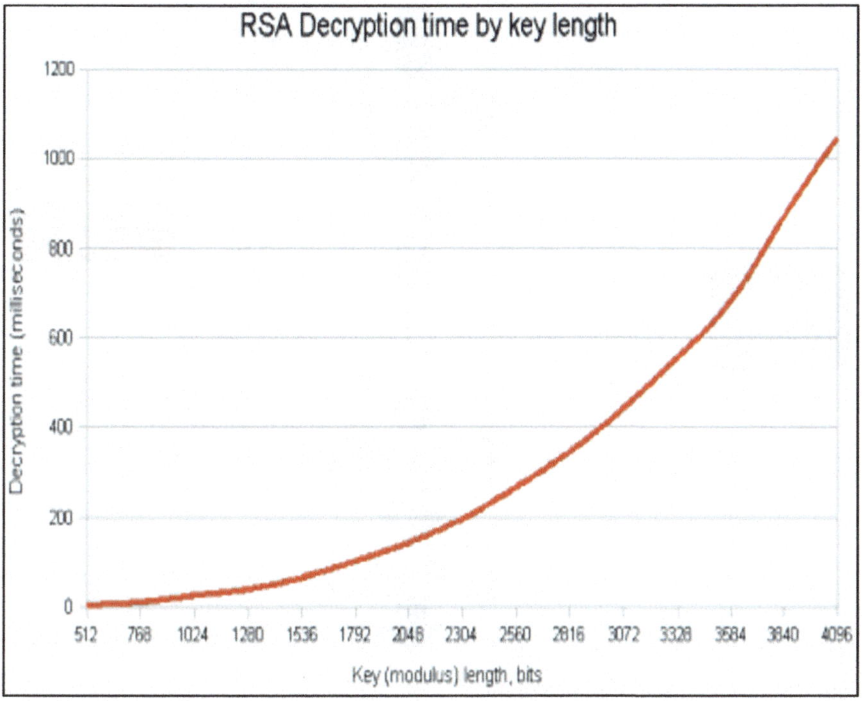

Fig. 2 RSA decryption time by key length

5.2 Symmetric Encryption

Symmetric-key algorithms are a class of algorithms for cryptography that use the same cryptographic keys for both encryption of plaintext and decryption of cipher text. The keys may be identical or there may be a simple transformation to go between the two keys. The keys, in practice, represent a shared secret between two or more parties that can be used to maintain a private information link. This requirement that both parties have access to the secret key is one of the main drawbacks of symmetric key encryption, in comparison to public-key encryption.

- **AES Algorithm**

AES is a variant of Rijndael which has a fixed block size of 128 bits, and a key size of 128, 192, or 256 bits. The key size used for an AES cipher specifies the number of repetitions of transformations rounds. The number of cycles of repetition is as follows (see fig. 2):

- 10 cycles of repetition for 128-bit keys.
- 12 cycles of repetition for 192-bit keys.
- 14 cycles of repetition for 256-bit keys.

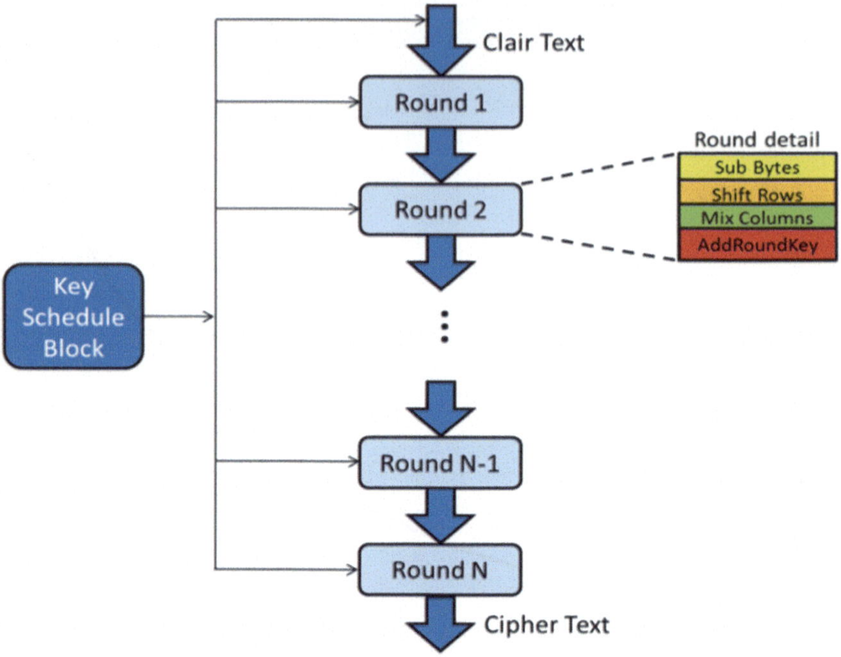

Fig. 3 Illustration of the AES Algorithm

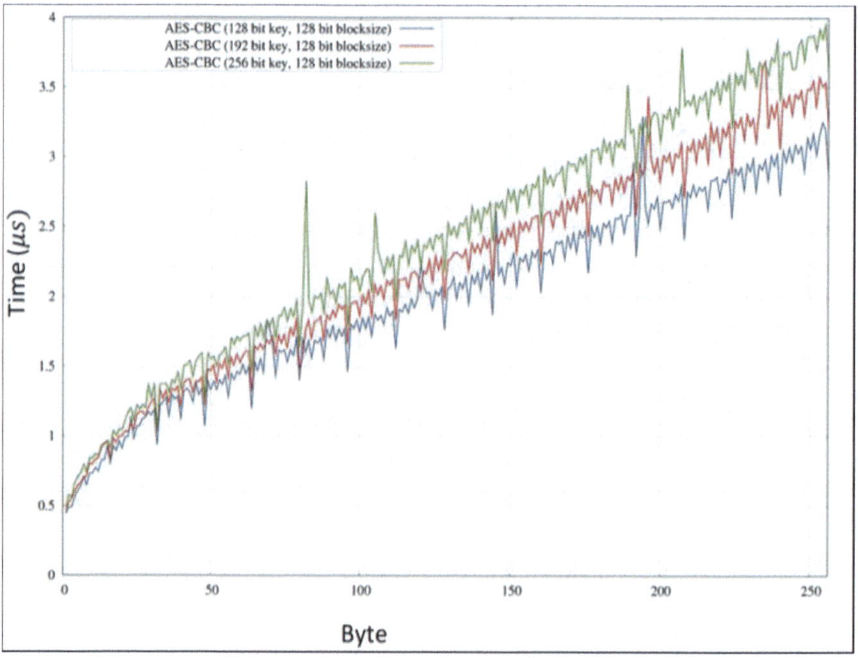

Fig. 4 AES speed at 128, 192 and 256-bit key sizes [12]

The advantages of AES are many. AES is not susceptible to any attack but Brute Force attack. However, Brute Force attack is not an easy job even for a super computer. This is because the encryption key size used by AES algorithm is of the order 128, 192 or 256 bits which results in billions of permutations and combinations. High speed [12] (see fig 4) and low RAM requirements were criteria of the AES selection process. Thus AES performs well on a wide variety of hardware; from 8-bit smart cards to high-performance computers. AES is also much faster than the traditional algorithms; therefore in our work AES is adopted [15]. Recently Compact AES S-box is developed to be more efficient [16].

5.3 Related Work

Security storage in cloud computing has been the object of several researches. In [13], they have addressed the security issues associated in cloud data storage and have explored many of them. Whenever a data vulnerability is perceived during the storage process, a precision verification across the distributed servers are ensured by simultaneous identification of the misbehaving nodes through analysis in term of security malfunctioning. It is proved that their scheme is effective to handle certain failures, malicious data modification attack, and even server colluding attacks.

In [12], the proposed technique emphasizes classical encryption techniques by integrating substitution cipher and transposition cipher. Both substitution and transposition techniques have used alphabet for cipher text.

In [14], it suggests the SPKS scheme for cloud storage services to allow users to efficiently access files containing certain keywords in a cloud anytime and anywhere using any device.

6 Proposed Algorithm

The above problems motivate us to provide a correct, safe and efficient algorithm for securing data saved in cloud storage. This algorithm suggests the encryption of the files to be uploaded on the cloud. The integrity and confidentiality of the data uploaded by the user is ensured doubly by not only encrypting it but also providing access to the data only on successful authentication. The existed file on the device will be encrypted using AES algorithm. To enhance security; AES key will be encrypted using RSA algorithm and will be stored in intern server. The authorized user can also download any of the uploaded encrypted files and read it on the system (see Fig. 4).

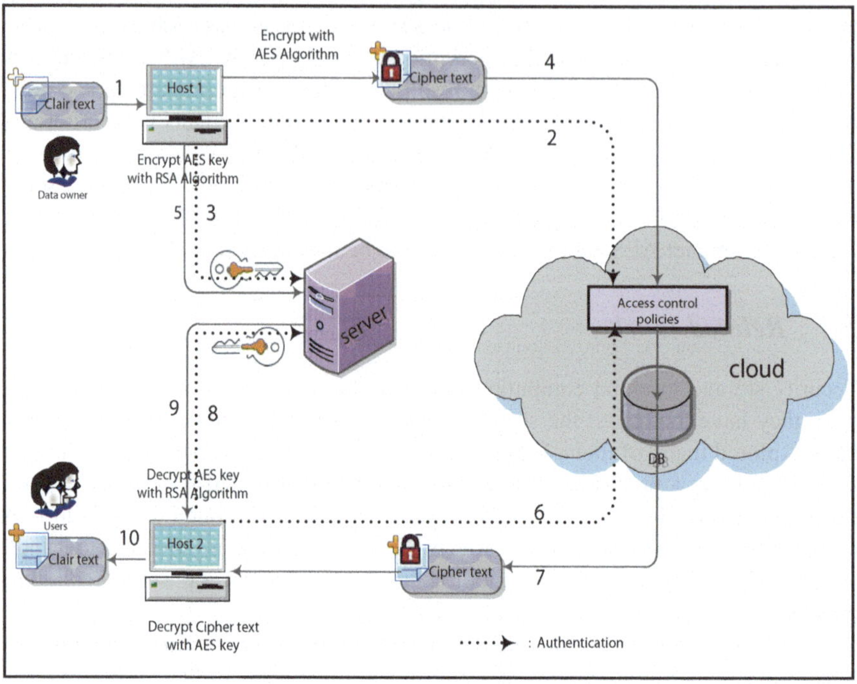

Fig. 5 Proposed model of data storage in cloud computing

6.1 File Upload

This algorithm got two phases. In the first phase, the algorithm encrypts Clair text with AES Algorithm. In the second phase, we encrypt AES key using RSA-1024 algorithm.

Our algorithm uses a set of the following functions:

NumberOfBlock(F) : It returns the number of block in the file F.
ENC_AES (B,K) : It encrypts the block B using AES Algorithm with key K.
send_to_cloud(F') : It permits to send the encrypted file F in Cloud storage.
ENC_RSA(k) : It encrypts k using RSA Algorithm.
Save_in_server(K') : It permits to save K' in the server.

Algorithm 1.0: FILE_UPLOAD

```
Encrypt_file (F) {
/* algorithm to encrypt file onto cloud storage */
/* to transform Clair text in file F into Cipher text in file F' */

/* Phase 1: Encrypt Clair text with AES Algorithm 6. */
For B←1 to numberOfBlock(F) do
 {
     B'=ENC_AES(B,K)
 }
 send_to_cloud(F')
/* Phase 2: Encrypt AES key with RSA Algorithm */
For k←1 to SizeOf(K) do
 {
     k'=ENC_RSA(k)
 }.
 Save_in_server(K')
}.
```

6.2 File Download

This algorithm got also two phases. In the first phase, the algorithm decrypts AES key using RSA Algorithm. In the second phase, it decrypts cipher text using AES key retrieved from the server.

Our algorithm uses a set of the following functions:

NumberOfBlock(F) : It returns the number of block in the file F.
DEC_RSA(k') : It decrypts k' using RSA Algorithm.
DEC_AES(B',K) : It decrypts the block B' using AES Algorithm with key K.

Algorithm 1.0: FILE_DOWNLOAD

```
Decrypt_file (F') {
/* algorithm to decrypt file downloaded from cloud storage */
/* to transform Cipher text in file F' into Clair text in file F */

/* Phase 1: Decrypt AES Key with RSA Algorithm */

For k'←1 to SizeOf(K') do
   {
       k=DEC_RSA(k')
   }
 return(K)

/* Phase 2: Decrypt Cipher text with AES Algorithm */
```

```
For B'←1 to numberOfBlock(F') do
  {
     B=DEC_AES(B',K)
  }.
return(F)
}.
```

6.3 *Implement Results and Analysis*

The implementation of results in this section highlights the time of execution in upload and in download of files with different sizes. Our application is developed in java7. The result obtained is in Fig 6 using a PC hp Compaq dc 5800 with the following specifications: Intel (R) Core (TM) 2 Duo CPU E6550 @ 2.33GHz (2 CPUs), with 3072MB of RAM. The download time is greater than the upload time. This is explained by the addition of key recovery time on server.

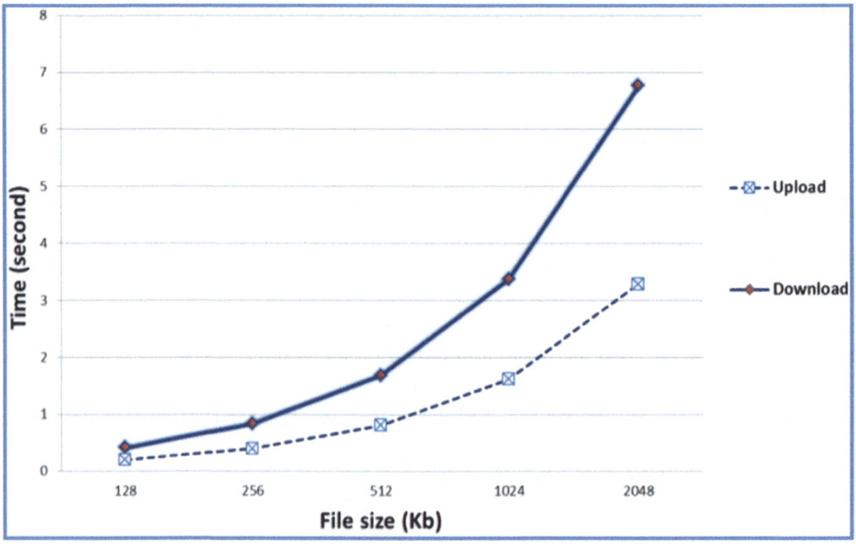

Fig. 6 Execution Time for Encryption and decryption

Our algorithm has the following advantages and strengths:

- The data sent to cloud is encrypted from the source machine to the destination machine and the decryption key does not exist in the cloud.
- AES algorithm used is a safe, fast symmetric algorithm and is one of the most secure encryption algorithms. It has not been broken to date. It means that our algorithm is fast in both directions: upload and download.
- Ability to change the symmetric key frequently to enhance security.

- The AES key used for encryption of the data is encrypted by RSA-1024 algorithm robust and has never been broken.
- The decryption of data requires double authentication. The user must have access rights to the company's server and cloud storage.

7 Conclusion and Future Work

Although Cloud storage has many advantages, there are still many actual problems concerning security that need to be solved. If we can eliminate or master this weakness of security, the future is going to be Cloud storage solutions for large as well as small companies. In this paper, we have suggested a solution that allows storage of data in an open cloud. Data security is provided by implementing our algorithm. Only the authorized user can access the data. Even if an intruder (unauthorized user) gets the data accidentally or intentionally, he can't decrypt it and needs two keys coming from two different locations. As perspectives, we will focus on several possible directions in this area, especially in the homomorphic encryption.

References

1. Kolodner, E.K., Tal, S., Kyriazis, D., et al.: A cloud environment for data-intensive storage services. In: 2011 IEEE Third International Conference on Cloud Computing Technology and Science (CloudCom), pp. 357–366. IEEE (2011)
2. Arora, P., Wadhawan, R.C., Ahuja, E.S.P.: Cloud Computing Security Issues in Infrastructure as a Service. Int. J. Adv. Res. Comput. Sci. Softw. Eng. **2**(1) (2012)
3. Kumar, M.D.K., Rao, G.V., Rao, G.S.: Cloud Computing: An Analysis of Its Challenges & Cloud Computing: An Analysis of Its Challenges & Security Issues
4. Mehta, N., Gupta, V. K.: A survey on use of SaaS of cloud in education. In: International Conference on Cloud, Big Data and Trust (2013)
5. Gong, C., Liu, J., Zhang, Q., Chen, H., Gong, Z.: The characteristics of cloud computing. In: 39th International Conference on Parallel Processing Workshops (2010)
6. Bezemer, C.-P., Zaidman, A., Platzbeecker, B., et al.: Enabling multi-tenancy: an industrial experience report. In: 2010 IEEE International Conference on Software Maintenance (ICSM), pp. 1–8. IEEE (2010)
7. Kamal Idrissi, H., Kartit, A., El Marraki, M.: Foremost security apprehensions in cloud computing. Journal of Theoretical and Applied Information Technology **59**(3), January 31, 2014
8. Kuyoro, S.O., Ibikunle, F., Awodele, O.: Cloud Computing Security Issues and Challenges. International Journal of Computer Networks (IJCN) **3**(5) (2011)
9. Gong, C., Liu, J., Zhang, Q., Chen, H., Gongng, H.: The characteristics of cloud computing. In: 2010 39th International Conference on Parallel Processing Workshopse Brazilian Computer Society (2010)
10. A Platform Computing Whitepaper. Enterprise Cloud Computing: Transforming IT. Platform Computing, p. 6 (2010)

11. Arockiam, L., Monikandan, S.: Data Security and Privacy in Cloud Storage using Hybrid Symmetric Encryption Algorithm. International Journal of Advanced Research in Computer and Communication Engineering **2**(8), August 2013
12. http://en.wikipedia.org/wiki/AES_implementations
13. Wang, C., Wang, Q., Ren, K.: Ensuring Data Storage Security in Cloud Computing. IEEE (2009). 978-1-4244-3876-1
14. Liu, Q., Wang, G., Wu, J.: Secure and privacy preserving keyword searching for cloud storage services. Journal of Network and Computer Applications **35**(3), 927–933 (2012)
15. Nagendra, M., Sekhar, M.C.: Performance Improvement of Advanced Encryption Algorithm using Parallel Imputation. International Journal of Software Engineering and Its Applications **8**(2), 287–296 (2014)
16. Zhang, X., Wu, N., Yan, G., et al.: Hardware Implementation of Compact AES S-box. IAENG International Journal of Computer Science **42**(2) (2015)

A New Shared and Comprehensive Tool of Cloud Computing Security Risk Assessment

Saadia Drissi, Siham Benhadou and Hicham Medromi

Abstract The cloud computing is a new trending paradigm that presents several benefits in achieving rapid and scalable resource provisioning capabilities to their users. Despite the fact that cloud computing offers many cost benefits for their cloud users, number of security risk are emerging in association with cloud usage that need to be assessed. Assessing risk in Cloud computing environment remains an open research issue. This paper presents a comprehensive and shared risk assessment method for cloud computing that will add a great help and assistance to both cloud consumers and cloud providers, which is also in compliance with all the specific characteristics of the Cloud Computing. An experimental result will be showed at the end to demonstrate the effectiveness of this new risk assessment model.

Keywords Cloud computing · Security risk · Comprehensive and shared risk assessment method · Cloud consumers · Cloud providers

1 Introduction

The cloud computing is a new model for enabling ubiquitous, convenient, on-demand network access to a shared pool of configurable computing resources that introduces several changes in technology (shared resources, multi-domain, multi-tenancy) [6]. Cloud computing introduces several challenges regarding security risk assessment. These include the assessment of several cloud actors in cloud computing environment, as well as an unknown risk profile that is affected by new tenants and originates from multiple points (e.g., the provider, Cloud users, cloud organization, the technology itself, other cloud actors, etc.). In spite of the advancement in cloud technologies, cloud computing being a novel technology

S. Drissi(✉)· S. Benhadou · H. Medromi
National High School of Electricity and Mechanics, ENSEM, 8118 Casablanca, Morocco
e-mail: Saadia.drissi@gmail.com

E. Sabir et al. (eds.), *The International Symposium on Ubiquitous Networking*,
Lecture Notes in Electrical Engineering 366,
DOI: 10.1007/978-981-287-990-5_13

introduces new security risks that need to be assessed [1]. Therefore, assessment of security risks is essential [2].

The current risk assessment methods (EBIOS, OCTAVE, and MEHARI [3], [4], [5], have not been designed specifically for cloud computing environments. In traditional IT environments, everyone in the business has to go to the IT department to obtain IT related services. However, for cloud computing, the risk assessment becomes more complex; cloud computing environment is multi-location environment in which each location can use different security and potentially employ various mechanisms. In addition of these classical methodologies, several risk assessment The framework defines risk as a combination of the probability of a security threat event and its severity, measured as its Impact. methods have been suggested in the literature and also the CSA and ENISA lead a number of ongoing research initiatives (security guidance, CCM and STAR). Despite all these methodologies and initiatives, currently no complete and concise methodology exists for assessing security risks of cloud based solutions. Thus, the adoption of cloud solutions in a number of industries is stopped. Most of the studies view the problem of assessing security risks either from cloud customer or cloud provider perspectives.

As consequence, there is a need of new risk assessment model which considers all relevant aspects of information security risk assessment and mainly the specific characteristics of Cloud computing environment. This paper proposes a new tool of the cloud computing security risk assessment on the basis of the previous researches.

The following section discusses the risk assessment for cloud computing in the literature. In section 3, we describe our proposed risk assessment model for cloud computing. In section 4 an experimental result will be showed. Finally, some concluding remarks are given at the end.

2 Related Work

Several risk assessment approaches have been discussed in the literature. However, none of them takes into account the relevant and specific characteristic nature of cloud computing [12], [13], [14], [15].

2.1 Risk Assessment in the Literature

Security as a service solutions have been suggested to provide and support security assessments in which a hosted cloud solution will make assessments and stores the resulting data. Actually, several tools for a number of security assessment areas have been implemented using the delivery model SecaaS [22], [24]. In the provision of SecaaS model, cloud consumers get the typical advantages of using cloud computing such as scalability and service on demand. [25] In [23], the risk assessment as a service is presented as a new paradigm for measuring real time risk by one or more entities in the cloud. A cloud provider can perform continuous self-assessments as a best practice by assessing its own execution environment. However, this work has not implemented such a service but rather offer it as a paradigm to be pursued.

Risk assessment has analyzed security risk by using qualitative or quantitative or the both approach. In [12], a quantitative risk and impact assessment framework (QUIRC) is introduced to assess associated six key categories of security objectives (SO) (i.e., confidentiality, integrity, availability, multi- party trust, mutual audit ability and usability) in a cloud computing platform. The impact is determined by Subject Matter Experts, the knowledgeable about the impact of threats on their particular type of business. In [26], a SEmi-quantitative BLO-driven Cloud Risk Assessment (SEBCRA) prioritizes and categorizes cloud risks according to their impact on different Business Level objectives in a given organization. The approach is designed for a Cloud Service Provider (CSP) to improve the achievement of a BLO, i.e., profit maximization, by managing, assessing, and treating Cloud risks. In an exemplary experimentation, the risk assessment approach demonstrates that it enables a CSP to maximize its profit by transferring risks of provisioning its private Cloud to third party providers of cloud infrastructures. However, a simple method for qualitative or quantitative analysis will lead to the inaccuracy and one-sidedness of the evaluation results. Therefore, several studies used an integrated method of qualitative and quantitative analysis to assess risk in cloud environment [27], [13], [28], [26].

Graphs and mathematical models can be used to address and calculate security risk in clouds by simulating attacker possibilities. In [29] they presented a mathematical model for threats that considers communication in order to identify security risk for individual entities, and then calculates it for a whole enterprise. The model is built by representing communications as a directed graph and then established a matrix to discover the risk. Furthermore, in [28] a hybrid risk-analysis method based on decision tree analysis (quantities) and risk matrix (qualitative) is proposed for risk assessment. In this method, risk factor from a user's viewpoint is systematically extracted with the Risk Breakdown Structure (RBS) method then analyzed and evaluated. A detailed countermeasure and proposal are produced on the basis of these results. The risk matrix method is used to classify risk into four kinds (Risk Avoidance, Risk Mitigation, Risk Acceptance, and Risk Transference) in accordance with the generation frequency and degree of incidence.

In [13] a security risk assessment method has been introduced based on an Analytic Hierarchy Process (AHP) model. The assessment is carried out using the principles of: decomposition, pairwise comparison, and synthesis of weights. Thus, AHP has three layers of decomposition: formulating the problem of assessing cloud security risk in a hierarchical structure is the first step in AHP. Then, in level two, 8 major factors were identified for assessing. In level three, 39 factors were identified corresponding to higher levels and specific local conditions. The evaluation module uses the constructed AHP tree to assess the system with the help of the judgment matrix that is filled by the cloud's experts.

Finally calculating the weighted vectors and getting the final risk order. In [30], a hierarchical framework is built to analyze the risk and set the goal for the assessment. After that, an indicator system is built under each principle and sub indicators are introduced for assessment. For example, the first indicator could be risk of cloud computing platform, risk of cloud storage, risk of cloud security and so on. Secondary indicators of cloud platform risk could then be risk of operating system, risk of application software and risk of availability.

In [31], Trust Matrix is used for security risk analysis in cloud environments. Two variables, namely "data cost" and "provider's history" are considered. In "data cost" users can assign a cost to data based on the data's criticality whereas "Provider's history" includes the record of the past services provided by the provider to consumers. Additionally, Cloud Control Matrix (CCM) has been released by CSA, as a baseline security control framework designed to help enterprises assess the risks associated with a cloud provider. The CCM has included a risk management domain to ensure that formal risk assessments are aligned with the enterprise-wide framework, planned and scheduled at regular intervals determining the likelihood and impact of identified risks, using qualitative and quantitative methods. Thereby, it facilities transparency and increase trust level between the cloud customer and the cloud in order to make cloud a secure environment to the future of business [32].

2.2 Discussion

After reviewing the literature, several risk assessment methodologies and frameworks have been reviewed and suggested. The risk assessment methods have been classified into five categories: assessment as a service, quantitative and qualitative, hierarchal, graph analysis and security matrix assessment. In addition to the risk assessment methods that have been reviewed, the CSA and ENISA lead a number of ongoing research initiatives (security guidance, CCM and STAR). Despite all these methodologies and initiatives, currently no complete and concise methodology exists for analyzing and evaluating security risks of cloud based solutions. A cloud-specific threats, vulnerabilities and risks have already been identified or assessed by numerous sources, but it still remains unclear how to assess risks basing on Information Risk Management frameworks or methods in the context of the Cloud. Thus, the adoption of cloud solutions in a number of industries is stopped. Most of the studies view the problem of assessing security risks either from cloud customer or cloud provider perspectives. The need for a comprehensive, shared, collaborative and intelligent risk assessment methodology that considers both customer and provider is recommended. Such as shared assessment enables the cloud provider to prove how the security risks have been managed and mitigated, as well as enabling the cloud consumer to determine the risk tolerance and define security requirements accordingly.

3 Research Methodology

The risk assessment is not a very easy task to do regarding cloud computing, because of its complex nature. Therefore, in this part we will describe how to assess risk in cloud computing environment.

Below the proposed risk assessment architecture that would be explained in detail in the next section.

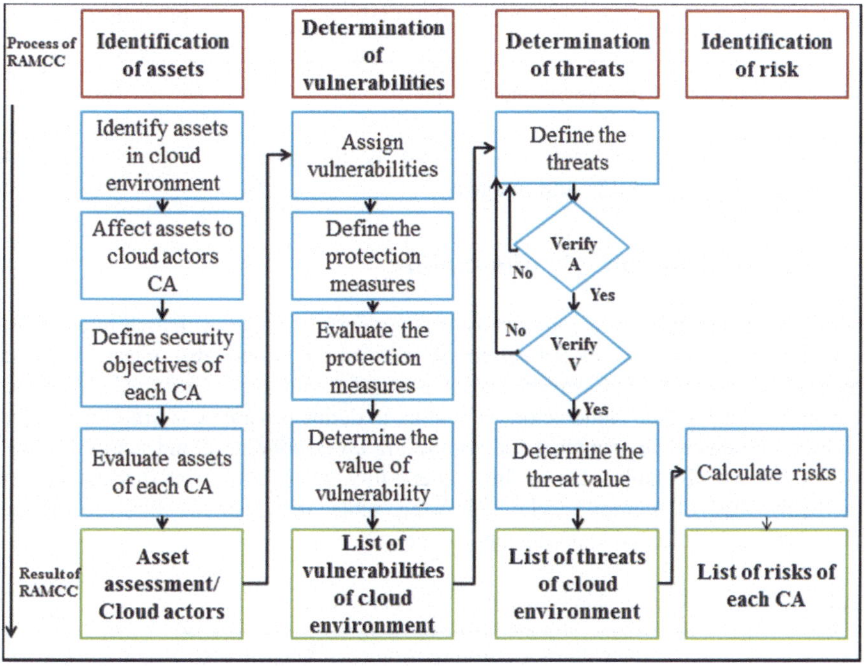

Fig. 1 Architecture of risk assessment model (RAMCC)

3.1 Identification of Asset Value

In cloud computing environment, there are several cloud actors and each one of them has their one security objectives and the risk assessment is purely based on decision making. Therefore, there is a need of model capable of dealing with such type of problems. To solve this problem, we will show up AHP in our work [17]-[18]-[19]. Such an asset assessment methodology incorporates a level of flexibility on the notion that several cloud actors can define their asset value in the same time. Thus, this paradigm can ensure the effectiveness, the flexibility and the automation to our proposed risk assessment model [21].

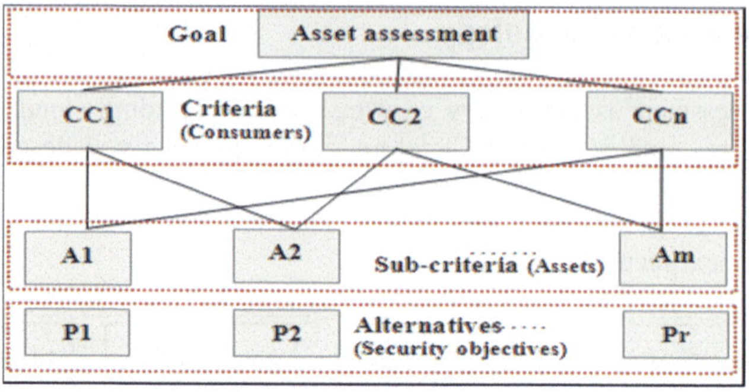

Fig. 2 Decision tree of asset assesmment [20]

The asset value of each cloud actor is the average of the weight of confidentiality, availability and integrity.

3.2 *Determination of Vulnerability*

Outsourcing services to cloud means been exposed to new vulnerabilities, thus, resulting in a modified identification of vulnerabilities and also means that the methodology used for conventional systems will be hard to use for cloud. To define vulnerabilities for each cloud actor, we require to define possible vulnerabilities in the cloud environment environment, determine the corresponding cloud actors to each vulnerability and finally, define the vulnerability value basing on the absence or ineffectiveness of controls. The following table demonstrates how to define vulnerabilities in cloud computing environment:

Table 1 Vulnerabilities in cloud computing environment

Vulnerabilities	Corresponding cloud actors	Vulnerability value
V1	Ac1 Ac2 Ac4	v1
V2	Ac2 Ac4 Ac6	v2
V3	Ac1 Ac8 Ac3	v3
V4	Ac7	v4
A5	Ac7 Ac8 Ac5	v5
……	…	….
Vn	Ac5 Ac4 Ac3	vh

3.3 *Identification of Threats*

The main reason that threats are important elements of the information security risk assessment is that they help to determine the scope of the vulnerabilities of the system being assessed. In the cloud environment, each threats is mapped to indicative number of vulnerabilities and assets as mentionned in [33]. The following table demonstrates how to define threats in the cloud environment:

Table 2 Threats in cloud computing environment

Threat	Corresponding assets	Corresponding vulnerabilities	Threats value
T(R1)	A1 A5 A6 A7	V13 V46 V31	t1
T(R2)	A1 A2 A3 A5 A6	V34 V35 V25	t2
T(R3)	A20	V25	t3
T(R4)	A1 A5 A6 A7 A9	V6 V7 V5	t4
T(5)	A1 A2 A3 A9 A10	V31 V46 V47	t5
……	…	….	….
T(Rn)	…	….	tn

The threat can be defined if there is at least one corresponding asset and vulnerability, and the threat value is product of probability of occurrence and the impact.

3.4 Determination of Risks

For the fourth process of risk assessment, the measure of an IT risk can be determined as a product of threat, vulnerability and asset values:

$$\text{Risk} = \text{Threat} * \text{Vulnerability} * \text{Asset} \quad (1)$$

$$\text{Threat} = \text{Impact} * \text{Probability} \quad (2)$$

$$\text{Asset} = \max(A_i) + 0.05(m_1 - 1) + 0.04(m_2 - 1) + 0.03(m_3 - 1) + 0.02(m_4 - 1) + 0.01(m_5 - 1) \quad (3)$$

Which m_5 is the amounts of assets when the corresponding assets value between 0 and 0.2

Which m_4 is the amounts of assets when the corresponding assets value between 0.2 and 0.4

Which m_3 is the amounts of assets when the corresponding assets value between 0.4 and 0.6

Which m_2 is the amounts of assets when the corresponding assets value between 0.6 and 0.8

Which m_1 is the amounts of assets when the corresponding assets value between 0.8 and 1

$$\text{Vulnerability} = \max(V_j) + 0.5(h_1-1) + 0.4(h_2-1) + 0.3(h_3-1) + 0.2(h_4-1) + 0.1(h_5-1) \quad (4) \quad \mathbf{(2)}$$

Which h_1 is the amounts of vulnerabilities when the corresponding vulnerabilities is 5
Which h_2 is the amounts of vulnerabilities when corresponding vulnerabilities is 4
Which h_3 is the amounts of vulnerabilities when corresponding vulnerabilities is 3
Which h_4 is the amounts of vulnerabilities when corresponding vulnerabilities is 2
Which h_5 is the amounts of vulnerabilities when corresponding vulnerabilities is 1

At the end each cloud actors will be aware of the risks and vulnerabilities present in the cloud computing environment.

4 Experimentations

We suppose the following scenario:

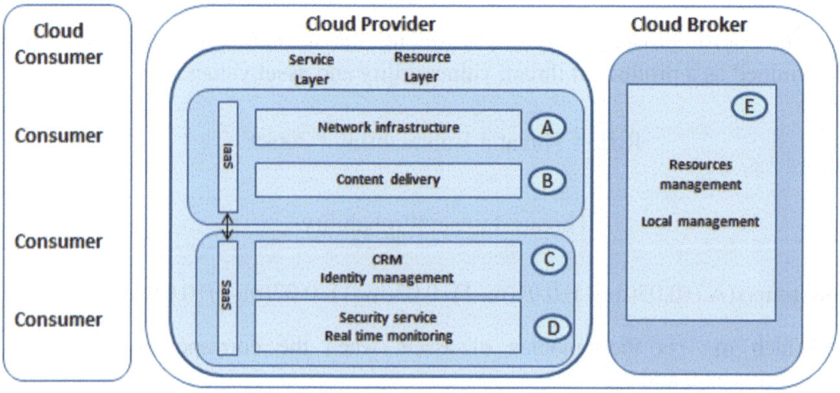

Fig. 3 Cloud computing scenario

4.1 Asset Value Identification

The asset identification table (Fig. 4 and Fig. 5) is as follow basing on preliminary analysis and assignment, in which each asset corresponds to one asset level and each defined asset can be assigned to one or more cloud actors:

Code	Asset	Confidentiality	Integrity	Availability	Asset Value
A20	Certification	4	5	5	0.213
A19	CP Application	4	5	3	0.141
A18	Physical buildings	3	4	3	0.098
A17	Physical hardware	5	4	5	0.11
A16	Network	5	4	5	0.093
A15	Management interface APIs	4	3	4	0.064
A14	Cloud service management interface	4	3	5	0.058
A12	Credentials	5	5	5	0.06
A11	Access control / authentication / authorization	4	4	5	0.044

Fig. 4 Part of asset identification for the SaaS

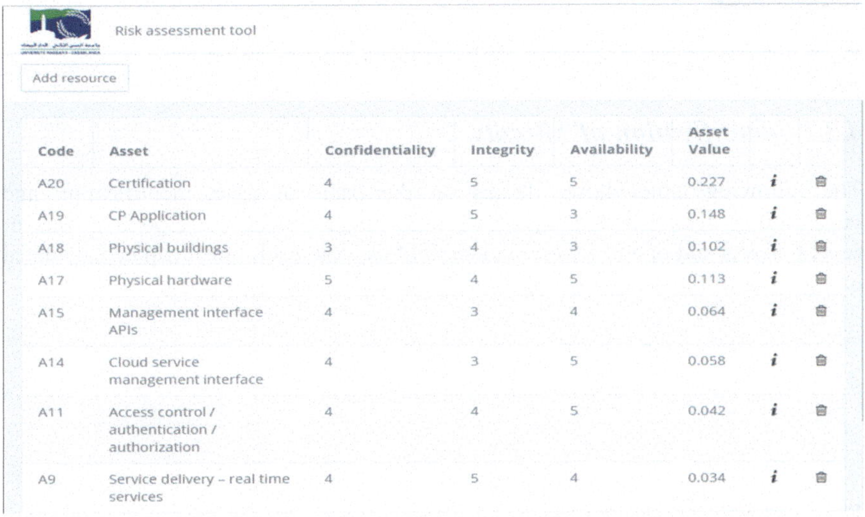

Code	Asset	Confidentiality	Integrity	Availability	Asset Value
A20	Certification	4	5	5	0.227
A19	CP Application	4	5	3	0.148
A18	Physical buildings	3	4	3	0.102
A17	Physical hardware	5	4	5	0.113
A15	Management interface APIs	4	3	4	0.064
A14	Cloud service management interface	4	3	5	0.058
A11	Access control / authentication / authorization	4	4	5	0.042
A9	Service delivery – real time services	4	5	4	0.034

Fig. 5 Part of asset identification for the IaaS

4.2 *Determination of Vulnerability*

The following figure shows a part of cloud security management vulnerabilities of cloud organization, in which each vulnerability corresponds to one vulnerability level and each defined vulnerability can be assigned to one or more cloud actors.

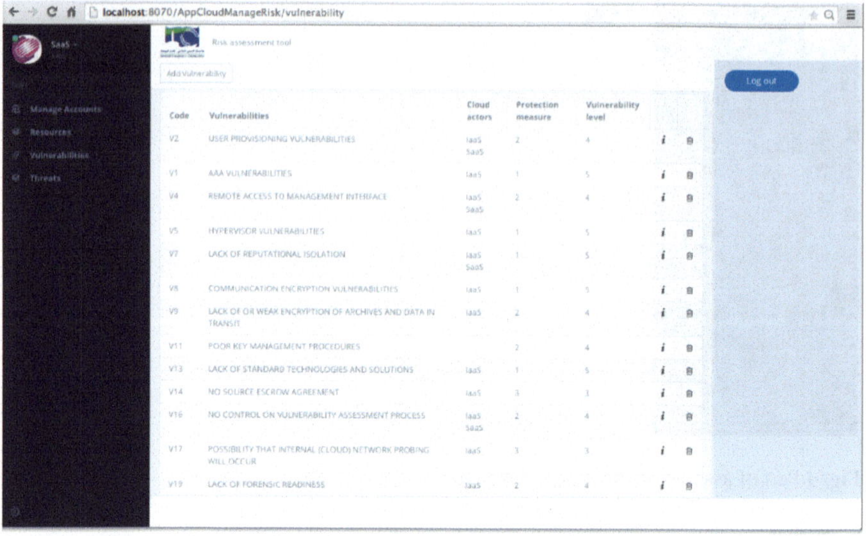

Fig. 6 Vulnerabilities Determination in cloud computing environment

4.3 Identification of Threats

The following table shows threats on the basis of asset identification and vulnerability determination, in which each threat can be correspond to one or several assets and one or several vulnerabilities. And each threat can be calculated on the basis of the occurrence possibility and their possible impacts.

Risk	Threat	Vulnerabilities	Assets	Probability	Impact	Th value
R29	T(R29)	[V2, V32]	[A5, A11, A12]	0.5	4.0	2.0
R28	T(R28)	[V2, V36]	[A5, A11]	0.25	4.0	1.0
R27	T(R27)	[V16]	[A5, A9]	0.25	4.0	1.0

Fig. 7 Part of threats identification

4.4 Identification of Risks

The following table shows all risks the present risk on cloud computing environment on the basis of the identification threats.

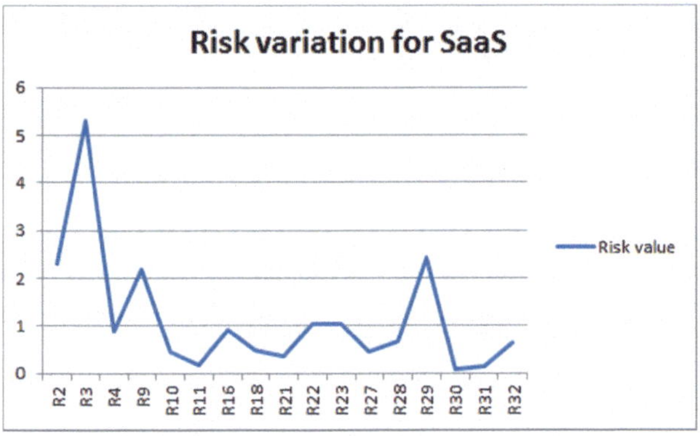

Fig. 8 Risk variation for SaaS

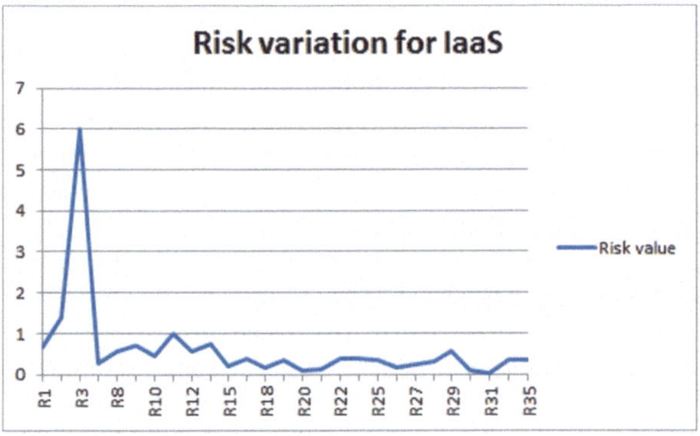

Fig. 9 Risk variation for IaaS

At the end each cloud actor will be aware of their corresponding risks as showen in the above figures (fig. 8 and fig. 9), the both figures are different, because each actor has their corresponding assets, their security objectives and their corresponding vulnerabilities.

5 Conclusion

In this paper, an use case have been performed in detail on the basis of the proposed web based solution, to demonstrate the effectiveness of this new comprehensive and shared risk assessment method for cloud computing that will add a great assistance and help to both cloud consumers and cloud providers. As consequence, with such an approach, the cloud consumers can be guaranteed data security and the cloud providers can win the trust of their consumers.

As future work, the authors will show how they can benefit from multi-agent systems to improve the architecture and consolidate the security risk assessment for cloud computing. In addition, the authors will give primordial improvements of the proposed risk assessment method.

References

1. Cloud Security Alliance (CSA): Top threats to cloud computing, version 1.0, March 2010. http://www.cloudsecurityalliance.org/topthreats/csathreats.v1.0.pdf
2. Kaliski Jr., B.S., Pauley, W.: Toward Risk Assessment as a Service in Cloud Environment. EMC Corporation, Hopkinton (2010)
3. EBIOS, Central Directorate for Information Systems Security, Version 2010 website. http://www.ssi.gouv.fr
4. Operationally Critical Threat, Asset and Vulnerability Evaluation (OCTAVE), Carnegie Mellon - Software Engineering Institute (1999)
5. Method Harmonized Risk Analysis (MEHARI) Principles and mechanisms CLUSIF, issue 3, October 2004
6. Mell, P., Grance, T.: Perspectives on cloud computing and standards. National Institute of Standards and Technology (NIST). Information Technology Laboratory (2009)
7. CSS, White paper on software and service architectures, Infrastructures and Engineering – Action Paper on the area for the future EU competitiveness: Background information, Version 1.3, vol. 2 (retrieved: 15.08.2010). http://www.euecss.eu/contents/documentation/volume%20two_ECSS%20White%20Paper.pdf
8. Miller, M.: Cloud computing: Web-based applications that change the way you work and collaborate online. Indianapolis (2008)
9. Van Scoy, R.L.: Software Development Risk: Opportunity, Not Problem
10. Farrell, R.: Securing the cloud-governance, risk and compliance issues reign supreme. Information Security Journal: A Global Perspective (2010)
11. Sayouti, A., Medromi, H.: Les Systèmes Multi-Agents: Application au Contrôle sur Internet, Auteurs Éditions universitaires européennes, Août 2012
12. Saripalli, P., Walters, B.: QUIRC: a quantitative impact and risk assessment framework for cloud security. In: The Proceedings of the IEEE 3rd International Conference on Cloud Computing, pp. 280–288 (2010)
13. Peiyu, L., Dong, L.: The New Risk Assessment Model for Information System in Cloud Computing Environment. Procedia Engineering **15**, 3200–3204 (2011)
14. Xuan, Z., Wuwong, N., et al.: Information security risk management framework for the cloud computing environments. In: 2010 IEEE 10th International Conference on Computer and Information Technology (CIT) (2010)
15. Sangroya, A., Kumar, S., Dhok, J., Varma, V.: Towards analyzing data security risks in cloud computing environments. In: International Conference on Information Systems, Technology, and Management (ICISTM), Bangkok, Thailand (2010)
16. Drissi, S., Houmani, H., Medromi, H.: Survey: risk assessment for cloud computing. International Journal of Advanced Computer Science and Applications, pp. 143–148 (2013)
17. Altuzarra, A., Moreno-Jimnez, J.M., Salvador, M.: A Bayesian prioritization procedure for AHP-group decision making. European Journal of Operational Research **182**(1), 367–382 (2007)

18. Ramanathan, R., Ganesh, L.S.: Group preference aggregation methods employed in AHP: An evaluation and an intrinsic process for deriving members' weightages. European Journal of Operational Research **79**(2), 249–265 (1994)
19. Dyer, R.F., Forman, E.H.: Group decision support with the analytic hierarchy process. Decision Support Systems **8**(2), 99–124 (1992)
20. Lichtenstein, S.: Factors in the selection of a risk assessment method. Information Management & Computer Security **4**(4), 20–25 (1996)
21. Drissi, S., Medromi, H.: A new risk assessment approach for cloud consumer. Journal of Communication and Computer **11**, 52–58 (2014)
22. Free Security Assessment by Trend Micro, Security Assessment Tool
23. Onwudebelu, U., Chukuka, B.: Will adoption of cloud computing put the enterprise at risk? In: 2012 IEEE 4th International Conference on Adaptive Science & Technology (ICAST), October 25–27, pp. 82–85 (2012)
24. Security Risk Assessment for Cloud and Web. Cenzic Cloud
25. SecaaS Category 5 Security Assessments Implementation Guidance. Cloud Security Alliance, September 2012
26. Fito, J.O., Macias, M., Guitart, J.: Toward business-driven risk management for cloud computing. In: 2010 International Conference on Network and Service Management (CNSM), October 25–29, pp. 238–241 (2010)
27. Djemame, K., et al.: A risk assessment framework and software toolkit for cloud service ecosystems. In: Cloud Computing 2011, The Second International Conference on Cloud Computing, GRIDs, and Virtualization (2011)
28. Tanimoto, S., Hiramoto, M., Iwashita, M., Sato, H., Kanai, A.: Risk management on the security problem in cloud computing. In: 2011 First ACIS/JNU International Conference on Computers Networks, Systems and Industrial Engineering (CNSI), May 23–25, pp. 147–152 (2011)
29. Leitold, F., Hadarics, K.: Measuring security risk in the cloud-enabled enterprise. In: 2012 7th International Conference on Malicious and Unwanted Software (MALWARE), October 16–18, pp. 62–66 (2012)
30. Zhang, J., Sun, D., Zhai, D.: A research on the indicator system of cloud computing security risk assessment. In: 2012 International Conference on Quality, Reliability, Risk, Maintenance, and Safety Engineering (ICQR2MSE), June 15–18, pp. 121–123 (2012)
31. Chandran, S., Angepat, M.: Cloud computing: analyzing the risk involved in cloud computing environments. In: Proceedings of Natural Sciences and Engineering, Sweden, pp. 2–4 (2010)
32. Cloud Security Alliance, Cloud Control Matri, September 26, 2013
33. Catteddu, D., Hogben, G. (eds.): Cloud Computing: Benefits, risks and recommendations or information security. ENISA (2009)

Part III
Advances in Physical Layer Characterization and Smart Antennas

The Behavior of CPW-Fed Slotted Cantor Set Fractal Antenna

Abdelati Reha, Abdelkebir El Amri, Othmane Benhmammouch and Ahmed Oulad Said

Abstract This paper presents the behavior of three iterations of a CoPlanar Waveguide-fed (CPW) CANTOR Set fractal antenna. This kind of antennas allows having an Ultra Wide Band (UWB) behavior and important gains. Also, the setup of slots allows having more lower resonant frequencies and therefore designing miniaturized antennas with good performances.
All the simulations were performed with FEKO 6.3.

Keywords Cantor set fractal antenna · Printed antennas · Antenna design

1 Introduction

In high-performance systems such as aircrafts, satellites, cars, Smartphones and other embedded systems, where size, weight, cost, performances, ease of installation, and aerodynamic profile are constraints, low-profile antennas may be required. To achieve this objective, microstrip antennas can be used. These antennas are low profile, conformable to planar surfaces, inexpensive and simple to manufacture using printed-circuit technology [1][2][3]. Also, we can have more resonant frequencies or miniaturizing antennas by adding loads between the patch and the ground or slots on the radiating elements [4][5].

A. Reha(✉) · A. El Amri
RITM Laboratory, ESTC, Casablanca, Morocco
e-mail: reha.abdelati@gmail.com

O. Benhmammouch
Mundiapolis University, Casablanca, Nouaceur, Morocco

A. Oulad Said
Royal Air Academy, Marrakech, Morocco

E. Sabir et al. (eds.), *The International Symposium on Ubiquitous Networking*,
Lecture Notes in Electrical Engineering 366,
DOI: 10.1007/978-981-287-990-5_14

In recent years, fractal geometry has been used in the electromagnetic, and especially in the design of antennas. Several studies have adopted fractal structures and showed that this technique can improve the performances of the antenna and it is one of the techniques to design antennas with multi-band and broad-band behavior [6][7][8][9].

In this paper, a CPW-Fed Slotted Cantor Set Fractal Antenna is designed. First we describe the antenna design methodology, after that, we study the behavior of the antenna when we increase the number of iterations and finally, we discuss the behavior of the studied antenna when we add slots.

2 The Patch Antenna Design Methodology

2.1 The Design of Miscrostrip Patch Antenna

The microstrip patch antennas can be analyzed in various methods, the most common are:

- Transmission-line method
- Cavity method
- Full-wave method

Although the transmission line model has the least accuracy, it is the easiest method to implement and gives good physical insight [10]. According to Balanis [1], the transmission-line model represents the microstrip antenna by two slots with a width of W and separated by a transmission line of length L (Fig. 1).

For the microstrip line shown in Fig. 2 (a), the field lines are inside the substrate and some of them are extended to outer space (Fig. 2 (b)). For this, an effective dielectric constant (ε_{reff}) is introduced to account for fringing and the wave propagation in the line (Fig. 2 (c)).

ε_{reff} can be calculated from [1] by (1)

$$\varepsilon_{reff} = \frac{\varepsilon_r + 1}{2} + \frac{\varepsilon_r - 1}{2} \times \left[[1 + 12\frac{h}{W}] \right]^{-\frac{1}{2}} \tag{1}$$

Where , W/h >1

ε_{reff} : Effective dielectric constant

ε_r : Dielectric constant of the substrate

W : Width of the radiating patch

h : Height of the substrate

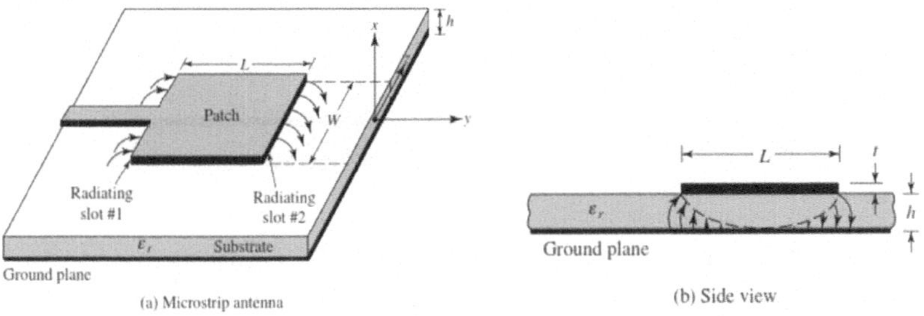

Fig. 1 Microstrip antenna [1]

As shown in Fig. 3, fringing effects looks greater than the microstrip patch dimensions. For the principal E-plane (xy-plane), the dimensions of the patch along its length have been extended on each end by a distance ΔL, which is a function of ε_{reff} and W/h given from[1] by (2)

$$\frac{\Delta L}{h} = 0.412 \times \frac{(\varepsilon_{reff} + 0.3) \times (\frac{W}{h} + 0.264)}{(\varepsilon_{reff} - 0.258) \times (\frac{W}{h} + 0.8)} \tag{2}$$

The effective length of the patch is given by (3)

$$L_{eff} = L + 2.\Delta L \tag{3}$$

It is also given by (4)

$$L_{eff} = \frac{\lambda}{2\sqrt{\varepsilon_{reff}}} \tag{4}$$

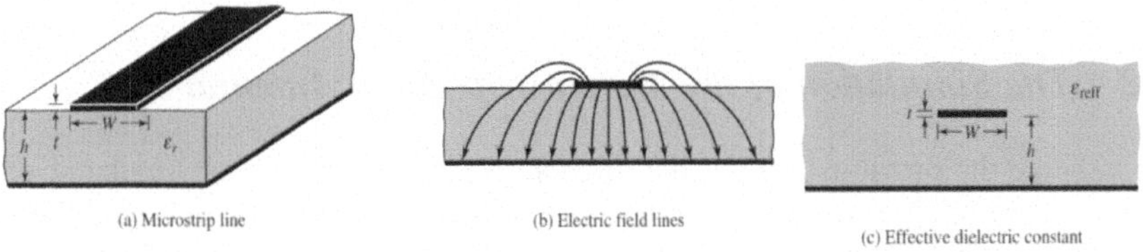

Fig. 2 Microstrip line and its electric field lines, and effective dielectric constant geometry [1]

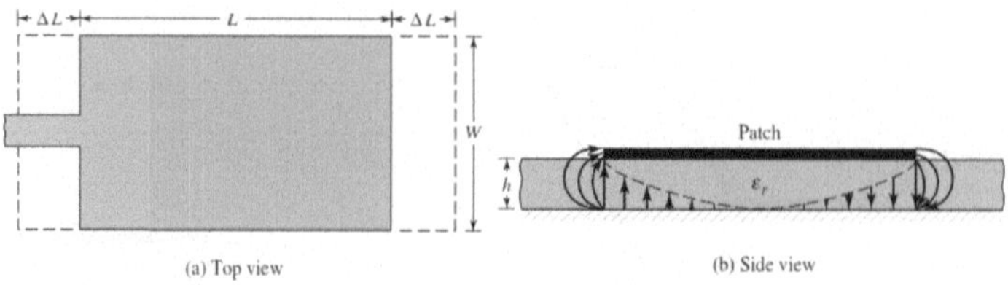

Fig. 3 Physical and effective lengths of rectangular microstrip patch[1]

The width of the patch is given from [1] by (5)

$$W = \frac{\lambda}{2}\sqrt{\frac{2}{\varepsilon_r + 1}} \tag{5}$$

Where, λ is the wavelength given by (6)

$$\lambda = \frac{c}{f} \tag{6}$$

The characteristic impedance of microstrip line (Z_m) is given from [11] by (7)

$$Z_m = \frac{120.\pi}{\sqrt{\varepsilon_{reff}}} \quad \frac{W_f}{H} + 1.393 + 0.667\ln\left(\frac{W_f}{H} + 1.444\right)^{-1} \tag{7}$$

With

W_f: the width of the microstrip line

To design a microstrip patch antenna operating in the frequency of 2.6GHz with the parameters: h=1.6mm and $\varepsilon_r = 4.4$ we follow the previous steps.

The results are:

W= 34.9mm, ε_{reff}= 4.06, L=27.1mm

To adjust Z_m=50Ω, the value of W_f should be 3.3mm.

2.2 The Simulation of the Miscrostrip Patch Antenna

To validate the previous design, the microstrip patch antenna was simulated with FEKO 6.3 witch based on the Method of the Moments (MoM), one of the more accurate methods [1][9]. As shown in Fig. 4, the first resonant frequency is 2.55GHz, 0.05GHz lower than the resonant frequency found by the Transmission Line Model.

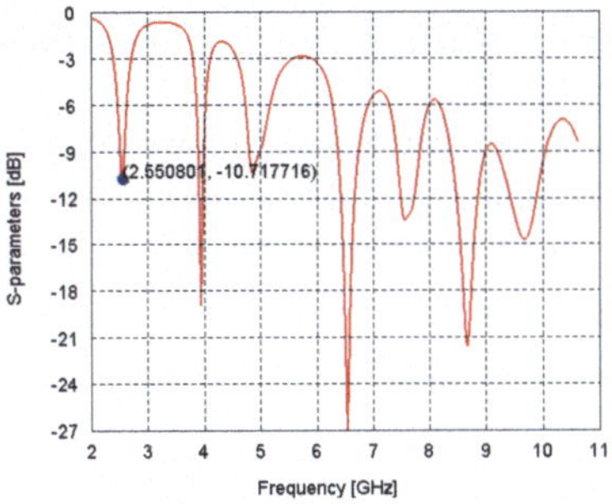

Fig. 4 The S_{11} parameter versus the frequency

2.3 *The Modification of the Ground Plane*

To achieve an ultra wideband behavior and a good efficiency, the ground plane were placed on the same plane as the resonant element as shown in Fig. 5. Several studies used this technique to design a broad-band antennas with high gains [4] [7][12]. Fig. 6 (a and b) shows respectively, the comparison between the S_{11} and the efficiency versus frequencies for the two configurations. In the following, the second configuration is adopted in order to design antennas with an ultra wide-band behavior and a good efficiency.

(a) : the first configuration (b) : the second configuration

Fig. 5 The two configurations of the ground plane

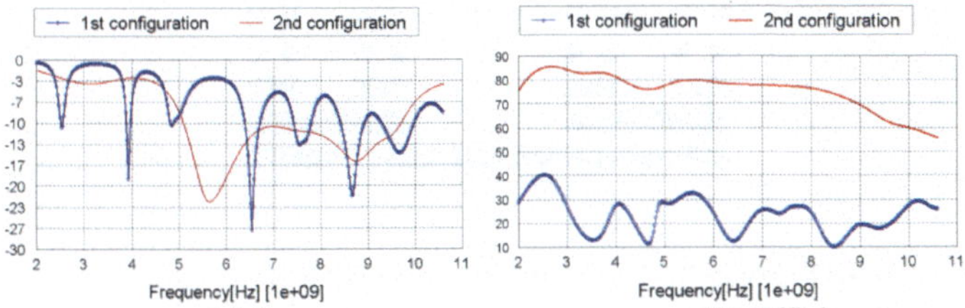

(a) S_{11} (b) Efficiency

Fig. 6 The S_{11} and Efficiency parameters versus frequencies for the two configurations

2.4 *The Modification of the Length of the Patch Antenna*

Initially, the dimensions of our patch antenna are: W= 34.9mm and L=27.1mm. The length of this patch is modified and becomes shortest $W_{\text{modified patch}}$ = 11.7mm. As shown in Fig. 7, the -10dB bandwidths of the two antennas are the same, but a difference of maximum gains is observed, the modified antenna have a better total gain in the bands of 5 – 6.6GHz and 6.9 – 7.9Ghz.

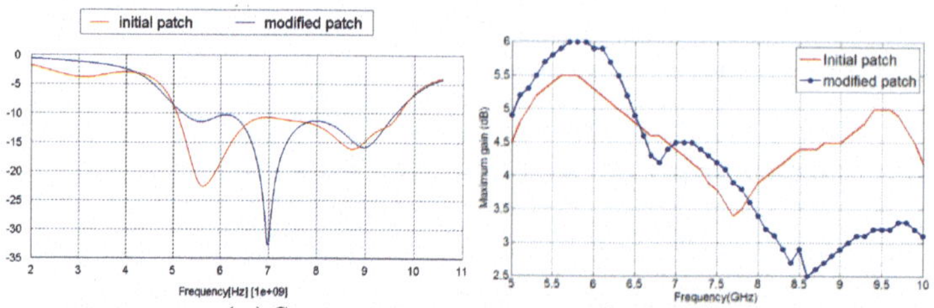

(a) S_{11} (b) Maximum Gain

Fig. 7 The S_{11} and maximum gain parameters versus frequencies for the two antennas

3 The Cantor Set Fractal Antenna Design

3.1 *The CANTOR SET Fractal Geometry*

CANTOR Set was invented by the German mathematician Georg CANTOR. It is built iteratively from the segment [0, T] by removing a central portion (a third for example); then the operation is repeated on the remaining two segments, and so on. Fig. 8 shows the 6 first iterations of the Cantor set structure [6].

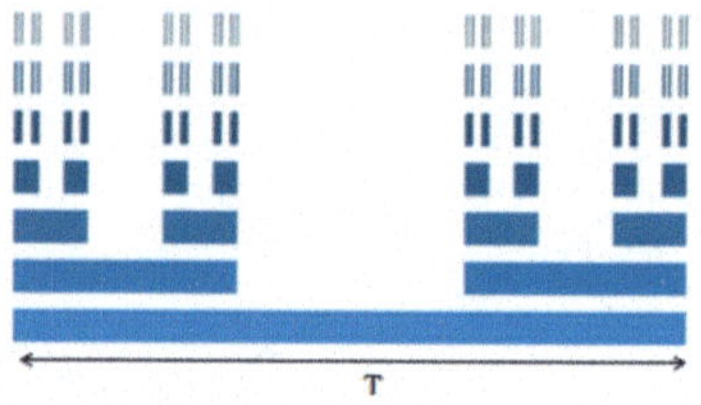

Fig. 8 The 6 first iterations of the CANTOR SET structure [6]

The HAUSDORFF dimension of the CANTOR Set fractal is given by (8) [6][13].

$$d = \frac{\ln(n)}{\ln(h)} = \frac{\ln(2)}{\ln(h)} \tag{8}$$

With: "n" is the number of copies from the iteration N to iteration N+1 and "h" is the reduction factor. If h=3, d=0.6.

3.2 *The Use of CANTOR SET Fractal Geometry in the Antenna Design*

Yingsong LI designed some antennas based on the modified Cantor Set Fractal structure. In 2011, he has designed a miniaturized and modified patch antenna fed by a microstrip line having a radiating element in the form of a CANTOR set. This structure is operational for Ultra Wide Band (UWB) applications and having an important gain except for the 5 - 6.3GHz applications[6] [14]. Fig. 9 shows the proposed structure, its return loss, and its gain.

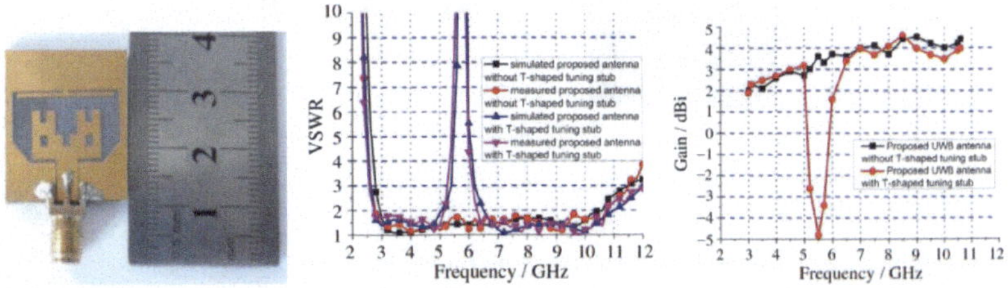

(a) (b) (c)

Fig. 9 the proposed structure by LI (a), its return loss (b), and its gain (c) [6][14]

In 2012, he has designed a printed diversity Cantor Set fractal antenna for UWB Applications[15]. Fig. 10 shows the proposed structure and its return loss.

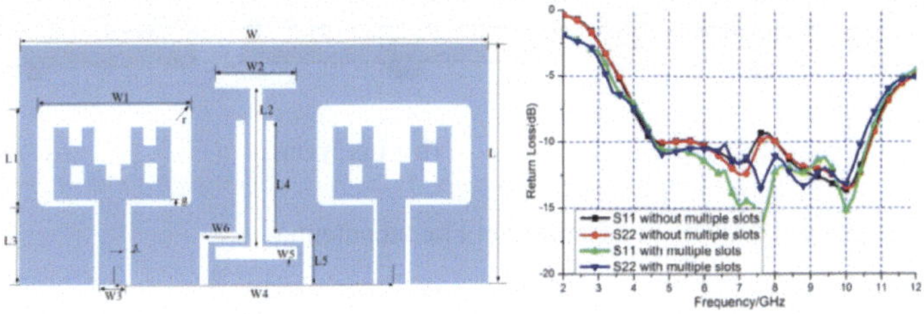

(a) (b)

Fig. 10 the proposed structure by LI (a) and its return loss [15]

In 2012, he has designed a Cantor Set fractal antennas for switchable UWB Applications[16]. Fig. 11 shows the proposed structure and its return loss.

3.3 *The Design of CPW-Fed Slotted CANTOR SET Fractal Antenna*

Fig. 12 illustrates the tree iterations of the CPW-Fed Cantor Set fractal antennas. The proposed structures are printed on a substrate with relative permittivity of 4.4, a loss tangent of 0.0025 and a thickness of h=1.6mm. The size of the antenna is 50×42mm^2 (L_s×W_s). The other parameters are: W_p=34.9mm, L_p=11.7mm, s=0.5 mm, L_g= 5mm and W_f=3.3mm.

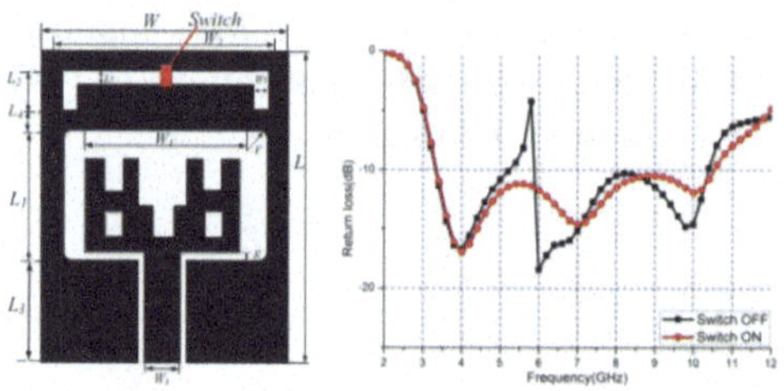

(a) (b)

Fig. 11 The proposed structure by LI (a) and its return loss [16]

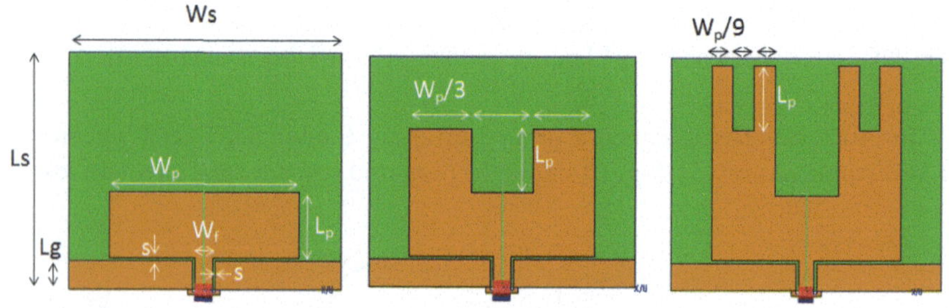

(a) Iteration 1 (b) Iteration 2 (c) Iteration 3

Fig. 12 The proposed CPW-Fed Cantor Set Fractal antenna

Through numerical simulations, the three iterations of the Cantor Set Fractal antenna demonstrate a return loss below -10dB over a wide frequency ranges (Fig.13). Also, we observe that the maximum gain of the antennas varies between 2.6 and 6dB for the first iteration, between 4.4 and 5.9dB for the second iteration and between 4.4 and 6.2 for the third iteration (Fig. 14). Table 1 summarizes for each iteration, the resonant frequencies, the S_{11} value on the resonant frequencies, the -10dB bandwidths and the maximum gains on the -10dB bandwidths.

3.4 *The Setup of the Slots on the CPW-Fed Slotted CANTOR SET Fractal Antenna (3rd Iteration)*

To create other resonance frequencies, a rectangular slot is setup on the 3rd iteration of the proposed antenna. The position of the slot and their dimension are Y_{slot}, W_{slot} and L_{slot} (Fig.15). A parametric study is based on the variation of W_{slot} was performed and Fig. 16 shows the variation of the S_{11} parameter versus frequencies and versus W_{slot}. The other parameters are: L_{slot} = 1mm and y_{slot} = 7.5mm.

We observe that the new resonance frequency decrease when the W_{slot} increase.

Table 1 Simulated resonant frequencies and bandwidths for the three iterations of the Cantor Set Fractal antenna

Iteration	Resonant frequencies (GHz)/S_{11}(dB)	(-10dB) bandwidths (GHz) : from → to	*Gains(dB) : Min - Max
First	5.6 / -11.5 7 / -32.7 9 / -15.9	4.42 : 5.2 → 9.62	2.5 - 6
Second	5.95 / -27.8 8.4 / -17.5 9.42 / -15.7	1.81 : 5.09 → 6.9 2.16 : 7.96 → 10.12	4 – 5.4 4.4 - 5.6
Third	5.3 / -39.5 6.03 / -25.4 9.27 / -13.6	1.82 : 4.98 → 6.8 2.4 : 7.6 → 10	4.4 – 4.8 3.7 – 6.2

*The gain is simulated on the (-10 dB) bandwidth

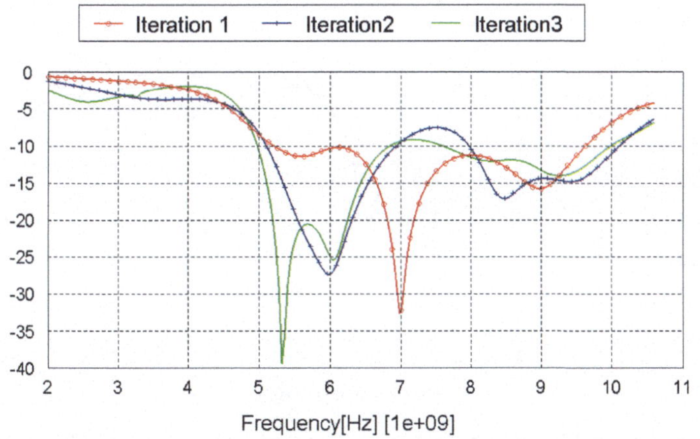

Fig. 13 The simulated S_{11} parameter versus frequencies for the three iterations

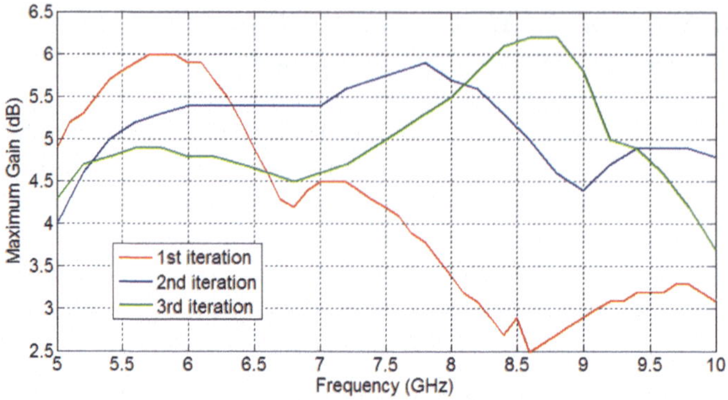

Fig. 14 The simulated maximum gains of the three iterations

Fig. 17 shows the 3D- total gain pattern when W_{slot} = 33mm for some frequencies (2.6, 5.3, 8 and 10GHz).

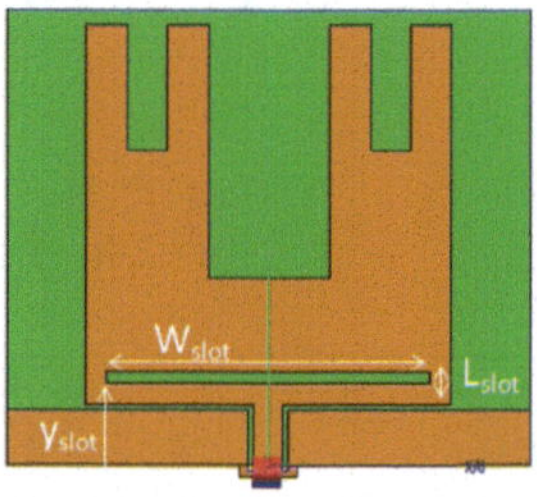

Fig. 15 The slotted CPW-Fed Cantor Set Fractal antenna (3rd iteration)

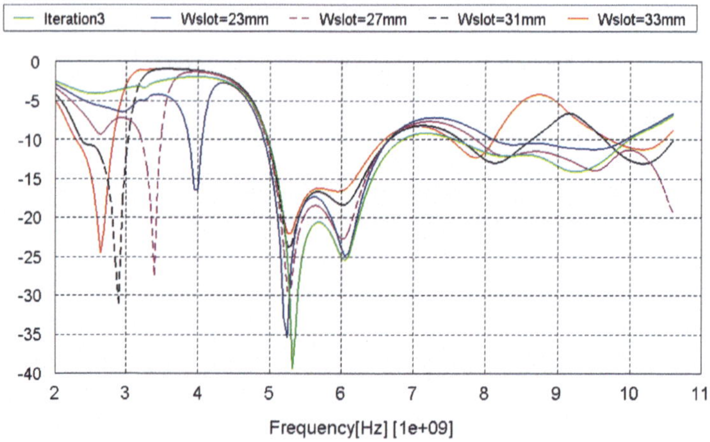

Fig. 16 The simulated S_{11} parameter versus frequencies and versus the W_{slot} for the 3rd iteration CANTOR SET fractal antenna

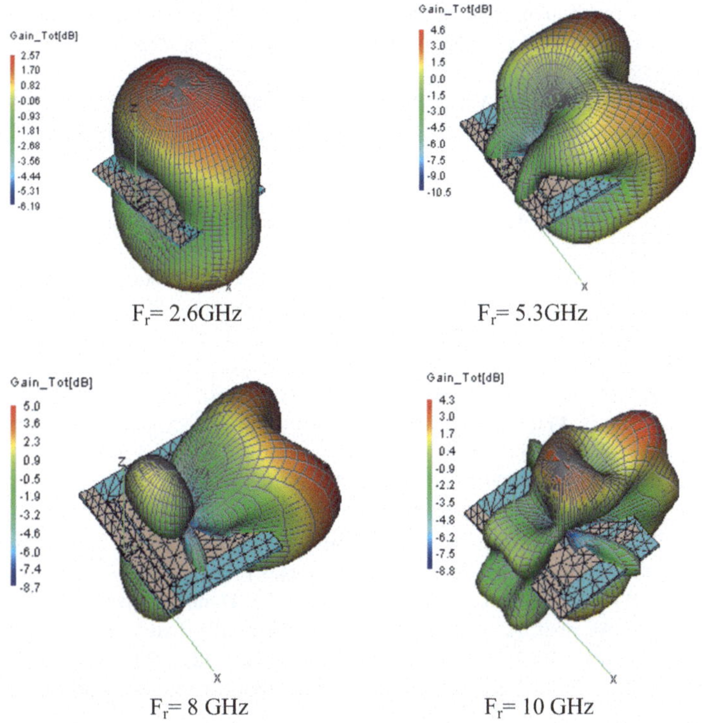

Fig. 17 The simulated 3D-total gain pattern for the frequencies 2.6, 5.3, 8 and 10GHz when W_{slot}=33mm.

4 Conclusion

The fractal concept is a one of the better solutions to design a simple, low profile and miniaturized antennas; the use of the CPW-Fed slotted CANTOR Set Fractal antenna allows having a broadband antenna with a lot of resonant frequencies and good efficiency.

Also, further dimensions and iterations can be done to obtain antennas with another sizes, more Ultra Wide Bands and better antenna performances.

References

1. Balanis, C.A.: Antenna theory: analysis and design, 3rd edn. John Wiley, Hoboken (2005)
2. Fang, D.G.: Antenna theory and microstrip antennas. CRC Press/Taylor & Francis, Boca Raton (2010)
3. Gross, F.B.: Frontiers in antennas next generation design & engineering. McGraw-Hill, New York (2011)

4. Shagar, A.C., Wahidabanu, S.D.: Novel wideband slot antenna having notch-band function for 2.4 GHz WLAN and UWB applications. Int. J. Microw. Wirel. Technol. **3**(04), 451–458 (2011)
5. Reha, A., Bouchouirbat, M.: A Dual-Band Rectangular CPW Folded Slot Antenna for GNSS Applications. Int. J. Adv. Res. Electr. Electron. Instrum. Eng. **3**(8), 11055–11061 (2014)
6. Reha, A., El Amri, A., Benhmammouch, O., Oulad Said, A.: Fractal Antennas: A Novel Miniaturization Technique for wireless Networks. Trans. Netw. Commun. **2**(5) (2014)
7. Reha, A., El Amri, A., Benhmammouch, O., Oulad Said, A., El Ouadih, A., Bouchouirbat, M.: CPW-fed H-tree fractal antenna for WLAN, WIMAX, RFID, C-band, HiperLAN, and UWB applications. Int. J. Microw. Wirel. Technol., 1–8 (2015)
8. Reha, A., El Amri, A., Benhmammouch, O., Oulad Said, A.: CPW-Fed KOCH SNOWFLAKE Fractal Antenna for UWB Wireless Applications. Trans. Netw. Commun. **2**(4) (2014)
9. Reha, A., Said, A.O.: Tri-Band Fractal Antennas for RFID Applications. Wirel. Eng. Technol. **04**(04), 171–176 (2013)
10. Van Lil, E., Van De Capelle, A.: Transmission line model for mutual coupling between microstrip antennas. IEEE Trans. Antennas Propag. **32**(8), 816–821 (1984)
11. Combes, P.F.: Micro-ondes. cours et exercices 1, 1. Dunod, Paris (1996)
12. Fallahi, H., Atlasbaf, Z.: Study of a Class of UWB CPW-Fed Monopole Antenna With Fractal Elements. IEEE Antennas Wirel. Propag. Lett. **12**, 1484–1487 (2013)
13. Falconer, K.J.: Fractal geometry: mathematical foundations and applications, 2nd edn. Wiley, Chichester (2003)
14. Li, Y.-S., Yang, X.-D., Liu, C.-Y., Jiang, T.: Analysis and Investigation of A Cantor Set Fractal UWB Antenna with A Notch-Band Characteristic. Prog. Electromagn. Res. B **33**, 99–114 (2011)
15. Li, Y., Li, W., Liu, C., Jiang, T.: A printed diversity Cantor set fractal antenna for ultra wideband communication applications, pp. 34–38 (2012)
16. Li, Y., Li, W., Liu, C., Jiang, T.: Cantor set fractal antennas for switchable ultra wideband communication applications, pp. 1–4 (2012)

Directive Beam-Steering Patch Antenna Using Adjustable Metamaterial Superstrate

Hayat Errifi, Abdennaceur Baghdad, Abdelmajid Badri and Aicha Sahel

Abstract Beam-steering antenna is the ideal solution for a variety of system applications, it is most commonly achieved through using phased array, where phase shifters are used to control the relative of the main-beam. However this technique cannot be used in the mobile terminals which use only single radiation element. In this work we present a high directivity and beam steering antenna using simple printed patch covered by adjustable left handed metamaterial superstrate. The design and simulations are carried out using HFSS software.

The adjustable metamaterial is based on double omega shaped unit cell. The antenna used is a probe feed rectangular microstrip patch which operates at X band. An appropriate loading of the metamaterial unit cell, through the use of a varactor diodes, enables to vary the medium index of the superstrate within an interesting range, therefore ± 30° deflection of the antenna beam is obtained. Furthermore, using metamaterial superstrate, the antenna performance has been improved significantly, indicating that the directivity can achieve 11.31 dBi and the HPBW is about 23° for all beam-steering angles.

Keywords Patch antenna · Directivity · Beam-steering · Metamaterial

1 Introduction

Antenna is an important component that affects the performance of wireless communication systems. Antennas with low profile, low manufacturing costs, and high gain are more desirable for the system. To satisfy the requirements, microstrip antenna is a good candidate for the antenna design. However it is difficult to obtain a high gain and high directivity using a normal microstrip patch

H. Errifi(✉) · A. Baghdad · A. Badri · A. Sahel
Electrical Engineering Department, EEA&TI Laboratory, Faculty of Sciences and Technology, Hassan II Casablanca University, 28800 Mohammedia, Morocco
e-mail: errifi.hayat@live.fr

E. Sabir et al. (eds.), *The International Symposium on Ubiquitous Networking*, Lecture Notes in Electrical Engineering 366,
DOI: 10.1007/978-981-287-990-5_15

antenna. To resolve this issue, some traditional technologies are used such as reflectors, directors, dielectric superstrates [1] [2], or array techniques [3], [4]. In recent years, electromagnetic bandgap (EBG) structures [5] and metamaterials [6] have been demonstrated to enhance the antenna performances and to steer the radiated beam in several directions.

Metamaterials denote artificial constructed materials that may not be found in nature. Metamaterials has negative permittivity (ε < 0) and/or negative permeability (μ < 0). The MTM is called double negative material (DNM) or left handed material (LHM) when it has double-negative ε and μ. With the same incident wave, the reflected wave through a LHM is in opposition to the reflected wave through a positive permittivity and permeability material [7]. LHM acts like lenses to focus wave in the same direction; thus it is usually placed above an antenna to increase its gain and to steer its radiated beam. In general, LHM uses a periodic structure and is modeled as an infinite array of MTM unit-cells.

For best quality of service, it is necessary to dynamically steer the radiation of the antennas in privileged directions and to present nulls in all the other directions for interference minimization. For this purpose, phased arrays are a well proven technology [8]; however it doesn't really fit in mobile terminals due to limited space availability. In fact, adjustable metamaterials have found applications in fabricating reconfigurable directive antennas and beam steering devices, especially for radar and communication systems: radar absorbent materials, reflectors, electromagnetic interference shielding, and terrestrial and satellite communications. Many of the efforts towards achieving beam steering focus on developing electrically and electronically adjustable metamaterial [9] [10].

In this paper, we propose to develop the design and the characterization of an actual adjustable metamaterial superstrate based on double omega shaped unit cell for beam-steering of patch antenna. The tuning of the metamaterial resonators is obtained by varying the gap impedance of the omega cell by an appropriate loading through a varactor diode. In the second part, we present the design and simulation of the probe feed rectangular patch antenna used. The details of the metamaterial design are presented in the third part of this paper. In the firth part, we propose the adjustable metamaterial as a superstrate for the designed patch antenna. The obtained results in terms of return loss, gain, directivity, HPBW and beam-steering capability are analyzed and discussed. Finally, we close with a conclusion and future works.

2 Probe Feed Patch Antenna Design

Fig. 1 shows the geometry of the proposed probe-feed rectangular patch antenna, it consists of three layers, the dielectric substrate is placed between a ground plane (lower layer) and radiating metallic patch (top layer).The dimensions of radiating patch are calculated by appropriate equations depending on the patch shape (rectangular, circular or triangular).

The formulas to determine the rectangular patch dimension are as follows [11]:
The width of the patch can be written as:

$$W = \frac{c}{2fr\sqrt{\frac{2}{r+1}}} \tag{1}$$

The length of the antenna becomes

$$L = L_{eff}\text{-}2\Delta L \tag{2}$$

Where

$$\frac{\Delta L}{h} = 0.412\ \frac{(\varepsilon eff+0.3)\left(\frac{W}{h}+0.264\right)^{1}}{(\varepsilon eff-0.258)\left(\frac{W}{h}+0.8\right)^{1}} \tag{3}$$

And

$$\varepsilon eff = \frac{r+1}{2} + \frac{r-1}{2}\left(1 + \frac{12h}{W}\right)^{-1/2} \tag{4}$$

Where fr is the resonant frequency; h is the thickness of the substrate. The rectangular patch antenna has been designed to operate at the frequency of 10 GHz with input impedance of 50 Ω, using RT-DURROID substrate (εr = 2.2). The Performance of the microstrip antenna depends on its dimension, the operating frequency, radiation efficiency, directivity, return loss and other related parameters are also influenced. For an efficient radiation, the patch antenna has been designed with the optimized parameters according to the table 1 [12]:

Table 1 Design parameters of probe feed patch antenna

Symbol	Parameter	Value (mm)
Ls	Substrate dimension Y	27
Ws	Substrate dimension X	35
h	Substrate thickness	0.79
W	Patch dimension X	11.86
L	Patch dimension Y	9.31
Y_0	Feed point	2.3

Below, we will present the simulation results in terms of the computed radiation patterns, return loss and directivity of the proposed antenna. We use HFSS, which is 3D High Frequency Structure Simulator software [13]. Fig. 2 shows the return loss simulated for the proposed rectangular patch antenna. This antenna resonates at the frequency of 10 GHz, with return loss -18.74dB.

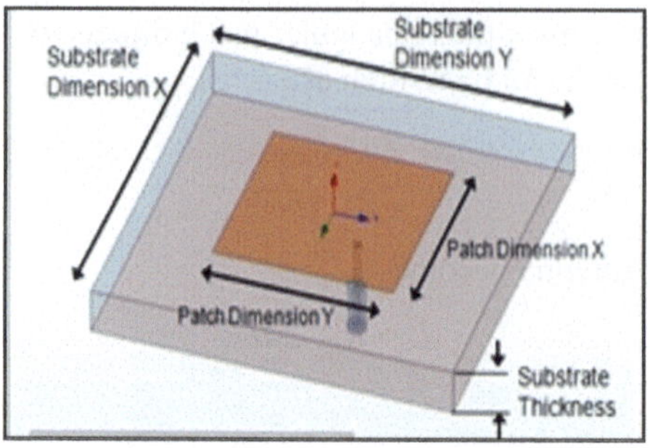

Fig. 1 Probe feed microstrip patch antenna

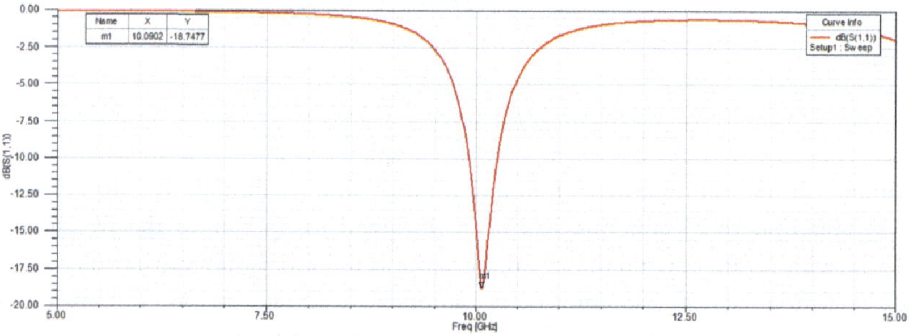

Fig. 2 Return loss of the proposed probe feed patch antenna

The simulated directivity of the antenna, according to Fig. 3, is 8.5dBi at φ = 0_ for the operating frequency. The HPBW is about 65°.

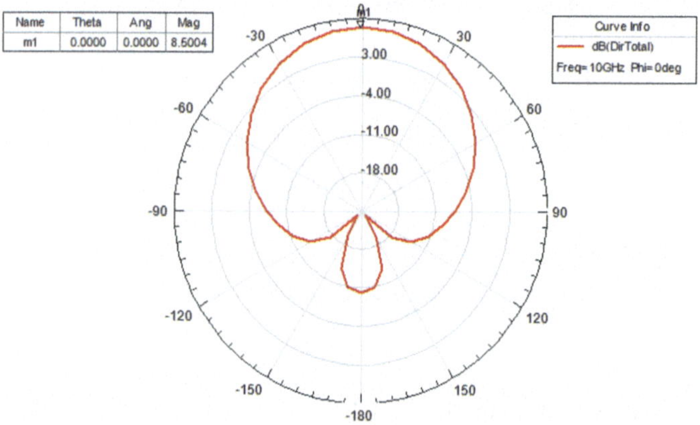

Fig. 3 Radiation pattern of the proposed probe feed patch antenna

From these results we can say that the proposed patch antenna don't exploit well the X-band.

3 Design and Characterization of Adjustable Metamaterial Unit Cell

The proposed metamaterial is based on double omega shaped unit cell, which was first suggested in 1992 by Saadoun and Engheta as single omega [14]. In 1997, Simovski et al. Presented "Ω" shaped LHM unit-cell in 3D for antenna gain enhancement [15], they also designed the "Ω" shaped LHM unit-cell in 2D [16]. This metamaterial unit cell exhibits a double negative behavior, as the magnetic and the electric resonators operate in the same frequency band. Compared to usual structures such as split ring resonators SRR, omega cells present the advantage to reach a large index medium, thanks essentially to its magnetic contribution.

Figure 4 shows the omega cell geometry and its dimensions: the magnetic resonator is formed by the metallic split ring and the electric resonator is formed by the metallic split strip. To realize this omega cell, both resonators are printed on RT/Duroid substrate characterized by $\varepsilon_r = 2.2$, $\tan\delta = 0.001$ and $h = 0.79$ mm. The dimensions of the unit cell are optimized in order to operate at the resonance frequency of the proposed patch antenna which is 10 GHz, where the global size is around $\lambda_0/8.5$. The double omega unit cell is designed and optimized by means of the finite element method (FEM) based HFSS software, respecting the conditions of the homogenization method [17].

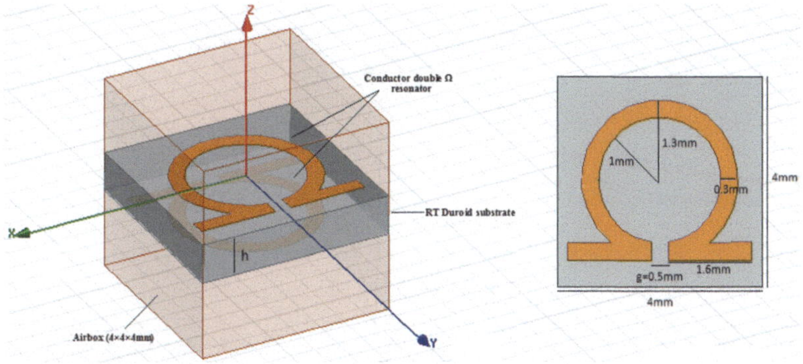

Fig. 4 Geometry of the proposed double omega unit cell

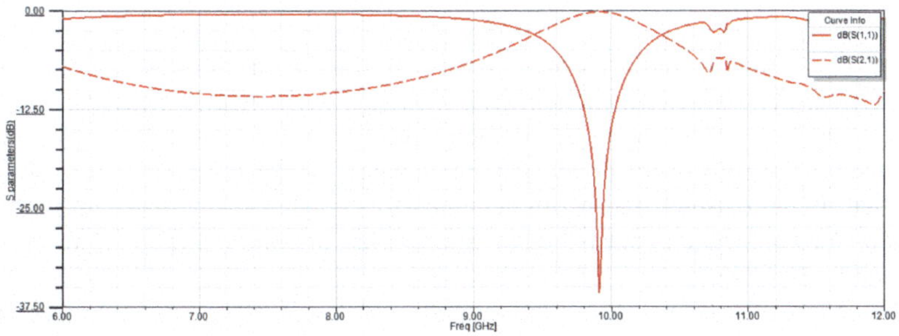

Fig. 5 Simulated S parameters of the double omega unit cell

The intrinsic parameters of the effective medium (ε_{eff} and μ_{eff}) are computed from the transmission and reflection coefficients presented in Figure 5 using Nicholson, Ross and Weir (NRW) approach [18]. Initially, the basic equations used are shown below:

$$\varepsilon_{\text{eff}} = \frac{2}{jkh} \times \frac{1-V1}{1+V1} \tag{5}$$

$$\mu_{\text{eff}} = \frac{2}{jkh} \times \frac{1-V2}{1+V2} \tag{6}$$

Where:

$V_1 = S_{21} + S_{11}$,
$V_2 = S_{21} - S_{11}$,
$k = \frac{\omega}{C}$,

$$n_{\text{eff}} = \sqrt{\varepsilon\text{eff} \times \mu\text{eff}} \tag{7}$$

ω: Radian frequency
C=Speed of light

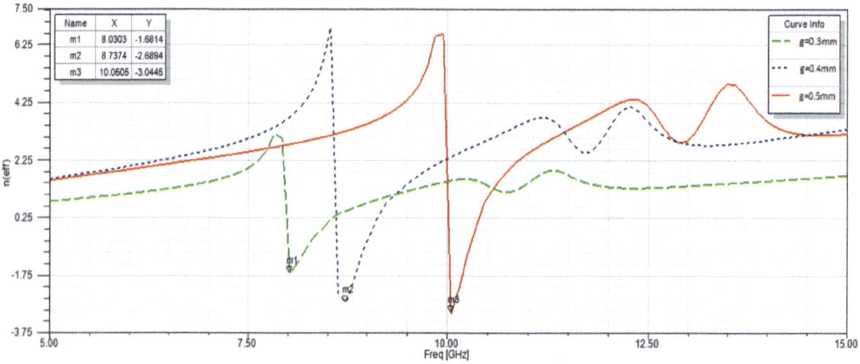

Fig. 6 Variation of the double omega unit cell effective medium index for different gap width values

The simulation result presented in Fig. 6 shows for the case of g=0.3mm, double negative metamaterial behavior around 8 GHz with a negative medium index of neff = −1.The gap width is then varied from 0.3 mm to 0.5 mm in order to illustrate the sensitivity of the gap impedance on the medium index. We can note that the increase of gap width accounts for the shift of the medium index, as much as the resonance towards higher frequencies, attaining neff =−3.5 for g = 0.5 mm and f =10 GHz.

The observable differences in the medium index and in the operating frequency are related to the small frequency shift between electric and magnetic resonances happening in the metamaterial cell during the control of the gap impedance.

This shift is very important since it will help to control the radiated beam direction of the patch antenna.

The propagation in the evanescent frequency range is only possible if the medium is left handed, thus confirming the metamaterial effect of the cell. The tuning of the gap width permits also to control the frequency where this metamaterial effect arises. In a second stage of the tuning study, we have loaded the gap of the omega cell by an equivalent complex impedance Z_d corresponding to the varactor diode model as it has been shown that the gap is truly the most sensitive position for an appropriate control of the resonating structure through the insertion of the variable load, Fig. 7.

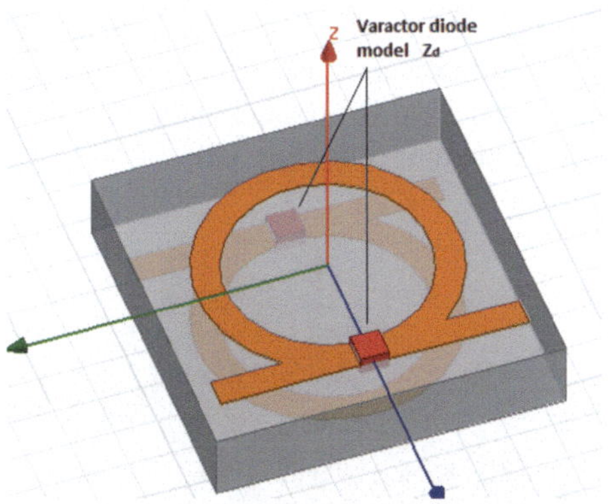

Fig. 7 Double Omega unit cell loading with a varactor diode

By a closer electromagnetic analysis, we find that a satisfactory control could be obtained with a varactor diode having a variation of 0.1–3 pF for a voltage biasing range of 1–20 V. Figure 8 shows the intrinsic parameters obtained by the voltage biasing of the chosen varactor diode. At the biasing of 10 V, Re (μeff) =-5 and Re (εeff) =−20.

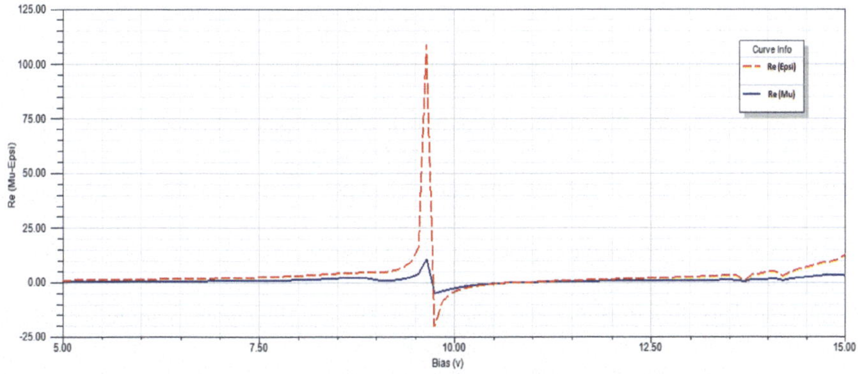

Fig. 8 Intrinsic properties versus the voltage biasing applied to the varactor diode at $f = 10$GHz

4 Patch Antenna Covered by the Metamaterial Superstrate – Results & Discussion

Figure 9 shows a configuration of the proposed adjustable metamaterial superstrate. It is formed by 7×7 active cells in the *XY*-plane and 1 cell along the *Z* axis. The dimensions of the superstrate (L_X= 28 mm, L_Y= 14 mm, L_Z= 4 mm) and the air gap between the layers (a=2 mm) were optimized in order to operate around the resonance frequency f =10 GHz. The proposed probe fed patch antenna is adopted as the exciting source. The orientation of the superstrate is chosen in a particular way in order to match the electromagnetic excitation requirements of the unit cells with the near field radiating of the primary source. The electrical near field is oriented along the *X* axis and is collinear to the dipole strips of the unit cells. The magnetic near field is oriented along the *Y* axis and is collinear to the axis of the split ring of the unit cells (Fig. 10).

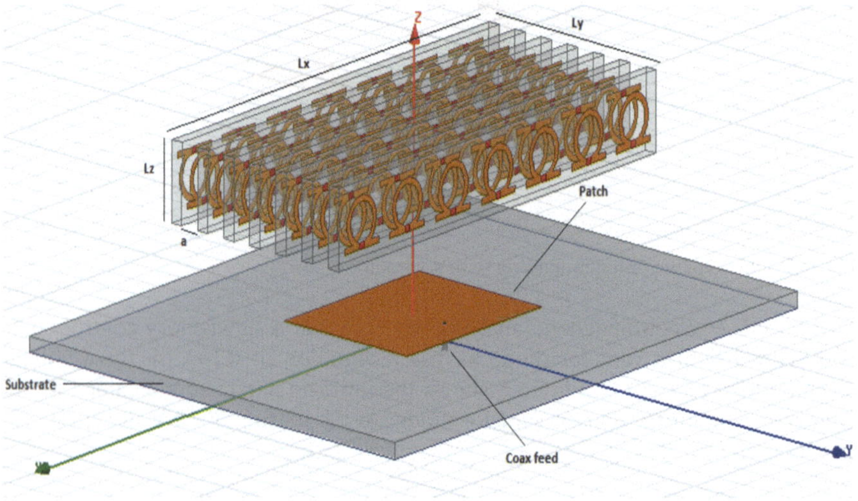

Fig. 9 Association of the adjustable metamaterial superstrate with a probe feed patch antenna

Figure 11 shows the radiation patterns of the patch antenna with MTM superstrate at f = 10 GHz. For the case of MTM without varactor diodes the beam is normal to the plane of the antenna and shows no deflection, which confirms the prediction on the constant phase metamaterial. However, according to the variation of the capacitive values of the varactor diodes, we can observe a deflection of the antenna beam either in the forward (clockwise) or backward (anticlockwise) direction. For the case of C = 0.1 pF, the steering angle is around θt =+30°. On the other hand, considering the load C = 3 pF, we observe a beam steering around θt =−30°. These results prove the feasibility of the beam steering through the adjustable metamaterial superstrate added to the simple patch antenna.

Furthermore the use of MTM superstrate increase the patch antenna directivity from 8.5 dBi to 11.31 dBi for all beam-steering angles.

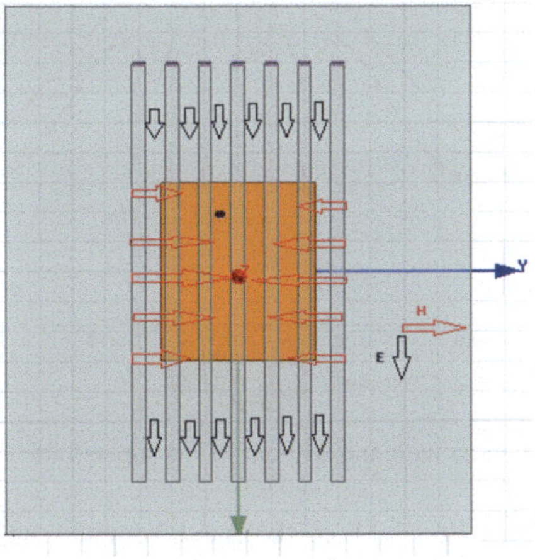

Fig. 10 Top view presenting orientation of the Metamaterial slabs according to the patch antenna *E* and *H* near fields

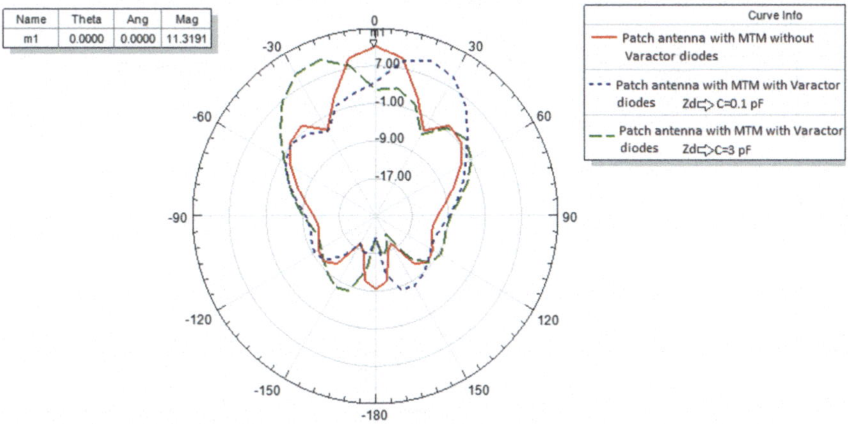

Fig. 11 Directivity radiation pattern of the patch antenna with MTM superstrate for different impedance varactor loading at f = 10 GHz

The gain in the direction of main beam (highest radiation) was increased by 2.58 dB after using the MTM superstrate (Fig. 12). Furthermore, the 3-dB beam-width for the patch antenna covered with the MTM superstrate is only 23° for all beam-steering angles compared with 65° for the patch antenna only. The reduction in the 3-dB beam-width is an indication of the improved directivity and the sharpening of the radiation pattern's lobe.

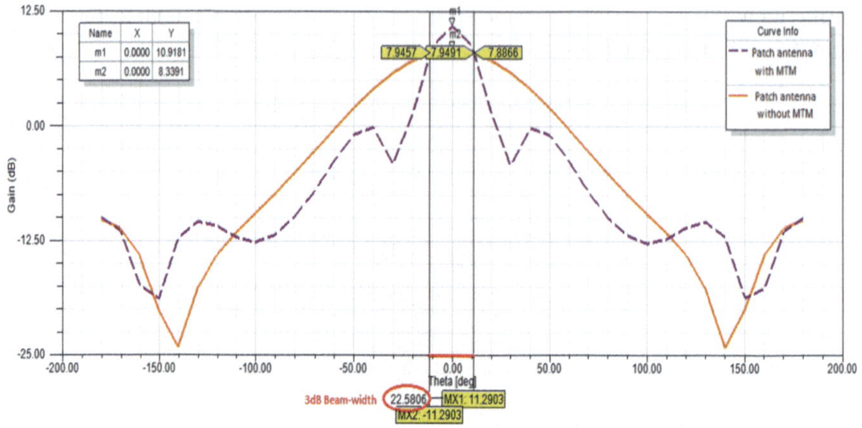

Fig. 12 Gain of patch antenna with and without MTM superstrate

The simulated return loss S11 of the patch antenna with and without MTM superstrate is as shown in Fig. 13. The return loss seems to shift to a higher region when the antenna is incorporated with the MTM, but it still shows a good agreement, where in both conditions, the antenna still operates around 10 GHz. The bandwidth of the antenna increases after the incorporation with MTM by 5%.

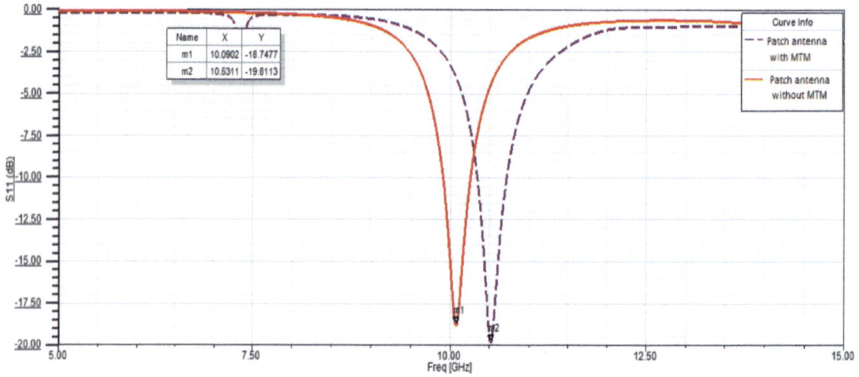

Fig. 13 Return loss S_{11} of the patch antenna with and without MTM superstrate

5 Conclusion

In this work, the design and the characterization of an adjustable metamaterial based on the double omega unit cell has been presented. The numerical results have confirmed the tunable double negative behavior with respect to the gap width, the loading impedance and the chosen varactor diode capacitance. The near field association of a simple rectangular patch antenna and the proposed adjustable metamaterial superstrate has been simulated. The results show that the antenna performance has been improved significantly, indicating that the directivity can achieve 11.31 dBi for all beam-steering angles. The radiating pattern of this smart and low profile antenna can be steered thanks to the tuning of the superstrate over a range of (θt =+/−30°) in the elevation plane. The results of this work can provide significant guidance for multi-functional reconfigurable antenna designers.

Acknowledgement This work falls within the scope of telecommunication projects. Our sincere thanks to the Faculty of Sciences and Technology, Hassan II University, Casablanca, Morocco, for providing us an opportunity to carry out our work in a well-equipped laboratory (EEA&TI). We are also thankful to all our colleagues who helped us while working on this project.

References

1. Errifi, H., Baghdad, A., Badri, A., Sahel, A.: Directivity enhancement of aperture coupled microstrip patch antenna using two layers dielectric superstrate. In: 14th Mediterranean Microwave Symposium, MMS-2014, Marrakech, Morocco, December 12–14, 2014
2. Errifi, H., Baghdad, A., Badri, A., Sahel, A.: Enhancement of Inset Feed Microstrip Semicircular Patch Antenna Directivity using Dielectric Superstrate. International Journal of Computer and Information Technology (IJCIT) **04**(02) (2015). (ISSN: 2279 – 0764)
3. Errifi, H., Baghdad, A., Badri, A., Sahel, A.: Design and Analysis of Directive Microstrip Patch Array Antennas with Series, Corporate and Series-Corporate Feed Network. International Journal of Electronics and Electrical Engineering (IJEEE) (2015). (article in press)
4. Errifi, H., Baghdad, A., Badri, A., Sahel, A.: Design and simulation of microstrip patch array antenna with high directivity for 10 GHz applications. In: International Symposium on Signal Image Video and Communications, ISIVC-2014, Marrakech, Morocco, November 19–21, 2014
5. Errifi, H., Baghdad, A., Badri, A., Sahel, A.: Improving Microstrip Patch Antenna Directivity using EBG Superstrate. American Journal of Engineering Research (AJER) **03**(11), 125–130 (2014). e-ISSN: 2320-0847 p-ISSN
6. Le, M.T., Nguyen, Q.C., Vuong, T.P.: Design of High-Gain and Beam Steering Antennas Using a New Planar Folded-Line Metamaterial Structure. International Journal of Antennas and Propagation (2014). Article ID 302580
7. Veselago, V.G.: The electrodynamics of substances with simultaneously negative values of ε and μ. Soviet PhysicsUspekhi **10**(4), 509–514 (1968)
8. Errifi, H., Baghdad, A., Badri, A., Sahel, A.: Directive beam-steering patch array antenna using simple phase shifter. In: Mediterranean Conference on Information & Communication Technologies, MedICT 2015, pp.7–9, Mai, Saïdia, Morocco (2015).(in press)
9. Sievenpiper, D., Schaffner, J.: Beam steering microwave reflector based on electrically tunable impedance surface. Electronics Letters **38**(21), 1237–1238 (2002)
10. Higgins, J.A., Xin, H., Sailer, A., Rosker, M.: Ka-bandwaveguide phase shifter using tunable electromagnetic crystal sidewalls. IEEE Transactions on Microwave Theory and Techniques **51**(4), 1281–1288 (2003)

11. Balanis, C.A.: Antenna Theory: Analysis and Design, 3rd ed. John Wiley and Sons (2005)
12. Errifi, H., Baghdad, A., Badri, A.: Effect of change in feedpoint on the antenna performance in edge, probe and inset-feed microstrip patch antenna for 10GHz. International Journal of Emerging Trends in Engineering and Development **1**(4), January 2014
13. HFSS software user guide. Ansoft Corporation (2005)
14. Saadoun, M.M.I., Engheta, N.: A reciprocal phase shifter using novel pseudochiral or ω medium. Microwave and Optical Technology Letters. **5**(4), 184–188 (1992)
15. Simovski, C.R., Tretyakov, S.A., Sochava, A.A., Sauviac, B., Mariotte, F., Kharina, T.G.: Antenna model for conductive omega particles. Journal of Electromagnetic Waves and Applications **11**(11), 1509–1530 (1997)
16. Simovski, C.R.: Plane-wave reflection and transmission by grids of conducting Ω-particles and dispersion of Ω electromagnetic crystals. AEU-International Journal of Electronics and Communications **57**(5), 358–364 (2003)
17. Smith, D.R., Vier, D.C., Kroll, N., Schultz, S.: Determination of effective permittivity and permeability of metamaterials from reflection and transmission coefficients. Physical Review B **65**(19) (2002)
18. Ziolkowski, R.W.: Design, fabrication, and testing of double negative metamaterials. IEEE Transactions on Antennas and Wireless Propagation **51**(7) (2003)

High Sensitive and Efficient Circular Polarized Rectenna Design for RF Energy Harvesting at 5.8 GHz

Mohamed Adel Sennouni, Jamal Zbitou, Benaissa Abboud, Abdelwahed Tribak, Hamid Bennis and Mohamed Latrach

Abstract In this paper a planar rectenna (rectifier + antenna) design which is the crucial part in a microwave power transmission system is presented. The developed structure is suitable to harvest RF energy at 5.8GHz ISM frequency band for powering devices involving low power consumption, located in unreachable area or needing expensive maintenance costs. The proposed design incorporate a 2 × 2 circular polarized antenna array with an enhanced directivity of about 10.2 dBi and an overall size of 14.6 × 8 cm^2 built and carried out by using CST Microwave Studio software. The receiving antenna is associated with a microstrip rectifier that combines series and shunt configuration employed Schottky diode HSMS2852 and HSMS2850. The RF-to-DC rectifier circuit is designed and simulated by using schematic simulation in ADS software. An output DC voltage of 2.2 V and a RF-to-DC conversion efficiency of more than 70% were achieved with an optimum load resistance of 6 kΩ for 0 dBm microwave incident power level.

Keywords Planar rectenna · Microwave power transmission · RF energy · ISM band · Antenna array · Microstrip rectifier · Conversion efficiency · CST-Microwave studio · ADS

M.A. Sennouni(✉) · B. Abboud
LITEN laboratory FSTS, University of Hassan the 1ST, Settat, Morocco
e-mail: adelsennouni@gmail.com

J. Zbitou · H. Bennis
LMEET laboratory FSTS, University of Hassan the 1ST, Settat, Morocco

A. Tribak
National Institute of Post and Telecommunication (INPT), Rabat, Morocco

M. Latrach
RF & Hyper group ESEO, Angers, France

E. Sabir et al. (eds.), *The International Symposium on Ubiquitous Networking*,
Lecture Notes in Electrical Engineering 366,
DOI: 10.1007/978-981-287-990-5_16

1 Introduction

Over 100 years ago, the concept of wireless power transmission was introduced and demonstrated by Tesla [1], [2], he described a method of "utilizing effects transmitted through natural media". This technology was advanced from 1960's by W. C. Brown [3] the pioneer in wireless power transmission technology, who has designed, developed a unit and demonstrated how power can be transferred through free space by microwaves. Consequently wireless power transmission technology (WPT) via microwave has received special attention recently for the implementation of low-cost and low-power battery-less operated sensors.

Due to the availability of different ambient sources of RF energy, it is now become an attractive research topic to harvest RF energy. Then many wireless systems have been developed and widely used around the world like GSM, WI-FI, Radio and TV broadcasting systems which radiate electromagnetic waves/energy into the air but a large amount of the energy is actually wasted. So harvested energy can be used as an alternative to use renewable energy; it is now possible to feed by low power devices without contact. One of the most promising methods to harvest the wireless energy from this RF sources is to use a rectenna (rectifier +antenna) [4]. In fact, a rectenna representing the key element for EM "Electromagnetic" energy harvesting and Wireless Power Transmission "WPT" applications, thus device is capable of capturing the microwave energy from surrounding environments sources and converting it into DC energy with a high performance over a large spectrum of frequencies. It's composed at least from two essential components. The first one is a receiving antenna to collect the Radio Frequency (RF)/microwave EM energy. The second component is a rectifying circuit (the rectifier) that can generate enough DC power from the incident electromagnetic waves. The basic architecture of a rectenna system is illustrated by a block diagram in Fig. 1. As highlighted in this circuit, two blocks were added before and after diode which is a low pass filter and DC filter respectively in order to increase the DC voltage and the power conversion efficiency. Further the good features of a rectenna for an electrical device is extending the device's lifetime and can assure a good mobility and integrity for wireless mobile devices such as cellular phones, RFID systems, wireless sensors, access control equipment monitoring and even personal identification. Recently, several rectenna devices have been proposed in the literature, investigating several operating frequencies and different input powers [5],[6],[7] and [8]. Moreover components of microwave power transmission have traditionally been focused on 2.45GHz and recently moving up to 5.8 GHz, which has a smaller antenna aperture area than that of 2.45 GHz. Also both frequencies can reported high conversion efficiency and have comparably low atmospheric loss.

In the literature many kinds of topologies of rectennas have been proposed such as arbitrary polarization [9], dual-polarization [10], linear polarization (LP) [11], [12], [13] and [14], Circular Polarization (CP) [15], [16], [17], [18], [19],[20] and wideband [21]. Eventually antennas with CP have more fading resistant than linear polarized antennas and the rotational orientation of the transmitter and receiver does not degrade the polarization loss [22]. Then the primary aim for a rectenna system is to increase efficiency and reduce conversion loss, therefore this is only possible by using an antenna arrays.

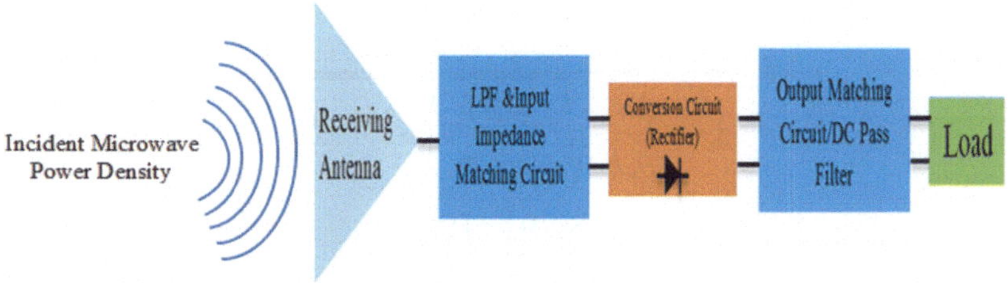

Fig. 1 The Block diagram of the Rectenna

In this paper we propose a new rectenna design with high performances for RF energy harvesting. The developed rectenna is aimed to scavenge the RF energy from the free space at the ISM frequency band of 5.8 GHz at low power densities. The rectenna design incorporate a series-fed microstrip antenna array with circular polarization (CP) providing an enhanced directivity of about 10.2 dBi, associated with a RF-to-DC microstrip rectifier which is formed by a combination of series and shunt topologies using a single Zero-Bias Schottky diode HSMS2852 and HSMS2850 for low power applications at low power densities. An output DC voltage of more than 2V with a RF-to-DC conversion efficiency of about 70% can be achieved for an input RF power level of 0 dBm with an optimum load of 6 kΩ.

2 Receiving Antenna Array Design

For rectenna systems it's recommended to meet demands to transmit power through wireless links for a long distance, so it's necessary to design antennas with very high directive characteristics, which can be only accomplished by increasing the electrical size of the receiving antenna. We present in this section the antenna array design which consisting of a 2×2 circular microstrip patch incorporating two antenna array types: the first one is a series feed array type that is formed by interconnecting two successive patch elements that are matched by using quarter wavelength transformer. Since, the series feed arrangement is compact, the line losses associated with this type of array are lower than those of the second type have been used which is the corporate feed array [23]. The combination of series feed and corporate feed are frequently used for array antennas to get benefits of both feeding networks. Then in this design, it is considered that the antenna array is printed on FR4 substrate (ε_r= 4.4, thickness = 1.58 mm, tan δ = 0.025), where in each antenna element we have introduced an inclined slot to achieve CP properties which is suitable for rectenna systems to prevent the power interruption across the supplied devices where the transmitter or the receiver changes its direction in the space, this technique could give reduction in the array sizes and avoid the degradation of the antenna array performances.

In order to achieve the proposed design, we have started from a single elementary antenna modeled theoretically. Firstly the radius, *(a)* and effective radius, (a_e) of the elementary circular patch antenna are determined by using the given formulas.

$$a = \frac{F}{1 + \frac{2h}{\pi \varepsilon_r F}\sqrt{In\left(\frac{\pi F}{2h}\right) + 1.7726}} \tag{1}$$

$$F = \frac{8.791 \times 10^9}{f_r \sqrt{\varepsilon_r}} \tag{2}$$

$$a_e = a\sqrt{\frac{2h}{\pi \varepsilon_r a}\left[In\left(\frac{\pi a}{2h}\right) + 1.7726\right]} \tag{3}$$

Where a is the radius of circular patch, F is the fringing field, h is the substrate height, ε_r the relative dielectric constant and a_e is the effective radius. Then the resonance frequency is given by (4) with v_0 is a speed of light.

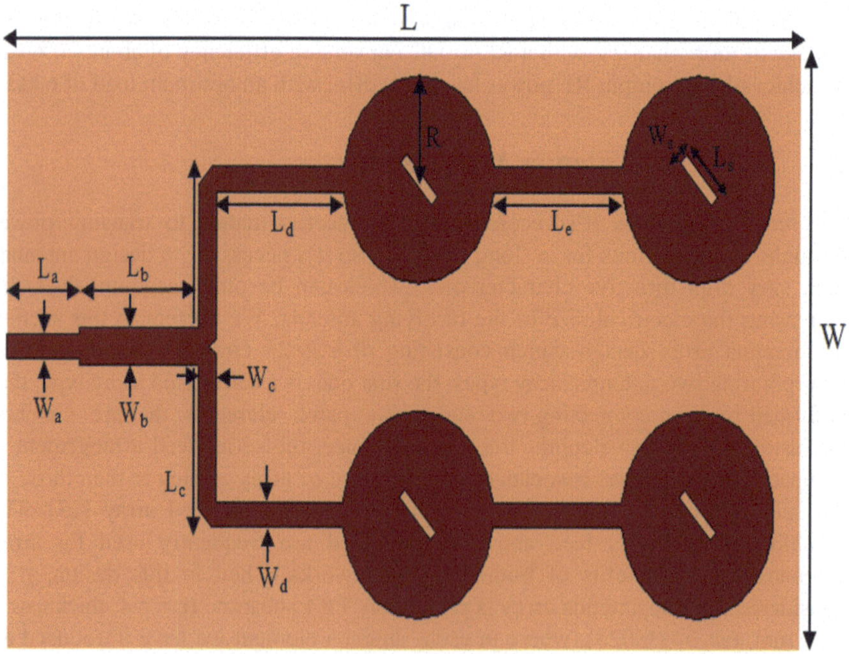

Fig. 2 Geometry of the proposed 2×2 Antenna Array

$$\left(f_r\right)_{110} = \frac{1.8412 \upsilon_0}{2\pi a_e \sqrt{\varepsilon_r}} \tag{4}$$

The developed antenna array has an overall size of about 14.6x8 cm^2 with the geometry along the parameters of the antenna array are shown in Fig. 2. Further several optimization processes were applied by using the full-wave simulator CST Microwave Studio [24] until we get the desired performance of the antenna array. The Table 1 below shows the various optimized parameters of the proposed antenna array.

Table 1 Physical Dimensions of the CP 2×2 Antenna Array

Parameter	Value (mm)	Parameter	Value (mm)
L	146	**W**	80
L_a	13.6	W_a	3.4
L_b	22	W_b	5.18
L_c	48.4	W_c	3.4
L_d	23.1	W_d	3.4
L_e	23.56	W_s	1.5
L_s	8	**R**	13.969

The performances of the proposed 4-elements CP antenna array, such as return loss, the voltage standing wave ratio (VSWR) also the axial ratio, the antenna radiation efficiency and the directivity of the developed array are evaluated by simulations. It can be noticed that at the resonant frequency this performances are greatly affected by the patch radius, the distance between the various patches, also the feed networks chosen to match all elements, further by the slots size. Figures 3 and 4 presents the simulated return loss and VSWR of the proposed CP antenna array.

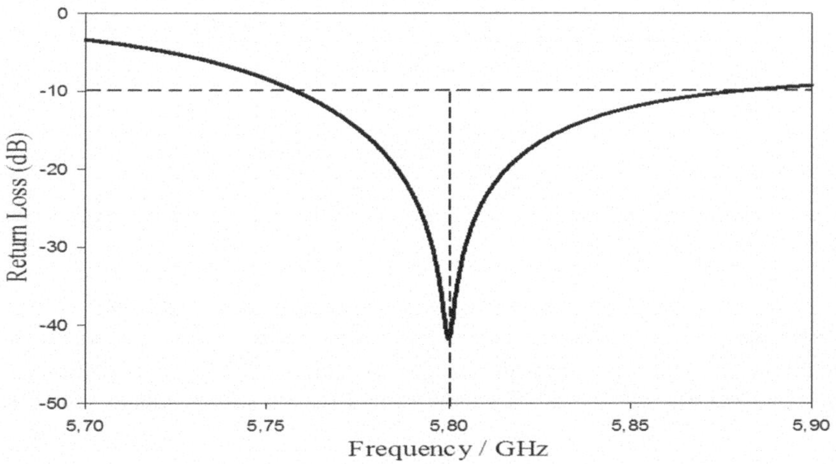

Fig. 3 Simulated return loss @ 5.8 GHz of the 2×2 CP antenna array

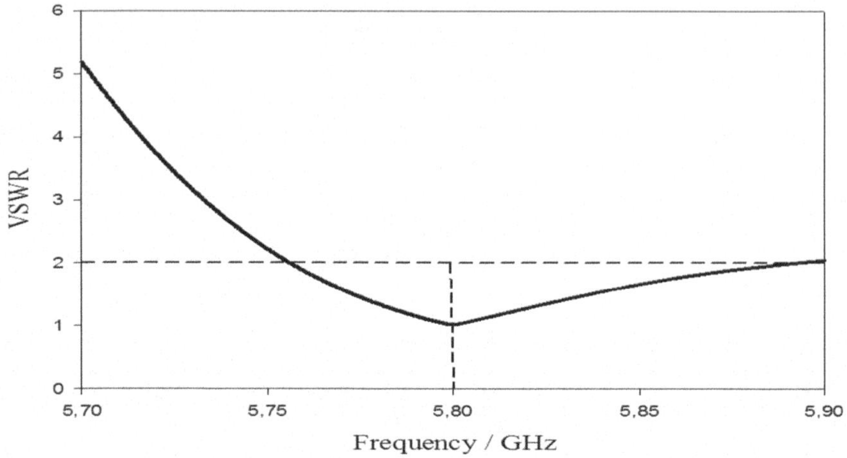

Fig. 4 Simulated VSWR @ 5.8 GHz of the 2×2 CP antenna array

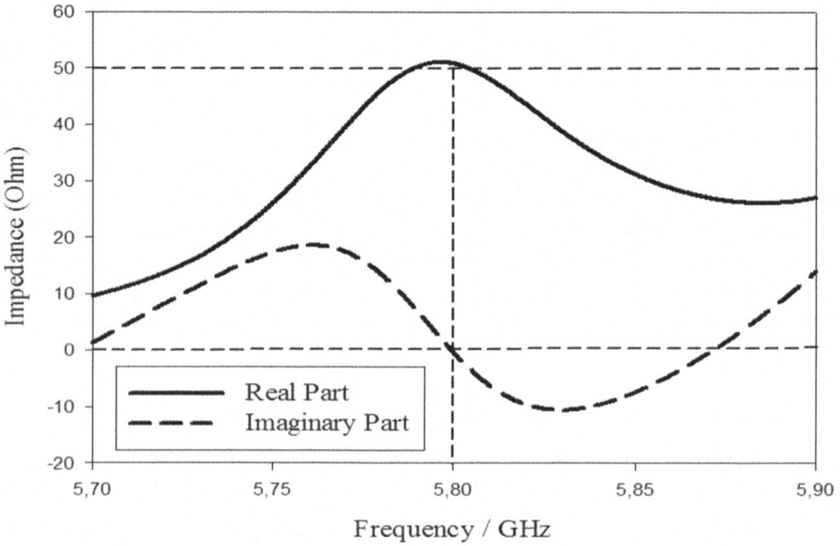

Fig. 5 Simulated antenna impedance (Ohm) vs. frequency of the proposed 2×2 CP antenna array

From the graphs it's can be seen that the reflection coefficient (|S11|) (dB) of the proposed antenna array provides a good matching input impedance with a good return loss of about -41.82 dB with a bandwidth of about 122MHz which is from 5.75 GHz to 5.88 GHz, also it exhibits a VSWR of 1.016 with a bandwidth of 136.6MHz.

Further the simulated variation of the input impedance versus frequency of the 2×2 antenna array is displayed in Fig. 5 where the 2×2 antenna array shows an impedance of Z_{in}=50.78-j0.21 Ω at 5.8GHz. These results indicate clearly that the developed antenna array is well matched to 50 Ω impedance at operating frequency of 5.8 GHz. Moreover the axial ratio (AR) is important criteria for CP operation which has to be kept below 3 dB. Eventually with the optimum values of the antenna array parameters a good axial ratio with a minimum of 2.07 dB is achieved at operation frequency of 5.8 GHz as shown in Fig. 6.

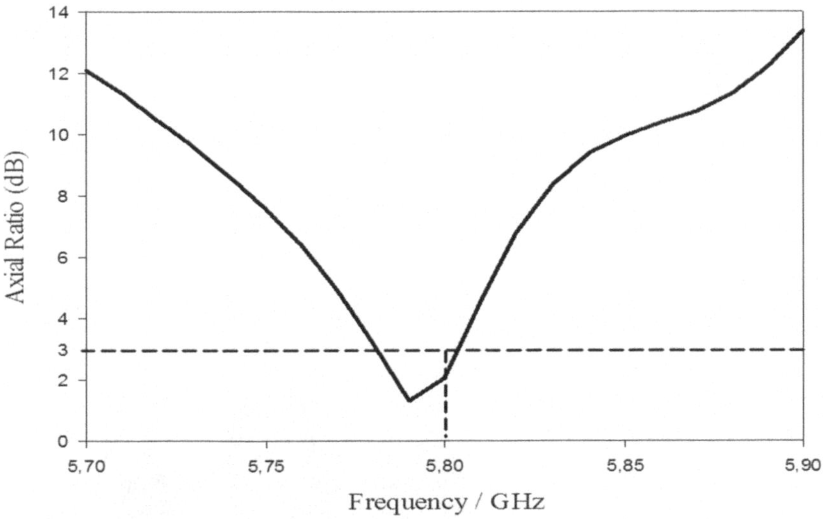

Fig. 6 Simulated Axial Ratio (AR) @ 5.8 GHz of the proposed 2×2 CP antenna array

Hence in practice, this antenna can capture the RF energy at this frequency band and transmit it into the rectifier circuit to produce DC power. Then the radiation efficiency is an important criterion to predict the ability of the developed design to harvest microwave power densities. For this reason we have evaluated this parameter for the proposed antenna array. Fig. 7 presents the simulated radiation efficiency versus frequency. From the graph it's clear that the antenna array radiation efficiency has a peak efficiency of about 97.89% at 5.8 GHz, which means that the previous developed antenna array has the required potential to be investigated in a rectenna system with a minimum power loss, which can boost the power levels at the input of the rectifier circuit.

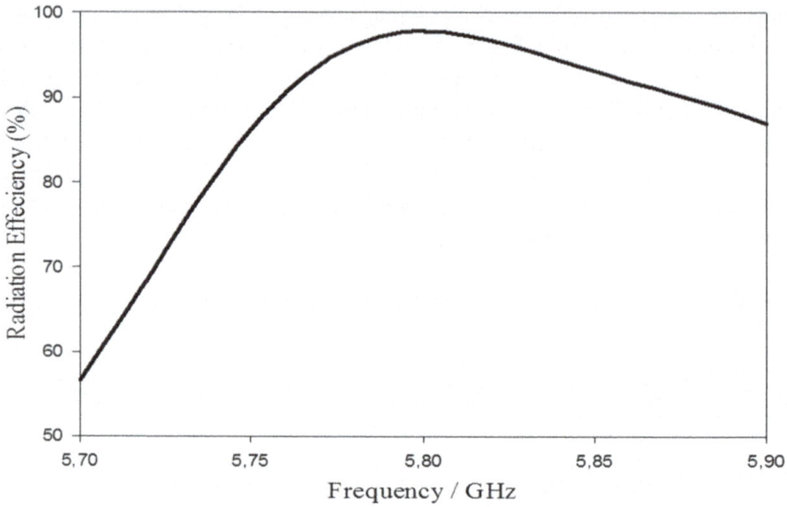

Fig. 7 Simulated Antenna Radiation Efficiency vs. frequency of the proposed 2×2 CP antenna array

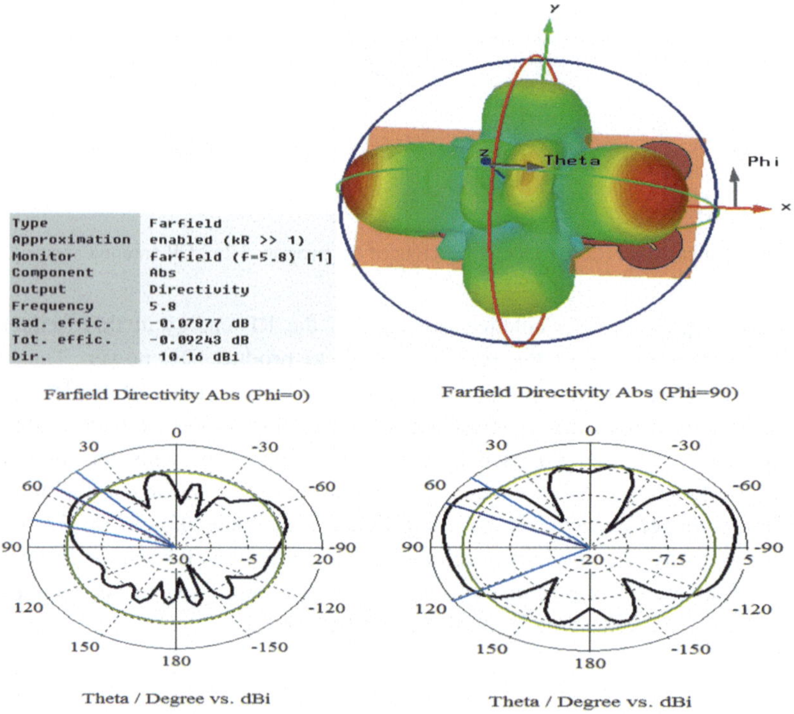

(a) (b)

Fig. 8 Three and Two-dimensional radiation pattern (a): E- plane, (b): H-plane of the 2×2 antenna array @5.8GHz

In order to verify the ability of the proposed antenna array to be used in a RF energy harvester system at microwave frequencies, the radiation pattern of the developed array is important criterion must take into account. Fig. 8 presents a 3-D and 2-D view of the simulated radiation pattern for the developed 2×2 CP antenna array at operating frequency of 5.8 GHz. From these results it is clear the antenna array provides a directional behavior with a half power beam width (HPBW) of 32.6° and 72.6° for E-plane and H-plane respectively with an enhanced directivity of about 10.2 dBi is obtained. On the other hand as the feed arrangement chosen is compact, the proposed corporate-series feed antenna array can control the beam shape in both planes and provides more directivity and radiation efficiency.

In this way it is clear from the above results have been presented and discussed that the proposed design has the required performances and a good capability due to its good input matching and its important bandwidth, furthermore due to its enhanced directivity and its important radiation efficiency to be investigated for a rectenna system to harvest microwaves power densities at free ISM band frequency of 5.8 GHz.

3 5.8 GHz RF-to-DC Microwave Rectifier Design

The rectifier is important component in a rectenna system which is able to convert the collected RF microwave power by the receiving antenna to a useful DC power. In the former work we focus on high efficiency at low incident power (small signal). We propose a rectifier model circuit which is a combination of shunt and series configuration in order to merge the advantages of both topologies using the packaged Schottky diode HSMS-2852 and HSMS2850 [25]. These diodes have relatively low barrier height and high saturation current compared to externally biased detector diodes. These features could provide higher output voltage at low power densities levels. The rectifier circuit is etched on the same substrate used to build the presented receiving antenna array. As we seen in the layout of the proposed rectifier in Fig. 9, the input filter are designed for penetrating only the f_0 frequency from the receiving antenna to the diode and suppressing the others harmonics rejected by the diode also to match correctly the input impedance between the circuit and the receiving antenna to 50 Ω impedance. Then, the output filter using three fan shaped stubs is optimized to assure only the DC voltage obtained over the load and block all the other harmonic frequencies produced in rectifying from 5.8 GHz as shown in Fig.10. Furthermore the developed circuit doesn't need neither any via-hole connections no bypass capacitor that can decrease its performances. Consequently the developed rectifier has a simple profile and it is easily fabricated for certain resonant frequency.

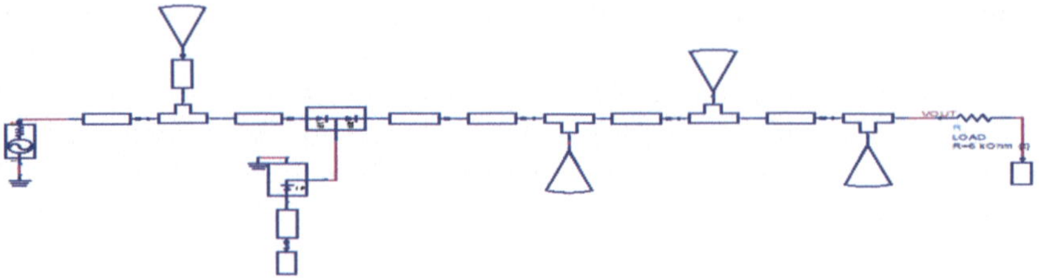

(a)

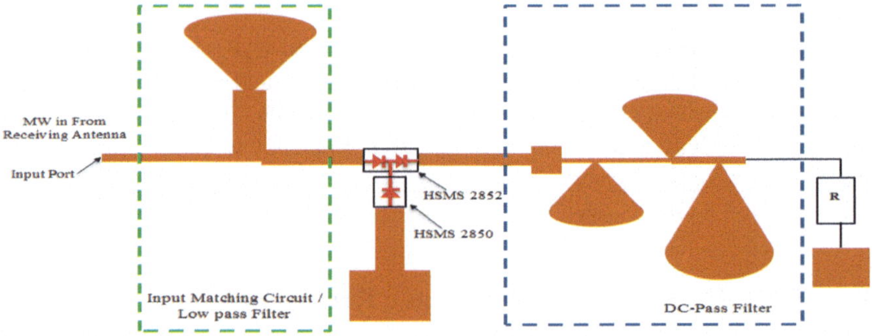

(b)

Fig. 9 Proposed rectifying circuit design: (a) schematic model with distributed elements, (b): Layout model of the RF-to-DC rectifier

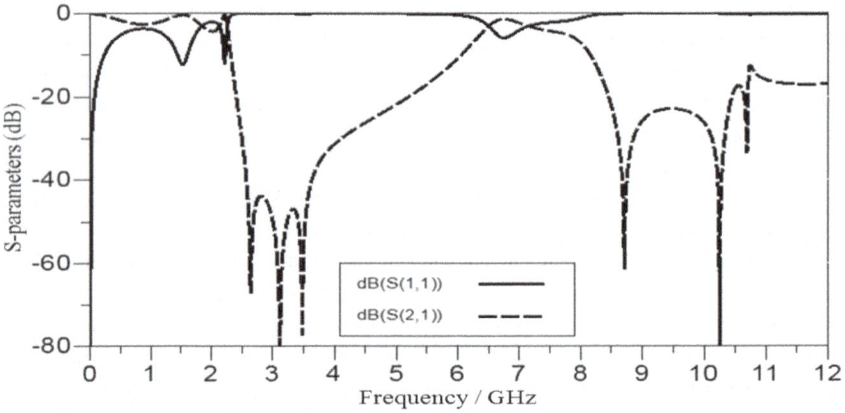

Fig. 10 Simulated Results of S-Parameter responses of the DC-pass Filter obtained by using Agilent ADS

In order to accurately predict the circuit performances an analysis approach by using ADS software is applied into the schematic simulation, by using Harmonic Balance Simulation and Large Signal Scattering Simulation in ADS. The input impedance changes with the variation of input power levels due to the nonlinear performance of Schottky diodes, According to ADS simulation; the corresponding simulated reflection coefficient versus input power is shown in Fig. 11. The displayed results shows that the proposed rectifier circuit is well matched at 5.8GHz with a minimum reflected power around -26 dBm input power (S11< -50 dB) is observed. Also Fig. 12 exhibits the output DC voltage of the proposed rectifier, versus input power. The proposed circuit occurs an output DC voltage of more than 2V where the input power is around 0 dBm and about 0.5 V at -10 dBm with an optimum load of 6 KΩ.

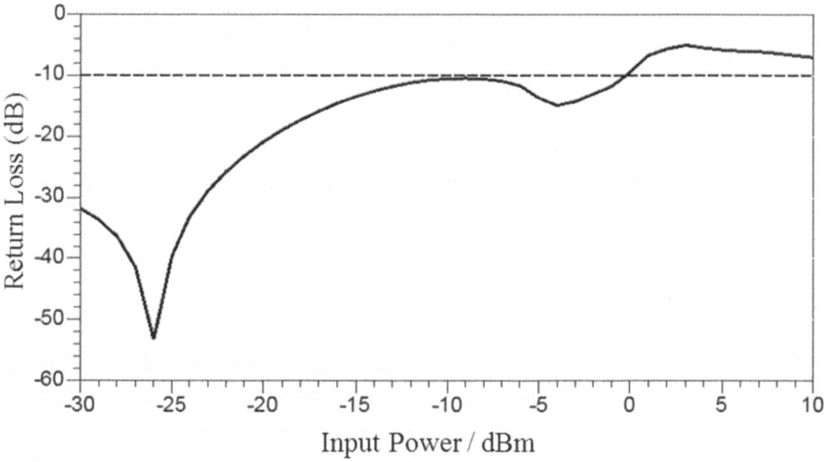

(a)

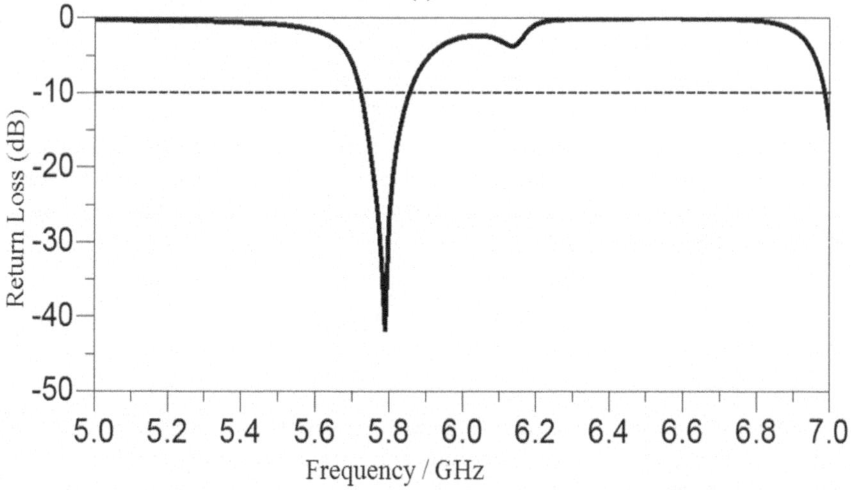

(b)

Fig. 11 Simulated return loss of the proposed Rectifier: (a) versus input Power (HB simulation), (b) versus frequency

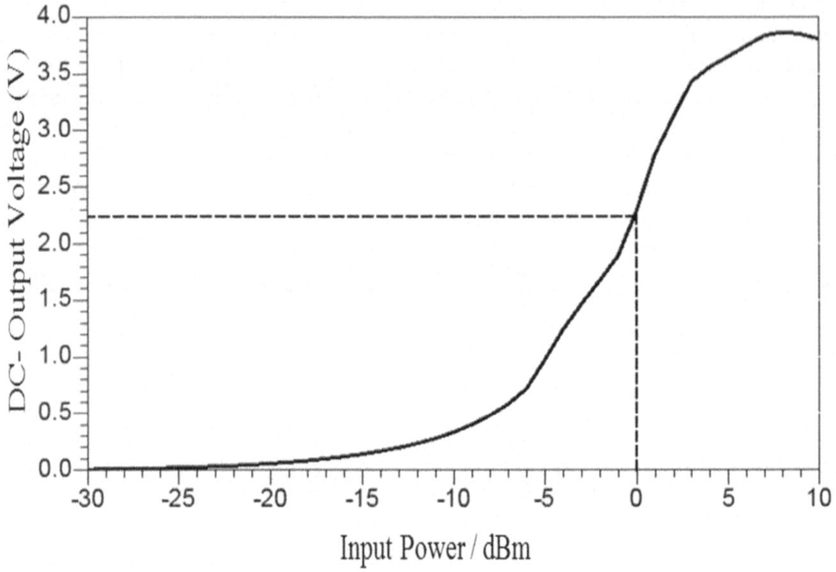

Fig. 12 Simulation result of DC Output Voltage: (a) versus input Power

Furthermore the RF-to-DC conversion efficiency is an important criterion must take in consideration for rectenna system to predict its ability to be used for microwave power conversion. Due to the non-linear characteristics of the diode the input power level is one of the main factors that can affect the rectifier conversion efficiency which (η) is defined as follows:

$$\eta_{RF-DC} = \frac{P_{DC}}{P_{RF}} = \frac{V_{DC}^2}{R_{Load} P_{RF}} \times 100\% \tag{5}$$

Where P_{RF} is power received by the receiving antenna that can be calculated using the Friis transmission equation (Equation 2). P_{DC} is the DC power produced at the load resistance R_{Load} of the rectenna and V_{DC} the output DC voltage delivered to the resistive load R_{Load}.

$$P_{\text{Received}} = \left(\frac{\lambda}{4\pi r}\right)^2 P_t G_t G_r \tag{6}$$

With G_r is the gain of the receiving antenna, λ is the operating wavelength, G_t and P_t are the gain and transmitted power of the transmitting antenna, r is the propagation distance.

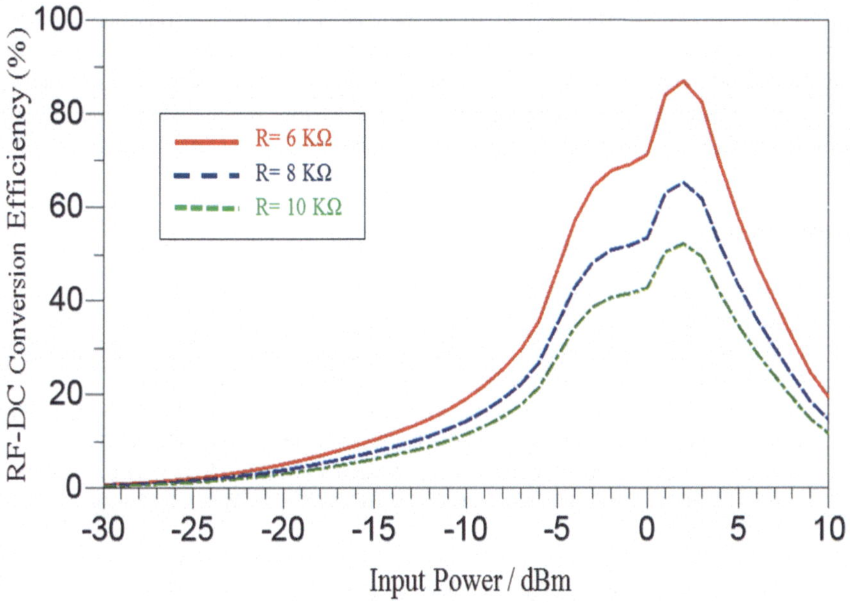

Fig. 13 Simulated result of rectifier conversion efficiency versus input Power

The simulated conversion efficiency versus the input power is demonstrated in Fig. 13. It can be seen from the obtained results that the circuit provides a maximum conversion efficiency of more than 70% and 46% at input power level of 0 dBm and -5 dBm respectively when the load is 6 KΩ, while it decrease when the value of the resistive load increase. From these results it's clear that the proposed rectifier circuit presents high sensitivity among low input power levels with important conversion efficiency and DC output voltage with a minimum power loss.

4 Conclusion

A planar, low profile and cost rectenna for RF power harvesting at 5.8 GHz ISM band was designed and evaluated. The rectenna structure incorporate a receiving antenna that is a circular polarized 2×2 antenna array combining tow feeding networks types and providing an enhanced directivity of about 10.2dBi also an important radiation efficiency of 97.89% with a reduced size of $14.6 \times 8 cm^2$, this antenna array was associated to a RF-to-DC microstrip rectifier combining series and shunt configuration in order to improve the performance of the circuit such as sensivity, output DC voltage and conversion efficiency at low power densities. The developed circuit exhibits an output DC voltage of more than 2 V where the input power level is around 0 dBm and 0.5V at -10 dBm with an optimum load of 6KΩ. The rectifier circuit shows a good sensivity at low power densities and occurs a

RF-to-Dc conversion efficiency of more than 70% at 0 dBm. Consequently from These features and the above results and analysis, the developed design shows the capability and the potential to be investigated using microwave power transmission at low power densities for power supplied wireless sensors and actuators. Future work will be focused to fabricate this structure and make it under experimentation tests.

References

1. Tesla, N.: Experiments with Alternate Current of High Potential and High Frequency, McGraw (1904)
2. Tesla, N.: The transmission of electric energy without wires. The thirteenth Anniversary Number of the Electrical World and Engineer (1904)
3. Brown, W.C.: The history of Power transmission by radio waves, IEEE transaction on microwave theory and techniques. IEEE Trans. MTT **32**, 1230–1242 (1984)
4. Akkermans, J.A.G., van Beurden, M.C., Doodeman, G.J.N., Visser, H.J.: Analytical models for low power rectenna design. IEEE Antennas Wireless Propag. Lett. **4**, 187–190 (2005)
5. Koert, P., Cha, J., Machina, M.: 35 and 94 GHz rectifying antenna systems. In: SPS 91-Power from Space Dig., Paris, France, pp. 541–547, August 1991
6. Yoo, T.W., Chang, K.: Theoretical and experimental development of 10 and 35 GHz rectennas. IEEE Trans. Microw. Theory Tech. **40**(6), 1259–1266 (1992)
7. Tu, W.-H., Hsu, S.-H., Chang, K.: Compact 5.8-GHz rectenna using stepped-impedance dipole antenna. IEEE Antenna and Wireless Propagation Letters **6**, 282–284 (2007)
8. Ren, Y.-J., Li, M.-Y., Chang, K.: 35 GHz rectifying antenna for wireless power transmission. Electronics Letters 24th **43**(11), May 2007
9. Hagerty, J.A., Popovic, Z.: An experimental and theoretical characterization of a broadband arbitrarily-polarized rectenna array. In: IEEE Microw. Symp. Dig., vol. 3, pp. 1855–1858, May 2001
10. McSpadden, J.O., Chang, K.: A dual polarized circular patch rectifying antenna at 2.45GHz for microwave power conversion and detection. In: IEEE Microw. Symp. Dig., vol. 3, pp. 1749–1752 (1994)
11. Zbitou, J., Latrach, M., Toutain, S.: Hybrid rectenna and monolithic integrated zero-bias microwave rectifier. IEEE Trans. Microw. Theory Tech. **54**(1), 147–152 (2006)
12. Huang, W., Zhang, B., Chen, X., Huang, K., Liu, C.: Study on an S-band rectenna array for wireless microwave power transmission. Progress in Electromagnetics Research **135**, 747–758 (2013)
13. Vera, G., Georgiadis, A., Collado, A., Via, S.: Design of a 2.45 GHz rectenna for electromagnetic (EM) energy scavenging. In: IEEE Radio Wireless Symp., pp. 61–64 (2010)
14. Falkenstein, E., Roberg, M., Popovi, Z.: Low-Power Wireless Power Delivery. IEEE Transactions on Microwave Theory and Techniques, 0018–9480 (2012)
15. Rasshofer, R.H., Thieme, M.O., Biebl, E.M.: Circularly polarized millimeter-wave rectenna on silicon substrate. IEEE Trans. Microw. Theory Tech. **46**(5), 715–718 (1998)

16. Heikkinen, J., Kivikoski, M.: Low-profile circularly polarized rectifying antenna for wireless power transmission at 5.8 GHz. IEEE Microw. And Wirel. Compon. Lett. **14**(4), April 2004
17. Ali, M., Yang, G., Dougal, R.: A new circularly polarized rectenna for wireless power transmission and data communication. IEEE Antenn. Wirel. Propag. Lett. **4**, 205–208 (2005)
18. Wong, K.-L., Huang, C.-C., Chen, W.-S.: Printed ring slot antenna for circular polarization. IEEE Trans. Antennas Propagation **50**, 75–77 (2002)
19. Fries, M.K., Vahldieck, R.: Uniplanar circularly polarized slot-ring antenna architectures. Radio Sci. **38**(2), VIC5/1–VIC5/10, March/April 2003
20. Strassner, B., Chang, K.: 5.8-GHz circularly polarized rectifying antenna for wireless microwave power transmission. IEEE Trans. Microwave Theory Tech. **50**(8), 1870–1876 (2002)
21. Slavova, A., Omar, A.S.: Wideband rectenna for energy recycling. In: Proc. IEEE Antennas and Propag. Society Int. Symp., vol. 3, pp. 954–957, June 2003
22. Soliman, E.A., De Raedt, W., Vandenbosch, G.A.E.: Circularly polarized slot antenna dual-fed with microstrip lines. Journal of Electromag. Waves & Appl. **22**, 2259–2267 (2008)
23. Garg, R.: Microstrip antenna design handbook. Boston, Mass.[u.a.]: Artech House (2001)
24. Computer Simulation Technology. www.cst.com
25. Avago technologies: HSMS-285x Series Surface Mount Zero Bias Schottky Detector Diodes data sheet. http://www.avagotech.com

Miniaturized Wideband Flexible CPW Antenna with Hexagonal Ring Slots for Early Breast Cancer Detection

Afyf Amal, Bellarbi Larbi, Achour Anouar, Riouch Fatima and Errachid Abdelhamid

Abstract This paper presents a miniaturized flexible microstrip antenna designed using a Liquid crystal polymer (LCP) material. The developed design consists on a hexagon radiator with two ring slots excited by a CPW feed line, providing an operating frequency in S-band at 3GHz with an important bandwidth of 600MHz. This structure offers a thin thickness (1.6mm) with an overall size of 30×20 mm^2 that's can assure an easy integration into clothes as wearable antennas used for early breast cancer detection. Modeling and performances evaluation of the proposed antenna in terms of return loss, voltage standing wave ratio, radiation pattern, and current distribution have been carried out using CST-MW STUDIO simulator.

Keywords Miniaturized flexible · CPW-fed · UWB antenna · LCP · Wearable antennas · Early breast cancer detection · Ring hexagonal slots · CST-MW STUDIO

A. Amal(✉) · B. Larbi · A. Anouar
Electrical Engineering Laboratory (LGE),
Higher National School of Technical Education (ENSET), Rabat, Morocco
e-mail: Afyf.amal@gmail.com, l.bellarbi@um5s.net.ma

R. Fatima
SRTS Laboratory, National Institute of Post and Telecommunications (INPT),
Rabat, Morocco

E. Abdelhamid
Institute of Analytic Sciences (ISA), University of Lyon, Lyon, France

E. Sabir et al. (eds.), *The International Symposium on Ubiquitous Networking*,
Lecture Notes in Electrical Engineering 366,
DOI: 10.1007/978-981-287-990-5_17

1 Introduction

Recent years have witnessed a great deal of interest from both academia and industry in the field of flexible electronics. According to market analysis, the revenue of flexible electronics is estimated to be 30 billion USD in 2017 and over 300 billion USD in 2028 [1]. Their light weight, low-cost manufacturing, ease of fabrication, and the availability of inexpensive flexible substrates (i.e.: papers, textiles, and plastics), make flexible electronics an appealing candidate for the next generation of consumer electronics [2].

Ultra wideband (UWB) is an emerging wireless technology, recently approved by the Federal Communications Commission (FCC) [3]. In low/medium data-rate applications, like wearable computing, UWB offers low-power operation and extremely low radiated power, thus being very attractive for body-worn devices. Wearable computing is a new, fast growing field in application-oriented research [4], [5], and [6]. Steadily progressing miniaturization in microelectronics along with other new technologies enables wearable computing to integrate functionality in clothing allowing entirely new applications. In this sense, wearable computing are highly attractive as future solutions for a wide ranging applications including, military, health care, sport entertainment and many others have been categorized into two main areas by the IEEE 802.15.6 standard [7], [8], and [9]. As wearable computing is developing; there is an increasing need for a wearable system in which antennas playing a decisive role. It is necessary to mention here that antennas for such applications should possess certain properties like lightweight, conformal design, low cost, easy system implant etc.

Breast cancer is a significant health issue affects one in every seven women [10]. Early diagnosis and treatment are the hot keys to survive from breast cancer. The present "golden" standard screening technology for detecting early-stage breast cancer is X-rays mammography. However, it has several limitations [11], [12]. There is a need for complementary, safe, and reasonably priced method [13]. Microwave thermography for breast cancer detection has been introduced as a complementary non-invasive method for breast cancer detection.

Moreover it is reported in literature that abnormal both tissues and it surrounding area are warmer than normal's [14], which mean a self radiation with a weak power. This report overviews the design and optimization of a novel microwave receiver antenna which must serve as an element in a radiometry system for early breast cancer detection (Fig. 1). According to Nyquiest low (Equation 1) and for getting the emitted difference of temperature generated by a tissue located at a depth of about 3cm, the microwave radiometer can indicate a previous anomaly at operating frequency around 3GHz [15].

$$T = \frac{P}{K \times \Delta F} \tag{1}$$

Where P is the power available at antenna port, K is the Boltzman constant, T is the tissue temperature and ΔF is the antenna bandwidth.

In this paper, we present a new UWB flexible antenna for early breast cancer detection. It's organized as follows: Section 2 and 3 respectively introduce the design procedure and obtained simulation results. Conclusion is shown in Section 4. As described in subsequent sections, the presented antenna structure was built and carried out by using CST Microwave Studio.

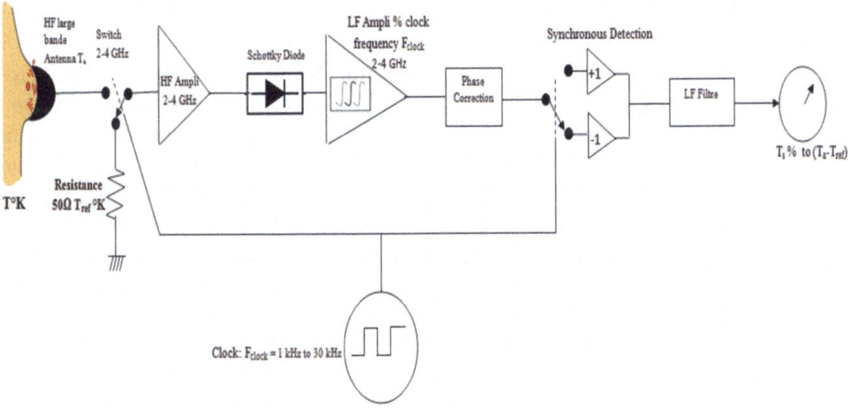

Fig. 1 Microwave radiometry system including antenna as the first circuit stage

2 Antenna Design

2.1 Choice of Antenna Substrate

In order to choose a substrate we have to consider its various features like the dielectric constant [16], the dielectric loss tangent, the cost of the material, the dimensional stability with time, the surface adhesion properties for the conductor coatings, the manufacturability (ease of cutting, drilling and shaping) [17]. In this research Liquid Crystal Polymer (LCP) is used as dielectric material. LCP is an emerging dielectric material that has gained attention in recent few years as a potential high performance microwave flexible substrate and packaging material [18], [19], and [20]. It offers an excellent combination of electronic, thermal, mechanical and chemical properties that make it as a promising substrate for electronics packaging [21].

2.2 Choice of Feed Technique

It is worth mentioning that there are several techniques used to characterize the electromagnetic properties of thin and flexible films/substrates such as: the near field microscopy, Coplanar Waveguide CPW approach, differential open resonator method, and goniometric time domain spectroscopy method [22],[23]. However, in this work, Co-Planar Waveguide (CPW) is preferred over other feeding techniques since no via holes or shorting pins are involved, in addition to several useful characteristics such as: low radiation losses, larger bandwidth, improved impedance matching, and more importantly, both radiating element and ground plane are printed on the same side of the substrate, which promotes low fabrication cost and complexity.

2.3 Antenna Parameters

The proposed antenna was printed on a 1.6mm thick LCP substrate (dielectric constant εr=3.1, tangent loss tan δ=0.002). The copper layer was only 0.035 mm thick and it was sputtered onto the LCP substrate. Gap between feed line and ground plane, which is used to create the CPW line, is labeled as G=0.412 mm. The compact CPW-fed hexagonal antenna design is presented in Fig. 1. The overall size is very small. The length of the board is L=30 mm and the width is W=20 mm. It is feed by a CPW line where the central conductor width W_F is 4 mm. Through full-wave EM simulation, we have found that, in particular, a slot with the similar shape of the antenna patch can generate a stronger resonance than any other shape due to the current distribution is concentrated at the edge of the patch. In this case, the antenna performs a narrower and stronger band-notched property. Consequently, hexagonal slots have been chosen as a basic resonant element to be etched on the patch, as shown in Fig. 2. Table 1 presents the various parameters of the proposed antenna shown in Fig. 2.

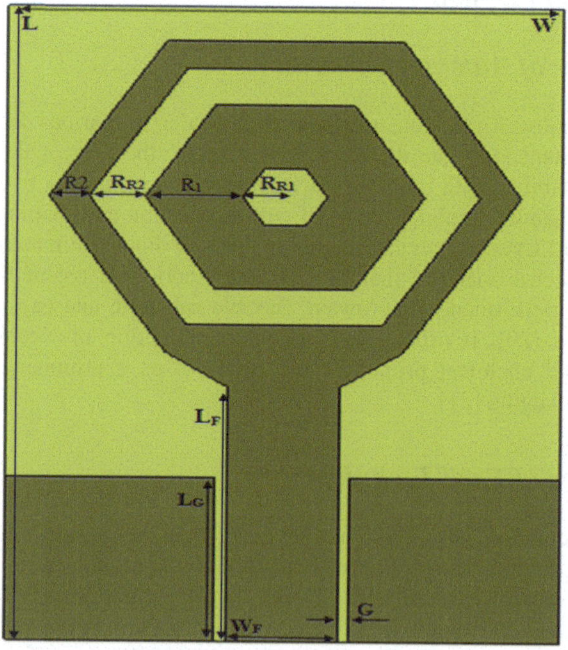

Fig. 2 The proposed flexible CPW antenna

Table 1 Physicals dimensions of the proposed antenna

Parameters	Values(mm)	Parameters	Values(mm)
W	20	R	$R_1+R_2+R_{R1}+R_{R2}$
L	30	R_1	3.5
W_F	4	R_2	1.5
L_F	14	R_{R1}	1.5
L_G	7.75	R_{R2}	2
H_S	1.6	G	0.412

3 Results and Discussion

Hence the main goal is to design an antenna with UWB radiation, we need to apply techniques that will improve the impedance bandwidth performance of the antenna. We have started from a simple circular patch (Antenna a), which is modeled to operate at 3 GHz according to the equation (2), and by changing the patch segmentation from 0 to 6 we obtained (hexagonal antenna, b), (antennas c and d) are achieved by adding respectively two hexagonal shaped slots, as the last technical step, both patch sides are tapered linearly down (Antenna c), Fig. 3.

$$R = \frac{87}{f_r \sqrt{\varepsilon_r}} \tag{2}$$

Where R is the patch radius, f_r is the operating frequency, and ε_r is the dielectric constant of substrate.

From Fig. 4 it is seen that the simple circular antenna (a) has an return loss (S11) of about -13dB at 3.6GHz, and by changing the patch segmentation from 0 to 6 we obtained (hexagonal antenna, b) that has a S11 of -15 at 3.7GHz, the third antenna is obtained by adding, for the first time, an hexagonal shaped slot (c) and another centric slot (d), we find respectively -25dB at 3.3GHz and -23dB at 2.9GHz, the last technical step is the antenna (e), where both patch sides are tapered linearly down, (e) has a good impedance matching with a return loss of about −27dB at the operating frequency of 3GHz. Moreover, from Fig. 4 it's clearly seen that the proposed antenna presents an ultra wideband propriety with an impedance bandwidth of about 600MHz which is from 2.711GHz to 3.3121GHz.

Further Fig. 5 shows the obtained simulated voltage standing wave ratio (VSWR). From the graph it's clear that at 3GHz the antenna provides a minimum VSWR of about 1 (less than 2), which is within the recommended range. The obtained result indicates that transmitter and antenna are well matched and a maximum possible amount of energy is absorbed at the input terminal with a minimum reflected power.

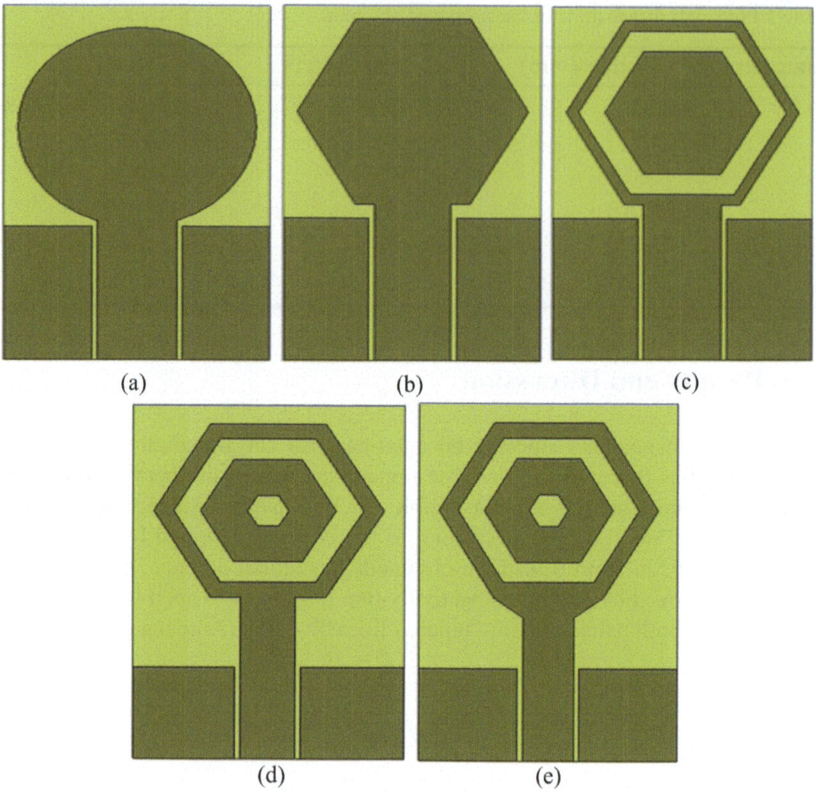

Fig. 3 The geometric shape of the proposed antenna design: (a) simple circular antenna, (b) hexagonal antenna (c) hexagonal antenna with a hexagonal ring slot, (d) hexagonal antenna with two hexagonal ring slots, and (e) proposed tapered path antenna.

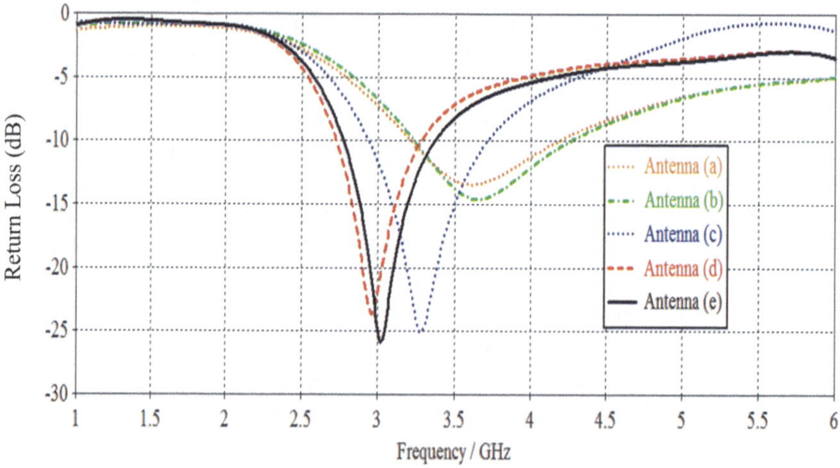

Fig. 4 Simulated return loss of the different antenna structures

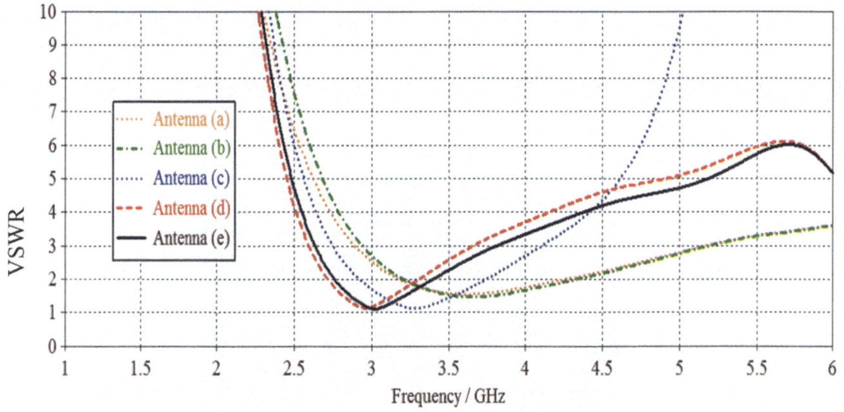

Fig. 5 The simulated VSWR of the three antenna structures

In Fig. 6, the simulated variation of the antenna input impedance versus frequency of the developed antenna can be seen. It is observed that the antenna is well matched to the 50 Ω impedance. Then at the operating frequency of 3GHz, the average value of the resistance (real part) is 50 Ohms, also the average value of the reactance (imaginary part) is 0-Ohms which provides the adequate input impedance matching at the desired resonant frequency.

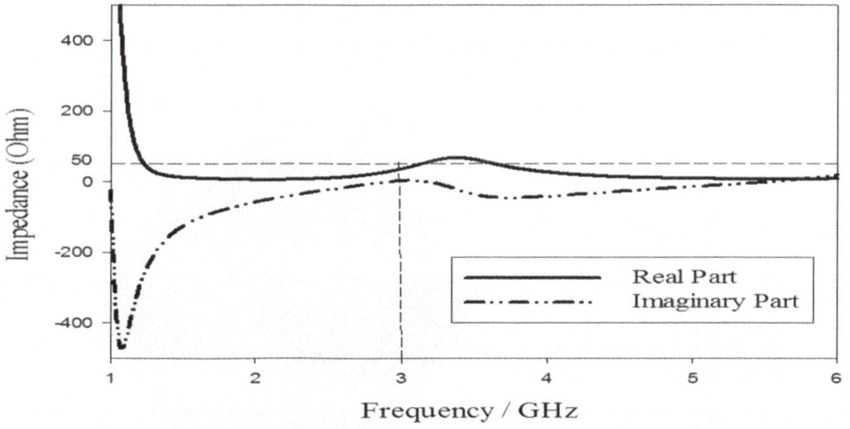

Fig. 6 Simulated antenna impedance (Ohm) vs. frequency of the proposed CPW antenna

Current distribution determines how the current flows on the patch antenna. Fig. 7 demonstrates these results. We observe a high strength of current radiates along the transmission line, the edges of the patch, and ring slots.

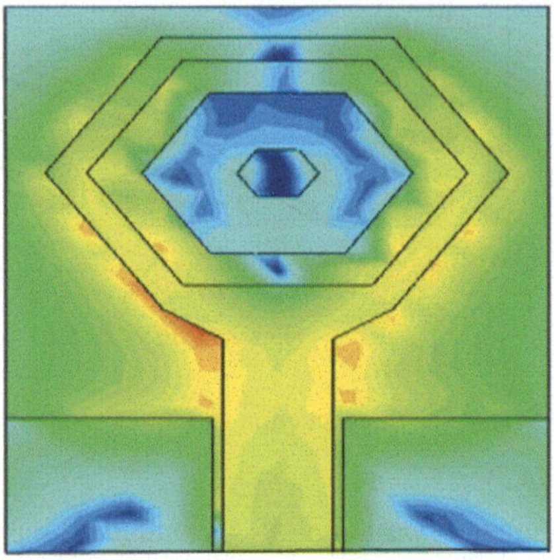

Fig. 7 Surface current density distribution at 3 GHz

The radiation pattern taken for the far-field at 3 GHz is indicated by the 3D view in Fig. 8. The directivity of the proposed antenna is about 2dBi. Further, Fig. 9 indicates that the antenna provides a directional behavior in E-plan (a) and a near omni-directional behavior in H-plan (b). Also Fig. 10 shows the gain variation over the operating frequency band. The graph shows that the proposed antenna provides a peak gain at 3GHz of about 2.37 dB.

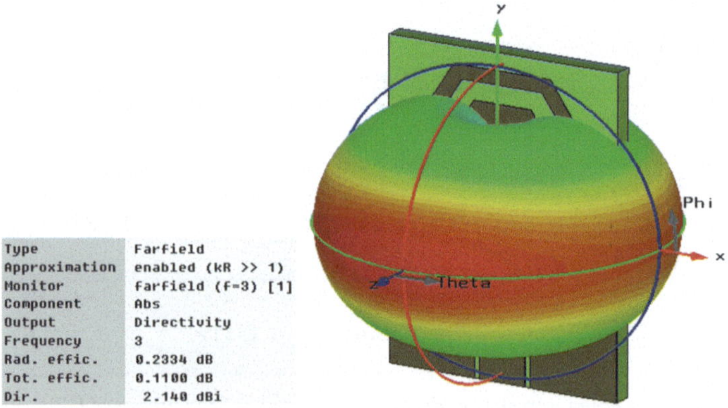

Fig. 8 3D radiation pattern at 3GHz

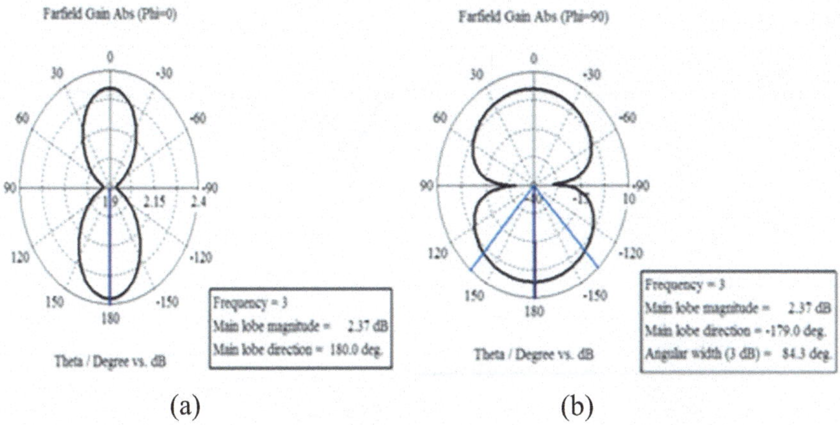

Fig. 9 2D polar radiation pattern at 3GHz for the antenna array; (a) E-Plan, and (b) H-Plan.

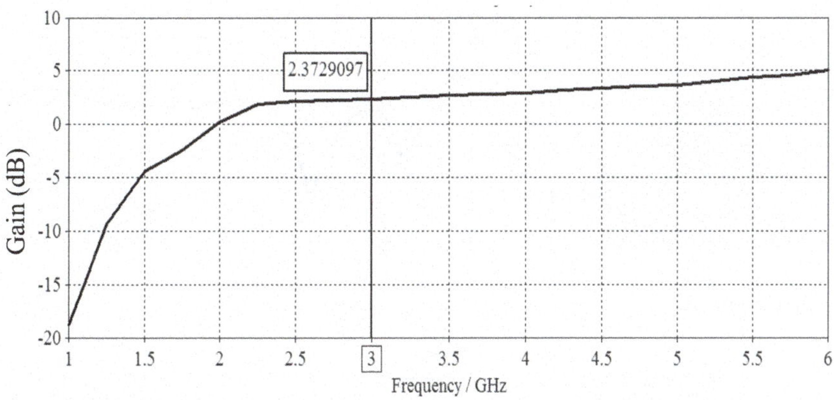

Fig. 10 Simulated antenna gain vs. Frequency

Fig. 11 presents a comparison between the proposed antenna and some recently developed miniaturized antennas printed on the same LCP substrate used in this work. Compared with the other structures the proposed antenna shows a wide impedance bandwidth around the operating frequency, compact size, and an important gain.

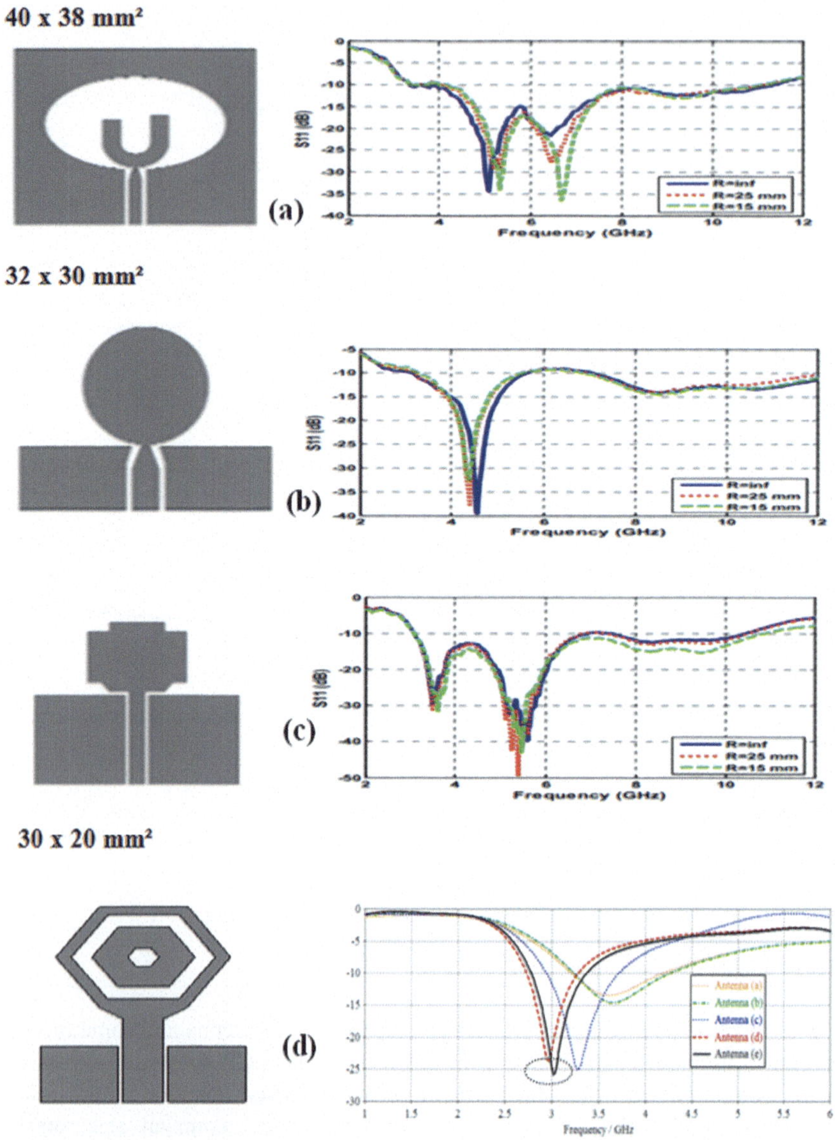

Fig. 11 (a), (b), and (c) respectively from [24], [25], and [26], (d) Our proposed antenna.

4 Conclusion

A new small CPW hexagonal antenna structure has been successfully designed and simulated via CST Microwave Studio. Effects of various antenna parameters on the bandwidth and the resonance characteristic were discussed. The performance criteria extracted from the software includes Return Loss, VSWR, Radiation Pattern, and Surface Currents, provide clear indication that the proposed design is suitable for smart clothes; due to its good matching input impedance at the operating frequency of 3GHz and its broadband propriety. Further, the miniaturized size of the developed antenna with the enhanced gain of 2dB that exhibits at 3GHz, are good features for such applications. Future work will be focused to explore the accuracy issues observed here for the inclusion of the body phantom in the simulation model before the realization stage and measurement tests.

Acknowledgment I acknowledge the contribution by all members of the GE Laboratory, ENSET/ENSIAS, Rabat, Morocco for providing facility and technical assistance in this research.

References

1. Hu, J.: Overview of flexible electronics from ITRIs viewpoint. In: VLSI Test Symposium (VTS), April 2010
2. Nathan, A., Chalamala, B.R.: Special issue on flexible electronics technology, part 1: systems and applications. Proceedings of the IEEE 93(7), 1235–1238 (2005)
3. Rhee, W., Xu, N., Zhou, B., Wang, Z.: Low power, non invasive UWB systems for WBAN and biomedical applications. In: ICTC 2010, Jeju, Korea (2010)
4. Starner, T.E.: Wearable computing for the developing world. IEEE Pervasive Comput. **4**, 87–91 (2005)
5. Salonen, P., Rantanen, J.: A dual band and wide-band antenna on flexible substrate for smart clothing. In: 27th Annual Conference of the IEEE, vol. 1 (2001)
6. Sanz-Izquierdo, B., Sobhy, M.I., Batchelor, J.C.: UWB wearable button antenna. In: European Conference on Antennas and Propagation EuCAP 2006, Nice, France, p. 131, November 2006
7. Heterogeneous electronics with low interfacial residual stress. IEEE Transactions on Components, Packaging and Manufacturing Technology **1**(9), 1368–1377 (2004)
8. Afyf, A., Bellarbi, L.: A Novel Miniaturized UWB antenna for microwave imaging. In: Proc. of ICMCS 2014. IEEE (2014)
9. Wellman, P.S., Dalton, E.P., Krag, D., Kern, K.A., Howe, R.D.: Tactile imaging of breast masses. Archives of Surgery **136**, 204–208 (2001)
10. Fear, E., Li, X., Hagness, S., Stuchly, M.: Confocal microwave imaging for breast tumor detection: localization in three dimensions. IEEE Trans. Antennas Propag. **49**(8), 812–822 (2002)
11. Zasowski, T., Althaus, F., Stäger, M., Wittneben, A., Tröster, G.: UWB for noninvasive wireless body area networks: channel measurements and results. At the IEEE Ultra Wideband Syst. Technol. Conf. (UWBST 2003), Reston, VA, November 2003
12. Langley, R.J., Ford, K.L., Lee, H.J.: Switchable on/offbody communication at 2.45 GHz using textile microstrip patch antenna on stripline. In: 6th European Conference on Antennas and Propagation (EUCAP), pp. 728–731 (2012)
13. Movassaghi, S., Abolhasan, M., Lipman, J.: Wireless body area networks: a survey. IEEE Communications Surveys & Tutorials (2014)

14. Ellis, H., Colborn, G.L., Skandalakis, J.E.: Surgical embryology and anatomy of breast and its related anatomic structures. Surgical Clinic North America **73**, 611–632 (1993)
15. Jacobi, C.E., Siegerink, B., Asperen, C.J.: Differences and similarities in breast cancer risk assessment models in clinical practice: which model to choose. Breast Cancer Res. Treat. **115**, 381–390 (2009)
16. Warren, L.S., Thiele, A.G.: Antenna Theory and Design. John Wiley & Sons Inc., New York (1998)
17. Kiani, I.G.: Coplanar Microstrip Active Integrated Antenna for Dual Band Low Noise Amplifier, M.Sc Thesis, Department of Electronic Engineering. GIK Institute of Engineering Sciences and Technology, Swabi, NWFP, Pakistan, May 2003
18. Thompson, C.D., Tantot, O., Jallageas, H., Ponchak, E.G., Tentzeris, M.M., Papapolymerou, J.: Characterization of liquid crystal polymer (LCP) material and transmission lines on LCP substrates from 30 to 110 GHz. IEEE Transactions on Microwave Theory and Techniques **52**(4), April 2004
19. Zou, G., Gronqvist, H., Starski, J.P., Liu, J.: Characterization of liquid crystal polymer for high frequency system-in-package applications. IEEE Transactions on Advanced Packaging (2002)
20. Karbassi, A., Ruf, D., Bettermann, A.D., Paulson, C.A., Weide, V.D., Daniel, W., Tanbakuchi, H., Stancliff, R.: Quantitative scanning nearfield microwave microscopy for thin film dielectric constant measurement. Review of Scientific Instruments **79**(9), 094706-094706-5 (2007)
21. Ming, L., Fortin, J., Kim, J.Y., Fox, G., Chu, F., Davenport, T., TohMing, L., XiCheng, Z.: Dielectric constant measurement of thin films using goniometric terahertz time-domain spectroscopy. Selected Topics in Quantum Electronics (2000)
22. Geise, A., Strohmaier, U., Jacob, A.F.: Investigations of transmission lines and resonant structures onflexed liquid crystal polymer (LCP) substrate sup to 67GHz. In: Proc. Eur. Microw. Conf., pp. 735–738, October 2009
23. Vyas, R., Rida, A., Bhattacharya, S., Tentzeris, M.M.: Liquid crystal polymer (LCP): the ultimate solution for low-cost RFflexible electronics and antennas. In: Proc. IEEE Antennas Propag. Soc. Int. Symp., pp. 1729–1732, June 2007
24. Pengcheng, L., Jianxin, L., Xiaodong, C.: Study of printed elliptical/circular slot antennas for ultrawideband applications. IEEE Trans. Antennas Propag. **54**, 1670–1675 (2006)
25. Liang, J., Guo, L., Chiau, C.C., Chen, X., Parini, G.C.: Study of CPW-fed circular disc monopole antenna for ultra wideband applications. IEEE Microwaves, Antennas and Propag. Proc., 520–526 (2005)
26. Nikolaou, S.: Design and implementation of compact reconfigurable antennasfor uwb and wlan applications, Thesis dissertation (2007)

Part IV
Advances in Provisioning QoS, QoE and QoC

Maximizing QoS in Heterogeneous Wireless Sensor Networks Using Game Theory and Learning Algorithms

Hajar El Hammouti, Loubna Echabbi and Yann Ben Maissa

Abstract A Wireless Sensor Network (WSN) is composed of sensor equipped devices that aim at sensing and processing information from the surrounding environment. Energy consumption is the major concern of WSNs. At the same time, quality of service is to be considered especially when dealing with critical WSNs.

In this paper, we present a game theory based approach to maximize quality of service, defined as the aggregate frame success rate, while optimizing power allocation. Game theory is designed to study interactions between players (e.g. chess players) who decide on a set of actions (e.g. the players moves) to reach the objective outcomes (e.g. to win the game). Here, we model the system as a potential game. We show that the optimal power allocation, crucial in a heterogeneous sensor network, is a Nash equilibrium of this game, and we discuss its uniqueness. For simulations, we present a fully distributed algorithm that drives the whole system to the optimal power allocation.

Keywords Game theory · Nash equilibrium · Global optimum · Potential game · Frame success rate · Distributed algorithms · Learning algorithms · Heterogeneous wireless sensor networks · Quality of service · Energy consumption

1 Introduction

1.1 Context

Wireless Sensor Networks (WSNs) are a set of sensor equipped devices that collect data from the surrounding environment and send it to a centralized entity called

H. El Hammouti(✉) · L. Echabbi · Y. Ben Maissa
Telecommunications Systems, Networks and Services Laboratories,
National Institute of Posts and Telecommunications, Rabat, Morocco
e-mail: {elhammouti,echabbi,benmaissa}@inpt.ac.ma

E. Sabir et al. (eds.), *The International Symposium on Ubiquitous Networking*,
Lecture Notes in Electrical Engineering 366,
DOI: 10.1007/978-981-287-990-5_18

the base station or the sink node. Sensor devices are tiny and resource limited in terms of batteries. Sensors are, most of the time, deployed over an inaccessible area. Therefore, the replacement or recharge of their batteries is mostly impossible. This makes the energy consumption the major concern of WSNs[1].

When dealing with WSNs, the limited resource is the most important preoccupation, sometimes more important than the quality of service(QoS). In order to prolong the lifetime [2] of a WSN, the energy consumed by batteries should be reduced [3]. This issue is to be considered either for homogeneous or heterogeneous WSNs. A Heterogeneous Wireless Sensor Network (HWSN) is composed of sensor devices that are heterogeneous in terms of sensors ability, processing capability, transmit power level, etc. For example, a network composed of MICAz nodes (MPR2400CB) that have a 7.37 MHz processor with 4 Kilo bytes of RAM (Random Access Memory) and LOTUS nodes (LPR2400) that have a 100 MHz processor with 64 Kilo byte of RAM.

1.2 Problem

The efficient use of transmission power resources contributes to lifetime maximization. The goal of the power allocation problem is to determine transmission power resources needed for the maximization of the overall system performance. From an optimization standpoint, optimizing resource allocation is an NP-hard problem [4]. A heuristic is therefore needed to find the exact solution or simply an approximate solution in a reasonable time. Many successful heuristic techniques have been used to solve this kind of problem. However, those using game theory are still rare.

In this work, we consider a *discrete* power allocation and we address the problem of the quality of service in HWSN. We focus mainly on maximizing the aggregate frame success rate function (FSR) while optimizing discrete transmission power allocation. FSR is particularly important for critical HWSNs that have a life-critical applications such as detecting a heart attack: in this case a frame that does not arrive in time just poses a risk to the life of the patient.

1.3 Contribution

We present a game theory based approach to study the problem of the global frame success rate (FSR) maximization. The novelty of this work is that we show that the game model we propose is a potential game. The property of this kind of games is that the global optimum is a Nash equilibrium (Nash Equilibrium is a situation where no player has the incentive to move unilaterally). Consequently, to reach the global optimum, existing algorithms that find Nash equilibrium can be used. We emphasize on the discrete case, where the power space is discretized, and we use a distributed *learning* algorithm that reaches to a Nash equilibrium. It is not guaranteed that the proposed algorithm converges to the Nash equilibrium that coincides with the global optimum, yet simulation experiments show that the proposed algorithm solves the maximization problem.

1.4 Structure

The remainder of the paper is organized as follows. In the next section, we discuss related works. Section 2 presents the system model. In section 3 we give the game formulation and prove that the studied game is a *potential game* . Section 4 describes a learning algorithm that converges to the global optimum for the numerical experiments we have conducted. Finally, in the last section, we make some concluding remarks and discuss various extensions to this work.

2 Related Work

One of the key success factors of game theory based approach is to build a good model that reflects the players' behaviors and best describes their preferences[5]. In the last decades, game theory has been increasingly used to describe sensors' behaviors and study challenges related to WSNs in general [6][7]. It has also been used to design heterogeneous WSNs.

Recent researches have also been interested in designing algorithms that solve the resource allocation problem. These algorithms are either centralized and need a controlling entity or decentralized and thus allow more autonomy to the nodes. The work proposed in [8] addresses a distributed algorithm that sets the transmission power independently for each node in order to increase the data recovery of the network. The work of authors in [3] goes in the same direction. They present the ant colony algorithm (a heuristic algorithm based on ants behavior) for solving the power and channel allocation problems. The scheduling problem was also tackled by Mahani et al. in [9]. Authors propose a scheduling algorithm based on node collaboration in sensing and transmitting phase. All of these works do not use game theory to model the resource allocation problem.

Algorithms based on game theory have been presented in some works to find specific solution concepts such as Nash Equilibrium (NE). In [7], authors use a distributed game based channel allocation algorithm to find the NE of the proposed game while taking into account both network topology and routing information. The work in [10] adopts a game theory based approach as well. Authors study the utility function for nodes to achieve desirable power efficiency and node degree while minimizing the power consumption.

As quality of service is important for HWSN especially for critical HWSN, we have chosen to work, in this paper, on the frame success rate as an objective function in its discrete form (using discrete transmission powers). To maximize the frame success rate minimize the bit error rate which is needed in critical WSN. The same function was studied in a previous work while using a continuous transmission power [11]. The transmitter can choose any continuous value of the power space. Meshkatiet al. study the FSR function and show that it is a quasi-concave function. This implies the existence of a Nash equilibrium. Hence, authors prove that the FSR has a unique NE which is the maximizer of the FSR. However, these properties (existence and uniqueness resulting from the quasi-concavity) may be lost in the discrete form. The reason behind this is that the quasi-concavity in the discrete case is an extension of the usual definition of concavity for continuous functions. Consequently, the restriction of any continuous concave function to a discrete space does not necessarily imply a discretely concave function. Therefore, the discussion of the existence and uniqueness of NE has a real sense when it comes to a discrete system.

When the potential function is continuous, it is sufficient to check that it satisfies the Moulin dominance solvability condition or the Rosen condition [12]. When these conditions are satisfied NE exists and it is unique. To find Nash equilibrium in the continuous case, many best-response learning algorithms (procedures by which agents learn NE) are designed to find NE. A best-response learning algorithm is described in [13].

To solve the problem of the uniqueness of Nash equilibrium in the discrete case, Song et al. in [14] propose an equilibrium selective algorithm which retains the favorable NE. However, it is shown in [15] that this algorithm achieves a very slow convergence. Consequently, this algorithm could be inappropriate in the context of WSNs. Actually, when using an equilibrium selective algorithm that converges slowly, a large number of control messages are needed. Thus, sensors spend their energy on sending control messages rather than performing measurements.Therefore, the existence and uniqueness of NE for the problem of FSR maximization worth to be discussed, and an algorithm that converges to the global optimum in a reasonable time is to be considered.

3 The System Model

3.1 Description

We work on a hierarchical wireless sensor network. We suppose that once the network is deployed, the nodes are organized in a hierarchical model. Hierarchical clusters are then formed (the clustering techniques are out of scope here). Each cluster consists of a large number of sensors with low transmission powers and a single sensor node with high transmission powers. The node transmitting with high power levels is considered as the cluster head (CH). To simplify, we will not consider all the network's clusters, but rather focus on one cluster and work on balancing the transmission power allocation in order to improve the network performance. The CH has a communication link with the base station.

3.2 The Model

We assume a set $\mathcal{S}$ of K sensors that form a cluster. The cluster consists of a node (the cluster head) transmitting with high discrete powers $\mathcal{P}^{high} = \{P_1^{high}, \ldots, P_{max}^{high}\}$, overlaid with nodes that transmit with lower discrete transmission powers $\mathcal{P}^{low} = \{P_1^{low}, \ldots, P_{max}^{low}\}$. The sensors are supposed fixed and the CH is always connected to the sink node (the gateway that collects data from all nodes and communicates it to the processing center). We denote by P_i the transmission power of node i. This power is transmitted to a node j through a radio channel. The quality of the radio channel depends on the distance between the transmitter and the receiver. When a frame is transmitted by i with P_i, it is received at j with $d_{ij}^{-\nu} P_i$, where d_{ij} is the distance between nodes i and j, and ν is the parameter of fading that describes the path loss. As the distance between nodes and the fading parameter are fixed, the only way to increase the receiving power is by increasing the transmitting power. However, this also increases interferences and, thus, decreases the quality of wireless sensor connections that is measured by the signal-to-interference-plus-noise ratio (SINR) as follows:

$$\gamma_{ij}(P) = \frac{d_{ij}^{-\nu} P_i}{\eta + \sum\limits_{k \neq i} d_{kj}^{-\nu} P_k}$$

3.3 Frame Success Rate

The primary concern of power control problem is to achieve a maximum frame success rate while including minimal power consumption [16]. The Frame Success Rate (FSR) is the probability of correct reception of a frame at the receiver [17]. Thus, the FSR depends on the Bit Error Rate (BER) which is the number of bit errors divided by the total number of transferred bits during a studied time interval. Let us denote by M the total number of bits per frame, the FSR $f(\gamma)$ could be written:

$$f_{ij}(P) = (1 - BER(\gamma_{ij}(P)))^M$$

Actually, the BER depends on the SINR $\gamma_{ij}(P)$ but also on the modulation scheme. The table below shows the BER as a function of SINR for various modulation schemes [17].

Some modulation Schemes	Value of BER
BPSK	$erfc(\sqrt{2\gamma})$
DPSK	$0.5e^{-\gamma}$
Coherent FSK	$erfc(\sqrt{\gamma})$
Non-coherent FSK	$0.5e^{\frac{-\gamma}{2}}$

Table 1 Formal game elements

Game Element	Meaning
$\mathscr{S}$	The set of players which are of sensor nodes
$\underbrace{\mathscr{P}^{low} \times \cdots \times \mathscr{P}^{low}}_{k-1} \times \mathscr{P}^{high}$	Space of actions of sensors
$\{U_i\}_{i\in\mathscr{S}}$	Function utilities which are, for each sensor, its FSR and FSRs of its neighbors (see section 4.2)

In this work we consider a non-coherent PSK (Phase Shift Keying) modulation scheme (which is a simpler scheme modulation to implement), and use $BER(\gamma_{ij}(P)) = 0.5e^{\frac{-\gamma_{ij}(P)}{2}}$.

4 Game Formulation

We consider a game G where sensor nodes are seen as players. The actions of these players are the decisions they make to choose their transmission powers. The elements of the game are defined in the table below:

4.1 *Sensors Neighborhood*

The notion of neighborhood is fundamental to distributed learning algorithms. The players learn how to reach the global optimum in a distributed fashion. Each player has limited knowledge about the status of the other players. It actually gathers information from neighboring players and decides locally its strategy. Thus, for this work, we define the neighborhood of each sensor as follows: two sensors are *neighbors* if one of them is within the range of the other. We denote by N_i the neighborhood of sensor i. We suppose that the neighborhood remains unchanged after changing the nodes power. Mathematically, we write:

$$N_i = \{j \in \mathscr{S}\backslash\{i\}, d_{ij}^{-\nu} P_i \succ \theta \text{ or } d_{ji}^{-\nu} P_j \succ \theta\} \tag{1}$$

With θ the received power threshold. According to equation 1, all nodes are neighbors of the CH.

4.2 *Objective Function*

Let us define at first the utility of nodes. Each node has an aggregate FSR wich is the sum of FSRs provided to nodes at its range. Let us use $\mathscr{S}_i$ to define the set of nodes that are within the range of node i:

$$\mathscr{S}_i = \{j \in \mathscr{S} \backslash \{i\}, d_{ij}^{-\nu} P_i \succ \theta\} \tag{2}$$

We define the utility of a node i as the sum of its aggregate FSR and the aggregate FSR of its neighbors. We denote by F_i the aggregate FSR of node i. Thus:

$$F_i(P) = \sum_{j \in \mathscr{S}_i} f_{ij}(P) \tag{3}$$

Hence, the utility function of a sensor node i is written as follows:

$$U_i(P) = F_i(P) + \sum_{j \in N_i} F_j(P) \tag{4}$$

Now, let us define the objective function of the game NG, which is the sum of the aggregate FSR received at each node i:

$$U(P) = \sum_{i \in \mathscr{S}} F_i(P) = \sum_i \sum_j (1 - 0.5e^{-0.5 \frac{d_{ij}^{-\nu} P_i}{\eta + \sum_{k \neq i} d_{kj}^{-\nu} P_k}})^M \tag{5}$$

As described in 5, the nodes cannot only increase their power to increase the frame success rate of the network, otherwise they would increase interferences as well and thus decreases the FSR. Therefore, we formulate the problem of power allocation as the maximization of the overall frame success rate of the network subject to the power state allowed to each node:

$$\begin{aligned} &\text{maximize} \quad U(P) = \sum_{i \in \mathscr{S}} F_i(P) \\ &\text{subject to} \quad \forall i \text{ sensor } P_i \in \mathscr{P}^{low}, P_{CH} \in \mathscr{P}^{high} \end{aligned} \tag{6}$$

4.3 Potential Game

Definition 1 A game G is a potential game if there exists a potential function U : $X \longrightarrow (R) \cup \{-\infty\}$ satisfying $U(x_i', x_{-i}) - U(x_i, x_{-i}) = U_i(x_i', x_{-i}) - U_i(x_i, x_{-i})$ for all x_i, x_i' in X_i and x_{-i} in X_{-i}, i in $\mathbb{N}$ [18].

A special characteristic of a potential game is that if a strategy profile X maximizes the potential function U, then X is a Nash equilibrium. Actually, if X maximizes U, then $U(x_i, x_{-i}) \geq U(x_i', x_{-i})$ for all x_i' in X_i, so $U(x_i, x_{-i}) - U(x_i', x_{-i}) = U_i(x_i, x_{-i}) - U_i(x_i', x_{-i}) \geq 0$, thus $U_i(x_i, x_{-i}) \geq U_i(x_i', x_{-i})$ for all x_i' in X_i, and this for all $i \in \mathbb{N}$, which is exactly the definition of a Nash equilibrium.

Corollary. *In a potential game, the global optimum of the potential function is a Nash equilibrium.*

Proposition 1 *The game G is a potential game with the global FSR as a potential function.*

Proof Now, let us prove that the game we consider is a potential game with the global FSR as a potential function.

$$U(\tilde{P}_i, P_{-i}) - U(\tilde{P}_i', P_{-i}) = U_i(\tilde{P}_i, P_{-i}) + \sum_{j\in\mathcal{S}\setminus\{N_i\cup\{i\}\}} F_j(\tilde{P}_i, P_{-i}) - U_i(\tilde{P}_i', P_{-i}) - \sum_{j\in\mathcal{S}\setminus\{N_i\cup\{i\}\}} F_j(\tilde{P}_i', P_{-i}) \quad (7)$$

Notice that $\sum_{j\in\mathcal{S}\setminus\{N_i\cup i\}} F_j(\tilde{P}_i, P_{-i})$ is independent of the strategy of i since none of the terms of the sum is related to i. Notice also, that $\mathcal{S}\setminus\{N_i\cup\{i\}$ remains unchanged by hypothesis (since we suppose that the neighborhood remains unchanged after changing the nodes power, see 4.1). Thus:

$$\sum_{j\in\mathcal{S}\setminus\{N_i\cup i\}} F_j(\tilde{P}_i', P_{-i}) = \sum_{j\in\mathcal{S}\setminus\{N_i\cup i\}} F_j(\tilde{P}_i, P_{-i}) \quad (8)$$

Consequently:

$$U(\tilde{P}_i, P_{-i}) - U(\tilde{P}_i', P_{-i}) = U_i(\tilde{P}_i, P_{-i}) - U_i(\tilde{P}_i', P_{-i}) \quad (9)$$

Which is equivalent to U is a potential function.

4.4 On the Uniqueness of Nash Equilibrium

To prove the uniqueness of the Nash equilibrium, it is sufficient to prove that the studied function is discretely quasi-concave [18]. Thus the aggregate FSR should satisfy: $\forall\lambda \in [0, 1]$, $\forall P$, $P' \in (\mathcal{P}^{low})^{K-1} \times \mathcal{P}^{high}$

$$\lambda U(P) + (1-\lambda)U(P') \leq \max_{z\in N(\lambda P+(1-\lambda)P')} (U(z)) \quad (10)$$

Where $N(x)$ is the set of power vectors neighbors of vector x, which is defined as:

$$N(x) = \{z \in (\mathcal{P}^{low})^{K-1} \times \mathcal{P}^{high} \text{ , } \|x - z\|_1 \leq 1\} \quad (11)$$

and

$$\|x - z\|_1 = \max_{1 \le i \le K} (|x_i - z_i|)$$

At the time of writing this paper, the proof of quasi-convexity of the aggregate FSR function has not been established yet. We will focus on this proof in our future work. However, simulation results of the learning algorithm used in section 5 show that the resulting NE is the unique maximizer of the aggregate FSR function.

5 Simulation Results

In this section, we present a selection of representative numerical experiments that we have conducted in order to validate our approach. To solve the problem described in 6, we use the learning algorithm proposed in [15], [19] and [20]. It is shown in [19] that the algorithm converges to a Nash equilibrium. Nash equilibrium may not be unique, yet simulations in our case show that the chosen algorithm reaches the Nash equilibrium that coincides with the global optimum. To refer to this algorithm we will use the same name used in [15]: FLAPH (Fully Learning Algorithm for Power allocation in Heterogeneous Networks).

5.1 The Proposed Algorithm

Learning algorithms are heuristic algorithms that allow to agents to build up new knowledge based on their accumulated experience.

In reinforcement learning, nodes should update their strategies following a probability distribution in order to learn. In the algorithm we propose probabilities are updated with respect to the response of the environment (function of the local normalized utility). Thus the successive choice of the power depends on the environment response, which is a property of a learning algorithm We denote by p_{ik} the probability that node i chooses the power P_k. Stochastic strategies, utilities and power strategies will be suffixed by t to refer to their values at time step t. The probabilities are calculated as follows:

$$p_{ik}^{t+1} = \begin{cases} p_{ik}^t - lU_i^t p_{ik}^t & \text{if } P_i^t \neq P_k \\ p_{ik}^t + lU_b^t \sum_{k' \neq k} p_{ik'}^t & \text{otherwise} \end{cases} \tag{12}$$

Where: U_i^t is the normalized local utility at the time t, such that:

$$U_i^t = \frac{U_i(P^t)}{\max_{t' \le t}(U_i(P^{t'}))} \tag{13}$$

And l a parameter in [0, 1], it is the step size of the probability updating rule. The results of the algorithm are more accurate when l is close to zero (see [19]).

The qualitative properties of this algorithm such as its distributed implementation, asynchronous performance and limited computational efforts, are shown in [15].

5.2 *Numerical Results*

We present two experiments with different sizes of HWSN: 5 and 12 nodes in the cluster respectively for each experience.

We suppose $\mathscr{P}^{low} = \{30\mu W, 60\mu W\}$, and $P_{high} = \{2000\mu W, 4000\mu W\}$. The power spectral density of the background noise is supposed equal to $4.0039e{-}15mW$. We take the step size of the probability updating rule $l = 0.008$. The figures 1 and 2 show the overall FSR of the network evolution and indicate the convergence of the FLPAH to the global optimum after a number of iterations. In Fig.1, the initial power allocation is $P = [303030302000]$, after the FLAPH convergence, the optimal power allocation is $P^* = [606030304000]$. Besides, in Fig.2, the initial power allocation is $P = [3030303030303030303030302000]$ while the optimal result given by the FLAPH is $P* = [60303030303030303060604000]$. Figures 3 4 5 and 6 show the network configuration and sensors links before and after the convergence of the FLAPH. The arrows in these figures designate communication links between sensors. The arrow show that the sensor at the origin of the arrow could reach the sensor at its end (the sensor is in its range). Links that are newly established after the convergence of the FLAPH are highlighted in blue color.

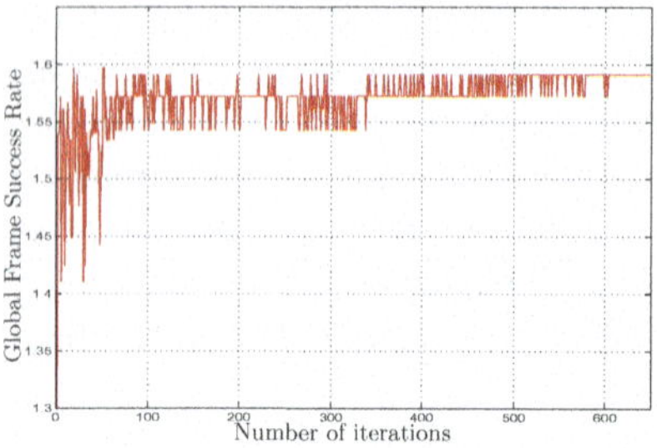

Fig. 1 The convergence of the FLAPH for $K = 5$ sensors

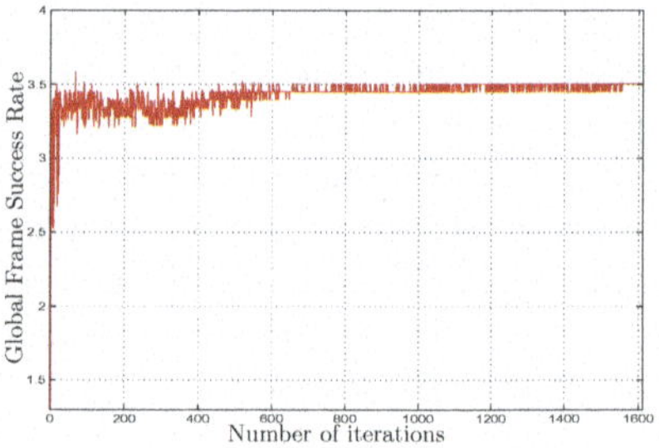

Fig. 2 The convergence of the FLAPH for $K = 12$ sensors

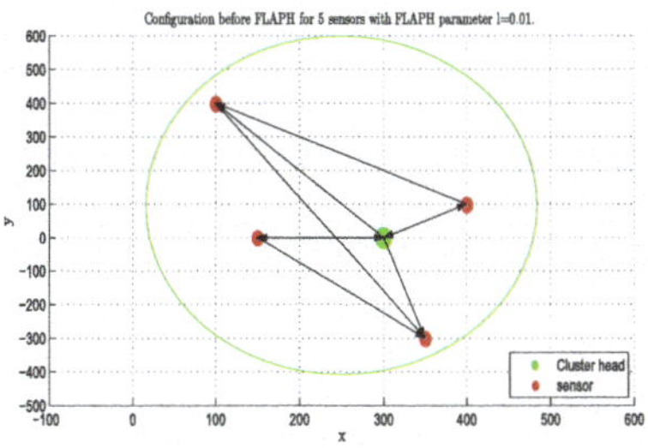

Fig. 3 Network Configuration for $K = 5$ sensors before the FLAPH

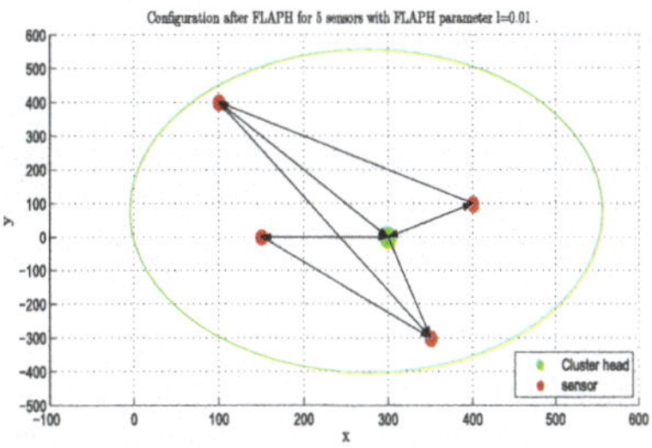

Fig. 4 Network Configuration for $K = 5$ sensors after the FLAPH

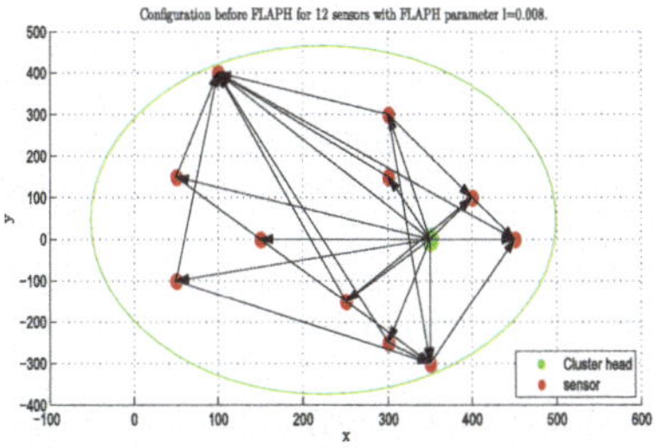

Fig. 5 Network Configuration for $K = 12$ sensors before the FLAPH

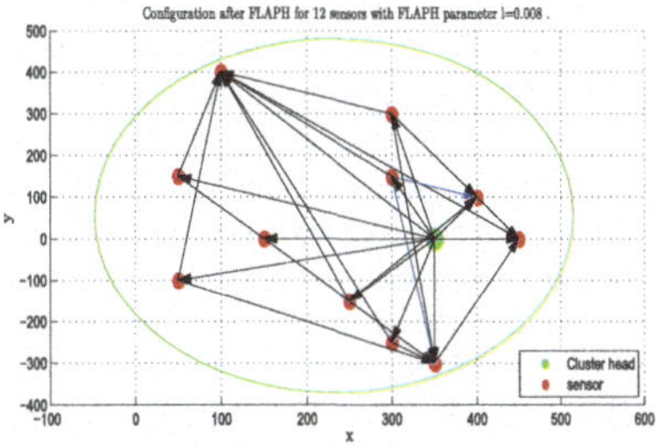

Fig. 6 Network Configuration for $K = 12$ sensors after the FLAPH

6 Conclusion

WSNs are one of the main trends in ubiquitous computing by their multiple applications and their ability to provide environmental data with high spatio temporal resolution. Energy consumption is the main constraint, which motivated our choice. In this work, we used game theory and learning algorithms in order to maximize Heterogeneous WSNs QoS depending on energy consumption constraints. We presented a game theory based approach to solve the problem of FSR maximization while optimizing discrete power allocation. We considered a potential game and showed that the optimal power allocation is a NE. The uniqueness of NE has not been proved yet. However simulation results show that the proposed algorithm converges to the NE that coincides with the global optimum.

In ongoing work, we will take into account other parameters that count while studying WSNs lifetime such as the coverage and connectivity of WSNs.

References

1. Tembine, H.: Distributed Strategic Learning for Wireless Engineers, 1st edn. CRC Press, May 2012
2. Dietritch, I., Dressler, F.: On the lifetime of wireless sensor networks. ACM Transactions on Sensor Networks (TOSN) (2009)
3. Aung, A.: Distributed Algorithms for Improving Wireless Sensor Network Lifetimewith Adjustable Sensing Range. Thesis, Georgia State University (2007)
4. Cheng, X., Narahari, B., Simha, R., Cheng, M.X., Liu, D.: Strong minimum energy topology in WSNs: NPCompleteness and heuristics. IEEE Transactions on Mobile Computing, 248–256 (2003)
5. Watson, J.: Strategy: An Introduction to Game Theory, 3rd edn. W. W. Norton Company (2013)
6. Shi, H.Y., Wang, W.L., Kwok, N.M., Chen, S.Y.: Game Theory for Wireless Sensor Networks: A Survey. Sensors **12**(7), 9055–9097 (2012)
7. Sengupta, S.: A Game Theoretic Framework for Power Control in Wireless Sensor Networks. IEEE Transactions on Computers **59**(2), 231–242 (2010)
8. Wazid, M., Gupta, N., Singh, D.P., Goudar, R.H.: Data recovery with energy efficient task allocation in wireless sensor networks. In: IEEE Colombian Conference on Communications and Computing (COLCOM), pp. 1–5, June 2014
9. Mahani, A., Farahmand, E., Kavian, Y.S., Rashvand, H.F.: A distributed sensing and medium access algorithm to prolong the lifetime of WSN. In: IET Conference on Wireless Sensor Systems (WSS), pp. 1–5 (2012)
10. Hongliang, R., Meng, M.Q.H.: Game-Theoretic Modeling of Joint Topology Control and Power Scheduling for Wireless Heterogeneous Sensor Networks. IEEE Transactions on Automaton Science and Engineering **6**(4), October 2009
11. Meshkati, F., Poor, H.V., Schwartz, S.C., Mandayam, N.B.: An Energy Efficient Approach to Power Control and Receiver Design in Wireless Data Networks. IEEE Transactions on Communications **53**(11), November 2005
12. Basslam, M., Elazouzi, R., Essabir, S., Echabbi, L.: Market share game with adversarial access providers: a neutral and a non-neutral network analysis. In: International Conference on NETwork Games, COntrol and OPtimization (NETGCOOP) (2011)
13. Weinberg, M., Rosenschein, J.S.: Best-response multiagent learning in non-stationary environments. In: The Third International Joint Conference on Autonomous Agents and Multiagent Systems AAMAS, pp. 22–23, July 2004
14. Song, Y., Wong, S.H.Y., Lee, K.: A game theoretical approach to gateway selections in multi-domain wireless networks. In: The Fifth Annual Conference of the International Technology Alliance (ACITA 2011), College Park, MD, September 2011
15. Elhammouti, H., Echabbi, L., Elazouzi, R.: A fully distributed algorithm for power allocation in heterogeneous networks. In: Proceedings of the International Conference of Networked Systems (NETYS), pp. 13–14, May 2015
16. Valli, R., Dananjayan, P.: A Non-Cooperative Game Theoretical Approach for Power Control in Virtual MIMO Wireless Sensor Network. International Journal of UbiComp (IJU) **1**(3), July 2010
17. Saraydar, U.C., Narayan, B.M., Goodman, D.J.: Efficient Power Control via Pricing in Wireless Data Networks. Transactions on Communications **50**(2), February 2002
18. Takashi, U.: A Note on Discrete Convexity and Local Optimality. Japan Journal of Industrial Applied Mathemathics, 21–29 (2006)
19. Barth, D., Echabbi, L., Hamlaoui, C.: Optimal Transit Price Negotiation: The Distributed Learning Perspective. Journal of Universal Computer Science **14**(5) (2008)
20. Xing, Y., Chandramouli, R.: Stochastic Learning Solution for Distributed Discrete Power Control Game in Wireless Data Networks. IEEE/ACM Transactions on Networking **16**(4), March 2008

Traffic Congestion Manager, a Cost-Effective Approach

Ahmed Adnane, Mohamed Salim Lmimouni, Maha Rezzai and Hicham Medromi

Abstract Traffic congestion is a worldwide problem which challenges both scientific community and governments. This paper proposes a novel cost-effective approach which aims to predict congestion levels of road segments and thus helps preventing high traffic congestions. Unlike other approaches based on cooperative vehicular communication, this work adopts a cooperation-without-communication approach inspired by social insects. In fact, vehicles cooperate and share their experiences through RSU (Road Side Unit) controllers placed in the traffic lights of the studied area. Each vehicle is equipped with a navigation device which notifies the driver about congestion levels of road segments. Hence, vehicles' distribution over the road network tends towards homogeneity. An evolutionary algorithm which optimizes traffic reports exchange between vehicles and RSU controllers is described.

Keywords Congestion detection · Congestion prevention · Traffic data optimization · Cooperation without communication · Collective intelligence

1 Introduction

Nowadays, cities suffer from traffic congestion. This is due to several factors, such as rapid population growth, poor road infrastructure, rapid increase in the number of cars, etc. [1]. We call congestion a situation on road networks where citizen's demand in terms of road utilization exceeds the network capacity. This undesirable event has many harmful effects [2] such as longer travel times, lower security level [3] and increased fuel consumption.

A. Adnane(✉) · M.S. Lmimouni · M. Rezzai · H. Medromi
Systems Architecture Team, Laboratory of Computer System and Renewable Energy, University Hassan II Casablanca, ENSEM, Casablanca, Morocco
e-mail: {ah.adnane,mohamed.salim.lmimouni,maha.rezzai}@gmail.com, hmedromi@yahoo.fr

E. Sabir et al. (eds.), *The International Symposium on Ubiquitous Networking*, Lecture Notes in Electrical Engineering 366,
DOI: 10.1007/978-981-287-990-5_19

In recent years, a lot of research has been done on the issue of congestion detection. Most of these studies rely on VANET (Vehicular Ad-hoc NETwork) approaches which mainly include two types of communication: V2I (vehicle-to-infrastructure) communication and V2V (vehicle-to-vehicle) communication. Therefore, vehicles are seen as mobile sensors which gather and disseminate traffic information such as speed, location, direction, etc. within their transmission range. Although borrowing WMSNs (Wireless Mobile Sensor Networks) techniques to V2V communication seems to be attractive, V2V communication suffers from many network issues such as bandwidth saturation and redundant data [4], late and wrong congestion level evaluation [5], and reliability problems [6].

In this paper, we propose a novel cost-effective congestion manager protocol which aims to predict and prevent congestion situations before they happen. In our approach, direct communication between vehicles does not exist anymore. Otherwise, vehicles coordinate and share traffic reports through indirect communication. This principle, which takes inspiration from social insects, considers the environment as a medium through which communications are performed. In fact, through RSU (Road Side Unit) controllers installed in traffic control signals, vehicles share their experiences and learning about road segments' congestion levels. In this way, the whole system converges to the real-time traffic map of the studied area. Equipped with a navigation system and a digital map, drivers are notified about road segments where congestion levels are reported to be high to avoid them. Hence, high level congestions are prevented and a smooth vehicles distribution is guaranteed.

This paper tries to expose theoretical background behind our approach, mainly focused on the calculation formula for detecting and evaluating congestion situations, and algorithms used to optimize and share traffic reports between vehicles and RSU controllers. Further work related to performance evaluation is under study and will be published in the near future.

The remainder of this paper is organized as follows: Section 2 highlights related work on the subject of traffic congestion. The vehicle controller device is discussed in Section 3. The latter is followed by a detailed description of the traffic lights RSU controller in section 4. Section 5 tackles the vehicle congestion manager algorithm. Section 6 focuses on the RSU congestion manager algorithm. Finally, Section 7 is dedicated to conclusion and future work.

2 Related Work

Communication between vehicles has been studied by many researchers in order to reduce traffic congestion. Varaiya [7] introduced a concept called 'platoon' where vehicles have sufficient intelligence to keep a minimum safe distance to the vehicle in front. The advantage of this approach is that the use of the road is improved, but this concept cannot prevent or unveil high level congestion segments.

In addition, Mamei and al [8] have conceived a navigation aid algorithm. Vehicles autonomously calculate less congested paths to their destinations based on traffic information provided by RSUs. Besides, Shah and al [2] discuss the vehicle scheduling problem. It tries to schedule vehicles trips so that the average travel time is minimized. However, both approaches are not adapted to high traffic density scenarios.

Moreover, Dimitrakopoulos and al [3] introduce cognitive management functionalities based on V2V communication using WMSNs mechanisms. The goal of the approach is to issue directives to drivers as well as to the overall transportation infrastructure to prevent congestion situations. As mentioned in the introduction, V2V communication causes many network problems such as bandwidth saturation and redundant data problem. In addition, its high setup cost makes it difficult to enter markets.

Finally, Bauza and al [9] present and evaluate a technique called CoTEC (CoOperative Traffic congestion detection) based on V2V communication. Although results are interesting in case of high rates of V2V penetration, CoTEC suffers from the same V2V communication drawbacks cited above.

3 Vehicle Controller Device

Each vehicle within the studied area is equipped with a controller device which allows two main functions: 1) visualizing high level traffic congestion segments in a road map to the driver and 2) performing calculations related to its congestion manager algorithm (see next section).

Therefore, the vehicle controller device consists of the following components:

- A GPS receiver, a digital map, and a navigation system for the first function,
- An IEEE802.11 transceiver, ultrasonic sensors and an embedded system for the second function.

The vehicle controller monitors its environment to predict any potential future congestion. To do so, it continuously calculates the congestion level of its location based on ultrasonic sensors installed in the four sides of the vehicle, as shown in Fig. 1. The sensors' task is to detect objects and measure the distance to them.

The congestion level (CL) is calculated based on distances generated by the sensors. The calculation formula is established as the weighted average of distances delivered by sensors. Weights are attributed to sensors depending on their installation position and thus on their influence in determining a congestion situation.

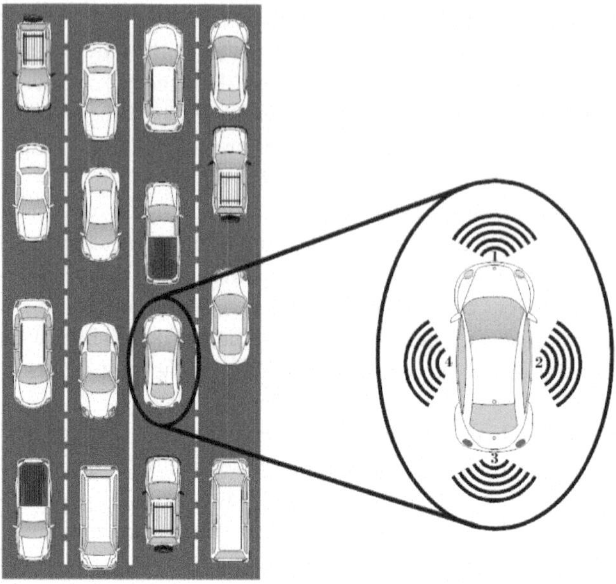

Fig. 1 Vehicle ultrasonic sensors

Let's consider a vehicle equipped with n ultrasonic sensors ($US_0, \ldots, US_i, \ldots, US_n$).

Since the environment is moving, it wouldn't be of great benefit to calculate instantaneous distance to any encountered obstacle. Instead, we define a damping time parameter called DT (Damping Time) during which average distances are calculated. If we set the DT to 60s for instance, then an US_i will continuously calculate the average distance to encountered obstacles from current time to last 60 seconds. The CL function is defined as follows:

$$CL(t) = \frac{1}{\sum_{i=0}^{n} \propto_i} \sum_{i=0}^{n} \propto_i d_{DT}(t, i)$$

Where $d_{DT}(t, i)$ is the average distance returned by the US_i from t time to last DT seconds and $\propto_i$ is the weight attributed to the US_i. Note that d_{DT} function has an upper limit d_{max} and a lower limit $d_{min} > 0$.

In practice, the DT parameter as well as USs' weights are determined with regard to the road network wherein the proposed approach will be deployed: in downtown scenarios, the DT is defined bigger than it is defined in highways scenarios since the average distance travelled during a DT is smaller in the downtown than it is in the highway, remember that a congestion level is reported according to a location, at most to a limited road segment (see Section 5).

Moreover, left-hand traffic (LHT) or right-hand traffic (RHT) scenarios directly affect USs weights. Since USs placed on the driver side are the most efficient sensors in terms of unveiling congestion situations, they should be given the biggest weights. Note that the passenger side road is the first to be congested.

In which case overtaking may be considered, but if the driver side road is not free as well, even overtaking would not possible, thus congestion level is more likely to be within the highest rates.

To illustrate the idea behind our work, let's consider a left-hand drive vehicle with 4 USs in a downtown scenario, each sensor is placed in the middle of a different side (See Fig. 1). The left sensor is more unveiling than the right one and the rear sensor is more unveiling than front one. So, according to this rule, USs weights could be defined as follows: weights 2 are given to the left and rear sensors, and weights 1 are given to the front and right sensors. We can set d_{max} to 5 meters, d_{min} to 1 meter and the damping time to 60s. USs are indexed as shown in Fig. 1. Therefore, the congestion level function becomes:

$$CL(t) = \frac{1}{6}(d_{60}(t,1) + d_{60}(t,2) + 2d_{60}(t,3) + 2d_{60}(t,4))$$

According to this function, we can define typical levels of congestion:

- High: nearby obstacles are detected at least at 3 sides: the front, the left and the rear sides, so the calculated threshold is expressed as: $CLT_H = 1/(\frac{1+5+2+2}{6}) = 0.6$. Therefore, congestion levels >=0.6 are considered to be high.
- Medium: nearby obstacles are detected at least at 2 sides: the front and the rear sides, so the calculated threshold is expressed as: $CLT_M = 1/(\frac{1+5+2+2*5}{6}) \approx 0.33$. Therefore, congestion levels within [0.33, 0.6[interval are considered to be medium.
- Low: nearby obstacles are detected at least at 1 side: the front one, so the calculated threshold is expressed as: $CLT_L = 1/(\frac{1+5+2*5+2*5}{6}) \approx 0.23$. Therefore, congestion levels within [0.23, 0.33[interval are considered to be low.

Other rules in expressing typical congestion levels or attributing sensors' weights may be considered as well. The purpose of the above-cited example is to clarify and show the flexibility potential of the approach.

In the remainder of this article, a location is considered to be relatively congested if the CL exceeds a predefined threshold called CLT (Congestion Level Threshold).

4 Traffic Lights RSU Controller

RSU controllers are installed in traffic lights. Their role is to offer a communication medium to vehicle controllers in order to share knowledge about road segments' traffic levels (see Fig. 2).

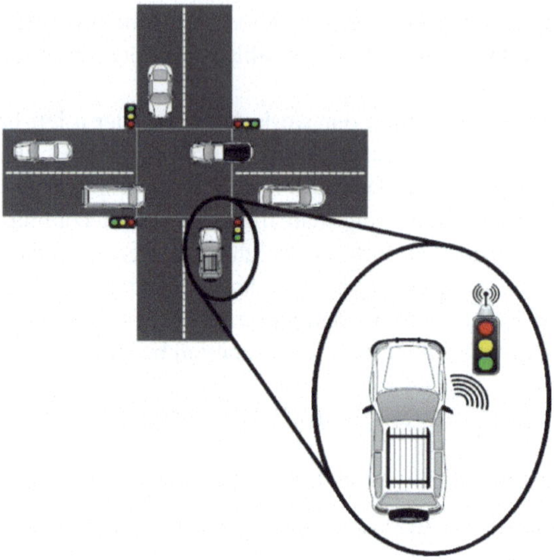

Fig. 2 Vehicle/RSU synchronization

RSUs are composed of an IEEE802.11 transceiver and an embedded system. When a vehicle is within the transmission range of a RSU, it established a connection to it and synchronizes its KB (knowledge base) with the RSU shared memory. The latter contains processed data about every vehicle KB which has run within its transmission range. In this way, vehicle controllers KBs are kept updated about the whole traffic map of the studied area.

5 Vehicle Congestion Manager Algorithm

A vehicle controller monitors its environment for two main events: 1) CLT reached and 2) a RSU detected, as shown in Fig. 3.

If the CLT is reached, which means that the vehicle's location is relatively congested and more likely to get congested in the near future, then the controller retrieves its GPS location and current datetime. This information is passed to the VCB (Vehicle Congestion Boxer) function to be processed, see algorithm1.

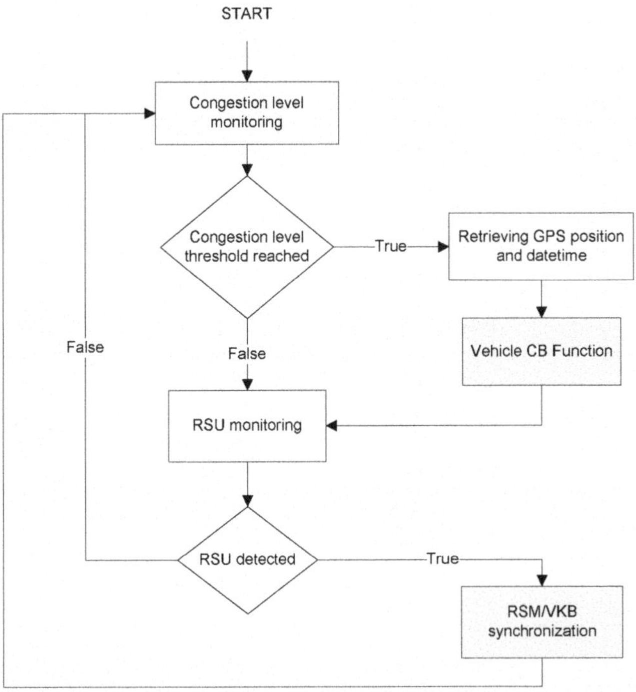

Fig. 3 Vehicle congestion manager flowchart

```
Algorithm 1. Vehicle CB function
```

```
Function VCB(position, datetime)
Begin
  iBox←generateInitialBoxPolygon(position)
  for each entry in VKB Do
    if isObsolete(entry) then
      VKB.delete(entry)
    else
      if areMergeable(iBox,entry.box) then
        entry.box←mergeBoxes(iBox,entry.box)
        entry.datetime←datetime
        return
      end if
    end if
  end for
  VKB.add(new Entry(iBox,datetime))
End Function
```

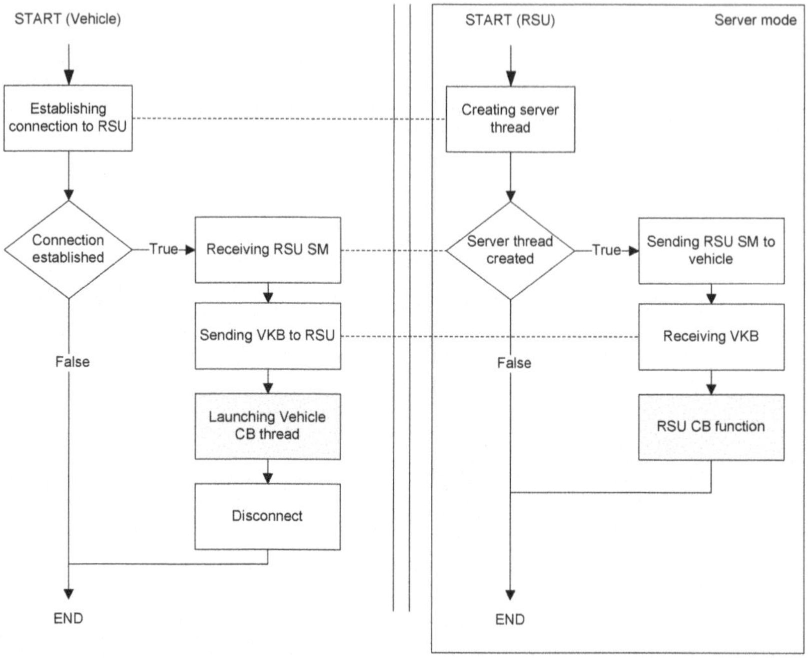

Fig. 4 Vehicle/RSU synchronization flowchart

A main drawback of several contributions in this field is redundant location data which rapidly turns into flooding bandwidth if it is not carefully investigated. To overcome such a problem, we have come up with a new concept which allows storing a huge amount of location data in small-sized blocks. The idea behind is that a geographically bounded area, which can be stored as locations of its edges, contains infinite locations which cannot be stored any way. So, instead of storing locations, storing areas is a much more efficient alternative.

So, the VCB function consists in generating an initial box including the GPS location. A box is a geographical rectangle which represents a relatively congested segment and whose size may grow up to cover other boxes in its vicinity. In this way, storing space is optimized by updating existing boxes instead of storing new ones.

So, after generating an initial box, the VCB function brows all entries within the vehicle KB. A KB contains information about vehicles experiences, i.e. high level congestion segments. It is composed of entries; each entry contains a box which represents a relatively congested road segment combined with the most recent datetime the congestion level was reported.

While browsing entries of the KB, the VCB function checks each entry for obsolescence. If it has been some time limit since the entry (mainly its datetime) was not updated, it is considered to be obsolete and removed from the KB. If it is not obsolete, then it is compared with the initial box through the areMergeable function. The latter consists in checking if the initial box can merge with an existing box. Cases where two boxes can merge into one box are the followings:

- Either one box includes or intersects with the other
- The two boxes are geographically close

If the two boxes meet one of the above conditions they are merged into one box. The latter overwrites the former box in the KB and its datetime is updated. If they cannot merge, then the function iterates to the next entry and repeats the same logic. After browsing all entries and no convenient box is found, the function creates a new entry in the KB where it stores the initial box.

If a RSU is detected, then the vehicle controller connects to it and launches a synchronization process as described in Fig. 4.

The synchronization process consists in synchronizing the RSM (RSU Shared Memory) and the VKB (Vehicle KB).

The RSM is a memory contained in the RSU which contains information about high level congestion segments. It is kept updated by vehicles as they drive within its transmission range. The RSM has the same data structure as the VKB.

So the synchronization process succeeds, connection time between the vehicle controller and the RSU should be as short as possible. This is why data processing is done asynchronously by both units after exchanging their KBs (see figure 4).

When a vehicle controller receives a RSM, it launches a CB (Congestion Boxer) thread as described in algorithm 2.

Algorithm 2. Vehicle CB thread

```
Thread VCB (RSM)
Begin
  for each rEntry in RSM do
    flag←true
    for each vEntry in VKB do
      if areMergeable(rEntry.box,vEntry.box) then
        vEntry.box←mergeBoxes(rEntry.box,vEntry.box)
        vEntry.datetime←(vEntry.datetime>rEntry.datetime)?
vEntry.datetime:rEntry.datetime
        flag ← false
        break
      end if
    end for
    if flag then
      VKB.add(rEntry)
    end if
  end for
End Thread
```

The CB thread consists in updating the vehicle KB based on the received RSM. It brows all entries within the RSM and checks whether they can be merged with any existing entries in the vehicle KB. If no entries are convenient then the concerned entries are added to the vehicle KB. It should be noted that this treatment may be time consuming; this is why it is performed by a background thread.

6 RSU Congestion Manager Algorithm

The RSU controller serves as a server to vehicles' controllers. When a vehicle controller is within the transmission range of a RSU, it tries to connect to it. Therefore, the RSU creates a server thread to perform all operations requested by the vehicle controller. When receiving a vehicle KB, the RSU server thread updates the RSM based on the received KB through the RCB (RSU Congestion Boxer) function (see algorithm 3).

```
Algorithm 3. RSU CB function

Function RCB (VKB)
Begin
  writeLock(RSM)
  for each vEntry in VKB Do
    flag←true
    for each rEntry in RSM Do
      if areMergeable(vEntry.box,rEntry.box) then
        rEntry.box←mergeBoxes(vEntry.box, rEntry.box)
        rEntry.datetime←(vEntry.datetime>rEntry.datetime)?
vEntry.datetime: rEntry.datetime
        flag←false
        break
      end if
    end for
    if flag then
      RSM.add(vEntry)
    end if
  end for
  releaseWriteLock(RSM)
End Function
```

The mechanism of synchronization is the same as that of the vehicle controller, except for one feature: during the updating process, the RCB function locks the RSM for writing. In this way, data coherence is guaranteed and concurrent accesses are managed. Vehicle controllers can read the RSM content but cannot update it until the lock for writing is released.

While the RSU doesn't exceed a certain predefined threshold based on the number of simultaneously connected vehicles, it periodically launches the CSB (Congestion Self Boxer) background thread as shown in Fig. 5.

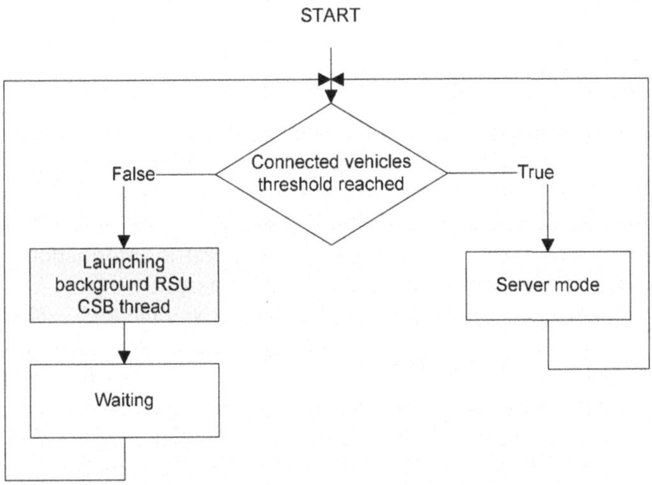

Fig. 5 RSU congestion manager flowchart

```
Algorithm 4. RSU self-congestion boxer thread
```

```
Thread RSCB()
Begin
  writeLock(RSM)
    for each entry in RSM Do
      if isObsolete(entry) then
        RSM.delete(entry)
      end if
    end for
  for each r1Entry in RSM Do
    for each r2Entry in RSM Do
      if areMergeable(r1Entry.box,r2Entry.box) then
        r1Entry.box←mergeBoxes(r1Entry.box, r2Entry.box)
        r1Entry.datetime←(r1Entry.datetime>
r2Entry.datetime)? r1Entry.datetime: r2Entry.datetime
        RSM.delete(r2Entry)
        break
      end if
    end for
  end for
  releaseWriteLock(RSM)
End Thread
```

The role of the CSB thread is to maintain the RSM optimal and cleaned from any obsolete entries. Therefore, it browses all RSM entries looking for any possible merging between them (see algorithm 4). In this way, space storage is optimized by reducing the number of RSM entries as well as its processing time during synchronizations.

7 Conclusion

The purpose of this article is to propose an efficient cost-effective approach which aims to avoid high traffic congestion situations. V2V communication, which faces many technical and cost-related challenges, is abandoned. Instead, the proposed approach employs indirect communication between vehicles. In fact, RSU controllers placed in traffic lights are used to exchange traffic information reports between vehicles and to synchronize their knowledge bases.

A key problem facing such an approach is data storage and exchange which theoretically should be optimized as much as possible. Therefore, optimization algorithms based on computational geometry are proposed and described.

Interactions between vehicles' controllers through the environment give birth to a collective intelligence which notifies drivers about congestion levels of road segments.

The proposed approach is still under investigation. Simulation results and performance evaluation in NS-2 platform will be published in the near future.

References

1. Rezzai, M., Dachry, W., Moutaouakkil, F., Medromi, H.: Designing an intelligent system for traffic management. In: Presented at the Mediterranean Conference on Information & Communication Technologies, Saadia, Morocco (2015)
2. Shah, N., Kumar, S., Bastani, F., Yen, I.-L.: Optimization models for assessing the peak capacity utilization of intelligent transportation systems. Eur. J. Oper. Res. **216**(1), 239–251 (2012)
3. Dimitrakopoulos, G., Demestichas, P.: Intelligent Transportation Systems. IEEE Veh. Technol. Mag. **5**(1), 77–84 (2010)
4. Schunemann, B., Wedel, J., Radusch, I.: V2x-based traffic congestion recognition and avoidance. Tamkang J. Sci. Eng. **13**(1), 63–70 (2010)
5. Zhong, T., Xu, B., Wolfson, O.: Disseminating real-time traffic information in vehicular ad-hoc networks. In: IEEE Intelligent Vehicles Symposium, pp. 1056–1061 (2008)
6. Korkmaz, G., Ekici, E., Özgüner, F., Özgüner, Ü.: Urban multi-hop broadcast protocol for inter-vehicle communication systems. In: Proceedings of the 1st ACM International Workshop on Vehicular Ad Hoc Networks, pp. 76–85 (2004)
7. Varaiya, P.: Smart cars on smart roads: problems of control. IEEE Trans. Autom. Control **38**(2), 195–207 (1993)
8. Mamei, M., Zambonelli, F., Leonardi, L.: Distributed motion coordination with Co-Fields: a case study in urban traffic management. IEEE Comput. Soc., 63–70 (2003)
9. Bauza, R., Gozalvez, J.: Traffic congestion detection in large-scale scenarios using vehicle-to-vehicle communications. J. Netw. Comput. Appl. **36**(5), 1295–1307 (2013)

Automated SLA Negotiation: A Novel Approach for Optimizing Cloud Data Overload

Latifa Maftahi, Said Rakrak and Said Raghay

Abstract Cloud computing is a set of processes in which a large group of remote servers communicating together so as to allow sharing data-processing tasks, data storage and online access to compute resources. Managing the drastically increasing number of requests presents the most challenging part of cloud computing. This paper discusses a novel hierarchy for optimizing load to achieve a high user satisfaction. By preserving data about the best cloud service providers, this hierarchy enhances the end user's satisfaction. In other words, as it is proposed in this hierarchy, suppliers with a maximum number of satisfied requests are more likely to be selected as soon as they have available resources/services. On the other hand, when no provider is capable to accomplish the requested tasks, a possibility of cooperation between two or more cloud suppliers is presented as a second option to ensure 100% of user's satisfaction. In order to apply such an approach, a huge amount of resources are wasted when the in-between communication is taking place. Therefore, an optimized method of SLA contract establishment is tackled in this paper.

1 Introduction

As organizations, over the globe, have begun to share and store digital information that are accessed by the internet, it is more suitable for them to focus on their daily business goals without having to worry about the installation of their computer applications. Cloud computing [1], which consist of renting virtual capacity of resources, enables achieving business goals more efficiently than acquiring the needed physical equipment. However, the growing need of such a technology results in growing number of problems related to traffic load, users associated

L. Maftahi(✉) · S. Rakrak · S. Raghay
Applied Mathematics and Computer Science Laboratory,
Cadi Ayyad University, Marrakech, Morocco
e-mail: {latifa.maftahi,srakrak}@gmail.com, s.raghay@uca.ma

E. Sabir et al. (eds.), *The International Symposium on Ubiquitous Networking*,
Lecture Notes in Electrical Engineering 366,
DOI: 10.1007/978-981-287-990-5_20

requirements and privilege management, and synchronization. Hence, contemporary research papers become more deviated to discussing different techniques to solve the preciously introduced problems related to Cloud Computing.

The general concept of our research focuses on the management of Quality of Service (QoS) [2][3] and Service Level Agreement (SLA) [4]. Given the need to ensure a reliable and efficient interaction between the composite services the concept of service contracts or service level agreement has taken importance. In cloud computing environments, the cloud provider and cloud consumer (enterprise, business, or government agency) need to establish an SLA. The SLA is a document written between two or more parties regarding service quality, responsibilities, and priorities. The main advantage of SLA is the understanding of the responsibilities of cloud customer and cloud provider. Before writing an SLA, both sides negotiate each term of the SLA and finalize it. In addition, the challenge of cloud provider is to ensure that the capacity of its resources used responds to the demand of its user's continuously during the threshold of requests. In other words, cloud provider searches to maintain the QoS in order to guarantee the SLAs. While it is important to design a cloud computing framework, that facilitates the automation of SLA negotiation [5].

In such context, cloud computing meets several difficulties about load management. Data overload is one of the major of these problems. This cloud receives millions providers requests of services [6] [7], in the same time, implying the need for a management of data overload.

Hence, this paper proposes a cloud computing framework to optimize cloud data overload. Using the proposed system a cloud broker [8] can establish the SLA regarding consumer preferences and provider resources through an automated SLA negotiation scheme and a coordinator can facilitate automatic selection of load between providers. The coordinator takes into account the resource allocation or resource provisioning and schedules the task in distributed environment. The proposed framework allows reducing request overloaded. Also, it enables to cooperate between two or more providers for the purpose to reduce SLA violation. Moreover, the technique enables to manage and negotiate SLAs with provider on behalf of the client without human intervention.

The rest of this paper is organized as follow: Section 2 gives a review of related work. Section 3 presents our architecture proposal and provide a use case based on a real world cloud usage scenario. In section 4 we present descriptive example based game theory. We conclude the paper with conclusion and future work of our proposal.

2 Related Work

Various data overload approach has been proposed for cloud computing to provide efficient distribution of load among several service providers. Actually, in what concerns calibration and optimization methods, the following papers discuss the main ideas to overcome the already introduced problems.

In [9], Yousri Kouki proposed a CSLA calibration solution that solves the problem of SLA dependencies among various Cloud layers XaaS. Calibrate SLA template level SaaS depends on the resources allocated to the IaaS provider (resource contract). The SLAs may be expressed at all levels. We are talking about intra-dependence when the SLA belongs to the same layer n and for different level SLAs we talking about inter-dependence. SaaS service caliber and negotiates the agreements (SLAs) with customers on the basis of the contract (SLAr) with its supplier IaaS. The objective of SaaS is twofold, on the one hand, the generation of template SLA parameters to ensure that de quality of service guarantees offered are achievable compared to SLAr. On the other hand, the selection of type/number optimum resource for the template generated. The goal is to offer customers profitable SLAs. The proposed solution is exclusively operated by human operators (domain expert). We observe the absence of an automatic solution to define SLA between both contracted parts.

In [8], Patel proposed a novel approach of VM provisioning method to improve the profit and SLA violation of cloud service provider. This approach aims to allocate VM in order to balance the load among multiple datacenters in a federated cloud environment. The approach focused on both 1) users to reduce SLA violation and 2) cloud providers to balance the load. The increase of resource allocation and load balancing among the datacenter is the disadvantage of this approach.

In [10], Gulshan presented a Central Load Balancer (CLB) technique which tried to avoid the situation of ever loading and under loading of virtual machines. This technique is efficiently shares the load of user requests among various virtual machines. The central load balancer calculates the priorities of virtual machines based on their memory and CPU speed. But load distribution is still less dynamic and robust.

Benay, et al. in [5] presented a middleware based service broker architecture where his aim is to select proper resources for a given task and manage SLA between Service Provider (SP) and Service Consumer (SC). They have considered, service broker as a third party which deals with service provider on behalf of service consumer. The SC and SP can negotiate to draft SLA document after selection of compatible resource for a given task. In the context of cloud computing, current research provides a limited support for automated SLA negotiation. There has not been much work done in the automation negotiation between SC and SP in cloud computing. Our work is a new step in this direction.

3 Architecture

Figure 2 shows an overview of our proposed framework (automated SLA negotiation) that cooperates between two or more suppliers in order to perfectly put the requested services in the hands of the customer. It is a method that satisfies all consumers' expectations, regarding the insurance and quality of service, from the service provider perspective; therefore, an automatic selection and synchronization between providers should take place. In other words, the negotiation process

[6][11] will undergo some important balancing algorithms so as for the network load to get reduced as much as possible. In fact, this method has the following essential characteristics such as the way it reduces traffic load, decreases the peak of demands, and reaches clients' satisfaction with him/her sending one single request even in case of multiple providers' cooperation.

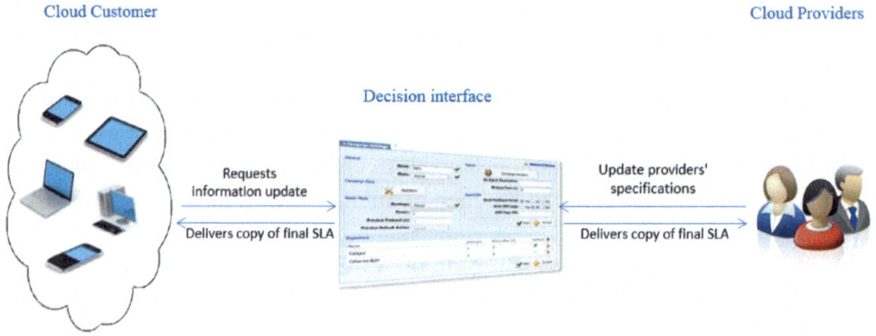

Fig. 1 Framework overview

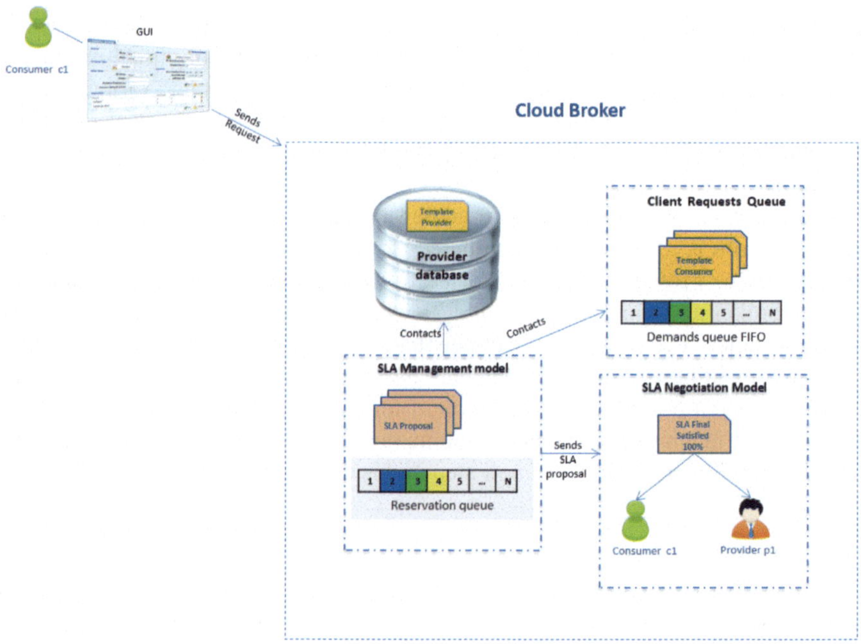

Fig. 2 Cloud broker architecture

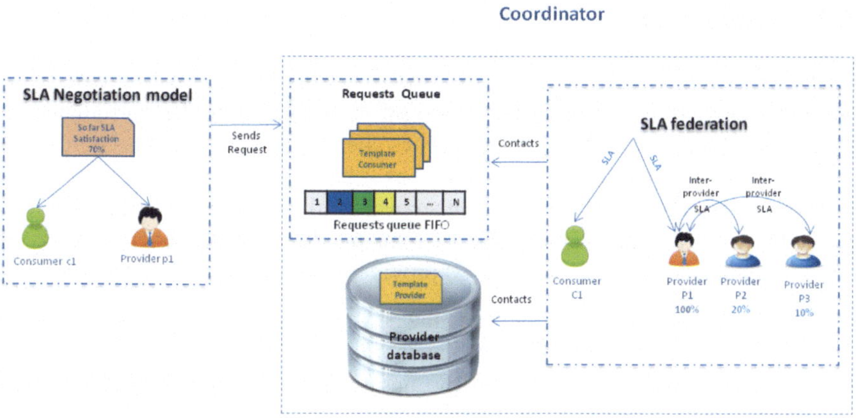

Fig. 3 Coordinator architecture

The process of selecting the most adequate parties in order to satisfy the incoming requests can be classified in two categories. One case (fig. 2) is when a request that can be fully satisfied by one single cloud service provider. The other case (fig. 3) is regarding requests that require cooperation between multiple providers. Consequently, our solution requires a third party that works as a decision point to link between the requesting client and one or multiple suppliers depending on its case. The main components of our framework are as follow:

3.1 Cloud Broker

As shown in fig. 2, the cloud broker provides a Graphical User Interface (GUI) to the client that hides the complexity and allows them to specify in a very simplified manner the required cloud services. It consists of four components: 1) provider database, 2) client requests Queue, 3) SLA manager, 4) SLA negotiation model.

- The provider database contains provider's template describing the service and its possible terms including the QoS. This template leaves several fields blank or modifiable in order to maintain the user specific needs.
- The client requests queue Fetches the template and fills it in with values which describe the planned resource usage. Some terms of the template may be modified (added, removed or changed).
- The SLA manager is based on the current resource availability (provider template) of the provider and customer policies (customer template), the SLA manager create SLA and sends it to SLA negotiation as a SLA proposal.
- The SLA negotiation model receives the SLA proposal. If the proposal accepted, it becomes an SLA officially between both parties (C1 and P1), and start to be a valid legal document.

3.2 Coordinator

As shown in figure 3, The coordinator precedes width cloud broker in case when the request cannot be fulfill by one cloud provider. Therefore the coordinator reduces cloud broker's traffic load and reaches client's satisfaction. This coordinator can be also considered as a second broker as it has the same role, but only works in case a cooperation is needed. In other words, the coordinator belongs to the regulator entity and has a main role of relaxing the main broker. It consists of three components: 1) requests queue, 2) provider database, 3) SLA federation.

- The SLA negotiation model sends the SLA of the selected provider with the biggest satisfaction percentage to the coordinator.
- The request queue reformulates the unsatisfied part of the SLA it receives from the SLA negotiation.
- The provider database as described in cloud broker, it contains provider's template. The coordinator proceeds with them to choose the most adequate providers for the request.
- The SLA federation assigns some satisfaction percentage for each cooperated providers and creates the inter-provider SLA between them defining the task achieved. In addition, the coordinator establishes the final SLA and sends copy to the concerned parties (C1 and P1).

In figure 4, we present our activity diagram. In this diagram we have explained in detail the interaction between the client and the cloud provider already participated in the negotiation process. Negotiation stages of our framework are as follows:

- The client creates his own request by describing the planned resource usage. Whereas it is essential for both a consumer and a provider to define its service level (price, execution time or response time) so that they can reach the mutual agreed level of requirements.
- To satisfy the incoming request, the cloud broker asking for the provider that gives the provisional answer based on the current resource availability and customer policies. The selection of the most adequate parties can be classified in two categories. One case is when a request can be fully satisfied by one provider. In this case a copy of SLA contract created and sent to both of the parties.

  The second case is regarding request that require cooperation between several providers. The SLA of the selected provider with the biggest satisfaction percentage sent to the coordinator. After receiving an incomplete SLA, the coordinator rechecks the list of providers excluding the selected one and compares the most adequate providers for the rest. We have considering two providers who can cooperate with the selected provider. In fact, the coordinator is responsible for doing this repetitive task until the request is fully satisfied. Lastly, the coordinator establishes the final SLA and sends copy to the concerned parties.

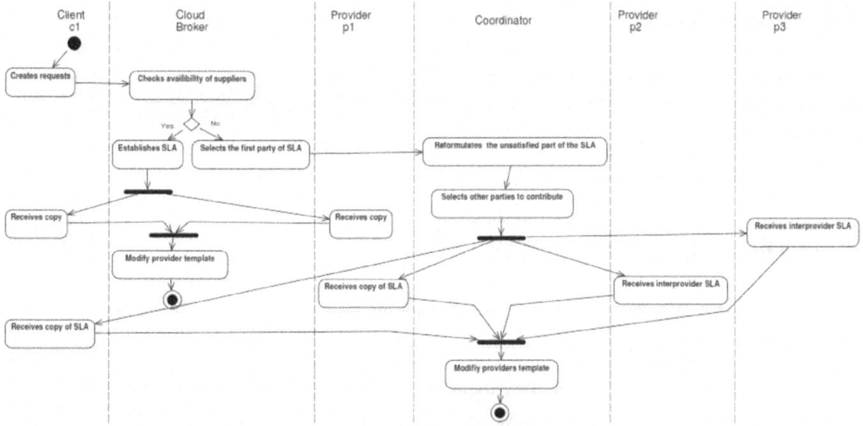

Fig. 4 The Interaction between client and provider during negotiation process

Table 1 Summary of intervening actors in our approach

Agent type	Function
Client/Customer	Requests and consumes services of cloud service provider.
Cloud service provider	Offers provisioned and metred computing resources (e.g, CPU, storage, memory, network), includes Infrastructure as-a-Service provider (IaaS), Platform as-a-Service provider (PaaS) and Software as-a-Service provider (SaaS)
Cloud broker	Manages SLAs between the user and the cloud service provider, negotiates SLA creation, handles fulfilement and violation, modify Provider information.
Coordinator	Cooperates between several providers, negotiates SLA creation, modify Provider information.

4 Descriptive Example

In this paper negotiation is viewed as common process between service provider and service consumer. Both parties often have a conflict of interest, it is difficult for them to reach an agreement. So cloud broker is used in negotiation, to minimize the complexity of the negotiation between them, his goal is to arrive at a mutually acceptable and beneficial agreement, and ensure there is no ambiguity between service provider and service consumer.

Cloud Broker still looking for a good service quality of service, therefore in case of failure to achieve users requirement, a cooperation between providers is needed to ensure 100% of user`s satisfaction. So game theory would be a very attracting tool to analyze this kind of interaction.

4.1 Game Theory Overview

Game theory is a branch of mathematical analysis developed to study decision making strategic situations in order to select the optimal strategy. This theory is a matrix representation of simultaneous set of one or more players of game theory. It consists of a set of players, and a set of strategies.

4.2 Description of the Game

A negotiation process has to be done between two or more parties. Our example may be have n-players in game theory where

P = $\{p_1, p_2, p_3, \ldots\ldots\ldots\}$ is a set of players.
S = $s_1 * s_2 * \ldots\ldots\ldots s_n$ is a set of strategy profiles defined by players

- Players:
  - p_1 is presented as principal provider which satisfy fully the task or the biggest percentage of the task.
  - $p_2 \ldots p_n$ the most adequate providers selected by the coordinator to accomplish rest of the task. Coordinator precedes width cloud broker in case when the task cannot be fulfill by one cloud provider.
- Set of strategies: coordinator tries to increase credibility of their suppliers. In case when no p_1 is capable to accomplish the requested tasks. The strategies are as follow:
  - s_1: Coordinator choose best providers p_n (has a good customer reputation and good quality of service) who could complete 25% of the requested task where p_1 satisfied 75%. In every task realization, coordinator note supplies and increase theirs percentages of credibility.
  - s_2: If no providers agree to participate in the cooperation, a draw of suppliers of same quality level would be mandatory and 20% of their credibility would be reduced.

One important part of our framework is that in any such type of request, the cloud broker and the coordinator are capable to fulfill the incoming request in order to perfectly put the requested service in the hands of the customer.

5 Conclusion and Future Work

As indicated in this paper, we see a very legitimate need for a methodology to handle data overload in the context of cloud computing for which this architecture is presented to enhance the end user's satisfaction. In our architecture we focused on 1) cloud broker to manage requests coming from the customers and to negotiate automatic SLA between concerned parties. In addition, we emphasized 2) coordinator to allow cooperation between two or more cloud services suppliers in case the task cannot be accomplished by a single provider.

In fact there are multiple paths to follow when improving this proposal such as doing some experimentation scenarios in order to evaluate our proposed architecture. Another one is to consider SLA violation penalties and how to apply them on the violating party. A third path is to implement a selection algorithm with a detailed list of criteria to enhance provider selection and percentage assignment. A fourth path can be to make decision whether the client is to be informed about the different participating providers at which level, broker or coordinator, to send the client a copy of SLA. In addition, the SLA can be reformed to enhance and guarantee the automatic SLA establishment. Finally the Cloud Broker can add another criteria to enhance the user`s satisfaction by submitting an email periodically after validation of SLA. This email will be the evaluation of each supplier`s quality of service.

References

1. Dialogic, Introduction to cloud computing (2010) (accessed February 2, 2015)
2. Praveen, S.P., Tulasi, U.: A Study on Qos Challenges in Cloud Computing. IJCC **2**(1) (2013)
3. Kritikos, K., et al.: A survey on service quality description. ACM Computing Surveys (CSUR) **46**(1), 1 (2013)
4. Kouki, Y.: Approche dirigée par les contrats de niveaux de service pour la gestion de l'élasticité du "nuage". Diss. Nantes, Ecole des Mines, 12–14 (2013)
5. Ray, B.K., Khatua, S., Roy, S.: Negotiation based service brokering using game theory. In: Applications and Innovations in Mobile Computing (AIMoC), pp. 1–8. IEEE, February 2014
6. Zulkernine, F.H., Martin, P.: An adaptive and intelligent SLA negotiation system for web services. IEEE Transactions on Services Computing **4**(1), 31–43 (2011)
7. Yan, J., et al.: Autonomous service level agreement negotiation for service composition provision. Future Generation Computer Systems **23**(6), 748–759 (2007)
8. Patel, K.S., Sarje, A.K.: VM provisioning method to improve the profit and SLA violation of cloud service providers. In: 2012 IEEE International Conference on Cloud Computing in Emerging Markets (CCEM). IEEE (2012)
9. Kouki, Y.: Approche dirigée par les contrats de niveaux de service pour la gestion de l'élasticité du "nuage". Diss. Nantes, Ecole des Mines, 96–104 (2013)
10. Soni, G., Kalra, M.: A novel approach for load balancing in cloud data center. In: 2014 IEEE International Advance Computing Conference (IACC), pp. 807–812. IEEE (2014)
11. Abdullah, R., Talib, A.M.: Towards integrating information of service level agreement and resources as a services (RaaS) for cloud computing environment. In: 2012 IEEE Conference on Open Systems (ICOS). IEEE (2012)

Frequency-Domain Analysis and Design of a TCP Flow Controller Using the AQM Mechanism

Khalid Lefrouni and Rachid Ellaia

Abstract In this article, we develop an Active Queue Management (AQM) mechanism for optimizing data traffic within the router, initially, we determine a linear model of TCP network around the operating point, secondly, based on a geometric method and taking into account the communication delay, we synthesize a state feedback controller acting on the probability of packet loss. Finally, we test the validity of our approach through simulations.

Keywords Frequency-Domain · Congestion control · TCP network · AQM mechanisms

1 Introduction

In this article, we are interested in the problem of router congestion within the communication network. In particular, we are focused on the implementation of a control law for optimizing data traffic in the TCP network.

Several studies have addressed this problem by using mainly the Active Queue Management (AQM) mechanisms, such as, the RED [1], the AVQ [2], however, the results were insufficient, especially when the number of network users is unpredictable and highly variable, it is the case, for example, of the Internet network.

In this context, many researchers have used the control theory, in particular, the state feedback control. For example the authors of [3,4] use a Lyapunov-Krasovskii functional, [5] use a Lyapunov-Razumikhin functional, others [6,7] consider less conservative conditions, by taking into consideration time-variable

K. Lefrouni(✉) · R. Ellaia
Mohammadia School of Engineers, Mohammed V University, BP. 765,
Ibn Sina avenue, Agdal, Rabat, Morocco
e-mail: {lefrouni,ellaia}@emi.ac.ma

E. Sabir et al. (eds.), *The International Symposium on Ubiquitous Networking*,
Lecture Notes in Electrical Engineering 366,
DOI: 10.1007/978-981-287-990-5_21

delays, the common point between these studies is that they rely on temporal criteria, also the simulation results are interesting, however the implementation is difficult.

In this paper, we also use the Active Queue Management (AQM) mechanism, through a state feedback controller acting on the probability of packet loss. However, we differ from previous research by using a geometric considerations (see [8,9,10]).

The remainder of this paper is structured as follows. In section 2, we present a mathematical model of TCP network. Then, in section 3, we expose the main results of this paper, which consists in developing a control law ensuring stabilization of the network and congestion avoidance. In Section 4, we validate the results obtained in the previous sections through simulations. Finally, we present our conclusions in section 5.

2 Mathematical Model of TCP Network

The control law developed in this article is based on a model found by analogy with fluid flow, first developed by Misra et al [11,12], this model is governed by the following equations:

$$\begin{cases} \dot{W}(t) = \dfrac{1}{R(t)} - \dfrac{W(t)W(t-R(t))}{2R(t-R(t))}p(t-R(t)) \\ \dot{q}(t) = \dfrac{W(t)}{R(t)}N(t) - C \\ R(t) = T_p + \dfrac{q(t)}{C} \end{cases} \tag{1}$$

Where R(t) is the Round-Trip Time (RTT), which is equal to the sum of the time taken by the packets to propagate in the communication lines T_p and the time required for processing and redirecting packets within the router $q(t)/C$, this latter quantity depends on the number of users, so it is variable in time.

After linearization around the operating point (W0, q0, p0) we find:

$$\begin{cases} \delta\dot{W}(t) = -\dfrac{2N}{R_0^2 C}\delta W(t) - \dfrac{R_0 C^2}{2N^2}\delta p(t-\tau_1(t)) \\ \delta\dot{q}(t) = \dfrac{N}{R_0}\delta W(t) - \dfrac{1}{R_0}\delta q(t) \end{cases} \tag{2}$$

with

$$W_0^2 p_0 = 2 \text{ and } W_0 = \frac{R_0 C}{N} \tag{3}$$

In the state-space, model (2) becomes:

$$\begin{cases} \dot{x}(t) &= A\,x(t)+B\,u(t-\tau_1(t)) \\ y(t) &= C\,x(t) \end{cases} \tag{4}$$

where $x(t)=[\delta W(t)\ \delta q(t)]^T$, $u(t)=p(t)$ and

$$A=\begin{bmatrix} a_1 & 0 \\ a_2 & a_3 \end{bmatrix},\ B=\begin{bmatrix} b_1 \\ 0 \end{bmatrix},\ C=\begin{bmatrix} 0 & 1 \end{bmatrix}$$

with

$$a_1=\frac{-2N}{R_0^2C},\quad a_2=\frac{N}{R_0},\quad a_3=\frac{-1}{R_0},\quad b_1=\frac{-R_0C^2}{2N^2}$$

3 Congestion Control: Geometric Considerations

Based on a geometric approach, we will develop in this section a state feedback control law acting on the packet loss rate and allowing the stabilization of the system (4).

3.1 State Feedback Control

The adopted control law has the following form:

$$u(t) = -K\,x(t) \tag{5}$$

With $K=[k_1\ \ k_2]$ is the state feedback gain, allowing to determine the rate of loss of packets $p(t)$, based on the internal state of the network $x(t)=[\delta W(t)\ \delta q(t)]^T$.

Introducing the Laplace transform, the equations (4) and (5) become:

$$\begin{cases} s\,x(s) &= A\,x(s)+B\,u(s)e^{-\tau_1 s} \\ y(s) &= C\,x(s) \\ u(s) &= -K\,x(s) \end{cases} \tag{6}$$

Therefore, the associated characteristic equation is given by:

$$H(s,k_1,\ k_2,\tau_1)=\det(sI_2-(A-BKe^{-\tau_1 s}))=0 \tag{7}$$

that can be written as:

$$H(s,k_1,\ k_2,\tau_1)=Q(s)+P(s)e^{-\tau_1 s}=0 \tag{8}$$

with
$$Q(s) = s^2 - (a_1 + a_3)s + a_1 a_3$$
$$P(s) = b_1(a_2 k_2 - a_3 k_1 + k_1 s) \tag{9}$$

A first phase of analysis of the system stability is to consider the ideal case corresponding to a zero-delay, in this case, we find the following equation:

$$s^2 - (a_1 + a_3 - b_1 k_1)s + a_1 a_3 + b_1(a_2 k_2 - a_3 k_1) = 0 \tag{10}$$

By applying the Routh criterion, we find the following conditions:

$$k_1 < \frac{a_1 + a_3}{b_1}, \quad k_2 < \frac{a_3}{a_2} k_1 - \frac{a_1 a_3}{b_1 a_2} \tag{11}$$

3.2 Analysis in the Controller Parameters Space

Now, Let us move to the real case, which corresponds to a non-zero communication delay. Indeed, the communication between the source and the router can not be instantaneous, it is at least equal to or greater than the signal propagation time in the transmission lines, which thus represents a lower limit of the delay τ_1.

In the same context, to determine the interval of delay values for which the stability of the system (4) is guaranteed, we will lead the system to the limit of the stability, which corresponds to the existence of a pure imaginary root of the characteristic equation (8), which translates into the following equation:

$$H(j\omega, k_1, k_2, \tau^*) = 0 \tag{12}$$

Which is equivalent to:

$$-\omega^2 - (a_1 + a_3) j\omega + a_1 a_3 + b_1(a_2 k_2 - a_3 k_1 + j\omega k_1) e^{-j\omega\tau^*} = 0$$

Separating the real and imaginary parts of this equation, we find

$$k_1 \alpha + k_2 a_2 b_1 \cos(\omega\tau^*) = \omega^2 - a_1 a_3$$
$$k_1 \beta - k_2 a_2 b_1 \sin(\omega\tau^*) = (a_1 + a_3)\omega$$

Thus, we find the following equations:

$$k_1 = \frac{a_1 + a_3}{\beta}\omega + \frac{\left[\omega^2 - \frac{\alpha}{\beta}(a_1 + a_3)\omega - a_1 a_3\right]\sin(\omega\tau^*)}{\alpha\sin(\omega\tau^*) + \beta\cos(\omega\tau^*)} \tag{13}$$

$$k_2 = \frac{\omega^2 - \frac{\alpha}{\beta}(a_1 + a_3)\omega - a_1 a_3}{a_2 b_1\left[\frac{\alpha}{\beta}\sin(\omega\tau^*) + \cos(\omega\tau^*)\right]} \tag{14}$$

with

$$\alpha = b_1 \omega \sin(\omega \tau^*) - b_1 a_3 \cos(\omega \tau^*)$$
$$\beta = b_1 a_3 \sin(\omega \tau^*) + \omega b_1 \cos(\omega \tau^*)$$

The two equations (13) and (14) define the set of pairs (k_1, k_2) which, by varying ω, allows us to find the crossings curves. So, we have find a second limit of the system stability regions in the space of controller parameters.

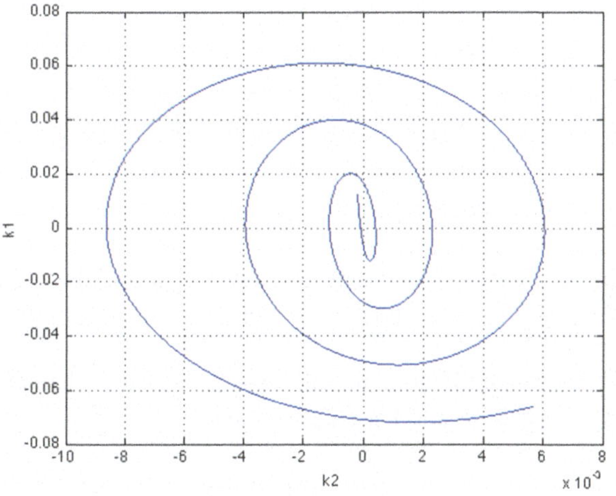

Fig. 1 Crossing curves in the (k_1, k_2) space with $\tau^* = 0.6$

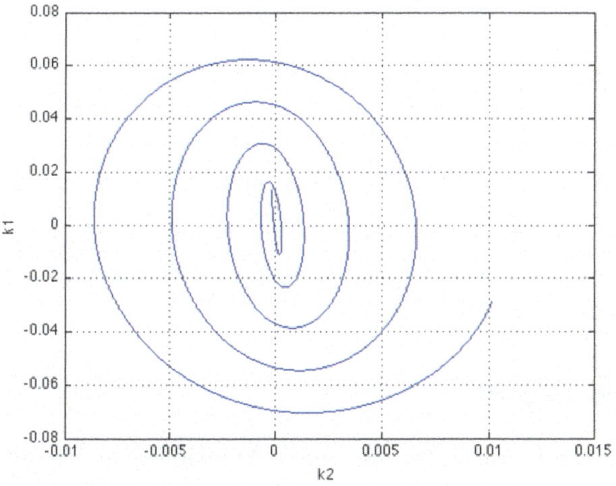

Fig. 2 Crossing curves in the (k_1, k_2) space with $\tau^* = 0.8$

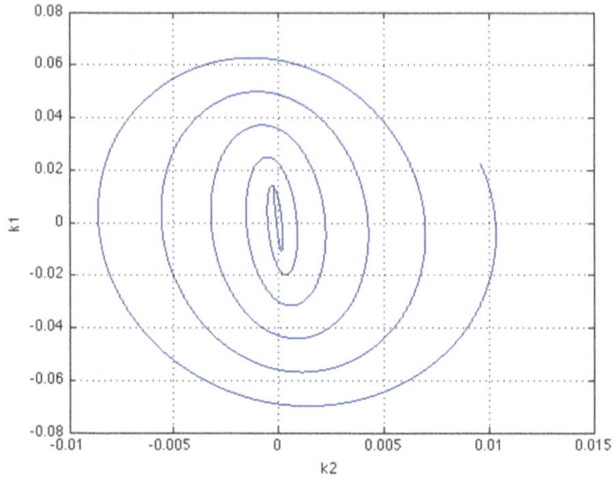

Fig. 3 Crossing curves in the (k_1, k_2) space with $\tau^* = 1$

In our case, using the same numerical values as in [3], and the following values of the delay {0.6. 0.8, 1}, we find respectively the crossings curves shown in Fig. 1, Fig. 2 and Fig. 3.

Now, we will determine the direction of crossing, which is characterized by the sign of $R_2I_1 - R_1I_2$, where R_i and I_i are defined by:

$$
\begin{aligned}
R_1 + jI_1 &= -\frac{1}{s}\frac{\partial H(s,k_1,k_2,\tau_1^*)}{\partial k_2}\bigg|_{s=j\omega} \\
R_2 + jI_2 &= -\frac{1}{s}\frac{\partial H(s,k_1,k_2,\tau_1^*)}{\partial k_1}\bigg|_{s=j\omega}
\end{aligned}
\tag{15}
$$

After determining the R_i and I_i, we find:

$$
R_2I_1 - R_1I_2 = \frac{-b_1^2 a_2}{\omega} < 0 \tag{16}
$$

Which meant that the crossing direction is to the left, i.e. when $(k_1,\ k_2)$ crosses the crossing curve from the left to the right, a pair of solution of (12) crosses the imaginary axis from the right to the left.

3.3 *Stability Region*

In this section, we will use all the results found previously to determine the stability region in the space defined by the controller parameters $(k_1,\ k_2)$.

So, we have demonstrated that the conditions (11) must be verified, and that the crossing direction is to the left. Thus, considering the crossing curves shown in Fig. 1, we find, in the case $\tau^* = 0.6$, the stability region shown in Fig. 4.

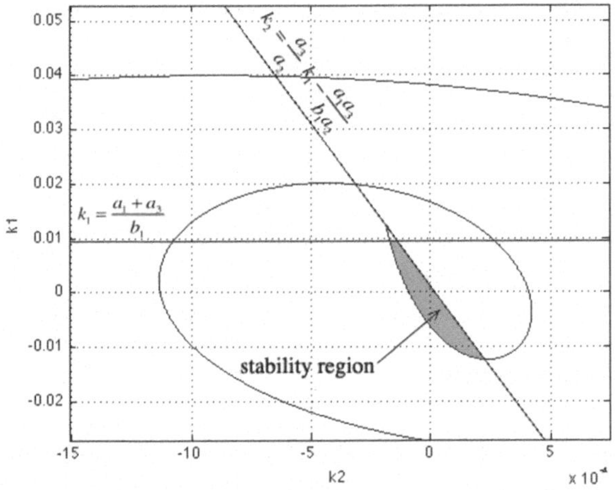

Fig. 4 Stability region in the (k_1, k_2) space with $\tau^* = 0.6$

Each point (k_1^*, k_2^*) belonging to this region defines a gain $K^* = [k_1^* \; k_2^*]$ for which the characteristic equation $H(s, k_1^*, k_2^*, \tau_1) = 0$ is Hurwitz for any delay $\tau_1 < \tau^*$.

Now, we will determine the critical delay τ^* beyond which, the selected pair (k_1^*, k_2^*) can no longer guarantee the stability of the system.

For this, we assume that the system is at the limit of the stability, i.e. the equation $H(s, k_1^*, k_2^*, \tau_1) = 0$ has a pure imaginary root $s = j\omega$, thus, we have:

$$Q(j\omega) + P(j\omega)e^{-j\omega\tau^*} = 0$$

with

$$P(j\omega) = b_1(a_2 k_2^* - a_3 k_1^* + j\omega k_1^*)$$
$$Q(j\omega) = -\omega^2 - (a_1 + a_3) j\omega + a_1 a_3$$

which is equivalent to

$$\cos(\omega\tau^*) - j\sin(\omega\tau^*) = -\frac{-\omega^2 - (a_1 + a_3) j\omega + a_1 a_3}{b_1(a_2 k_2^* - a_3 k_1^* + j\omega k_1^*)}$$

Taking the real parts of this equation gives:

$$\tau^* = \frac{1}{\omega} arc\cos\left[\frac{a_1(\omega^2 + a_3^2)k_1^* + a_2(\omega^2 - a_1a_3)k_2^*}{b_1(a_2k_2^* - a_3k_1^*)^2 + b_1(k_1^*\omega)^2}\right] \tag{17}$$

Similarly, by a direct resolution, we find the associated crossing frequency ω_0:

$$\omega_0 = \sqrt{\frac{\alpha_1 + \sqrt{\alpha_1^2 + 4\alpha_0}}{2}} \tag{18}$$

with

$$\alpha_0 = b_1^2(a_2k_2^* - a_3k_1^*)^2 - (a_1a_3)^2$$
$$\alpha_1 = (b_1k_1^*)^2 - a_1^2 - a_3^2$$

4 Simulation Results

In order to regulate the queue size of the router at q_0 = 175 *packets*, we consider the system presented in [3], where: T_p = 0.2 *s*, C = 3750 *packets/s* and for $N = 60$ TCP sessions, we have W_0 = 15 *packets*, p_0 = 0.008, R_0 = 0.246 and τ_1 = 0.6 s.

To test the validity of our approach, we will first choose three pairs (k_1^*, k_2^*) belonging to the stability region shown in Fig. 4. Then using the equations (17) and (18), we determine the critical delay and the associated crossing frequency. The results are shown in the following table.

Table 1 Critical delay and associated crossing frequency for each selected pair (k_1^*, k_2^*)

(k_1^*, k_2^*)	K^*	τ^*	ω_0
(-0.005, 0.00005)	10^{-3} [-5 0.05]	2.5187	0.9446
(-0.0075, 0.0001)	10^{-3} [-7.5 0.1]	2.9481	0.9215
(-0.01, 0.00016)	10^{-3} [-10 0.16]	1.1737	2.3580

Considering the pairs (k_1^*, k_2^*) indicated in the previous table, we obtained the temporal evolution of the queue, shown in Fig. 5, Fig. 6 and Fig. 7.

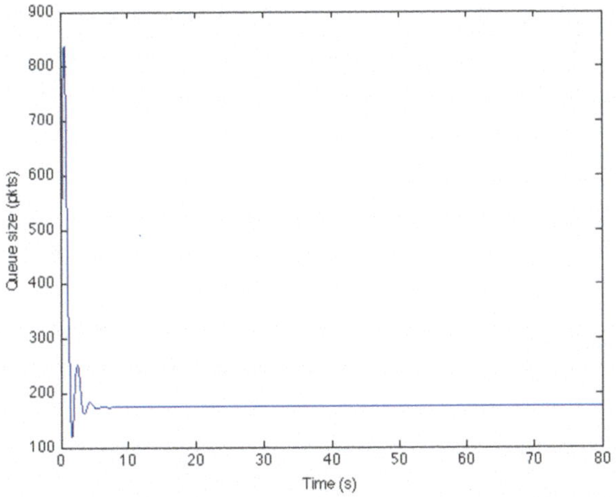

Fig. 5 Temporal evolution of the queue length with the state feedback gain $K^* = 10^{-3}$ [-5 0.05]

Analyzing the results, we see that all the selected state feedback gains can guarantee the system stability. However, we also note that an inappropriate choice can give rise to strong oscillations (see Fig. 7) and reduce the maximum allowable values τ^* of the delay τ_1 (see Table 1). Therefore, we recommend the development of an algorithm to ensure the desired behavior of the temporal evolution of the queue size, taking into account the requirements of the specifications.

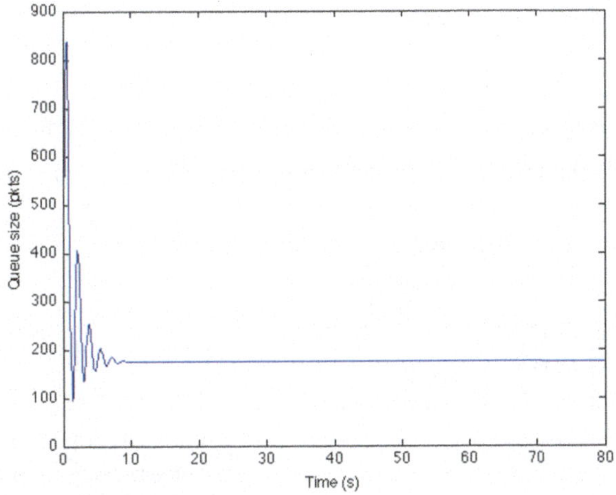

Fig. 6 Temporal evolution of the queue length with the state feedback gain $K^* = 10^{-3}$ [-7.5 0.1]

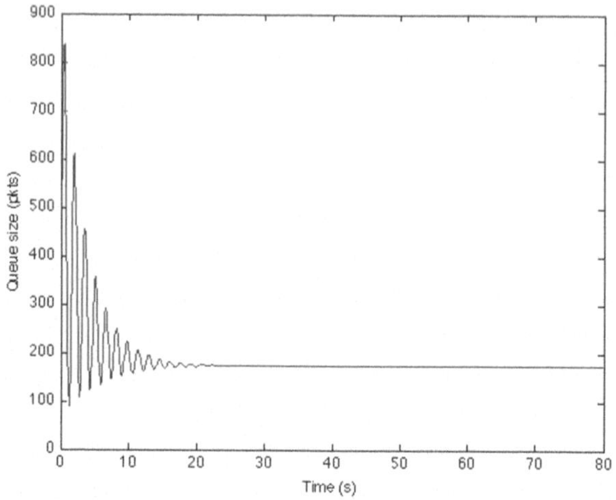

Fig. 7 Temporal evolution of the queue length with the state feedback gain $K^* = 10^{-3}$ [-10 0.16]

It should also be noted that the developed control law (5) is calculated from the system state $x(t)$, therefore, the effectiveness of this controller is conditioned by the exact knowledge of the system state. In practice, this state is not always accessible, and therefore it is necessary to implement an observer allowing both the reconstruction of the system state and the consideration of the measurement delay τ_2.

Under these conditions, the Luenberger observer defined by (19) is the most appropriate.

$$\begin{cases} \dot{\hat{x}}(t) = A\hat{x}(t) + Bu(t-\tau_1) + L\left[y(t-\tau_2) - \hat{y}(t-\tau_2)\right] \\ \hat{y}(t) = C\hat{x}(t) \end{cases} \tag{19}$$

Where $\hat{x}(t)$ is an estimate of the state vector $x(t)$ and L is the observer gain, ensuring a rapid convergence of the estimation error $e(t) = x(t) - \hat{x}(t)$ to zero.

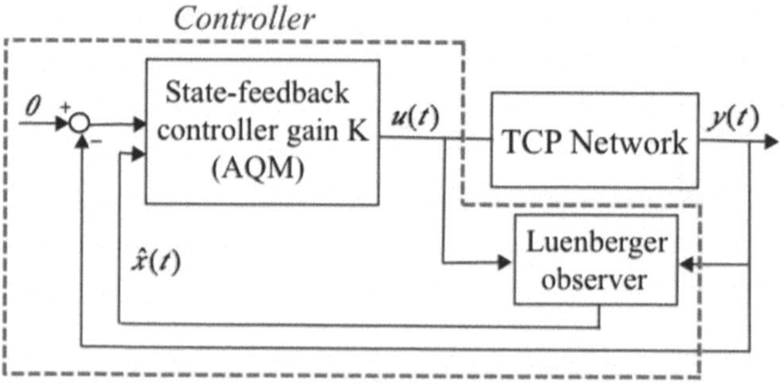

Fig. 8 The block diagram of the TCP/AQM Network

Based on the system model (4), we can therefore develop a monitoring mechanism, whose dynamics are governed by the equation (19). This mechanism, called in control theory: observer, will be implemented within the router, thus allowing to estimate, in real-time, the state of the system $x(t)$, i.e., the TCP window size $W(t)$ and the queue length $q(t)$. The figure Fig. 8 shows the closed loop structure of TCP/AQM network with the Lueunberger observer.

Remark 1. In comparison with classical algorithms such as Tahoe, SACK, New Reno or Vegas, which are based only on acknowledgments, the AQM mechanism anticipates the congestion and acts directly on the probability of packet loss $p(t)$, which allows to regulate the TCP flow and to avoid the strong oscillations in the routers queues. Thus, for efficient management of TCP traffic, it is recommended that all the routers are equipped with an algorithm based on the AQM.

5 Conclusion

In this paper we have proposed a geometric method for the synthesis of a state feedback controller in the frequency domain, also, through the control law, we have implemented an AQM mechanism acting on the probability of packet loss. Thus, as shown in section 4, this approach ensures the stabilization of the TCP flow with some flexibility in the choice of controller parameters. However, in practice, it is difficult to measure the state of the system, thus, as future work, we plan to increase the controller performance via the implementation of the Luenberger observer.

References

1. Floyd, S., Jacobson, V.: Random Early Detection gateways for congestion avoidance. IEEE/ACM Transactions on Networking **1**(4), 397–413 (1993)
2. Kunniyur, S., Srikant, R.: Analysis and design of an adaptive virtual queue (AVQ) algorithm for active queue management. In: Proceedings of ACM SIGCOMM, August, San Diego, CA, pp. 123–134 (2001)
3. Alami, N.E., Ellaia, R., Lefrouni, K.: State feedback controller and observer design for TCP network with consideration of UDP flow. Journal of Theoretical and Applied Information Technology **55**(1), 14 (2013)
4. Seuret, A., Michaut, F., Richard, J.-P., Divoux, T.: Networked control using gps synchronization. In: Proc. of ACC 2006, American Control Conf., June, Mineapolis, USA (2006)
5. Yu, M., Wang, L., Chu, T.: An LMI approach to network control systems with data packet dropout and transmission delays. In: MTNS 2004 Proc. of Mathematical Theory Networks and Systems, Leuven, Belgium (2004)
6. Witrant, E., Georges, D., Canudas-de-Wit, C., Alamir, M.: On the use of state predictors in networked control system. In: Applications of Time Delay Systems. Lecture Notes in Control and Information Sciences, vol. 352, pp. 17–35. Springer, New York (2007)

7. Ariba, Y., Labit, Y., Gouaisbaut, F.: Congestion control of a single router with an active queue management. International Journal on Advances in Internet Technology **2**(1) (2009)
8. Lefrouni, K., Ellaia, R.: State-feedback control in TCP network: Geometric approach. International Review of Automatic Control (IREACO) **8**(2) (2015)
9. Gu, K., Niculescu, S.-I., Chen, J.: On stability crossing curves for general systems with two delays. J. Math. Anal. Appl. **311**, 231–253 (2005)
10. Morărescu, C.-I., Barrios, C.-M., Niculescu, S.-I., Gu, K.: Stability crossing boundaries and fragility characterization of PID controllers for SISO systems with I/O delays. In: American Control Conference on O'Farrell Street, June 29-July 01, San Francisco, CA, USA (2011)
11. Hollot, C.V., Misra, V., Towsley, D., Gong, W.: A control theoretic analysis of red. In: IEEE INFOCOM, pp. 1510–1519, April 2001
12. Misra, V., Gong, W.-B., Towsley, D.: Fluid-based analysis of a nerwork of AQM routers supporting TCP flows with an application to red. In: Proc. ACM SIGCOMM, pp. 151–160 (2000)

Quality of Service Provision and Data Security in Communication Networks Based on Traffic Classification

Jeferson Wilian de Godoy Stênico and Lee Luan Ling

Abstract In this paper we present a new approach to admission control of new flows in communications networks. Initially, we introduce a traffic classification procedure based on a new construction approach of conservative multiplicative cascades, to ensure information security by blocking malicious attempts and creating priorities over well-intentioned traffic. The admission control process accepts or rejects then the request for a new call by evaluating the available bandwidth within a particular transmission interval for different pre-classified traffic types. The experimental investigation by simulation showed that the proposed method is capable of ensuring the efficient use of available network resources and at the same time providing consistent QoS provisions and data security for network traffic flows.

Keywords Multiplicative cascade · Traffic classification · Admission control

1 Introduction

The CAC (Connection Admission Control) is defined as a set of actions taken by the network during the call setup phase in order to determine whether a connection request can be accepted or rejected [1,2]. Call admission by no means can be considered as a robust procedure for preventing network congestion due to the fact that users are not always faithful to the maintenance of the negotiated connection parameters established during the call setup phase. Consequently the network needs to implement some additional control strategy imposing in some manner the compliance of the negotiated connection parameters on the traffic sources.

J.W. de Godoy Stênico(✉) and L.L. Ling
School of Electrical and Computer Engineering,
State University of Campinas – UNICAMP, Av. Albert Einstein 400, Campinas, SP, Brazil
e-mail: {jeferson,lee}@decom.fee.unicamp.br

E. Sabir et al. (eds.), *The International Symposium on Ubiquitous Networking*,
Lecture Notes in Electrical Engineering 366,
DOI: 10.1007/978-981-287-990-5_22

The use of multifractal characteristics observed in traffic to help to optimize connection admission control methods can be a relevant option. This is because the multifractal behavior greatly alters the aggregated traffic queue properties, provoking severe impacts on networks performance [3,4].

Network applications that require guaranteed quality of service (QoS) have increased considerably, especially over the internet including voice over IP and video conferencing. However, the current internet structure is still based on the best effort service strategy and has become precarious. Network congestions or serious outbreaks of traffic now occur frequently. This fact has motivated many studies and the surge of proposals for implementations of reliable QoS mechanisms.

Most of available call admission control approaches found in literature make accepting and rejecting decision is only based on the current availability of link bandwidth, without taking into account security issues. As a consequence, *malicious* traffic can easily access the network system and, therefore, cause irreparable damage to the network. To preventing such malicious attempts, we recommend making call admission control more robust by additionally implementing a traffic classification mechanism capable of blocking malicious traffic and even possibly established priorities to the well-intentioned traffic flows. The objective this work is to illustrate our approach for developing a new and robust admission control scheme and traffic classification based on the multifractal characteristics of input traffic flows.

The paper is organized as follows. In Section 2, we provide a brief description of the multifractal model (multiplicative cascade) for traffic representation and how multifractal traffic parameter being extracted from real network traffic traces. Section 3 devotes to the presentation of proposed traffic classification scheme based on multiplicative cascades. Section 4 shows our admission control mechanism which has incorporated the proposed traffic classification algorithm. Section 5 is dedicated for the presentation of obtained experimental results. Finally in Section 6, we present our conclusions.

2 Multiplicative Cascades

According to [5], a multiplicative cascade can be defined as follows.

Definition 1. A multiplicative cascade is an iterative process that fragments a given set into smaller and smaller pieces according to a geometric rule and, at the same time, distributes the total mass of the given set according to another scheme.

2.1 *The Proposed Binomial Multiplicative Cascade*

Assuming a unit total mass, the proposed multiplicative binomial cascade iteratively establishes the fragmented masses on sub-intervals according to the Newton Binomial weights, $\binom{2^N}{k}(x)^{2^N-k}(1-x)^k$, where N is a positive integer representing the total number of stages in the cascade and the stage number $k = 0,1,\dots,2^N-1$.

Without losing the generality, consider an initial interval $I = [0,1]$ and let x be a real-valued random variable uniformly distributed over the interval I. At the Nth stage of the cascade, there are $2^N - 1$ subintervals and the first subinterval has its mass equal to:

$$W_{\underbrace{00\ldots0}_{N\ digits}} = (x)^{2^N} + (1-x)^{2^N} \tag{1}$$

while for the remaining subintervals their masses are:

$$W_{b_1 b_2 \ldots b_N} = \binom{2^N}{i} (x)^{2^N - i} (1-x)^i |_{i=1,\ldots,2^N-1} \tag{2}$$

where $b_1 b_2 \ldots b_N$ is the binary representation of decimal numeral i, also used to denote the corresponding sub-interval position at the Nth stage of the cascade. As consequence, it is easy to see that the cascade is mass conservative in expectation.

The kth stage of the cascade is built based on the sub-interval information of the $(k-1)^{th}$. In other words, each subinterval of the $(k-1)^{th}$ stage is divided in the middle resulting in two equal length sub-intervals of the kth stage. Thus, at k^{th} stage of the cascade, the mass measure of the first interval $I_k = [0, 2^{-k}]$ is iteratively computed and equal to:

$$\mu[I_k] = \mu[0, 2^{-k}] = \mu[I_{k-1}] W_{\underbrace{00\ldots0}_{k\ digits}} = \mu[0, 2^{-k+1}] W_{\underbrace{00\ldots0}_{k\ digits}} =$$
$$\mu[0, 2^{-k+1}] \left[(x_{k-1})^{2k} + (1 - x_{k-1})^{2^k} \right] \tag{3}$$

For the other sub-intervals of the at k^{th} stage, the mass of the corresponding subinterval is:

$$\mu[I_k] = \mu[I_{k-1}] W_{b_1 b_2 \ldots b_k} =$$
$$\mu[I_{k-1}] \binom{2^k}{i} (x_{k-1})^{2^k - i} (1 - x_{k-1})^i |_{i=1,\ldots,2^k-1} \tag{4}$$

Notice that $x_1, x_2, x_3, \ldots$ are i.i.d. random variables uniformly distributed on $[0,1]$. Let Δt_k denote the length of each subinterval at the kth stage of the cascade. The mass measures of a multifractal process modeled by this binomial multiplicative cascade distributed over its dyadic intervals of length Δt_k localized at $t = 0.b_1 \ldots b_k = \sum_{i=1}^{k} b_i 2_i$ can be calculated as:

$$\mu(\Delta t_k) = R(b_1) R(b_1 b_2) \ldots R(b_{1,\ldots,} b_k) \tag{5}$$

where $R(b_{1,\ldots,} b_i)$ is the multiplier of the corresponding sub-interval at the stage i of the cascade.

2.2 Constructing Multiplicative Cascades for Network Traffic Modeling

The construction of a multiplicative cascade from real data means obtaining the $(k-1)^{st}$ stage of the cascade by iteratively aggregating masses of the kth stage.

For this end, the first step is to specify the random variable x of equation $\binom{2^N}{k}(x)^{2^N-k}(1-x)^k$ based on the real traffic trace data. For this, we used the following method:

We start at stage N and let $(X_i^N, i = 1, \ldots, 2^N)$ denote the given 2^N traffic samples. The data at stage $(N-1)$ is derived from aggregating the data localized in two adjacent intervals at stage N. The same procedure is repeated at other coarse stages $(N-j)$ by obtaining its sample value X_i^{N-j}, $(i = 1, \ldots, 2^{N-j}$ by grouping the sample data X_i^{N-j-1} of stage $(N-j)$ as follows:

$$X_i^{N-j-1} = X_{2i-1}^{N-j} - X_{2i}^{N-j} \tag{6}$$

for $i = 1, \ldots, 2^{N-j-1}$. The procedure for cascade construction ends with a unit interval [0, 1] and a single number. Figure 1 illustrates a part of this cascade construction procedure.

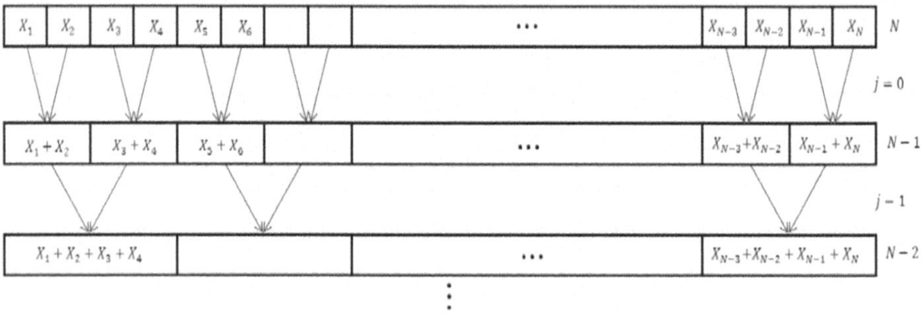

Fig. 1 Data Aggregation Process

A simple way to describe stage data sample X_i^{N-j-1} is by its statistical distribution. According to [6], the data sample holds a beta distribution having the following generic format:

$$f(w) = \frac{1}{(v-u)}\frac{\Gamma(\alpha+\beta)}{\Gamma(\alpha)\Gamma(\beta)}\left(\frac{w-u}{v-u}\right)^{\alpha-1}\left(1-\frac{w-u}{v-u}\right)^{\beta-1}, \tag{7}$$

where u and v are, respectively, the lower and upper bounds of the random variable, respectively, $\Gamma(.)$ is the Gamma function, α and β are Beta distribution parameters and w is the sample data value. For our modeling purpose, the random variable assumes values on the interval [0,1]. The normalization is carried out through the transformation [6]:

$$x_i = \frac{w-u}{v-u} \tag{8}$$

Therefore, the probability density function (7) now becomes

$$f(x_i) = \frac{\Gamma(\alpha+\beta)}{\Gamma(\alpha)\Gamma(\beta)}(x_i)^{\alpha-1}(1-x_i)^{\beta-1} \tag{9}$$

where $0 \leq x_i \leq 1$, for $\alpha > 1$ e $\beta > 1$.

For illustration purposes, Figures 2 and 3 show the histograms for two real traffic, represented by X_i^{N-j-1}, as well as the beta distribution approximation given by Equation (9). The first traffic trace, called "lbl-tcp-3", is extracted from a wired network of Digital Equipment Corporation [7] and available at [8]. The second traffic trace, denoted by "TrafIETF", is recorded from a large-scale wireless network during the 62nd Internet Engineering Task Force (IETF) meeting [9].

Using the data values of the constructed cascade, denoted by x_i^j, we complete the process of multifractally multiplicative cascade modeling by estimating the corresponding multipliers. An estimate of the multiplier the expand data of stage j to that of stage $j+1$ is obtained as follows. For the first two sub-intervals the multiplier estimates are obtained by using:

$$R_j^{i=1} = \frac{(x_i)^{2^{N-j}}+(1-x_i)^{2^{N-j}}}{(x_{i-1})^{2^{N-j-1}}+(1-x_{i-1})^{2^{N-j-1}}} \tag{10}$$

and

$$R_j^{i=2} = \frac{(x_i)^{2^{N-j}}+(1-x_i)^{2-j}}{\binom{2^{N-j-1}}{1}(x_{i-1})^{2^{N-j-1}}(1-x_{k-1})^{2^{N-j-1}}} \tag{11}$$

For the other subsequent sub-intervals,

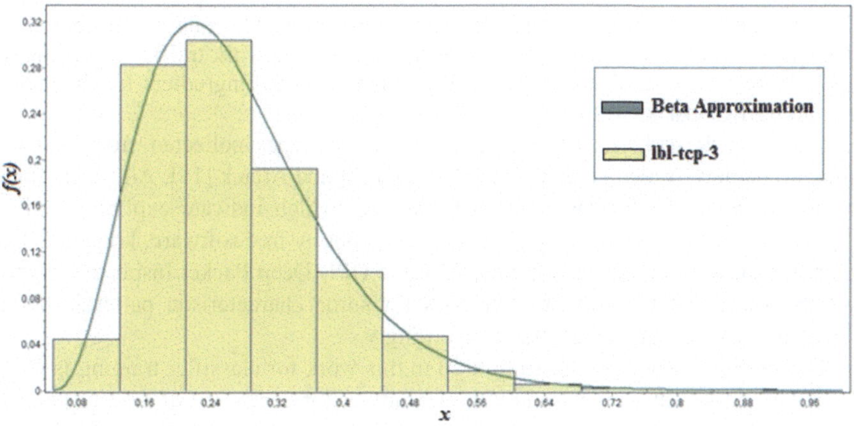

Fig. 2 Beta Distribution Approximation for lbl-tcp-3 traffic trace when N=6

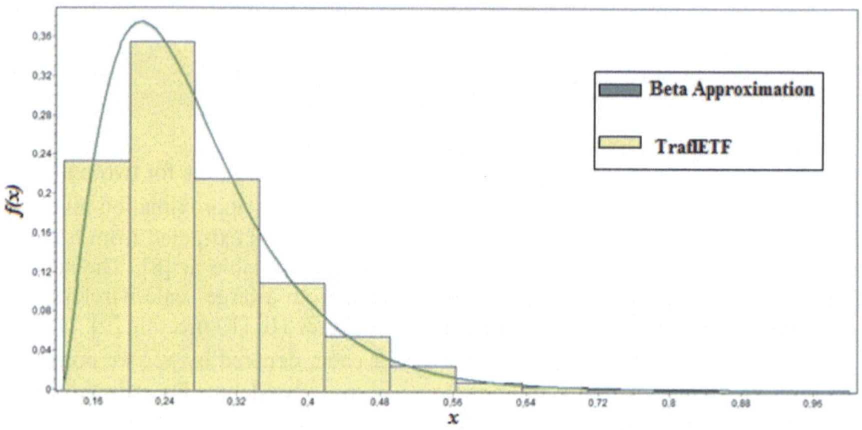

Fig. 3 Beta Distribution Approximation for TraflETF Traffic Trace when N = 7.

$$R_j^{i=3,\ldots,2^{N-j-1}} = \frac{\binom{2^{N-j}}{k}(x_i)^{2^{N-j}}+(1-x_i)^{2^{N-j}}}{\binom{2^{N-j-1}}{k}(x_{i-1})^{2^{N-j-1}}+(1-x_{i-1})^{2^{N-j-1}}} \tag{12}$$

where $j = 2,\ldots,2^N - 1$.

3 Traffic Classification

The compilation of a robust reference database is a critical step for pattern classification, especially via machine learning approaches. The quality of reference databases will dictate classification performance. For network traffic, according to [10], the reference database of traffic flows is an essential ingredient for the precision in traffic flow identification.

Our database consists of four different traffic types collected from distinct sources, namely, Video [11], Voice [12], Data [8] and Attack [13]. All traffic traces are found in the format of simple text files which indicate explicitly traffic types. The traffic flow labeling process carried out by free software, known as the L7-filter software [14]. The software is of the DPI (Deep Packet Inspection) type, which means that the software searches for some characteristic patterns in the payload of packets and labels them accordingly.

The machine learning algorithm used in this work for classifier training is C4.5 [15]. The classification is based on the feature vector composed of a set of parameters extracted from the constructed multiplicative cascades. In other words, the feature vector consists of the variances of cascade multiplier values estimated from real network traffic flows. Here, the size of a feature vector depends on the length of traffic flow or number of components used. Each component of a traffic flow contains the traffic data observed on the corresponding time interval. In this work, we only make use of the total number of traffic packets on the observed time interval. For instance, a traffic flow of 1024 components will result to a

10-stage multiplicative cascade and, as consequence, 10 distinct multipliers and 10 corresponding variances.

The variances obtained at each level of the cascade follow an exponential characteristic. This can be verified by Figure 4.

The training dataset consists a group of pairs {*feature vector*, *label*}, where *feature vector* consists of estimated multiplier variances and *label* is a name identifying the network application that generated the traffic flow.

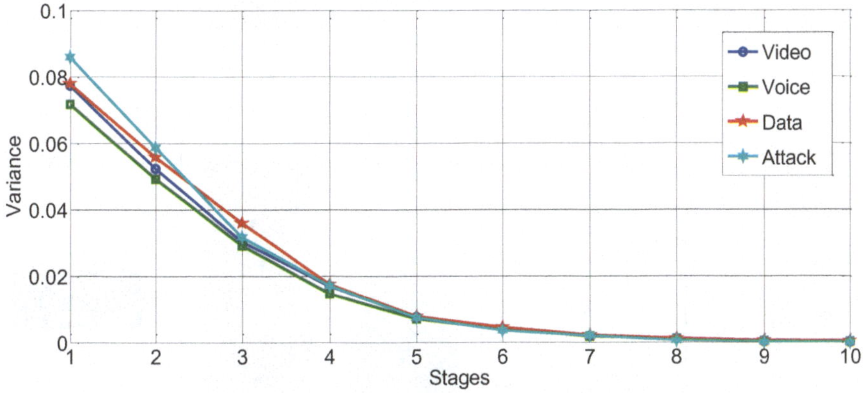

Fig. 4 Measured Variance

Figure 5 shows the functional blocks of the training process. The input reference data is labeled with the name of the application by an open source tool, L7-filter, and used for the cascade construction, multiplier estimation and variance computation.

The extracted numeric variance values are the feature vectors used by the algorithm C4.5 for the classifier training. The C4.5 algorithm is adopted in this work due to its attractive quality of high accuracy and small overheads compared to other machine learning techniques [15]. The decision procedure performed by the C4.5 algorithm consists of generating a decision tree from the input data and performing recursively data partitioning. The algorithm tests all possible combinations of input data and retains those combinations that grant the highest information gain.

In a block diagram, Figure 6 shows the classification scheme. Initially the statistical information necessary for later analysis is extracted from the input data of unclassified traffic flow. Then the input statistic data is used for the cascade construction, multiplier estimation and variance computation. Finally the decision model, previously constructed in the training phase, is applied to identify the class of the input traffic data.

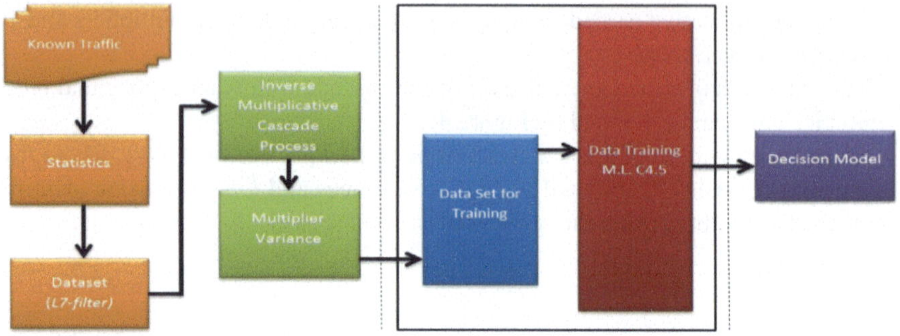

Fig. 5 Training Phases

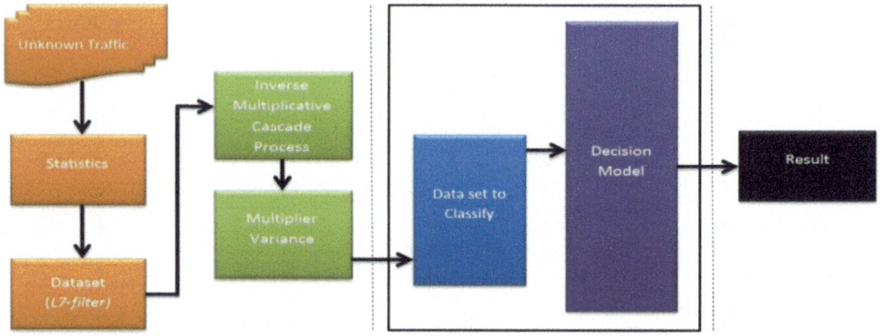

Fig. 6 Test Phases

4 Admission Control

The section is dedicated to the presentation of the new admission control algorithm that incorporate the traffic classification mechanism described in the previous section. A detected malicious traffic flow is classified into abnormal traffic type, therefore the admission control algorithm block its connection request. On the other hand, if the input traffic trace is not classified as malicious attempt, or a normal traffic type, it will be identified and further processed by the admission control algorithm.

The proposed admission control algorithm will allocate a bandwidth to a normal traffic flow according to the priority criterion pre-defined. Video traffic has the highest priority, voice traffic next and data traffic the lowest one. Figure 7 depicts in a block diagram the proposed admission control scheme. Next, we present in detail the proposed admission control algorithm. For this end, first of all, we need to define and provide suitable notations to the evolved variables and parameters.

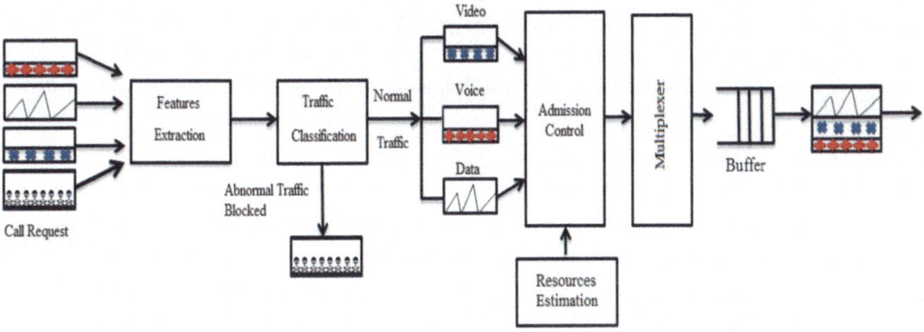

Fig. 7 The Proposed Admission Control Scheme

To obtain this goal, follows an outline of the proposed algorithm.

4.1 Notation of Variables and Parameters

The QoS parameters involved in connections (Video, Voice and Data) and the terminology used in this work are given below.

t: time (ms);
d: maximum delay allowed in a connection (ms);
m: d/t (an integer);
C_T: Total bandwidth allocation;
C_{Vd}^{max} : Maximum Allocated Bandwidth to Videos connections;
C_V^{max}: : Maximum Allocated Bandwidth to Voice connections;
C_d^{max}: Maximum Allocated Bandwidth to Data connections;
C_d^{min}: Minimum Allocated Bandwidth to Data connections;
C_V: Available bandwidth for Voice connections;
C_d: Available bandwidth for Data connections;
C_{alloc}: Allocated Bandwidth;
λ: The average data rate;
B: Buffer size (Kbits);
C_{REQ}: Bandwidth required for the connection. To satisfy the delay guarantees of the connections, we adopt the criterion equation proposed in [16]:

$$B \leq [(m-1)(1 + C_{NRT}/C_{Video})]\lambda t \tag{13}$$

where
C_{Video}: The total amount of bandwidth allocated to video connections (Real-time);
C_{NRT}: The total amount of bandwidth allocated for other connections (Non Real-time).

4.2 The Proposed Admission Control Algorithm

```
CAC for new Video connection
  If a new call request arrives at the server (Video)
    If (C_REQ ≤ (C_T − C_alloc)) → Check if there is available
bandwidth
```

```
      If Equation 13 is satisfied
          Admits the new connection
      Else
          Reject the connection
      End
    Else
% Perform degradation of data connections
    If (C_d > C_d^min)
      If Equation 13 is satisfied
          Admits the new connection
      Else
          Reject the connection
      End
    Else
% Perform degradation of Voice connections
    If (C_V > C_V^min)
      If Equation 13 is satisfied
          Admits the new connection
      Else
          Reject the connection
      End
    End
  End
CAC for new Voice connection
  If a new call request arrives at the server (Voice)
    If (C_REQ ≤ (C_T − C_alloc)) → Check if there is available
bandwidth
      If Equation 13 is satisfied
          Admits the new connection
      Else
          Reject the connection
      End
    Else
% Perform degradation of data connections
      If (C_d > C_d^min)
          If Equation 13 is satisfied
              Admits the new connection
          Else
              Reject the connection
          End
      End
  End
CAC for new Data connection
  If a new call request arrives at the server (Data)
    If (C_REQ ≤ (C_T − C_alloc)) → Check if there is available
bandwidth
      If Equation 13 is satisfied
          Admits the new connection
      Else
          Reject the connection
      End
    End
  End
```

5 Experimental Results

The effectiveness of the proposed admission control scheme that involving the traffic classification strategy is evaluated by simulation using the traffic traces: Video traffic trace available in [11], Voice traffic trace available in [12], Data traffic trace available in [8] and traffic traces considered as Attack (malicious traffic), extracted from [13].

Initially we compiled a database composed of real traffic traces (Videos, Voice, Data and Attack). Next we extracted the multifractal parameters via constructed multiplicative cascades as described in Section 3. Figure 8 shows the obtained correct classification rate (or precision metrics in literature [8, 10]) provided by the proposed traffic classification method for different sizes of used multifractal cascades. The traffic classification method proposed is able to obtain excellent estimates in the correct traffic classification, i.e. are obtained correct classification rate very high.

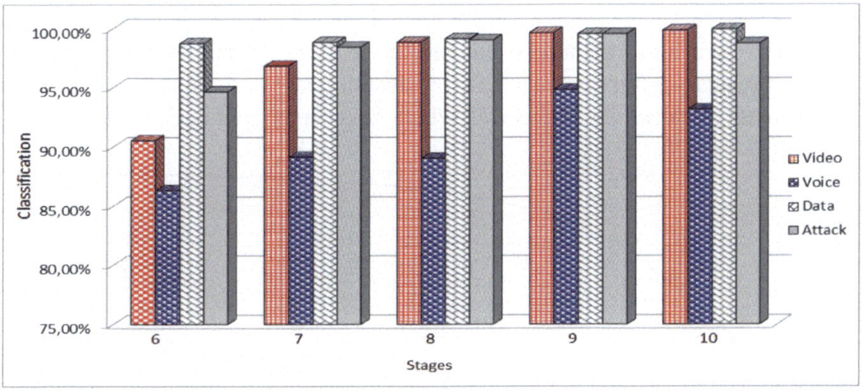

Fig. 8 Traffic Classification by Precision Metrics

The evaluation of the proposed admission control method is based on the performance measures: connection blocking probability and total bandwidth usage rate. Table I presents some relevant parameter values of the single server queuing system adopted in simulation of the single server queue used in the simulation. The total server capacity of the network was set to $C_T = 15000kbps$.

Table 1 System Configuration

Traffic Trace	Maximum Rate (Kbps)	Minimum Rate (Kbps)	*B* (Kbits)	*d* (ms)
Video	*9000*	*4000*	*10000*	*20*
Voice	*5000*	*3000*	*7000*	*30*
Data	*1000*	*100*	*600*	-

Figure 8 shows the call blocking probability in function of service requests. Notive that the video call has the lowest call blocking rate, followed by the voce call.

This result is intuitively plausible due to the fact that video traffic holds the highest connection priority and followed by voce traffic.

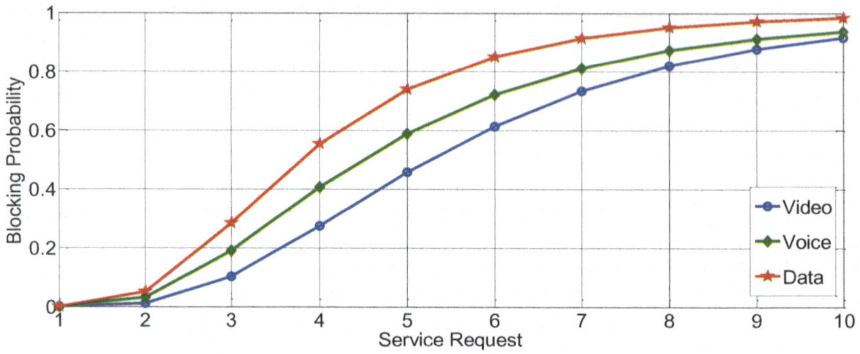

Fig. 9 Blocking Probability

Figure 10 will show the total bandwidth usage of the system. For large amounts of traffic requesting service, available bandwidth tends to the maximum, causing also increase the blocking of new call, which has been proven in Figure 9.

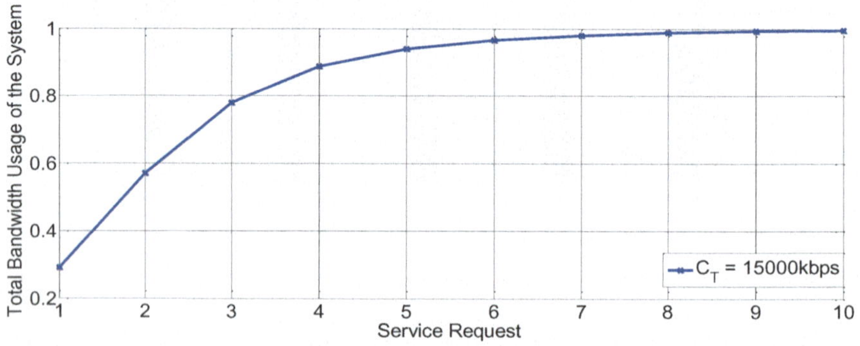

Fig. 10 Total Bandwidth Usage of the System

6 Conclusion

In this work we introduced a new admission control scheme that incorporates a new traffic classification scheme. The new approach to traffic classification considers that the networks flows can be represented by multiplicative cascade. Thus, using multifractal properties, it was possible to obtain relevant characteristics of several traffic. Through this procedure it is possible to distinguish the different traffic in communication networks and thereby block the traffic immediately considered Attack (*malicious* traffic). Based on this perspective the admission control scheme proposed will allocate a bandwidth to a normal traffic flow according to the priority criterion pre-defined. The proposed strategy aims to ensure QoS for higher priority calls (Video Traffic) and thereby reduces the transmission rate of traffic with lower priority. The simulations results showed that the scheme proposed is able to ensure the efficient use of available network resources while providing consistency in the QoS provisioning and data security in the system.

References

1. Newman, M.: Networks: An Introduction. Oxford University Press, Inc., NY (2010)
2. Tanembaum, A.S.: Computer Networks, 4th edn. Campus (2003)
3. Stênico, J.W.G., Lee, L.L., Vieira, F.H.T.: Queuing Modeling Applied to Admission Control of Network Traffic Flows Considering Multifractal Characteristics. IEEE Latin America Transactions **11**(2), 749–758 (2013)
4. Stênico, J.W.G., Lee, L.L.: A new binomial conservative multiplicative cascade approach for network traffic modeling. In: 27th IEEE International Conference on Advanced Information Networking and Applications–IEEE AINA 2013, Barcelona, Spain, vol. 1, pp. 794–801, Mach 25–28 (2013)
5. Waymire, E.C., Willians, S.: Multiplicative Cascades: Dimension Spectra and Dependence. Journal of Fourier Analysis and Applications, Special Issue in Honour of Kahane, J.-P., pp. 589–609 (1995)
6. Falls, L.W.: The Beta Distribution: A Statistical Model for World Cloud Cover, pp. 1–6. NASA Tecnical Memorandum, TMX–64714, Alabama (1973)
7. Erramilli, A., Narayan, O., Neidhardt, A., Saniee, I.: Performance Impacts of Multi-Scaling in Wide Area TCP/IP Traffic. Proceedings Infocom. **1**, 352–359 (2000)
8. http://ita.ee.lbl.gov/html/traces.html (retrieved: March, 2015)
9. Jardosh, A., Ramachandran K.N., Almeroth, K.C., Belding, E.: CRAWDAD Data Set UCSB / IETF–2005 out 2005 (2005). http://crawdad.cs.dartmouth.edu/ucsb/ietf2005 (retrieved: March, 2015)
10. Carela-Español, V., Barlet-Ros, P., Cabellos-Aparicio, A., Solé-Pareta, J.: Analysis of the Impact of Sampling on NetFlow Traffic Classification. Computer Networks **55**, 1083–1099 (2011)
11. Schulzrinne, H.: Internet Technical Notes and Resources (2008). http://www.cs.columbia.edu/~hgs/internet/traces.html (retrieved: March, 2015)
12. UPC.: Traffic Classification at the Universitat Politècnica de Catalunya (UPC). http://loadshedding.ccaba.upc.edu/trafic_classification (retrieved: March, 2015)
13. Mirkivic, J., Reiher, P., Prier, G., Michel, S., Li, J.: D-WARD: DDos Network Attack Recognition and Defense (2001). http://lever.cs.ucla.edu/ddos/traces (retrieved: March, 2015)
14. L7-filter.: Application Layer Packet Classifier. http://l7-flter.sourceforge.net/
15. Quinlan, J.R.: C4.5: Programs for Machine Learning. Morgan Kaufmann Publishers (1993)
16. Wongthavarawat, K., Ganz, A.: Packet scheduling for QoS support in IEEE 802.16 broadband wireless access systems. International Journal of Communication Systems **16**(1), 81–96 (2003)
17. Riedi, R.H., Crouse, M.S., Ribeiro, V.J., Baraniuk, R.G.: A Multifractal Wavelet Model with Application to Network Traffic. IEEE Transactions on Information Theory **45**(3), 992–1018 (1999)
18. Dehghani, F., Movahhedinia, N., Khayyambashi, M.R., Kianian, S.: Real-time traffic classification based on statistical and payload content features. In: 2nd International Workshop on Intelligent Systems and Applications (ISA), pp. 1–4 (2010)

Part V
Advances in Ad hoc Networking: MANET, VANET, WSN, DTN, etc.

A Survey on QoS for OLSR Routing Protocol in MANETS

Fatima Lakrami, N. Elkamoun and M. El Kamili

Abstract A mobile ad hoc network (MANET) is a decentralized type of wireless network, characterized by a dynamic topology. Supporting appropriate quality of service for mobile ad hoc networks is a complex and difficult task because of the bandwidth constraints and dynamic nature of the network. A routing protocol has a significant role in terms of the performance. It is used to discover and to establish correct and efficient route between a pair of source and destination nodes so that messages may be delivered in a timely manner. In this paper we have done the study of OLSR (Optimized Link State Routing) protocol. The key concept used in the protocol is that of multi-point relays (MPRs), which are selected nodes that forward broadcast messages during the flooding process. The objective of this paper is to examine QoS constraint in OLSR protocol. We present a state-of-the-art review and a comparison of typical representatives OLSR extensions, designed to enhance the quality of service in the original OLSR. The report aims to create a taxonomy of OLSR extension's with QoS support on the basis of the nature and the number of the metrics used to adapt the protocol to QoS requirements.

Keywords OLSR · QoS · Mobility · Manets · Vanets

1 Introduction

Ad hoc wireless networks inherit the traditional problems of wireless and mobile communications, such as bandwidth optimization, power control, and transmission

F. Lakrami(✉)
NTIC Laboratory, Institut Supérieur du Génie Appliqué, El Jadida, Morocco
e-mail: fatima.lakrami@gmail.com

N. Elkamoun
STIC Laboratory, Chouaib Doukkali University, El Jadida, Morocco

M. El Kamili
LIMS Laboratory, Sidi Mohammed Ben Abdellah University, Fez, Morocco

E. Sabir et al. (eds.), *The International Symposium on Ubiquitous Networking*,
Lecture Notes in Electrical Engineering 366,
DOI: 10.1007/978-981-287-990-5_23

quality enhancement [1]. In addition, the multi-hops nature and the lack of fixed infrastructure generate new research problems such as configuration advertising, discovery, and maintenance, as well as ad hoc addressing and self-routing. In mobile ad hoc networks, topology is highly dynamic and random. In addition, the distribution of nodes and, eventually, their capability of self-organizing play an important role. Quality of Service (QoS), as the name suggests, involves studying the level of user satisfaction in the services provided by a communication system. In computer networks, the goal of QoS support is to achieve a more deterministic communication behavior [3], so that information carried by the network can be better preserved and network resources can be better utilized. Efficient, dynamic routing is one of the key challenges in mobile ad hoc networks. Especially while introducing QoS constraint.

Nowadays, routing protocols are no longer limited to establishing routes, but become aware of user needs in term of reliability and availability of transmission resources. The QoS routing protocol [3] is now an integral part of any QoS solution since its function is to ascertain which nodes, if any, are able to serve applications requirement. QoS routing problem was addressed by many research efforts [2, 3], resulting in a large body of literature. We address this work to the study of QoS problem in OLSR routing protocol, a proactive protocol[6] conceived for ad hoc networks. This document offers an up-to-date survey of most major contributions for enhancing QoS in OLSR. A metric based classification is presented, in which the improvements and the cost of each extension is thus revealed. The remainder of this article is structured as follows. In Section 2 a brief presentation of OLSR is introduced, Section 3 describes the problem statement of QoS in OLSR routing protocol. In section 4, we discuss different Q-OLSR extensions and summarize their main points, including their strengths and weaknesses. Section 5 concludes the paper.

2 Optimized Link State Routing (OLSR)

The Optimized Link State Routing (OLSR) protocol [6] is a proactive, table-driven protocol that applies a multi-tiered approach with multi-point relays (MPRs). MPRs allow the network to apply scoped flooding, instead of full node-to-node flooding, with which the amount of exchanged control data can substantially be minimized.This is achieved by propagating the link state information about only the nodes chosed as MPR. Since the MPR approach is most suitable for large and dense ad hoc networks, in which the traffic is random and sporadic, OLSR [6] works best in this kind of an environment. The one great advantage of OLSR is that it immediately knows the status of the link and it is possibly to extend the quality of service(QoS) information to such protocol so that hosts know in advance the quality of a route [9]. A number of efforts have been made to concentrate on the comparative study of well known routing protocols in manets. However, it remains impossible to establish an assessment of the resulted works, since each work considers different simulation/experiment contexts [10, 11]. During the last few years, there were a large number of performance measurement studies on OLSR evaluating its reactivity in comparison with other standardized routing protocols, from other routing classes: proactive, reactive and hybrid. Authors in [11] compare OLSR, AODV and GRP, considering two mobility models, random way point model(RWP) and the vector mobility model (VMM). The collected results using OPNET simulator show that OLSR outperform the other protocols under mobility constraint.

Similar results have been obtained by [10], where authors evaluate OLSR, DSR and FSR using NS2 simulator. The study shows that OLSR performs better than the other protocols in terms of throughput and transmission delay. In [7], OLSR is compared to TORA and GRP, using different mobility models. In [12] three protocols were evaluated: OLSR, TORA and DSR. In [9], OLSR performance is evaluated in comparison with DSDV and AODV, OLSR provides better performance than AODV whose establishing and repairing routes delay as well as routing overload increase significantly as network conditions deteriorate (high mobility), on its part, DSDV generates a very important routing traffic, which becomes handicapping especially when network density increases in number.

3 QoS in OLSR Routing Protocol

Quality-of-service routing in an Ad-Hoc network [4] is difficult because the network topology may change constantly and the available state information for routing is inherently imprecise. To support QoS, the link state information such as delay, bandwidth, jitter, cost, loss ratio and error ratio in the network should be available and manageable. However, getting and managing the link state information in a MANET is by all means not trivial because quality of a wireless link changes with the surrounding circumstance. Compared to best-effort routing protocols, QoS routing has added costs, which may affect the performance of the routing protocol [16]. QoS routing protocols search for routes with sufficient resources to support various QoS requirements. However, finding a path subject to multiple constraints is inherently hard [17], considering such difficulties, together with the fact that node movements in ad-hoc networks make the problem even more complex, so polynomial-time algorithms for the problem may not exist[6]. In OLSR, the QoS must regards two fundamental functions: MPR selection and route calculation. The MPR technique optimizes the broadcast of control packets in the network[5], which reduces significantly the number of retransmissions in a flooding process, in plus it minimizes the size of control packets, as OLSR nodes will simply circulate the list of MPRs instead of the list containing all the neighbors. OLSR uses simple greedy algorithm for MPR selection, a node elects from its one-hop-neighbors, nodes that reach the maximum nodes in the two-hops-neighborhood. OLSR uses as routing algorithm: Dijkstra shortest-path[5], that considers hops count as routing metric. OLSR protocol was designed in such a way that it can be able to react to the mobility and the link failures of the network, while minimizing its control traffic overhead. OLSR can be extended to support quality of service. It should be equipped with additional mechanisms and make use of dynamic metrics to satisfy varied application requirements.

For making OLSR QoS aware, we could distinguish two main families of solutions. The first one seeks to modify the MPRs selection algorithm to became QoS aware while the second one focalizes on the choice of high QoS-cost paths for routing. A combination of both solutions is also feasible. We present in the next section a survey of different QoS solution for OLSR routing protocol. A metric based classification is adopted, we propose to overview QOLSR extensions referring to the used metric and the algorithmic part subject to change.

4 QoS Solutions for OLSR

4.1 Approaches Based on the Use of One QoS Metric

Energy. The energy metric is a concave function, which means that the representative energy of a given path is the minimum energy spotted at an intermediate node from the path:

$$E(C_1, C_2, ...C_n) = \min[E(C_1), E(C_2), ...E(C_n)] \quad (1)$$

A minimum energy routing protocol reduces the energy consumption of other nodes in a wireless ad hoc network by routing packets on routes that consume the minimum amount of energy, to get the packets to their destinations. Many energy-efficient routing protocols have been developed[9]. In OLSR, energy-aware solutions propose to choose either to use Min power consumption or Max lifetime algorithm.

The first category of solutions aim to make OLSR energy aware by modifying the MPR selection algorithm, in order to select nodes with high residual energy. In [10] authors proposed two MPR selection mechanisms named E-OLSR1 and E-OLSR2. They use the amount of residual energy and cost to send packets as metric in order to select MPR nodes. In [11], researchers integrated a modified MPR selection scheme of [10] with a path determination algorithm based on the residual energy level of each link. Researchers in [17], propose EE-OLSR: an extension of OLSR with a modification of the MPR selection mechanism, EE-OLSR propose to integrate the energy cost in the decision of the willingness: a parameter defined by the RFC3626 to express nodes ability to become MPRs. The evaluation of energy consumption and residual energy level, in each station, is performed locally before it is transmitted to the rest of the network, through the dissemination of the willingness value. they are two advantages of this approach, the use of the lifetime of a node which give a more accurate idea about node status, and the dissemination of the information through the willingness, that no control additional traffic is required. The problem encountered with all these extensions is that they don't take into account the maximum degree of nodes selected as MPRs. In fact, once a node manifests a high residual energy, according to the modified MPR algorithm, the node will be added to the MPR list, which may generate thereafter a high flooding traffic. In [18], a solution is proposed: RE-OLSR. It provides a new MPR selection mechanism that mainly considers residual energy of each node. RE-OLSR selects MPR nodes based not only on the residual energy of one-hop neighbors but also on their reachability and degree. The aim of RE-OLSR is to avoid selecting mobile nodes with small residual energy as MPRs, without exceeding the optimal MPR number. In [19], an improvement of [18] is proposed through the usage of a new system to resolve the willingness setting in OLSR at the basis of what is called the zero-Order Sugeno Fuzzy System. This technique enables a more precision of the willingness whose value is calculated on the base of the remaining energy (RE) and the expected residual lifetime(ERL). The specification of this solution remains more accurate compared to EE-OLSR. However, the mean local time for calculating metric value become more significant, specially when the task is performed in a very dense and mobile network. The problem encountered with all previous OLSR_energy_extensions, is that they don't take into account the transmission status of a node, which can lead to a rapid energetic depletion, even if for a high residual energy at the moment of the election. In [1], we proposed to combine mobility and energy lifetime in the decision of the willingness. Our algorithm selects stable nodes with high remaining lifetime to become MPR. The proposed idea is based on a simple mechanism of willingness selection with a dual optimization of mobility and energy. The combination of the two metrics enable a remarkable enhancement compared to other approaches considering energy as the only metric.

The second category of solutions chooses to take into account of the energy metric during routing process. In [25], authors proposed OLSRE, an enhanced version energy efficient routing based on OLSR. In OLSRE, the energy consumption is taken into account during packets routing by calculating the cost of packet transmission along a path. However, this protocol has a high overhead and it does not consider a node residual energy. OLSR_EA [15] measures and predicts per-interval energy consumptions using the well-known Auto-Regressive Integrated Moving Average time series method. Authors develop a composite energy cost, by considering transmission power consumption and residual energy of each node, and use this composite energy index as the routing metric. However, in the OLSR_EA, there is a higher chance of collision if a relay node has a longer transmission range than surrounded nodes and, therefore, it is not preferred in large networks.

In [15], authors proposed the OLSRM protocol, based on the standardized OLSR. They have tried to make it energy efficient by proposing an efficient neighbor selection based on nodes residual energy and drain rate. They have considered the multipath version of OLSR and the source routing concept for route selection. The proposed algorithm, modifies radically the original OLSR, on the cost of a high latency time during route establishment. In [17], a new multipath routing algorithm for OLSR is given, the new heuristic incorporates an energy optimization of the nodes in the network. it improves the number of nodes remaining actives for about 10% to 25% by always choosing paths with intermediate nodes having the highest level of residual energy. The test of the considered approach, shows that the improvement of

[17] remains insufficient compared with the investment cost, which is a high traffic overload.

An interesting approach was proposed by [16]: MBA-OLSR, based on the multipath extension of OLSRv2. It considers the remaining battery energy of the nodes for calculating the initial cost of the multiple links between source-destination pairs. The MBA-OLSR aims to construct energy efficient routes and extend the network lifetime to avoid network failure due to battery exhaustion. It succeeds to construct energy efficient routes and extend the network lifetime, first by deploying OLSRv2 (an enhanced version of OLSRv1) and second by using the multipath routing algorithm that enables a very effective load balancing.

Bandwidth. Bandwidth is a critical performance metric. Estimating the remaining bandwidth at a given time and in a given part of the network is tricky, as the medium is shared between close nodes in a wireless network[18]. Available bandwidth between two neighbor nodes is defined as the maximum throughput that can be transmitted between these two peers without disrupting any ongoing flow in the network. As for energy, the bandwidth is a concave metric. For a path, it is quantified by the minimum bandwidth of all the links constructing the path from the source to the destination:

$$Bw(C_1, C_2, ...C_n) = \min[Bw(C_1), Bw(C_2), ...Bw(C_n)] \qquad (2)$$

In [22], Y. Ge presents three revised MPR selection algorithms using the bandwidth metric: OLSR_R1 , OLSR_R2 and OLSR_R3. In OLSR_R1, MPR selection is almost the same as that of OLSR. However, when there are more than one 1-hop-neighbor covering the same number of uncovered 2-hop-neighbors, the one with the largest bandwidth link to the current node is selected as MPR. In OLSR_R2, the principle is to select the best bandwidth neighbors as MPRs until all the 2-hop-neighbors are covered. As for OLSR_R3, it is about selecting the MPRs in a way such that all the 2-hop-neighbors may have the optimal bandwidth path through the MPRs to the current node. In term of flooding optimization, OLSR_R3 is decidedly the best, the optimum number of MPRs is guaranteed. However, OLSR_R1 and R2 are indeed optimal regarding link reliability. OLSR_R2 has fewer overheads than OLSR_R3. Also, and compared to OLSR_R3, OLSR_R2 is simpler and more straightforward. Authors modify also the basic OLSR routing algorithm, for this end, two algorithms were tested: the maximum bandwidth spanning tree, and an extension of Bellman-Ford shortest path algorithm. The aim is to introduce bandwidth as a route selection criterion while changing routing cost. This modification addresses the problem of asymmetric links in wireless networks [22], knowing that natively, nodes selected as MPRs are the ones used to form the routes built later during routing process. The problem in most approaches proposing to use bandwidth as a criterion for QoS, is the complete neglect of the effect of the interference. Indeed, the authors often skip the part related to the method of estimating the capacity of the channel (passive, active or analytic)[23], thing that may affect widely the results and the performance obtained. Another constraint is revealed when estimating the bandwidth designated for each type of application, normally the proactive aspect of OLSR does not calculates routes on demand, so that usually the best effort and sensitive traffic will be able to borrow the same paths, previously established according to the QoS criterion defined by the protocol. In [24], a solution is proposed to overcome this problem, it is the source routing. Authors modify the design of the routing protocol to accommodate reservations of resources of an entered stream to its requirements. In fact, they take into account the superlative parts consumed by interference while estimating available bandwidth offered by a wireless link, but at the cost of a radical change in the basic algorithm of OLSR, which does not seem to be the rightful solution, as it does not allow to take advantage from the proactive nature of the protocol, in plus of loosing compatibility with the standardized version. On the other hand, authors of [25] consider that the only interference affecting the transmission are those of 3 hops, which is not true in practice. To resolve interference problem in bandwidth estimation, [26] proposes a formula of interference and a novel Link-disjoint Interference-Aware Multi-Path routing protocol (LIA-MPOLSR) , based on OLSR. The more difference between LIA-MPOLSR and the other multi-path routing protocols is that LIA-MPOLSR calculates interference by taking into account of the geographic distance between nodes instead of hop-by-hop.

Delay. With the emergence of real-time applications in wireless networks, delay guaranties are increasingly required. In order to provide support for delay sensitive traffic in such networks, an accurate evaluation of the delay is a necessary first step. Delay indicates the time to send a packet from a source to a destination node. Delay is an additive metric. Thus, the delay along a path is equal to the sum of the delays on the one-hop links of this path. With the use of IEEE 802.11 [25], the mean packet delay on a specific one-hop link, denoted by D, can be divided into three parts:

$$D = D_q + D_c + D_t \qquad (3)$$

where
D_q the mean queuing delay,
D_c the mean contention delay,
D_t the mean transmission delay.

The constraint of delay in OLSR is presented and modeled by [27], authors study analytically different estimation methods for probing the wireless channel. They point out the problem of synchronization between the sender and the receiver in an ad hoc architecture, especially when using broadcast packet to estimate the average transmission time. In OLSR, control packets are broadcasted with a jitter to avoid eventual collisions, due to synchronization problem. This jitter must be considered while using those packets to evaluate average one-hop transmission delay.

Authors of [28] propose OLSR_ETX, a new extension of OLSR that uses a new metric called expected transmission count (ETX). Each node independently measures the ETX of the link to each neighbor. The routing decisions are made such that the ETX of the route, which is the sum of the ETX of each link along this route, is minimized. Nodes have to use periodical link probe packets to measure the delivery ratios required for ETX calculation. Although ETX claims its design is independent of network load, but the delivery ratios it uses in calculation is actually affected by network load, also, additional traffic overhead of ETX is very significant. More over, ETX has the limitation of overestimating link delivery ratio when the data packet sizes are much larger than the link probe packet size. In a later paper [32], the same research group derived another routing metric, estimated transmission time (ETT), from ETX. In addition to link delivery ratio, ETT also handles varying nominal bandwidths of the links. ETT is designed to be used in a hybrid MANET routing scheme combining both link stake routing table computation and DSR-style on-demand querying. ETX and ETT are considered as complex metrics. To solve their complexity, authors in [33] investigate the use of the queuing delay as a routing metric, by using the neural network methods to predict delays. In the new extension named OLSR_NN, delay prediction/evaluation system is integrated with OLSR. In order to enable utilization of the predicted delays in routing table calculation, authors developed a node state based algorithm called TierUp, which is a light-weighted derivative of Dijkstra algorithm. To simplify queuing delay estimation, they consider a simplest queuing model assuming that all the nodes share a common omnidirectional radio channel with a fixed nominal bandwidth, and that node uses a simple FIFO queue for all the outbound packets, with the exemption of letting control packets having priority over data packet, which is not accurate in a real wireless transmission and can decrease OLSR_NN performances in a real implementation. In the same context, we can cite OLSR-MD[34], the main idea behind the Minimum delay is to measure the link delay between the nodes. It is calculated through the ad hoc network. Therefore, all calculations of routing tables are based upon each neighboring nodes. Therefore OLSR-MD is the protocol with the route selection between the current node and the other nodes in the network which have the lowest sum of different transmission delays of all the links along the path. In [35], H. Badis *et al.* propose to use the average travel time between two nodes for choosing the best path. During neighbors discovery, each node includes in its Hello message, the moment of its creation. When the Hello message is received by the neighboring node, the time between the sending node and the receiving node is calculated, the difference is added to the jitter, already included in HELLO packets, in order to solve the problem of synchronization. Similar approaches were proposed by [36] and [37]. Authors in [38] present a cross-layer framework for a delay estimation protocol as an extension to OLSR. Furthermore, the information from this protocol can be used to compute the route that satisfies the QoS delay requirements specified by a multimedia application. Despite of the huge modification of routing algorithm; they succeed to achieve a polynomial, Dijkstra algorithm.

Mobility. The mobility support remains one of the most difficult issues in ad hoc architectures, especially for routing protocols. The proactive nature of OLSR routing protocol makes it suitable for dense and less mobile networks. Authors in [48] propose fast-OLSR, an extension of OLSR, based on differentiating the behavior of mobile and fixed nodes. Mobile node will transit to a fast moving state, where the MPR selection algorithm is lightly rectified to manage efficiently nodes displacement. A new mechanism in [49] is proposed for predicting nodes mobility via elaborating a mathematical proceeding, it consists on calculating coordinates and velocities of nodes to construct a global view of neighbors mobility graph. Authors in [50] introduce the idea of Link Duration criterion as mobility metric for MPR selection, this approach has the advantage of using a simple modification with no need of any additional packet header. However, the estimation remains local, unless nodes are not informed about the mobile state of their neighbors. Authors in [50], modify the MPR selection process to enable selecting stable nodes; each node calculates locally its relative mobility rate and communicates it to its neighbors as a parameter of a QoS equation, aiming to select MPRs capable of assuring the relay function with certain optimality. In a similar way, the idea of [21] is based on changing the auxiliary functioning of OLSR, to make it mobility aware. Modifications consist mostly on including the mobility (stability) rate, exchanged between neighbors, in the MPR selection heuristic. Such propositions have the advantage of being dynamics and effectives. Since they use a number of mobility metrics, involving for their calculation two or more communication ends. The principal of such vision is to enable the determination of the link instability, on the basis of links status. However, their incompatibility with the standard OLSR, stands as an impediment in front of its practical deployment in real world, where heterogonous nodes may participate in a single communication scheme. Besides this, the dissemination of QoS information inside the network involves the addition of supplement traffic to control packets, of which the size increases proportionately to the neighborhood density. In [21][23], we propose a new solution scheme for implementing mobility in OLSR, the idea is based on the use of the willingness, whose value is mapped to the degree of stability of each node. For this purpose, we deploy two mobility metrics, the first is nodes speed while the other is based on the calcul of neighboring change. Nodes use those metrics to conclude if theirs 1-hop-neighbors are moving or not, and use the signal strength information to know the direction of the mobility. Compared to previous approaches, this one remains the simpler and the more accurate, in the context that there is no huge modification in OLSR, no additional traffic is needed and it outperforms other approaches in different simulation context.

4.2 Multiple QoS Metrics

To improve the quality of an ad hoc network service, it is appropriate to include parameters such as delay, bandwidth, cost of the link, packet loss and error rate. However, the optimal calculation of a route taking into account two or more heterogeneous metric is a NP-hard Problem [39]. To achieve a good adaptation to the constraints of mobile ad hoc networks, a series of optimization algorithms have been proposed by Costa *et al.* [40]. They propose an optimization of OLSR on the basis of combining three QoS metrics: bandwidth, propagation delay, and link loss probability. They propose simple algorithms that can scale instead of trying to find the optimal solution to the routing problem. The key idea of the developed SMM[40] is to prune links (or paths) that do not have enough bandwidth and then to run a modified version of classical routing algorithms based on a single mixed metric. This single mixed metric combines delay and loss probability, but uses the absolute value of the logarithmic transmission-success probability function (slog) instead of the loss probability to avoid complex composition. Badis *et al.*[41] proved that it is practically difficult to find the path that is at the same time the widest (maximum bandwidth) and the shortest (minimum delay), they tend via multiple mathematical solutions (Lagrangian relaxation) to achieve the best combination of metrics to optimize the cost function using bandwidth and delay. Another algorithm is proposed by[42], it proposes to combine multiple QoS metrics in one cost function, in the context of what we call multi-objective routing. Except for the definition of the cost function, this solution is identical to the previous one, with an extension to the use of three metrics instead of two. Authors in [40] propose a new technique for considering four metrics of QoS at once, the QoS algorithm is used only for MPR selection, while they preserve best-effort routing, to keep a low profile of complexity. It should be noted that algorithms, proposed to enhance routing part in OLSR, are based on the partial graph resulting from a neighboring view of the network topology. In [44], authors introduce an approach of using three QoS metrics; it is the bandwidth, delay, and loss rate. The proposed work introduced these criteria separately in the selection of MPRs. The main purpose is to improve the responsiveness of connected topological graph with 2-hop neighbors by adding specific weight on connectivity arcs. However, routing algorithm has not been modified to calculate routes with QoS, in fact authors assume that by constructing reliable connections with the second neighborhood, routes constructed using MPR nodes will therefore benefit from the same reliability, which is not the case, due to the asymmetric nature of wireless links.

Similarly, Zhihao *et al.*[45] propose to construct a cost function based on the combination of a multitude metrics. The goal is to design a multi-objective OLSR protocol. The special feature of this work is about the mathematical precision and accuracy of the estimation formulas, especially for the delay calculation. The proposed approach allows much better performance, despite a significant computational effort and a lack of efficiency in the dissemination of QoS values to the network. Authors in [47] propose to incorporate several QoS metrics, taking into consideration mobility and energy constraints, this approach aims to extend QoS while generating the minimum traffic overload considering two major aspects: the first by taking into account mobility and QoS metrics during the MPR selection stage and the second by minimizing the supplement traffic added to the control packets, to exchange QoS and mobility information between nodes. Authors choose to deploy an important number of constraints: Energy, Mobility, Bandwidth, Delay and Loss Rate. Different QoS approaches are resumed in (TABLE I) according to different selection criteria. The problem notified with the evaluation of different QoS solutions for OLSR routing protocol become clearly notable when measuring the Normalized Routing Load of OLSR. QoS-OLSR extensions manifest always highest values. Due to the bad dissemination of QoS metrics, which become handicapping when network density became consistent. In plus, the flooding problem is always neglected while evaluating an extension performance; except for [42], while the MPR set is revised many time to verify its optimality. In fact, QoS problem still always a challenge, it is not simply related to the choice of QoS metrics or their estimation process, but also on how to develop an efficient way to implement QoS algorithm in an existent protocol, while conserving its basic functionalities and also providing an optimal diffusion of QoS parameters in the network.

	Metrics						Traffic load for QOS			Mpr or routing		Mobility Support
	Energy		Bandwidth		delay		weak	medium	high	MPR selection	Routing	
	Residual-energy	Battery	Interference aware		Mean local delay	Transmission delay						
			yes	No								
EE-OLSR[18]		✓					✓			✓		
RE-OLSR[19]	✓						✓			✓		
EMOLSR[1]	✓									✓	✓	✓
OLSR_V1 & OLSR_V2[23]				✓				✓		✓		
Sondi & al[56]		✓		✓		✓			✓	✓		
Cost & al[46]				✓					✓		✓	
Badis & al[41]			✓			✓		✓		✓		
Lakrami & al[51]	✓			✓		✓		✓		✓	✓	✓
Guo & al[39]					✓	✓	✓					
Zahiho & al[49]		✓		✓	✓				✓		✓	

Table I: A Comparative table of OLSR Extensions with QoS support

5 Conclusion

Adaptive routing in MANETs is a very challenging issue. Dealing with the uncertainty of available routing information constitutes only one aspect of the problem. Wireless routing protocols, whether proactive or reactive, are intended to establish routes to route packets to their destinations while responding to changes in the topology and link failures. But in any case, they do not take into account the specific applications, namely the various bandwidth requirements, or delay disestablishing these roads. The particularity of the proactive protocols, including OLSR is that they establish paths in advance, without being aware of the special needs of the applications using the network. Efforts to introduce a support for mobility and quality of service in OLSR, are further multiplying. The goal is to take advantage of the protocol performances, in dense and, relatively, stable environments. The current deployment of OLSR in VANET sparked new ideas and new proposals. We proposed in this paper a survey of a number of most remarkable contributions, aiming to enhance quality of service support in OLSR. We tried to describe the principal, advantages and weakness of each proposition, in order to construct a formal synthesis for further optimizations, especially for our future works.

References

1. Lakrami, F., Elkamoun, N.: Energy and mobility in OLSR routing protocol. Cyber Journals: Multidisciplinary Journals in Science and Technology, Journal of Selected Areas in Telecommunications (JSAT) (2012)
2. Vashistha, S.: Improving the QoS in MANET by Enhancing the Routing Techniqueof AOMDV Protocol. In: ICT and Critical Infrastructure: Proceedings of the 48th Annual Convention of Computer Society of India, vol. I, pp. 381–392. Springer International Publishing, January 2014
3. Sarma, N., Nandi, S.: A multipath QoS routing with route stability for mobile ad hoc networks. IETE Technical Review **27**(5), 380 (2014)
4. Perkins, D., Hughes, D.: A survey on quality of service support for mobile ad hoc networks. Wireless Communications and Mobile Computing **2**(5), 503–513 (2002)
5. Waoo, A., Sharma, A.: Optimized Multipath Extension of AODV to Minimize Delay and Network Traffic with Reduce Packet Retransmission Attempts to Improve Performance. International Journal of Modern Computer Science **2**(4) (2014)
6. Clausen, T., Jacquet, P.: RFC 3626, Optimized link state routing protocol (OLSR) (2003)
7. Huhtonen, A.: Comparing AODV and OLSR routing protocols. Telecommunications Software and Multimedia, 1–9 (2004)
8. Ade, S., Tijare, A.: Performance comparison of AODV, DSDV, OLSR and DSR routing protocols in mobile ad hoc networks. International Journal of Information Technology and Knowledge Management **2**(2), 545–548 (2010)
9. Gangwar, P., Yadav, P., Tolani, M.: Performance analysis of different MANET protocols for different mobility of node. In: National Student Conference on Advances in Electrical Information Communication Technology, AEICT 2014 (2014)
10. Maurya, A.K., Singh, D., Kumar, A.: Performance Comparison of DSR, OLSR and FSR Routing Protocols in MANET Using Random Waypoint Mobility Model. International Journal of Information and Electronics Engineering **3**(5), September 2013
11. Pal, N., Dhir, R.: Analyze the Impact of Mobility on Performance of Routing Protocols in MANET Using OPNET Modeller. International Journal of Advanced Research in Computer Science and Software Engineering **3**(6), June 2013
12. Sharma, V., Khambra, S.: Performance Comparison of DSR, OLSR and TORA Routing Protocols. International Journal of Scientific Technology Research **3**(8), August 2014
13. Wang, Z., Crowcroft, J.: Quality-of-service routing for supporting multimedia applications. IEEE Journal on Selected Areas in Communications **14**(7), 1228–1234 (1996)
14. Yu, C., Lee, B., Youn, H.Y.: Energy efficient routing protocols for mobile ad hoc networks. Wireless Communications and Mobile Computing **3**(8), 959–973 (2010)
15. Ghanem, N., Boumerdassi, S., Renault, E.: New energy saving mechanisms for mobile ad-hoc networks using OLSR. In: Proceedings of the 2nd ACM International Workshop on Performance Evaluation of Wireless Ad Hoc, Sensor, and Ubiquitous Networks, pp. 273–274. ACM (2005)
16. Kunz, T.: Energy-Efficient Variations of OLSR. In: Proc. International Wireless Communiction and Mobile Computing Conference (IWCMC), Crete Island, pp. 517–522 (2008)
17. Rango, F.D., Fotino, M., Marano, S.: EE-OLSR: Energy Efficient OLSR Routing Protocol for Mobile Ad Hoc Networks. In: Proc. Military Communication Conference (MILCOM), San Diego, CA, pp. 1–7 (2008)
18. Hirata, K., Higami, Y., Kobayashi, S.Y.: RE-OLSR: Residual Energy-Based OLSR Protocol in Mobile Ad Hoc Networks. International Journal of Multimedia Technology
19. Chettibi, S., Chikhi, S.: FEA-OLSR: An Adaptive Energy Aware Routing Protocol for MANETs Using Zero-Order Sugeno Fuzzy System. IJCSI International Journal of Computer Science Issues **10**(1), March 2013
20. Guo, Z., Malakooti, B.: Energy aware proactive MANET routing with prediction on energy consumption. In: Wireless Algorithms, Systems and Applications (2007)

21. Nguyen, D., Minet, P.: Analysis of MPR Selection in the OLSR Protocol. In: Advanced Information Networking and Applications Workshops (2007)
22. Adoni, K.A., Joshi, R.D.: Optimization of energy consumption for OLSR routing protocol in MANET. International Journal of Wireless Mobile Networks (IJWMN) **4**, 251–262 (2012)
23. Jabbar, W., Ismail, M., Nordin, R.: MBA-OLSR: a multipath battery aware routing protocol for MANETs. In: 2014 Fifth International Conference on Intelligent Systems, Modelling and Simulation, Langkawi, Malaysia, pp. 630–635 (2014)
24. Adoni, K.A., Joshi, R.D.: Optimization of Energy Consumption for OLSR Routing Protocol in MANET. International Journal of Wireless Mobile Networks (IJWMN) **4**, 251–262 (2012)
25. Giles, R.R., Smith, P.G.: U.S. Patent No. 5,231,634. U.S. Patent and Trademark Office, Washington, DC (1993)
26. Chen, L., Heinzelman, W.B.: QoS-aware routing based on bandwidth estimation for mobile ad hoc networks. IEEE Journal on Selected Areas in Communications **23**(3), 561–572 (2005)
27. Ge, Y., Kunz, T., Lamont, L.: Quality of service routing in ad-hoc networks using OLSR. In: Proceedings of the 36th Annual Hawaii International Conference on System Sciences, p. 9. IEEE (2003)
28. Kim, K., Shin, K.G.: On accurate measurement of link quality in multi-hop wireless mesh networks. In: Proceedings of the 12th Annual International Conference on Mobile Computing and Networking, pp. 38–49 (2006)
29. Lakrami, F., Elkamoun, N.: Mobility Support in OLSR Routing Protocol. In: Network Computing and Information Security, pp. 804–812 (2012)
30. Chen, L.J., Sun, T., Yang, G., Sanadidi, M.Y., Gerla, M.: Ad hoc probe: path capacity probing in wireless ad hoc networks. In: Proceedings of the First International Conference on Wireless Internet, pp. 156–163. IEEE (2005)
31. Campista, M., Esposito, P., Moraes, I., Costa, L., Duarte, O., et al.: Routing metrics and protocols for wireless mesh networks. IEEE Network **22**(1), 6–12 (2008)
32. Bianchi, G.: Performance analysis of the IEEE 802.11 distributed coordination function. IEEE Journal on Selected Areas in Communications **18**(3), 535–547 (2000)
33. Naimi, A., Jacquet, P.: le contrôle du délai dans le protocole OLSR (2003)
34. Passos, D., Teixeira, V., Muchaluat-Saade, D.C., Magalhães, L.C.S., Albuquerque, C.: Mesh network performance measurements. In: International Information and Telecommunicatios Technologies Symposium (I2TS), pp. 48–55 (2006)
35. Sinky, H., Hamdaoui, B.: Implementation and performance measurement and analysis of OLSR protocol. In: Proceedings of the 6th International Wireless Communications and Mobile Computing Conference, pp. 286–290 (2010)
36. Guo, Z., Malakooti, B.: Predictive delay metric for OLSR using neural networks. In: Proceedings of the 3rd International Conference on Wireless Internet, p. 21. ICST (Institute for Computer Sciences, Social-Informatics and Telecommunications Engineering) (2007)
37. Cordeiro, W., Aguiar, E., Moreira, W., et al.: Providing quality of service for mesh networks using link delay measurements. In: Computer Communications and Networks (2007)
38. Munaretto, A., Badis, H., Al Agha, K., Pujolle, G.: QOLSR:Routage avec QoS dans OLSR. In: 5eme Rencontres Francophones sur les aspects Algorithmiques, AlgoTel 2003 (2003)
39. Munaretto, A., Badis, H., Al Agha, K., Pujolle, G.: A link-state QoS routing protocol for ad hoc networks. In: MWCN, pp. 222–226 (2002)
40. Ramrekha, T., Politis, C.: An adaptive QoS routing solution for manet based multimedia communications in emergency cases. In: Mobile Lightweight Wireless Systems, pp. 74–84. Springer, Heidelberg (2009)
41. Jacquet, P., Meraihi Naimi, A., Rodolakis, G.: Asymptotic delay analysis for cross-layer delay-based routing in ad hoc networks. Advances in Multimedia (2007)
42. Nguyen, D., Minet, P.: Analysis of MPR Selection in the OLSR Protocol. In: 21st International Conference on Advanced Information Networking and Applications Workshops, vol. 2, pp. 887–892. IEEE (2007)
43. Costa, L., Fdida, S., Duarte, O.: Developing scalable protocols for three-metric QoS routing. Computer Networks **39**(6), 713–727 (2002)

44. Badis, H., Al, K.: Agha, QOLSR, QoS routing for ad hoc wireless networks using OLSR. European Transactions on Telecommunications **16**(5), 427–442 (2005)
45. Lavén, A., Hjärtquist, P.: Multimetric OLSR and ETT. In: 5th OLSR Interop & Workshop (2009)
46. Guo, Z., Malakooti, S., Sheikh, S., Al-Najjar, C., Malakooti, K.: Multi-objective OLSR for proactive routing in MANET with delay, energy, and link lifetime predictions. Applied Mathematical Modelling **35**(3) (2011)
47. Obwang, S.: Le routage à qualité de service dans les réseaux mobiles ad hoc (Doctoral dissertation, Valenciennes) (2010)
48. Lakrami, F., Elkamoun, N.: Mobility and QoS Management in OLSR Routing Protocol. International Journal of Computer Networking, Wireless and Mobile Communications (IJCNWMC), **2**(4), December 2012. ISSN 2250-1568
49. Benzaid, M., Minet, P., Al Agha, K.: Integrating fast mobility in the OLSR routing protocol. In: IEEE International Conference on Vehicular Technology (2003)
50. Ghumaliya, D., Chauhan, S.: A Survey on Quality of Service for Optimized Linked State Routing protocol in Mobile Ad hoc Network. IJCSMC **3**(3), 89–93 (2014)

New Classification of Nodes Cooperation in Delay Tolerant Networks

Salah Eddine Loudari, Maria Benamar and Nabil Benamar

Abstract Delay and Disruptive Tolerant Networks (DTNs) is a concept related to environments characterized by very long delay paths and frequent network disruptions. DTN is nowadays a recognized area in networking and communications research, due to its suitability and practical experiences with mobile ad-hoc networks especially in situations where continuous end-to-end paths may not be always guaranteed. In DTNs, nodes store carry and forward messages, called bundles, to other nodes. The forwarding mechanism can occur opportunistically. However, some nodes may show some selfish or malicious behavior, which leads to less cooperation in the network. Thus, one of the main challenges in DTN is to ensure the security and confidentiality within the Network and assure cooperation among nodes. In this paper, we classify some of the threats that have been considered and treated by researchers in the field of DTN, and we propose a new classification based on the degree of cooperation of nodes. We describe different incentive mechanisms used to enhance cooperation among nodes in DTN environment focusing on the strengths of these mechanisms and also their limitations and drawbacks.

Keywords Delay Tolerant Networks · Cooperation · Bundle

1 Introduction

Delay Tolerant Network is a network of regional networks; it is an occasionally-connected network [1] that permits communication where connectivity issues like long and variable delay, high error rates, sparse and intermittent connectivity, highly asymmetric data rate and non-end-to-end connectivity exist [2]. The purpose of the DTN is to support interoperability among these underlying stressed regional networks. DTN concept was originally designed for communicating with spacecraft, to compensate for disconnections over interplanetary distances [3]. However, after several years of research in the field of DTN, numerous implementations and applications have appeared with a broad variety of performance and

S.E. Loudari(✉) · M. Benamar · N. Benamar
University Moulay Ismail, Meknes, Morocco
e-mail: {salaheddine.loudari,mariabenamar,benamar73}@gmail.com

E. Sabir et al. (eds.), *The International Symposium on Ubiquitous Networking*,
Lecture Notes in Electrical Engineering 366,
DOI: 10.1007/978-981-287-990-5_24

application domains [4] (inter-planetary networks [3], underwater networks [5], wildlife tracking networks [6], military tactile networks [7]...). We regroup in table 1 the different examples of DTN application. A new work group has been created in the IETF to tackle DTN issues and application in terrestrial context [8].

The unit of information exchange in Delay Tolerant Networks is a bundle [9][10]. A DTN node is an entity with a bundle layer. A node may be a host, router, or gateway acting as a source, destination, or forwarder of bundles [4].

Fig.1 shows the bundle layer and its position with respect to the network layers when compared to the Internet layers. The DTN bundled layer is common across all the DTN regions while the lower layers are all region specific.

DTN utilizes the concept of store, carry and forward [11][12][13] (Fig. 2) which is a long established form of postal system. Each node has a persistent storage system, which is used to store messages as backup just in case the network fails during transmission. The node stores bundle in its buffer for a certain time (can be minutes, hours or even days) and will be forwarded only when connection between the intermediate nodes is established.

Table 1 Example of DTN applications

DTN Applications	Purpose	DTN nodes	Delay
ZebraNet [6]	Track African zebras across large regions	Zebra, mobile base station	Hours or days
DakNet [14]	Digital communications for rural areas	Coaches, motorbikes, ox carts, kiosks, access points	Minutes or hours
KioskNet [15]	Digital communications for rural areas	Buses, kiosks, hand-held devices, desktop computers with a dial-up connection	Hours or days
Widernet [16]	To improve educational communication systems in Africa	Desktop computers with sufficient storage to store web sites with rich educational contents	Days or monhs

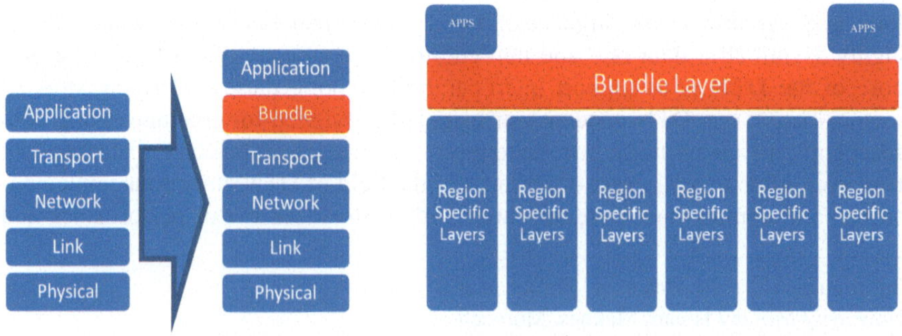

Fig. 1 Position of bundle layer

Fig. 2 Store-carry and forward paradigm

2 Security Threats: An Overview

2.1 Non-DTN Threats

The first series of threats to consider are those from the elements that are not directly part of the DTN, The network is an overlay network. Packets can traverse multiple underlying networks where the message might be edited or deleted.

An overlay network inherits all of the good and all of the bad of the underlying networks upon which it resides [17]. For example, if an overlay network passes over four or five different concatenated underlying networks, then the overlay network is vulnerable to all of the insecurities of any of the underlying networks. This makes overlay networks much more difficult to secure as one has to secure each underlying network in addition to applying proper security to the network overlay itself.

2.2 Resource Consumption

Due to resource scarcity that characterizes DTN, unauthorized use of resources – particularly battery power and storage - can be considered as threats on the infrastructure of DTN [18]. For example, if an unauthorized application could control certain DTN infrastructures (through an attack on the routing protocol) resource consumption could be catastrophic for the network, which reduces its performance.

Limited resources challenge requires an efficient protocol design. In other words, nodes must utilize their limited hardware resources such as CPU, memory and battery efficiently. For example, in WSNs [19], nodes can be located in an open environment for years before data are collected, and hence it requires nodes to carefully manage their energy usage. Additionally, a good routing protocol will leverage the resources of multiple nodes. For example, nodes may choose to shift some of their stored bundles to other nodes to free up memory or to reduce transmission cost.

2.3 *Routing Attacks*

Denial of Service

This type of attack must be considered in the DTN context because it is at the same position of other MANETs. So, all the problems with secure routing in ad-hoc networks exist for many DTNs too.

Denial of service is one of the major threats and among the toughest security issues in networks nowadays [20] because it attempts to limit access to a machine or service. The effect of Dos in Delay Tolerant Network is even more aggravated due to the scarcity of resources. Perpetrators of DOS attacks in DTN-like environments look beyond the objective of rendering a target node useless [21]. The aim of an attacker is to cause a network-wide degradation of resources, service and performance. This can easily be achieved by exhausting node or link resources and partitioning the network.

Table 2 Some solutions proposed to protect against attacks

Techniques	Objectives	Details
Bundle Authentication Block (BAB) [16]	Ensures bundle authenticity and integrity	Used to thwart DOS and to ensure routing information exchange between "neighboring" DTN nodes is authenticated.
Internet Security Association and Key Management Protocol (ISAKMP) [25]	Anti-clogging technique where a client is required to return a server generated cookie	Used to prove a client's identity and is verified by the server before any costly authentication protocol is triggered.
Tackling the junk mail [20]	Tackle the problem of connection depletion attacks	Use cryptographic puzzles where a sender is required to compute a puzzle for very message sent. The cost of this technique is negligible for normal users when compared to mass mailers.

Black-Hole

This attack aims to destroy the network services, causing a sharp decline in delivery rate [22]. Malicious nodes involved in launching the black-hole attack disseminate false probable information delivery to increase -or decrease- their chances to be selected and attract the maximum number of messages to delete later.

Gray-hole It is an extension of the Black-hole attack where malicious nodes delete certain packets (as opposed to black-hole that remove all packages) [23].

Worm-Hole
Malicious nodes build a tunnel between each other using a low latency link to convince victims that it is the best path to send packets [24]. These nodes can manipulate the routing algorithm and control information that is shared by the honest nodes which disrupts the packet delivery operation.

3 Cooperation in Delay Tolerant Networks

Due to the uncertainty of transmission opportunities between nodes, Delay Tolerant Networks adopts a store–carry–forward method [26]. This method requires nodes to store and forward messages in a cooperative way. However, threats mentioned in the previous chapter directly affect the cooperation between nodes in the DTN and implicitly affect the performance of DTN.

The authors in [27] define three types of nodes behavior: cooperative nodes, partly-cooperative and uncooperative.

Cooperative nodes: stock and forward messages to another node without restriction.

Partly-cooperative: Agree to forward the messages coming from other nodes, but on condition that transmits messages directly to the destination.

Uncooperative: DTN are resource-constrained networks. The selfish behaviors may occur among the nodes to preserve its own resources (energy, storage space, CPU…) by ignoring the packet from other nodes and will forward only its own message [28]. Another type of uncooperative nodes is that malicious nodes provide forged metrics to other nodes that come in contact with and attract packets from them [29]. After receiving these forwarded packets the malicious node can either drop or modify them [30].

On the basis of types mentioned above, we propose a classification of node behavior (Fig. 3):

Fig. 3 Classification of node behavior

3.1 *Impact of Uncooperative Behavior*

The existing research works, based on theoretical analysis models and simulations, demonstrate two characteristics of the performance degradation caused by uncooperative behavior, Firstly, the routing performance (i.e., delivery ratio, delivery cost and delivery latency) is seriously degraded, if a major portion of the nodes in the network is selfish [31]. Secondly, the impact on the routing performance is related to the uncooperative behavior (i.e., non-forwarding of messages and dropping of messages). The behavior of non-forwarding of messages reduces the delivery cost, while the behavior of dropping of messages increases the delivery cost. However, both of them decrease the delivery ratio, and prolong the delivery latency, even if messages are eventually delivered.

3.2 *Strategies for Preventing Uncooperative Behavior*

In order to reduce the impact of uncooperative behavior on routing performance, a number of studies focus on stimulating uncooperative nodes to be cooperative. The existing incentive strategies can be classified into three categories [27]: Barter based, Credit based and Reputation based, table 3 summarizes these strategies. In the following sections, we present a comprehensive discussion about these different strategies.

Barter-Based Strategies

Barter-based or pair wise Tit-For-Tat strategy is the simplest strategy and one of the most popular motivations to tackle the problem of uncooperative behavior [32]. It is based on the fact that every node forwards as much traffic for a neighbor as the neighbor forwards for it. In [33], the authors divide the message into two types: primary messages and secondary message. A message is a primary message for a given mobile node, if the mobile node is interested in the content of the message and secondary if the mobile node is not. It is worth for the users collecting messages even if they are not interested in them to exchange them later for messages that they are interested in.

However, the requirement of exchanging the same amount of messages is a two-edged sword. The problem is when the node has no enough messages to exchange; the message should not be delivered to destination even if the destination is in connection [32], which reduces routing performances.

Credit-Based Strategies

Due to disadvantages of barter-based, Credit-based strategies are proposed to avoid these disadvantages. This strategy encourages nodes to be cooperative by paying reward for cooperative nodes. The concept is that if a node cooperates to forward a message for others nodes, it receives an amount of credit as a reward that it can later explore for its own benefit.

Credit-based strategies can be subdivided into two models [31]: **Message Purse Model**: the source node pays credits to nodes which participate in delivering the message to the destination. **Message Trade Model**: the source node do not pay for the message forwarding, contrary to message purse model, the receiver pays credits to the sender of a message until the message reaches the destination, which finally pays for the message forwarding. Because the source nodes do not pay for the message forwarding, the network can be flooded by the source node. For this reason, most of the credit-based works utilize the message purse model.

Reputation-Based Strategies

In Reputation-based strategies, each node observes the behavior of other nodes and assigns each of them a reputation, which measures how well a node is behaving. Reputation is calculated from the opinion of neighbor. The routing of messages is done on the basis of the reputation: the lower the reputation the lower the probability that a node is chosen as next hop for forwarding a message [34].

Reputation-based strategies can work even if a major portion of the malicious nodes. However, this kind of strategy mistakenly considers the collaboration of intermediate nodes as selfish behavior, which results in the decrement of the delivery probability of the messages generated.

Table 3 Advantages and limitations of preventing strategies

Strategies	Advantages	Limitations
Barter-based	Requirement of exchanging the same amount of message	When the node has no enough messages to exchange
Credit-based	Encourage nodes to be cooperative by paying reward	Cannot work in an environment in which the nodes have a high probability of being selfish to other nodes
Reputation-based	Work well even if a major portion of the nodes are uncooperatives	Consider the collaboration of intermediate nodes as selfish behavior

4 Conclusion

In this paper, we presented an overview of the various security threats in Delay Tolerant Networks such as non-DTN threats, resource consumption and routing attacks. We focused on threats related to nodes behavior and cooperation and their effects on DTN performances. Thus, we proposed a new cooperation degree classification: Cooperative node (Fully cooperative and partly cooperative), and uncooperative node (Malicious, selfish and misbehaving). We pointed out the negative impact of uncooperative behavior on the network, and we presented the existing strategies aiming to solve or at least to reduce the impact of this issue. We discussed their strengths and we made a special focus on the limitations of these strategies and their drawbacks.

Due to the importance and the utility of Delay Tolerant Networks, especially in their terrestrial applications, further work should take into consideration security threats in order to solve the problems encountered in these networks and reduce the negative impact of threats.

References

1. Scott, K., Burleigh, S.: Bundle Protocol Specification. RFC 5050, November 2007
2. Pereira, P.R., Casaca, A., Rodrigues, J.J., Soares, V.N., Triay, J., Cervelló-Pastor, C.: From Delay-Tolerant Networks to Vehicular Delay-Tolerant Networks. IEEE Commun. Surveys Tuts. **PP**(99), 1–17 (2011)
3. Burleigh, S., Hooke, A., Torgerson, L., Durst, B., Scott, K., Corporation, T.M.: Delay-Tolerant Networking: An Approach to Interplanetary Internet. IEEE Communications Magazine, 128–136 (2003)
4. Voyiatzis, A.G.: A Survey of Delay- and Disruption-Tolerant Networking Applications. Journal of Internet Engineering **5**(1), 331–344 (2012)
5. Partan, J., Kurose, J., Levine, B.N.: A survey of practical issues in underwater networks. In: Proc. 1st ACM Int. Work. Underw. networks - WUWNet 2006, p. 17 (2006)
6. Juang, P., Oki, H., Wang, Y.: Energy-Efficient Computing for Wildlife Tracking: Design Tradeoffs and Early Experiences with ZebraNet. In: Proc. ASPLOS, October 2002
7. Krishnan, R., Basu, P., Mikkelson, J.M., Small, C., Ramanathan, R., Brown, D.W., Burgess, J.R., Caro, A.L., Condell, M., Goffee, N.C., Hain, R.R., Hansen, R.E., Jones, C.E., Kawadia, V., Mankins, D.P., Schwartz, B.I., Strayer, W.T., Ward, J.W., Wiggins, D.P., Polit, S.H.: The SPINDLE disruption-tolerant networking system. In: MILCOM 2007 - IEEE Mil. Commun. Conf., pp. 1–7, October 2007
8. DTN Work Group: http://datatracker.ietf.org/list/wg/
9. Benamar, N., Singh, K.D., Benamar, M., Bonnin, J., El Ouadghiri, D.: Routing protocols in Vehicular Delay Tolerant Networksm: A comprehensive survey. Computer Communication (Com Com) **48**, 141–158 (2014)
10. Benamar, N., Benamar, M., Bonnin, J.M.: Routing protocols for DTN in vehicular environment. In: 2012 International Conference on Multimedia Computing and Systems (ICMCS), pp. 589–593, May 10–12, 2012
11. Benamar, N., Benamar, M., Ahnana, S., Saiyari, F.Z., el Ouadghiri, M.D., Bonnin, J.-M.: Are VDTN routing protocols suitable for data collection in smart cities: a performance assessment. J. Theor. Appl. Inf. Technol. **58**(3), 589–600 (2013)
12. Benamar, M., Ahnana, S., Saiyari, F.Z., Benamar, N., El Ouadghiri, M.D., Bonnin, J.-M.: Study of VDTN routing protocols performances in sparse and dense traffic in the presence of relay nodes. J. Mob. Multimedia **10**(1&2), 78–93 (2014)
13. Benamar, M., Benamar, N., Singh, K.D., El Ouadghiri, D.: Recent study of routing protocols in VANET: survey and taxonomy. In: WVNT 2013: 1st International Workshop on Vehicular Networks and Telematics, May 2013, Marrakech, Morocco (2013)
14. http://www.unitedvillages.com/
15. Guo, S., Falaki, M.H., Oliver, E.A., Rahman, S.U., Seth, A., Zaharia, M.A., Keshav, S.: Very Low-Cost Internet Access Using KioskNet. ACM Computer Communication Review, October 2007
16. http://www.widernet.org/

17. Ivancic, W.D.: Security analysis of DTN architecture and bundle protocol specification for space-based networks. In: 2010 IEEE Aerosp. Conf., pp. 1–12, March 2010
18. DTN Research Group - www.dtnrg.org/
19. Prabhakar, T.V., Nambi Sn, A.U., Jamadagni, H.S., Swaroop, K., Prasad, R.V., Niemegeers, I.G.M.M.: A novel DTN based energy neutral transfer scheme for energy harvested WSN Gateways. ACM SIGMETRICS Perform. Eval. Rev. **38**(3), 71 (2011)
20. Ardra, P.S.: A Survey On Detection And Mitigation Of Misbehavior In Disruption Tolerant Networks, vol. 2, pp. 656–660, December 2012
21. Ansa, G., Cruickshank, H., Sun, Z.: A proactive DOS filter mechanism for delay tolerant networks. In: ICST PSATS, Conference, Malaga, Spain, February 2011
22. Jain, S.: Black Hole Attack in Delay Tolerant Networks: A Survey, no. 4, pp. 172–175 (2014)
23. Nogueira, M., Silva, H., Santos, A., Pujolle, G.: A Security Management Architecture for Supporting Routing Services on WANETs. IEEE Trans. Netw. Serv. Manag. **9**(2), 156–168 (2012)
24. Ren, Y., Chuah, M.C., Yang, J., Chen, Y.: Detecting wormhole attacks in delay tolerant networks. IEEE Wireless Commun. Mag. **17**(5), 36–42 (2010)
25. Onen, M., Molva, R.: Denial of service prevention in satellite networks. In: 2004 IEEE Int. Conf. Commun. (IEEE Cat. No. 04CH37577), vol. 7, pp. 4387–4391 (2004)
26. Mangrulkar, R.S., Atique, M.: Procedia Computer Science Heterogeneous Highly Dense Mobile Environment, pp. 1–13 (2013)
27. Keranen, A., Pitkanen, M., Vuori, M., Ott, J.: Effect of non-cooperative nodes in mobile DTNs. In: Proc. 2011 IEEE International Symposium on a World of Wireless, Mobile and Multimedia Networks (WoWMoM), pp. 1–7, June 2011
28. Mishra, D., Chawla, M.: Minimax Theory Based Scheme to Detect Selfish Node and Reduce Latency in Delay Tolerant Network, vol. 2013, no. Cac2s, pp. 78–82 (2013)
29. Ahnana, S., Saiyari, F.Z., Benamar, N., el Ouadghiri, M.D.: Study of DTN routing protocols in vehicular environment in the presence of misbehaving nodes. In: Proc. - WINCOM (2013)
30. Li, F., Wu, J., Srinivasan, A.: Thwarting blackhole attacks in distruption-tolerant networks using encounter tickets. In: Proc. - IEEE INFOCOM, pp. 2428–2436 (2009)
31. Miao, J., Hasan, O., Mokhtar, S.B., Brunie, L., Yim, K.: An Analysis of Strategies for Preventing Selfish Behavior in Mobile Delay Tolerant Networks (2012)
32. Liu, L.: A survey on barter-based incentive mechanism in opportunistic networks. In: 2013 2nd Int. Symp. Instrum. Meas. Sens. Netw. Autom., pp. 365–367, December 2013
33. Buttyan, L., Dora, L., Felegyhazi, M., Vajda, I.: Barter trade improves message delivery in opportunistic networks. Ad Hoc Networks **8**(1), 1–14 (2010)
34. Dini, G., Duca, A.L.: Ad Hoc Networks Towards a reputation-based routing protocol to contrast blackholes in a delay tolerant network. Ad Hoc Networks **10**(7), 1167–1178 (2012)

A Survey on Flat Routing Protocols in Wireless Sensor Networks

Hassan Echoukairi, Khalid Bourgba and Mohammed Ouzzif

Abstract The study of routing protocols in wireless sensor networks has been addressed to improve the usage of limited network resources. In this paper, we give in detailed survey of flat routing protocols (flooding, data centric and forwarding) based on network structure in wireless sensor networks, while providing the advantages and disadvantages of each routing protocol. By taking into account several parameters, some of routing protocols in this category will be compared. This paper also highlights the associated problem and application type designated for each protocol.

Keywords Wireless Sensor Network (WSN) · Network structure · Routing protocols · Flat routing protocols

1 Introduction

A wireless sensor network (WSN) is a collection of sensor nodes that cooperate and communicate by wireless links deployed randomly or deterministically in a geographical area in order to detect, collect and transmit data about a phenomenon observed during a sink node (base station). Depending on the number of sensor nodes and the extent of the battlefield, some nodes cannot transmit their messages to the collector node (sink) directly, which requires collaboration between network nodes to ensure that transmission. This call for a delivery of a source node to a destination node is called the routing process message. Taking into account the limited capacity (computing power, memory, scalability, data aggregation, etc) of the sensor nodes, communication with the sink should be carried out by broadcasting messages until they reach the collector without routing protocol.

H. Echoukairi(✉) · K. Bourgba · M. Ouzzif
Laboratory of Computer Networks, Telecommunications and Multimedia,
Higher School of Technology, Hassan II University, Casablanca, Morocco
e-mail: {echoukairi,bourgba2008,ouzzif}@gmail.com

E. Sabir et al. (eds.), *The International Symposium on Ubiquitous Networking*,
Lecture Notes in Electrical Engineering 366,
DOI: 10.1007/978-981-287-990-5_25

However, this simple solution causes significant disadvantages such as implosion and overlapping messages. In addition, sensor nodes do not take into account their resources to limit their operation issue (calculation). Thus, the implementation of a routing algorithm for a network becomes inevitable for effective sensors. The potential field of application is unlimited. Wireless sensor networks (WSNs) are becoming increasingly attractive for numerous application areas, such as military reconnaissance, disaster management, security surveillance, environmental monitoring, medical and health, industrial automation. In WSN, the data transmission to the sink may be continuous, event-driven or query-driven. Nevertheless, some nodes may stop working. So, the routing protocol becomes a challenging task, it must be able to adapt the topology changes and always find the optimal route between the source node and the destination node (sink). Routing is one of the critical technologies in WSNs because: the resources are constrained in terms of energy, processing, storage capacities and transmission bandwidth, it's difficult to design a global addressing scheme as Internet Protocol (IP) for a large number of sensor nodes as data collection by sensor nodes usually results in a high probability of data redundancy. Due to such differences, researchers have designed a number of different routing protocols to solve these problems in WSNs. In here we shall deal in detail with the routing protocols based on network structure in wireless sensor networks, as shown in Fig. 1:

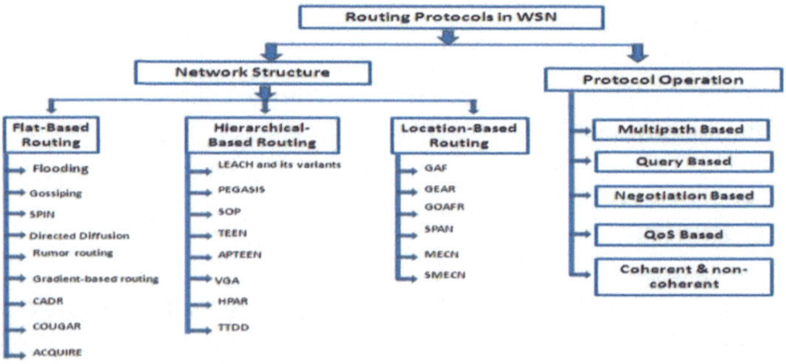

Fig. 1 Routing Protocols in WSN

In general, network structure routing in WSNs can be classified into three categories namely flat-based routing (flooding, data centric and forwarding), hierarchical-based routing and location-based routing. In this structure of routing, the protocols are classified into multipath-based, query-based, negotiation-based, QoS-based and coherent and non coherent routing techniques depending on the protocol operation. In flat-based routing, all nodes are treated equally and have the same functionalities in the network. In hierarchical-based routing, however, nodes will play different roles in the network. In location based routing, sensor nodes are addressed based on their location to route data in the network by using GPS or via coordination among the nodes.

This paper specially provides an overview of data-centric routing and dissemination protocols in flat-based networks. It is structured as follows: In section 2 we describe in detail the category of Flat-based Routing Protocols (Flooding and Gossiping, Sensor Protocols for Information via Negotiation (SPIN), Directed Diffusion (DD), Rumor routing (RR), Gradient-Based routing (GBR), Constrained Anisotropic Diffusion routing protocol (CADR), COUGAR, Active Query Forwarding in Sensor Networks (ACQUIRE)) with some advantages and disadvantages. The comparison of some flat routing protocols in WSNs with their characteristics, problems associated and application type is presented in section 3. Finally, the paper is concluded in section 4 with future perspectives.

2 Flat-Based Routing Protocols

This category of routing protocols have three types of flat routing schemes, namely, flooding, forwarding and data-centric based routing [1], [2], [3], [5]. In flat topology, all nodes in the sensor network have equal roles in gathering information. They all have the same information about the state of the network. In this type of network, it is not possible to assign a particular identification (ID) to each node due to the large number of sensor nodes. This leads to data-centric routing approach in which the sink sends query to a group of particular nodes in a region and waits for a response. The property of data is specified by an attribute-based naming.

2.1 Flooding and Gossiping

In this category, the authors present a family of adaptive protocols: Flooding and Gossiping [4], [5]. There are two classical mechanisms to relay data in sensor networks without using topology maintenance and complex routing algorithms. In Flooding as shown in Fig. 2, each node receiving a packet broadcasts it to all its neighbors without testing the capabilities of nodes. It will continue this process until the maximum number of hops for the packet is reached or until the packet reaches its destination.

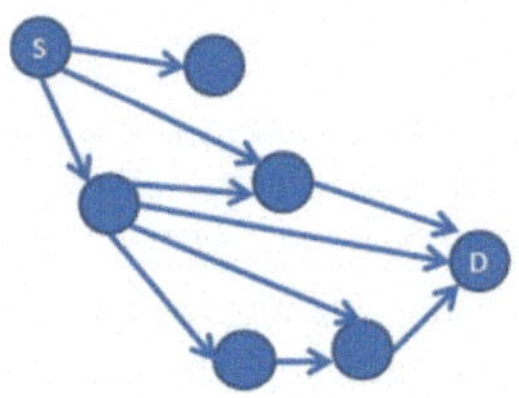

Fig. 2 Flooding Protocol Scenario

However, this protocol raises few drawbacks such as: impulsion, overlap and resource blindness. In gossiping, instead of the broadcasting each packet to all its neighboring (components) the packet is transmitted to a single neighbor chosen at random from a neighboring table after testing all the nodes capability, as shown in Fig. 3. The selected node picks another random neighbor to forward the packet to and so on.

Gossiping will help to reduce multiple copies of the same packet traversed in the network by selecting a random node for packet relaying. This avoids the problem of impulsion but the delay to reach destination may be large in some cases.

2.2 Sensor Protocols for Information via Negotiation (SPIN)

Sensor Protocols for Information via Negotiation (SPIN) is a family of Flat protocols [2], [5], [6], [7], [8]. It is a data centric routing protocol based on a negotiation model to propagate information in a sensor network. It is designed to address the deficiencies of flooding by negotiation and resource adaptation. This protocol disseminates all the information at each node to every node in the network treating all nodes as potential sink nodes. The SPIN family rests upon two basic ideas [7]:instead of sending all the data, sensor nodes negotiate with each other the description of the data that they have and that they still need to obtain. Sensors nodes use meta-data to describe the sensor data briefly. This ensures that there is no redundant data sent throughout the network. Each node monitors and adapts to changes in its own energy resource a fact, which is used to perform energy-aware decisions and extend the operating lifetime of the system. SPIN uses three stages for basic operations: ADV(Data Advertisement), REQ(Request For Data), DATA. The Fig. 4 illustrates this:

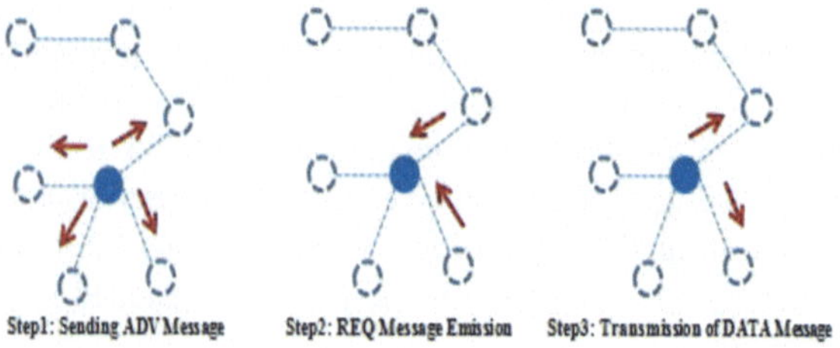

Fig. 3 Spin Operations and Steps

The neighbor sensor node then repeats this process with its neighbors. As a result, the entire sensor area will receive a copy of the data. The SPIN family of protocols is two protocols called SPIN-1 and SPIN-2 .The SPIN-1 is simple handshake protocol for the dissemination of data through a lossless network. SPIN-2 is an extension to SPIN-1, which incorporates threshold-based resource awareness mechanism in addition to negotiation. Other protocols of the SPIN family are: SPIN-BC, SPIN-PP, SPIN-EC, SPIN-RL.

Advantages

- Topological changes are localized since each node needs to know only its single-hop neighbors, i.e. does not affect the whole network.
- SPIN reduces energy consumption compared to flooding and metadata negotiation almost halves the redundant data.

Drawbacks

- SPINs data advertisement mechanism cannot guarantee the delivery of data, i.e it is not sure if the data will certainly reach the target.
- Not good for applications requiring reliable data delivery, for example intrusion detection (where data should be reliably reported over periodic intervals).

2.3 Directed Diffusion (DD)

Directed diffusion (DD) [9], [10], [11] is a popular data aggregation paradigm for WSNs proposed by C. Intanagonwiwat et al. in 2000. It's a data-centric routing protocol for data propagation and application-aware paradigm uses on-demand data query and send data for delivery to inquiring destination. All the data generated by sensor nodes is named by attribute-value pairs (objects, interval, duration and geographical area) used for the description of the information. The main objective is to save network energy and prolong its lifetime by keeping localized interactions, in terms of message exchanges between neighbors or nodes within limited network vicinity. DD can also achieve multiple paths for routing information. This protocol includes the following elements: interest, data, gradient, and reinforcement. In DD [13], the sink broadcasts interest packet which describes a task required to be done by the network. Sensor nodes set up a gradient to each neighbor from which they receive an interest and store the interest entry in its cache. Based on the detection of the corresponding event, the sensor generates a data packet and sends it to the sink via the neighbors for which it has a gradient. This process is repeated by each node receiving the data packet. These data packets are called exploratory packets, since they are sent to the sink along multiple paths. The sink reinforces the path from which for example it received the first exploratory data packet. Fig. 5 illustrates this process

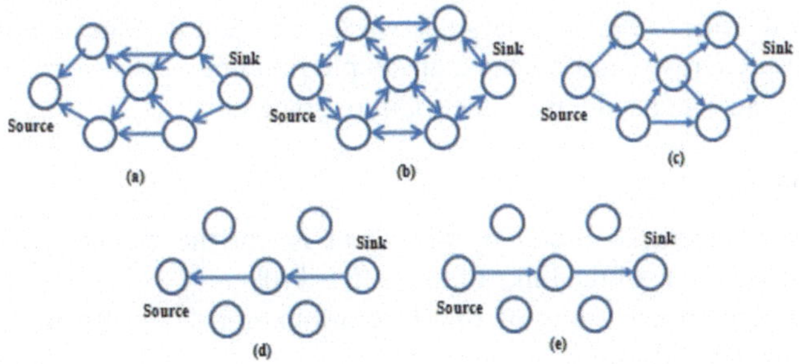

Fig. 4 a- Broadcasting interest b-Establishment of gradients c- Data propagation d- Selecting and reinforcing path e- Transmitting data

Advantages

- Directed diffusion allows on demand data queries while SPIN allows only interested nodes to query.
- Unlike SPIN, there is no need to maintain global network topology in directed diffusion.

Drawbacks

- Matching data to queries might require some extra overhead at the sensor nodes.
- There is limit memory storage for data caching inside the sensor node.
- Directed diffusion is not suitable for the applications where continuous flow of data is required (e.g., environmental monitoring).

2.4 Rumor Routing (RR)

Rumor Routing (RR) **[13]**, **[6]** is another variant of Directed diffusion approach proposed by Braginsky and Estrin in 2002. It is intended for network application which geographic routing is not suitable for. This routing technique will route queries to the nodes that have observed a particular event rather than flooding the entire network. In RR whenever any event occurs at the node, it will add the event to its local table called events table and create an agent, which is a long –lived packet and travels the network in order to propagate information about local events to distant nodes. On the other hand, this routing table will be updated if there's a shorter path to an event within the routing table of the node it is visiting. When a node generates a query for an event, the nodes that know the route, may respond to the query by inspecting its event table. Otherwise, it can be sent on to a random direction until it finds the route (path) to the event. RR maintains a single path from source to destination and reduces the communication cost because it doesn't need to flood the whole network. For large events, cost of an agent is also increasing and hence RR is preferred when there is small event.

Advantages

- Deliver queries to events in large networks with less average cumulative hops and lower energy requirements than simple flooding.
- The RR can also handle the node's failure gracefully.

Drawbacks

- The path sometimes found by the agent is not the shortest and could be unavailable if one of the links of the path is broken.
- The agent may carry lots of routing information of events and even some events would disappear.
- Rumor routing technique fails in case of large number of nodes since the cost of maintaining agents and event-tables in each node becomes infeasible.

2.5 Gradient-Based Routing (GBR)

In 2001**[14]**, Schurgers and Srivastava proposed an enhanced version of the Directed Diffusion called Gradient-Based Routing (GBR) that includes the number of hops (hop-number) when the interest is diffused to the network. The basic idea in GBR **[11]** is to choose the shortest route (minimum number of hops to sink) to reach the sink. This parameter is called the height of the node and it can be calculated by each node. The gradient is the difference between a node's height and that of its neighbor. The packet is forwarded to the link, which has the largest gradient. In addition, GBR uses traffic spreading techniques in order to uniformly divide the traffic over the network. When multiple routes pass through a node, which acts as a relay node, their data may be combined. Their different data spreading techniques have been discussed in GBR: Stochastic Scheme, Energy-based scheme and Stream-based scheme.

Advantages [5], [16]

- Simulation results of GBR have shown that GBR outperforms directed diffusion in terms of total communication energy.
- Nodes in GBR deliver the message in a point to point manner and do not use the broadcast nature of wireless networks.

Drawbacks [15], [16]

- One limitation of the multi-hop data propagation approach is that it seems inevitable that bottle-neck nodes (or bottle-neck regions) in the routing graph tend to be overused and run out of energy before others.
- The nodes in GBR which are near the sink will be overused and will die before others. The failure of these nodes leads to the failure of the entire network.

2.6 Constrained Anisotropic Diffusion Routing Protocol (CADR)

Constrained Anisotropic Diffusion routing [17], [18] is a variant data centric protocol proposed by M. Chu et al in 2002. CADR aims to be a general form of directed diffusion. The main idea of this protocol is to query the sensors and route data in a network in order to maximize the information gain, while the bandwidth and latency are minimized. This aim is reached by activating only the sensors that are close to a particular event and dynamically adjust data routes. The principal difference from directed diffusion is the consideration of information gain in addition to communication cost. Based on sensor values of neighboring nodes, CADR routes the query through a greedy search by creating a sequence of local decisions. There exist 3 modes of these decisions for to forward the query to the neighboring node with: highest objective function, steepest (local) gradient in the objective function and maximizes combination of the local gradient of the objective function and the distance improvement to the estimated optimum location.

Advantages

- This technique can minimize the energy needed to route the query to its destination by choosing the shortest path.
- It can also maximize the information gain by taking an irregular walk with more steps.

Drawbacks

This routing scheme is complemented by the technique information-driven sensor querying (IDSQ) that it will not specifically define how data and the queries are routed between sensors and the base stations.

2.7 *Cougar*

COUGAR [19],[20],[2] is data-centric routing protocol proposed by Y. Yao and J. Gehr in 2002, taking into consideration the sensor networks as a huge distributed database system. The main idea is to introduce a new query layer between the applications and the sensor network. This layer uses declarative queries in order to abstract query processing from the network layer functions such as selection of relevant sensors and utilize in-network data aggregation to save energy. In COUGAR, sensor nodes select a leader node to perform aggregation and send the data to the sink (destination). The sink generates a query plan which specifies the necessary information about the data flow and in-network computation for the incoming query and transmits it to the relevant nodes. The query plan also describes how to select a leader for the query. The procedure in the selection of a query leader is shown in Fig. 6:

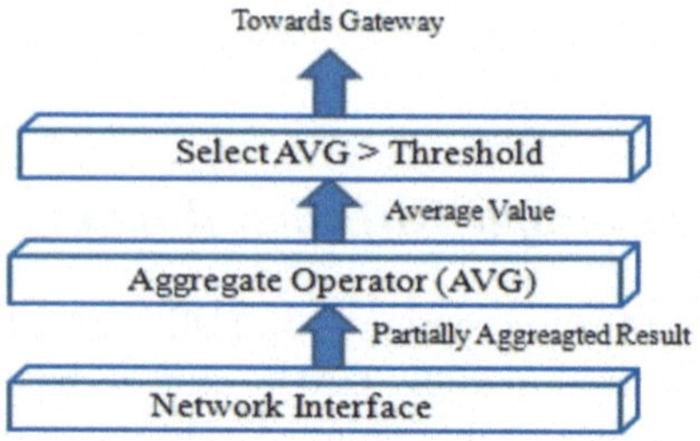

Fig. 5 Query plan at a leader node: The leader node gets all the readings, calculates the average if it is greater than a threshold sends it to the gateway (sink)

Advantages

- COUGAR provides energy savings especially when the generated data is huge.
- COUGAR provides network-layer independent methods for data query.

Drawbacks

- Extra overhead of energy consumption and memory storage by introducing additional query layer on each sensor node.
- Dynamic maintenance of a leader node to avoid failure.
- For successful in-network data computation, it requires synchronization.

2.8 Active Query Forwarding in Sensor Networks (ACQUIRE)

Active Query Forwarding in Sensor Networks (ACQUIRE) [21], [22] is mechanism for querying sensor nodes proposed by Sadagopan et al. in 2003. Similar to COUGAR, it views network as distributed database. This approach is well-suited for complex queries which consist of several sub queries can be resolved partially en route. The basic principle behind ACQUIRE is to consider the query as an active entity that is forwarded through the network (either randomly or in some directed manner) in search for the solution. ACQUIRE also incorporates a look-ahead parameter **d** in the following manner: intermediate nodes that handle the active query use information from all nodes within **d** hops in order to partially resolve the query. When the active query is fully resolved, a completed response is sent directly back to the querying node.

Advantages

- Energy efficiency may be controlled by varying a parameter and thus limiting the number of hops to include in the query processing.

Drawbacks

- Revealing the shortest path is often more energy consuming.
- Setting for the parameter d as maximum directed diffusion behaves flooding.

3 Comparison of Flat Routing Protocols and Discussion

The flat routing protocols compared in this study has certain limitations in terms of scalability, mobility, data aggregation, multipath, overhead (in terms of route discovery), energy efficiency, data delivery model and do not support quality of service of the sensor network.

By opposition of hierarchical and geographical protocols. Flooding and gossiping performs very limited in scalability because the nodes send the received data through every outgoing link, use a large amount of power and they suffer from reliability as well as information delay issues.

SPIN, DD, ACQUIRE, RR, GBR, and COUGAR protocols which data centric routing is used to control the redundancy of data because sensor node does not have global identification number.

All these protocols do not present the same characteristics and differ in many ways although they have been designed for the same underlying network.

Table 1 Describes the characteristics of flat routing protocols [6], [24], [25], [26], [27],[28]

Routing Protocols	*Classification*	*Scalability*	*Mobility*	*Negotiation Based*	*Data Aggregation*	*Multipath*	*QoS*	*Power Usage*	*Query Based*	*Over head*	*Data delivery model*
Flooding	Flat	No	Ltd	No	No	Simplest	No	Large	No	Hight	No
Gossiping	Flat	Ltd	Ltd	No	No	-------	No	Large	No	Mod	No
SPIN	Flat/DataCentric	Ltd	No	Yes	Yes	No	No	Ltd	Yes	Low	Event driven
DD	Flat/DataCentric	Ltd	No	Yes	Yes	Yes	No	Ltd	Yes	Low	Demand driven
RR	Flat	Good	Very Ltd.	No	Yes	No	No	Low	Yes	Low	Demand driven
GBR	Flat	Ltd	Ltd	No	Yes	No	No	Low	Yes	Low	Hybrid
CADR	Flat	Ltd	No	No	No	No	No	Low	Yes	Low	Continuously
COUGAR	Flat	Ltd	No	No	Yes	No	No	Ltd	Yes	Hight	Query driven
ACQUIRE	Flat/DataCentric	Ltd	Ltd	No	Yes	No	No	Low	Yes	Low (Ltd)	Complex query

No: Not existent, Ltd : limited

Table 2. describes the associated problem and the applications for each flat routing protocol in WSN [5], [6], [23], [24]:

Routing Protocols	*Problem associated*	*Application type*
Flooding	Implosion	Monitoring servers, Fire detection and Intrusion detection.
Gossiping	Overlap problem	
SPIN	implosion, overlapping and resource blindness are solved	Habit Monitoring
Directed Diffusion	Not fit for the application where continues flow of the data is required.	Environment Monitoring (heating, air-conditioning control)
Rumor Routing	Unable to handle large number of events	Habit/Environmental monitoring
GBR	Problem of data propagation (each node can find the minimum number of hops to the sink)	Health monitoring
CADR	*Problem* addressed here is how to perform queries and route data maximizing the information gain.	Tracking and identification *problems* in both civilian and military
COUGAR	• We have to dynamically maintain a leader node to avoid failure. • Extra overhead to sensor nodes by introducing additional query layer on each sensor node. • For data computation it requires synchronization.	Environment monitoring
ACQUIRE	If d is equal to network size, then the protocol behaves similar to flooding	

We can't say any particular protocol is better than the other since each protocol in this category has a specific application. For all applications, an important issues and challenges for routing protocols in WSN are:

- The treatment of node mobility and their self configuration.
- The density of sensors with large number for to prolong the lifetime of the network.
- Make secure routing protocols to ensure the transmission of messages between the nodes.

4 Conclusion

Routing in sensor networks is a new area of research. As sensor nodes are constrained by limited battery, backup, memory, computation capacity, scalability, data aggregation, etc .Many routing protocols have been proposed by many researchers to take into account these sensors nodes constraints. So, the routing protocols design for this type of networks is a crucial challenge to improve the usage of limited network resources. It can be divided into flat, hierarchical and location routing protocol. This paper gives in detail general overview about flat routing protocols based on network structure in wireless sensor networks with their comparisons and specifies the associated problems and applications designated for each protocol. It highlights the challenges in WSN and helps the researchers to understand the work done in this domain.

For future perspective, we aim to improve the contribution of one of the protocols in order to solve the challenges identified in the discussion by evaluating their performances based on metrics: packet delivery ratio (PDR), end-to-end delay, energy consumption, throughput and confirm the improvements in the simulator NS2 network.

References

1. Kanavall, A., et al.: A flat routing protocol for sensor networks. In: Proceeding of International Conference on Methods and Models in Computer Science, Delhi, pp. 1–5 (2009)
2. Akkaya, K., Younis, M.: A survey on routing protocols for wireless sensor networks. Ad Hoc Networks Journal **3**, 325–349 (2005)
3. Salman, H.M.: Survey of routing protocols in wireless sensor networks. International Journal of Sensors and Sensor Networks **2**(1), 1–6 (2014)
4. Kumar, A., Pahuja, S.: A Comparative Study of Flooding Protocol and Gossiping Protocol in WSN. Int. J. Computer Technology & Applications **5**(2), 797–800 (2014)
5. AL-Karaki, J.N., Kamal, A.E.: Routing techniques in wireless sensor networks: a survey. IEEE Wireless Communications **11**(6), 6–28 (2004)
6. Doohan, N.V., Tokekar, S.: A Survey on Routing Techniques of Data-Centric Wireless Sensor Networks. International Journal of Computer Applications **53**(16), 1–5 (2012)

7. Heinzelman, W., Kulik, J., Balakrishnan, H.: Adaptive protocols for information dissemination in wireless sensor networks. In: MobiCom 1999 Proceedings of the 5th Annual ACM/IEEE International Conference on Mobile Computing and Networking, pp. 174–185 (1999)
8. Kulik, J., Heinzelman, W., Balakrishnan, H.: Negotiation-Based Protocols for Disseminating Information in Wireless Sensor Networks. Wireless Networks **8**, 169–185 (2002)
9. Intanagonwiwat, C., Govindan, R., Estrin, D.: Directed diffusion: a scalable and robust communication paradigm for sensor networks. In: 6 thACM/IEEE International Conf. on MobiCom, Boston, pp. 56–67. ACM (2000)
10. Intanagonwiwat, C., Govindan, R., Estrin, D., Heidemann, J., Silva, F.: Directed Diffusion for WSN. IEEE/ACM **11**, February 2003
11. Dargie, W., Poellabauer, C.: Fundamentals of Wireless Sensor Networks: Theory and Practice. John Wiley and Sons, Ltd. (2010)
12. Xiaoqing, Z., Chunlin, L., Qiongfen, Q.: An Energy Efficient Directed Diffusion Model for Wireless Sensor Networks. IEEE Wireless Communications, 1–4 (2010)
13. Braginsky, D., Estrin, D.: Rumor routing algorithm for sensor networks. In: The Proceedings of the First Workshop on Sensor Networks and Applications (WSNA), pp. 22–31 (2002)
14. Schurgers, C., Srivastava, M.B.: Energy efficient routing in wireless sensor networks. In: The MILCOM Proceedings on Communications for Network-Centric Operations: Creating the Information Force, McLean, VA, pp. 357–361 (2001)
15. Powell, O., et al.: Gradient Based Routing in Wireless Sensor Networks: a Mixed Strategy, Arxiv preprint cs 051, pp. 10–13 (2005)
16. Miao, L., Djouani, K., Kurien, A., Noel, G.: Energy-efficient algorithm based on gradient based routing in wireless sensor networks. In: Proc. Southern Africa Telecommunication Networks and Applications Conference (SATNAC), Spier Estate, South Africa (2010)
17. Chu, M., Haussecker, H., Zhao, F.: Scalable Information-Driven Sensor Querying and Routing for ad hoc Heterogeneous Sensor Networks. The International Journal of High performance Computing Applications **16**(3), 293–313 (2002)
18. Krishnamachari, B.: Networking wireless sensors: deployment & configuration, localization, synchronization, wireless characteristics, sleep-oriented MAC, efficient routing, data-centric concepts, congestion control. Cambridge University Press (2005)
19. Yao, Y., Gehrke, J.: The cougar approach to in -network query processing in sensor networks. SIGMOD Record **31**(3), 9–18 (2002)
20. Chang, W.Y.: Network-Centric Service Oriented Enterprise, pp. 167–177 Springer-Verlag, New York Inc. (2007)
21. Sadagopan, N., Krishnamachari, B., Helmy, A.: The ACQUIRE mechanism for efficient querying in sensor networks. In: Proceedings of the First IEEE International Workshop on Sensor Network Protocols and Applications (SNPA), Anchorage, AK, pp. 149–155, May 2003
22. Rajon, S.A.A., Rahman, M.M.: Energy Efficient Routing in Wireless Sensor Network. IJCIT **03**(01), 12–19 (2012)
23. Biradar, R.V., et al.: Classification and Comparison of Routing Protocols in Wireless Sensor Networks. Special Issue on Computing Security Systems, UbiCC Journal **4**, 704–711 (2009)
24. Garg, P., Rani, R.: A Survey on Wireless Sensor Networks Routing Algorithms. ICFTEM, 38–42 (2014)

25. Patel, G., Raj, N.M.: Routing Protocols to provide Quality of Service in Wireless Sensor Networks. International Journal of Research in Advent Technology **2**(2), 201–209 (2014)
26. Abdullah, M., Ehsan, A.: Routing Protocols for Wireless Sensor Networks: Classifications and Challenges. Journal of Electronics and Communication Engineering Research **2**(2), 05–15 (2014)
27. Tyagi, M.P., Jain, M.S.: Comparative Study of Routing Protocols in Wireless Sensor Network. International Journal of Advanced Research in Computer Science and Software Engineering **2**(9), 134–140 (2012)
28. Krishnaveni, P., Sutha, J.: Analysis of routing protocols for wireless sensor networks. International Journal of Emerging Technology and Advanced Engineering (ijetae) **2**(11), 401–407 (2012)

An Energy-Aware Clustering Approach Based on the K-Means Method for Wireless Sensor Networks

Ridouane El Mezouary, Ali Choukri, Abdellatif Kobbane and Mohammed El Koutbi

Abstract This paper describes a new approach to build energy aware clusters for Wireless Sensor Networks (WSN). The algorithm behind is based on the k-means method which is well known as a clustering technique and widely used in several fields of engineering and research. K-means clustering tends to find clusters of comparable spatial extent (density clustering). We try to enhance the clustering process by selecting nodes as clusters that are centric and have a high level of energy. This will give the same QoS results as given by the K-means approach with a reduction of energy consumption and a prolongation of the lifetime of the sensor network. For the simulation purposes, we have implemented our approach on the OLSR routing protocol which has been selected by our research team as a test bed routing protocol. We have compared our new approach (OLSR-Kmeans-Energy) with a previous work we developed and where we use a Max-Min heuristic (OLSR-MaxMin-Energy) to enhance QoS parameters and the lifetime of the WSN. Our new approach seems to give better results than the MaxMin approach.

Keywords Clustering · K-means · Routing protocols · Energy aware routing protocols · Wireless Sensor Networks

1 Introduction

Sensors are small embedded sensing platforms with computing and communication capabilities, which combine low cost, flexible and fast deployment, resilient self-management and embedded intelligence for cooperatively delivered, value added services. Such nodes are called sensor nodes and are a type of a transducer that converts some physical phenomenon such as heat, light, and sound into electrical signals. Each sensor node is capable of only a limited amount of processing.

R. El Mezouary(✉) · A. Choukri · A. Kobbane · M. El Koutbi
SIME Lab, MIS Team, ENSIAS, Mohammed V University, Rabat, Morocco
e-mail: {ridouane.elemezouary,choukriali}@gmail.com, {kobbane,elkoutbi}@ensias.com

E. Sabir et al. (eds.), *The International Symposium on Ubiquitous Networking*,
Lecture Notes in Electrical Engineering 366,
DOI: 10.1007/978-981-287-990-5_26

But when coordinated with the information from a large number of other nodes, they have the ability to measure a given physical environment in great detail. Thus, WSNs can be described as a collection of sensor nodes which coordinate to perform some specific action. Unlike traditional networks, sensor networks depend on dense deployment and coordination to carry out their tasks.

Routing is a very important process in WSNs. The concept of routing is a set of procedures to ensure the establishment and maintenance of paths between the networks nodes. When increasing the density of WSNs, routing protocols may generate more control traffic and a heavy overload on sensor nodes. To solve this problem, several solutions have been proposed and some of them use a clustering approach (classification) to reduce control traffic and nodes overhead. The clustering approach can be also used to address the problem of scalability in case of large WSNs. Therefore, the routing process can be divided in two main functions: communication intra-cluster and communication inter-cluster.

In this paper, we aim to use some learning techniques to build intelligent and energy-aware clusters. Several supervised and unsupervised classification techniques exist in the literature. We have chosen the k-means approach which is one of the most popular in the field of clustering and datamining.

The remaining of this paper is organized as follow: Section 2 presents some of the works that have use k-means in the field of WSN; Section 3 gives more details about the algorithm based on k-means and that will be implemented in sensor nodes. Section 4 discusses some numerical and simulation results and the last section of this paper summarizes the work done and presents some future works.

2 Related Work

In this section, we present a few rare initiatives that have tried to exploit classification methods in mobile ad hoc networks.

The classification is an essential task for data analysis and interpretation. It is used to divide a set of data into groups or clusters [1]. The most popular method of mobile centers is undoubtedly the k-means method [11] which is designed to process massive data sets.The k-means algorithm [12] uses intra-class inertia based on a criterion called distance for classifying a set of objects described by variables.

The authors of [2] exploited the k-means method to calculate Meta-clusters by adapting it to a peer-to-peer environment. Each cluster is constructed as a weighted combination of source data. However, they had difficulty to specify beforehand the number of clusters. In [3] the authors proposed an approach that combines the content-based subscriptions in a compact data format (Bloom lters) and ODMRP (On-Demand Multicast Routing Protocol). In order to build a mesh of dynamic scattering optimized, the approach is extended to use the context of the middleware layer. Dynamic multicast groups are created by aggregating subscriptions. The grouping method used by authors is k-means. In Article [4] the authors based on the classification method k-means for study the strategy of a hidden Markov model

(HMM) for intrusion detection using a multivariate Gaussian model for observations that are then used to predict an attack that exists in the form of a hidden state. This model consists of a network self-organized for event classification, a classifier observation, a drift detector, a profile estimator, an accelerator GMM (Gaussian Mixture Model) and an HMM engine. The k-means method is also used by the authors of the article [5] to limit the flood search to a circular area around the local source in a MANET deploying RDMAR (Relative Distance Micro-discovery Ad Hoc Routing) by grouping nodes with similar characters in a same cluster. Remi Badonnel and al. [6] have proposed an approach for monitoring errors in ad hoc networks. This approach is based on a measure of information theory adapted to ad hoc nodes intermittent and can detect network failures by inference. They defined a distributed monitoring system with multiple detection methods of collaboration and a self-configuration mechanism based on k-means algorithm.

3 Algorithm Description

This work falls within the scope of research to conceive a strong WSN clustering algorithm that generates equitable, energy-aware and stable clusters. Thus this work represents an improvement of our previous contributions [7, 8, 9].

In the absence of any assumption about the distribution of nodes in a WSN, an unsupervised classification of nodes into classes is required. We propose a model based on geometric considerations. This proposal requests define a measure of the proximity between nodes.

3.1 Partition

C is a partition of X if and only if C satisfies the following properties:

- $C_a \subset X$ for all $C_a \subset C$
- $\cup_{a=1}^{t} C_a = X$
- $C_a \cap C_b = \phi \ \forall \ (a, b)$ such as $a \neq b$

The property (3) expresses that the clusters formed are disjoint; each object of X belongs only to a single cluster of C.

3.2 Similarity and Dissimilarity

To define the homogeneity of observations set, it is necessary to measure a similarity between two observations. Then we introduce the concepts of similarity and dissimilarity:

Dissimilarity is a function d which associates a value in IR^+ to each pair (x_i, x_j) such that:

$$d(x_i, x_j) = d(x_j, x_i) \geqslant 0$$
$$d(x_i, x_j) = 0 \Rightarrow x_i = x_j$$

Conversely, another possibility is to measure the resemblance between observations using a similarity:

Similarity is a function s to which associates a value in IR^+ to each pair (xi, xj) such that:

$$s(x_i, x_j) = s(x_j, x_i) \geqslant 0$$
$$s(x_i, x_i) \geqslant s(x_i, x_j)$$

3.3 Distance

The choice of the distance is a key issue for classification methods. To offer a relevant measure of similarity between elements, it is necessary to well use the available information at the nodes.

The Minkowski distance is the most used to determine the similarity between elements:

$$d(x_i, x_j) = \left(\sum_{k=1}^{P} | V_k(x_i) - V_k(x_j) |^l \right)^{\frac{1}{l}}$$

Where $V_k(x_i)$ is the value of the object x_i on the variable v_k. Depending on the values taken by the parameter l, we talk about:

- Euclidean distance ($l = 2$)
- Manhattan distance ($l = 1$)
- Chebychev distance ($l = \infty$)

We note that the metrics commonly used to analyze the WSN performances such as density, mobility and energy can be used to express distance.

3.4 Classification of Nodes in a WSN (Clustering)

Our objective is to divide the nodes into homogeneous and distinct clusters. To do this, we start from the definition of network inertia that can be modelled by a cloud of points.

Consider a network of n nodes $x_1, ..., x_n$ and u_G designates the centroid of nodes cloud:

$$u_G = \frac{1}{n}\sum_{i=1}^{n} x_i \tag{1}$$

The cluster inertia is by definition the sum of squares of distances from the center:

$$I = \sum_{i=1}^{n} d^2(x_i, u_G) \tag{2}$$

$$= \sum_{i=1}^{n} \| x_i - u_G \|^2 \tag{3}$$

Assume that the network consists of P distinct clusters $C_1, ..., C_p$ Each of these clusters having as centroid u_{C_k}.

We can then decompose the total inertia of the cloud of nodes as follows:

$$\begin{aligned} I &= \sum_{i=1}^{n} \| x_i - u_G \|^2 \\ &= \sum_{k=1}^{P}\sum_{i \in C_k} \| x_i - u_G \|^2 \\ &= \sum_{k=1}^{P}\sum_{i \in C_k} \| x_i - u_{C_k} + u_{C_k} - u_G \|^2 \\ &= \sum_{k=1}^{P}\sum_{i \in C_k} (\| x_i - u_{C_k} \|^2 + \| u_{C_k} - u_G \|^2) \\ &= \sum_{k=1}^{P}\sum_{i \in C_k} d^2(x_i, u_{C_k}) + \sum_{k=1}^{P} d^2(u_{C_k}, u_G) \end{aligned} \tag{4}$$

The first term called intra-cluster inertia calculates the sum of distances between nodes and their centroid. Low intra-cluster inertia indicates that the nodes in the same cluster are more nearer.

The second term called the inter-cluster inertia calculates the sum of distances between centroids of clusters and global centroid, that is to say the separation degree of clusters.

$$I = I_{intra} + I_{inter} \tag{5}$$

From a formal point of view, the optimal partition is:

- That minimizes the intra-class inertia.
- Or that maximizes the interclass inertia.

These two affirmations have the same meaning because:

- $I_{intra} + I_{inter} = I$
- I doesn't depend on the clusters.

Thus the optimal partition would be defined as follows:

$$C_k^{optimal} = Argmin_{C \in c_k} \sum_{k=1}^{P} \sum_{i \in C_k} d^2(x_i, U_{C_k}) \tag{6}$$

The objective of classification algorithms based on this principle is the search of optimal partition. In practice it is impossible to generate all patterns of clustering for evident reasons of complexity. We then seek a scheme sufficiently close to the optimum. This optimum is obtained in an iterative manner by improving an initial scheme randomly selected by reallocating objects around mobile centers.

In order to partition the nodes into clusters, we used this technique (iterative reallocation) based on k-means algorithm.

3.5 K-Means Method

The k-means algorithm [10] is a partitioning method widely used in various application areas. From P separate clusters, the k-means algorithm assigns iteratively objects $(x_1, .., x_n)$ at P centers of clusters $(u_1, .., u_P)$, followed by calculating the positions of the new centers. The stopping algorithm is a criterion fixed by the user and can be:

- Achieve a limited number of iterations;
- Clusters formed remain the same after two successive iterations;
- The inertia intra-cluster is not improving between two iterations.

We start from the following considerations at the end of a determined step j:

- Centers of clusters $(u_1^j, .., u_1^P)$ have been calculated;
- The classes $(C_1^j, .., C_1^P)$ were obtained by assigning at the center (C_k^j) the n_k^j nearest nodes.

We define the following quantity at the end of step j:

$$Q_i = \sum_{i=1}^{P} \sum_{i \in C_k^j} \| x_i - u_k^j \|^2$$

The redefinition sub-step of new centers and reassigning of nodes (next iteration) requires:

1. Recalculate centers of clusters (u_1^{j+1}, ..., u_P^{j+1}) based on points belong to each of clusters (C_1^{j+1}, ..., C_P^{j+1}) possessing respectively n_k^j elements. We have :

$$u_k^{j+1} = \frac{1}{n_k^j} \sum_{j \in C_k^j} x_i$$

And:

$$W_{j+1} = \sum_{k=1}^{P} \sum_{i \in C_k^j} \| x_i - u_k^{j+1} \|^2$$

Where u_k^{j+1} is the center of the cluster C_k^j and W_{j+1} is the inertia intra-clusters associated to clusters (C_1^j, ..., C_P^j). We will have:

$$\begin{aligned}
Q_j &= \sum_{K=1}^{P} \sum_{i \in C_k^j} \| x_i - u_G \|^2 \\
&= \sum_{K=1}^{P} \sum_{i \in C_k^j} \| x_i - u_k^{j+1} + u_k^{j+1} - u_G \|^2 \\
&= \sum_{K=1}^{P} \sum_{i \in C_k^j} \left(\| x_i - u_k^{j+1} \|^2 + \| u_k^{j+1} - u_k^j \|^2 + 2 < x_i - u_k^{j+1}, u_k^{j+1} - u_k^j > \right) \\
&= W_{j+1} + \sum_{K=1}^{P} \sum_{i \in C_k^j} \| u_k^{j+1} - u_k^j \|^2 + 2 \sum_{k=1}^{P} < \sum_{i \in C_k^j} (x_i - u_k^{j+1}), u_k^{j+1} - u_k^j > \\
&\geqslant W_{j+1}
\end{aligned}$$

The vector $\sum_{i \in C_k^j} \left(x_i - u_k^{j+1} \right)$ being the null vector by definition of u_k^{j+1}.

2. Reassign nodes to the nearest centers. We then obtain new clusters (C_1^{j+1}, ..., C_P^{j+1}) and we define:

$$Q_{j+1} = \sum_{k=1}^{P} \sum_{i \in C_k^{j+1}} \| x_i - u_k^{j+1} \|^2$$

After reassignment all distances decrease because each node x_i is assigned to the cluster center u_k^{j+1} minimizing thereby the gap: $\| x_i - u_k^{j+1} \|^2$. We therefore have:

$$Q_{j+1} \leqslant W_{j+1}$$

Thus, for each j we have proved the following inequality:

$$Q_{j+1} \leqslant W_{j+1} \leqslant Q_j$$

So we have in particular

$$W_{j+2} \leqslant Q_{j+1} \leqslant W_{j+1}$$

This shows that after each algorithm iteration, the improvement of nodes classification is effective within the meaning of the Intra criterion. Because the intra-cluster inertia of the optimal partition is the smaller, the margin of improvement is finite. This implies that the algorithm converges inevitably.

At the beginning each node represents its own cluster. Thereafter we make a series of ascending partitions by combining the nodes belonging to the same neighborhood into same cluster until reaching a stable number of clusters. We present below the adapted k-means algorithm:

Input: a WSN of n nodes.
Output: Network virtually partitioned into P clusters: $P = (C_1, .., C_P)$.
Distance (d): The density of nodes is the dissimilarity measure we have used.
Step 0:
1. Initialization with n centers $(u_1^0, .., u_n^0)$: Each node is a cluster.
2. Creation of an initial partition $P_0 = (C_1^0, ..., C_k^0)$:
a. Initialize l to 1 (l is the iteration index).
b. assign to u_l^0 its two-hop neighbors:

$$C_l^0 = (x_i \in network \mid d(x_i, u_1^0) \leqslant 2hops)$$

c. Remove from list of centers the nodes assigned to the center u_l^0;
d. Move to the center $l + n_{C_l}$.
e. Repeating steps b through d until all nodes are assigned.
f. Calculation of new centers of k clusters obtained $(u_1^1, .., u_k^1)$: nodes having the greatest density within their neighbors and whose energy is greater than or equal to the average of neighboring energies.
Step t:
Creation of a new partition $P_t = (C_1^t, ..., C_2^t)$ by assigning to each center U_l^t its two-hop neighbors:

$$C_l^t = (x_i \in network \mid d(x_i, u_1^t) \leqslant 2hops)$$

3. Centers assigned to other centers are removed from the list of centers.
4. The isolated nodes (unassigned) are added to the list of centers.
5. Calculate the centers of k clusters obtained:$(u_1^{t+1}, ..., u_k^{t+1})$.
6. Repeat steps 3 to 6 until that a stable partition is achieved (structure of partition P_t equals that of the partition $P_{(t+1)}$ or reach n iterations (n: number of nodes).

4 Simulations Results

We have begun our evaluation by the OLSRkmeansEnergy protocol then we have done some comparisons with a previous developed clustering approach based on the MaxMin heuristic [9] (OLSRMaxMinEnergy developed by our team research). Simulations have been carried out on NS-2 version 2.34. Table 1 provides all chosen simulation parameters:

Table 1 Simulation parameters

Parameter	**Value**
Routing Protocol	OLSRkmeansEnergy OLSRMaxMinEnergy
Simulation Time	500 s
Number of nodes	10, 20, 30, 50, 70, 90, 100
Environment Size	$500m \times 500m$
Traffic Type	CBR
Maximum Speeds	10m/s
Mobility Model	Random Waypoint

Performance comparison of the two algorithms has been conducted by analysing their stability and by examining the number of generated clusters during the simulation period.

4.1 Number of Generated Clusters

A reduced number of clusters means that each cluster contains a maximum number of nodes and each node is eccentric in its neighbourhood. Figure 1 shows the evolution of the number of generated clusters using OLSRkmeansEnergy and OLSRMaxMinEnergy according to the network density. We note that OLSRkmeansEnergy generates few clusters especially for denser networks. We can conclude that the clusters formed by OLSRkmeansEnergy are more stable and more durable.

4.2 End to End Delay

The figure 2 shows the behaviour of OLSRkmeansEnergy and OLSRMaxMinEnergy protocols in terms of the average time needed to route a packet from a source to a destination. We observe that the required time to transmit a packet using the OLSRkmeansEnergy protocol is very interesting in particular in dense networks. This can be interpreted by the fact that OLSRkmeansEnergy forms less clusters.

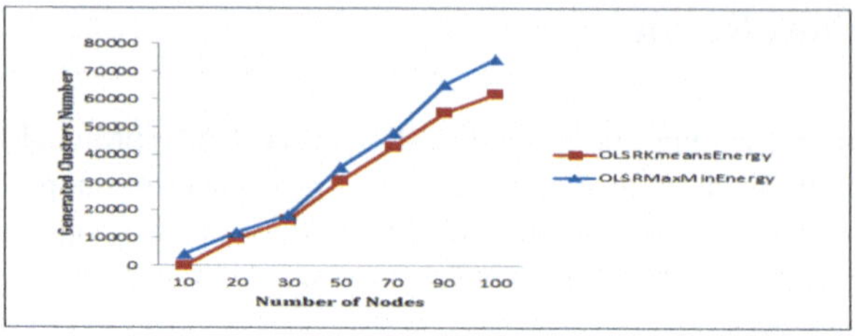

Fig. 1 Number of generated clusters vs. network density

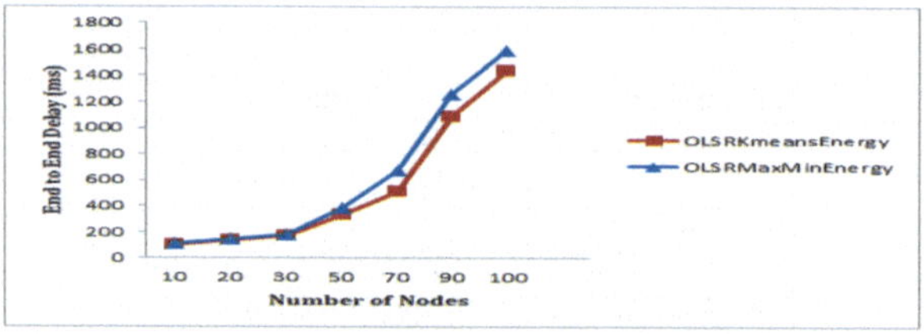

Fig. 2 Average end to end delay vs. network density

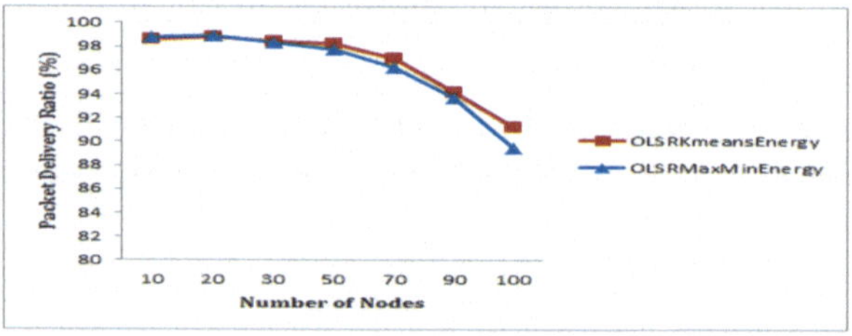

Fig. 3 Packet delivery ratio vs. network density

Inter-cluster interactions are limited to clusterheads exchanges, this helps to reduce the path length regarding of the number of hops and this consequently improves the average of the transmission delay.

4.3 Packet Delivery Ratio

In a WSN, several events may cause packet loss such as congestion, high error rate or broken link. This will have an impact on the packet delivery ratio.

The figure 3 shows the behaviour of OLSRkmeansEnergy and OLSRMaxMinEnergy protocols based on a CBR traffic in terms of packet delivery ratio according to network density. We observe that OLSRkmeansEnergy reduces the number of packets lost compared to OLSRMaxMinEnergy.

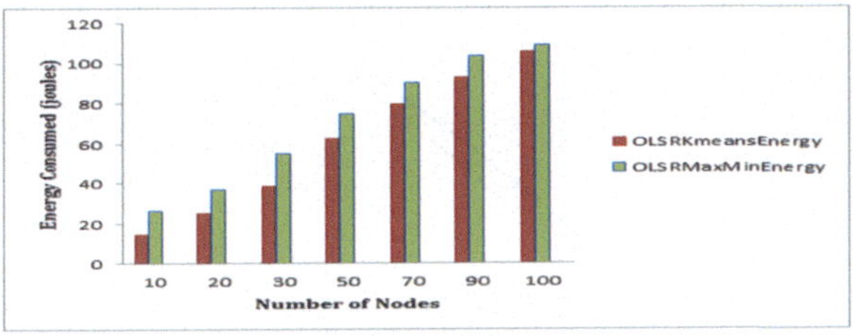

Fig. 4 Average Energy Consumption vs. network density

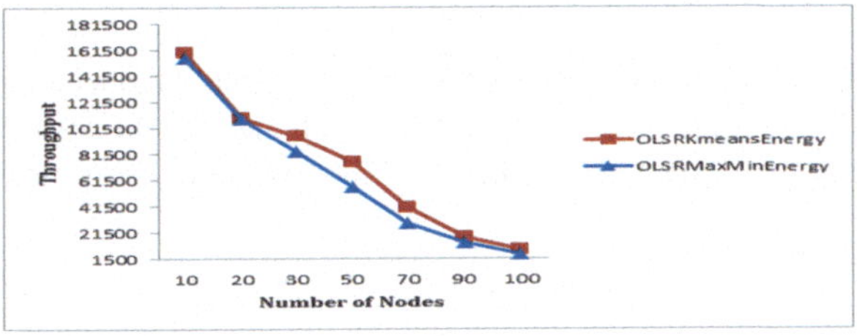

Fig. 5 Throughput vs. network density

4.4 Energy Consumption

Energy consumption is mainly impacted by transmission and reception of data packets. We have evaluated OLSRKmeansEnergy and OLSRMaxMinEnergy protocols regarding to the energy consumption when using a CBR traffic.

Figure 4 shows that OLSRKmeansEnergy is more efficient compared to OLSR-MaxMinEnergy in terms of energy consumption. This allows prolonging the network lifetime.

4.5 Throughput

The figure 5 shows the comparison of OLSRkmeansEnergy and OLSRMaxMinEnergy protocols in terms of throughput according to the network density. We observe that OLSRkmeansEnergy improves the throughput that is organically linked to packet losses.

4.6 Control Routing Overhead

Figure 6 shows the evolution of the number of control packets used in the evaluated protocols: OLSRMaxMinEnergy and by OLSRk-meansEnergy depending on

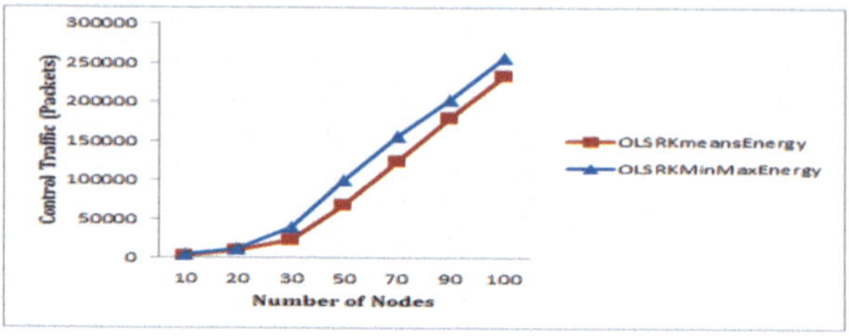

Fig. 6 Control traffic overhead vs. network density

network density. We notice that OLSRkmeansEnergy has a reduced number of control packets. This will preserve the link bandwidth and should positively influence other performance QoS parameters such as PDR (packet delivery ratio) and energy consumption.

5 Conclusion

In this paper, we have used the k-means classification method to enhance the process of clusters design. The elected cluster head will be eccentric and will have enough energy to support the routing operations. We have shown that an optimal partition exists (at least theoretically), and the algorithm converges inevitably. The fall of a clusterhead causes isolation of all members of the cluster with the rest of the network.

Simulations have shown that the results obtained by OLSRkmeansEnergy exceed those by OLSRMinMaxEnergy in terms of stability (number of nodes per cluster), end-to-end delay, packet delivery ratio, power consumption, throughput, Control traffic overhead.

As a future work, it would be interesting to consider some k-mean variants (c-means) to introduce incertanity in the decision process.

References

1. Larose, D.T.: Discovering Knowledge in Data: An Introduction to Data Mining. John Wiley and Sons, Inc., Hoboken (2005)
2. Dornbush, S., Joshi, A.: StreetSmart traffic: discovering and disseminating automobile congestion using VANETs. In: IEEE 65th Vehicular Technology Conference, VTC2007-Spring, pp. 11–15 (2007)
3. Yoneki, E., Bacon, J.: Distributed multicast grouping for publish/subscribe over mobile ad hoc networks. In: 2005 IEEE Wireless Communications and Networking Conference, vol. 4, pp. 2293–2299 (2005)
4. Khanna, R., Liu, H.: System approach to intrusion detection using hidden Markov model. In: IWCMC 2006 Proceedings of the 2006 International Conference on Wireless Communications and Mobile Computing, pp. 349–354. ACM (2006)

5. Hamad, O.F., Kang, M.-Y., Jeon, J.-H., Nam, J.-S.: Neural network's k-means distance-based nodes-clustering for enhanced RDMAR protocol in a MANET. In: IEEE 2008 Signal Processing and Information Technology, pp. 192–197 (2008)
6. Badonnel, R., State, R., Festor, O.: Self-configurable fault monitoring in ad-hoc networks. Ad Hoc Networks **6**(3), 458–473 (2008)
7. Choukri, A., Habbani, A., El Koutbi, M.: A hierarchical version of OLSR for MANET. World Applied Sciences Journal **21**(12), 1739 (2013)
8. Choukri, A., Habbani, A., El Koutbi, M.: Efficient heuristic based on clustering approach for OLSR. Journal of Computer Networks and Communications **2013** (2013)
9. Choukri, A., Habbani, A., El Koutbi, M.: An energy efficient clustering algorithm for MANETs. In: ICMCS 2014 (2014)
10. MacQueen, J.: Some methods for classification and analysis of multivariate observations. In: Proceedings of the Fifth Berkeley Symposium on Mathematical Statistics and Probability, pp. 281–297, Berkeley (1967)
11. Forgy, E.W.: Cluster Analysis of Multivariate Data: Eficiency versus Interpretability Models. Biometrics **61**(3), 768–769 (1965)
12. Benzecri, J.P.: L'analyse des donnes: la taxinomie, Tome 1. Dunod, Paris (1973)

An Efficient Key Establishment Protocol for Wireless Sensor Networks

Yassin Maleh and Abdellah Ezzati

Abstract Wireless sensor network represents a new perspective for many applications, including security and surveillance, control, actuation and maintenance of complex systems and fine-grain monitoring of indoor and outdoor environments. However, due to limited resources of sensors and hostile environments in which they could be deployed, this type of networks is vulnerable to several types of attacks similar to those occurring in ad hoc networks. Therefore, it is necessary to use effective mechanisms to protect this type of network. Cryptography can play an important role in detecting and preventing security attacks. Our contribution is intended for applications using the WSN nodes with low mobility and require a high level of security. Unlike most methods in the literature for specific topologies, our security scheme can cover both the needs of flat and hierarchical topologies. In this paper, we proposed a new lightweight cryptography algorithm based on LEAP+. Simulations results of the proposed model are presented using TOSSIM simulator and the complexity of the protocol is analyzed and compared with other symmetric schemes.

Keywords Wireless sensor network · Cryptography · Key management · Key establishment · LEAP+ · Performance evaluation

1 Introduction

Wireless Sensor networks (WSNs) are highly vulnerable against attacks, it is very important to have certain mechanisms that can protect the network from all kinds of attack. The different characteristics of wireless sensor networks (energy limited, low-power computing, use of radio waves, etc...) expose them to many security

Y. Maleh(✉) · A. Ezzati
LAVETE Laboratory, Faculty of Science and Technology, Hassan 1st University, Casablanca Road 3.5 KM Hassan 1st University, BP 539, Settat, Morocco
e-mail: {y.maleh,abdeezzati}@uhp.ac.ma

E. Sabir et al. (eds.), *The International Symposium on Ubiquitous Networking*,
Lecture Notes in Electrical Engineering 366,
DOI: 10.1007/978-981-287-990-5_27

threats. The most important tools that ensure security and its services are the security primitives [1] which are symmetric key encryption, public key cryptography and hash functions [2]. SKE and Hash functions are the primitives that can be called the building blocks which offer a basic protection of the information flow, because these assure the confidentiality and integrity of the channel. Public key cryptography assure protection from the participation of external entities and also eliminate the problem of a malicious insider which try to use more than one identity. PKC assures this by allowing authentication of the peers involved in the information exchange. Based on the primitives it is possible to create a better network services. It is also equally important to have a key management system for constructing a secure key infrastructure. Each primitive has different requirements in-terms of processing power and word size etc.

The key management protocols ensure safe routes within the sensor networks. The idea is to before opening the channel securely the nodes must share some security credentials. We term these security credentials as keys. These security keys itself needs a secure system which takes the responsibility of creating, maintaining and distributing security keys, which we called key management systems. The limitation introduced above discouraged us to use resource-intensive algorithm for most of the scenarios of sensor networks.

The paper is structured as following. In the next section, we present in related work. Section 3 proposes our security model for WSNs. Section 4 describes the algorithms and analyses defense methods against routing attacks. In the Section 5, we present performances analyses and discussion of our scheme. Finally, the paper ends with a conclusion and future works.

2 Related Works

At present, there are many security scheme designed for general ad hoc network, but very few for wireless sensor network specifically. In this section, we describe three typical security protocols for sensor network and summarize their features.

Perig et al [3] proposed SPINS protocol which is a suite of security building blocks proposed. It is optimized for resource constrained environments and wireless communication. SPINS has two secure building blocks: SNEP and μTESLA. SNEP uses a shared counter between the two communicating parties and applies the counter in calculating encryption and a message authentication code (MAC) to provides data confidentiality, semantic security, data integrity, two-party data authentication, replay protection, and weak message freshness. What's more, the protocol also has low communication overhead, for the counter state is kept at each end point and the protocol only adds 8 bytes per message. For applications requiring strong freshness, the sender creates a random nonce (an unpredictable 64-bit value) and includes it in the request message to the receiver. The receiver generates the response message and includes the nonce in the MAC computation. μTESLA provides authenticated broadcast for severely resource-constrained environments. μTESLA constructs authenticated broadcast from symmetric primitives, but introduces asymmetry with delayed key disclosure and one-way function key chains. SPINS realizes an authenticated routing application and a security two-party key agreement with SNEP and μTESLA separately with low storage, calculation and communication consumption. However, SPINS still have some underlying problems. Due to use the pairwise key pre-distribution scheme in the security routing protocol, SPINS rely on the base station excessively.

Deng et al. [4] suggested an Intrusion-tolerant routing protocol for wireless sensor networks (INSENS). This protocol leverages some concepts of SPINS to implement intrusion-tolerant multi-hop routing for WSNs. For example, we utilize keyed message authentication codes (MAC) similar to SNEP to verify the integrity of control packets. We also employ the concept of a one-way hash chain seen in μTESLA, but use the chain to provide one-way sequence numbers for loose authentication of the base station, rather than as the key release mechanism seen in SPINS. In addition, INSENS limits flooding of messages by allowing communication only between the base station and the sensor nodes, and by having sensor nodes drop duplicate messages. These techniques essentially limit the damage an intruder may cause. Together, these design choices ensure that an intruder may be able to take out a small part of the network, but cannot compromise the entire network. What's more, INSENS minimizes computation by performing all heavy-duty computations at the base station(s) and minimizing the role of sensor nodes in building routing tables, or dealing with security and intrusion-tolerance issues. Although INSENS limits the damage in the area around the invaders or the downstream area of the invaders, the damage range still will be very big when the internal attackers are around the base station.

Zhu et al proposed LEAP (Localized Encryption and Authentication Protocol) [5] protocol for sensor networks that is designed to support in-network processing, while at the same time restricting the security impact of a node compromise to the immediate network neighborhood of the compromised node. The design of the protocol is motivated by the observation that different types of messages exchanged between sensor nodes have different security requirements, and that a single keying mechanism is not suitable for meeting these different security requirements.

Hence, LEAP supports the establishment of four types of keys for each sensor node-an individual key shared with the base station, a pairwise key shared with another sensor node, a cluster key shared with multiple neighboring nodes, and a group key that is shared by all the nodes in the network. The protocol used for establishing and updating these keys is communication and energy efficient, and minimizes the involvement of the base station. LEAP also includes an efficient protocol for inter-node traffic authentication based on the use of one-way key chains. A salient feature of the protocol is that it supports source authentication without precluding in-network processing and passive participation. LEAP adopts different types of key to different network protocol message packages, and therefore can support multiple communication mode. What' more, this protocol weakens the role of base station. The main defect of the protocol is that the protocol increases overhead traffic, the failure of single point is serious, and the storage capacity is large. Besides, LEAP is inefficient in the formation and update of cluster key. What' more, the HELLO news in the network is written in plaintext rather than ciphertext, which may lead nodes to respond to invalid news and waste node resources.

The diagrams SPINS and LEAP use master keys in the key establishment. Which reduces the storage key in the memory nodes. However, resistance to attacks is low. Because the master key can be compromised at any time, set the key after deployment using this key can be compromised too. By adopting a symmetrical system, they are the most suitable and among the fastest in terms of calculation. Note that the symmetrical patterns are expensive in operations (if any) as key renewal and revocation using secret keys to exchange other secret keys. The problem is easier in asymmetrical patterns since the public key does not need to be secret.

3 The Proposed Model

In the previous chapters it has been tried to give a general description of some of the state of the art algorithms available in the literature. It can be concluded that sensor network nodes are mostly deployed in unattended adversarial environment for example battlefield. Therefore it is extremely important for the applications of many sensor networks to have a security mechanism which provide authentication and confidentiality. It has been also discussed briefly that providing security to sensor network is specifically challenging because of resource limitations of sensor nodes. The reason of using symmetric algorithm for encryption is that a sensor node consist of slow processor approximately 7.8 MHZ also they are low powered devices with memory up-to 4KB [6]. Keeping in view the above limitations different state of the art algorithms has been studied in depth for the proposal of a key-management scheme based on symmetric shared keys. The two base algorithms for the proposed schemes are "Leap+" and "improving key negotiation in transitory key schemes". A straight forward and practical approach for bootstrapping secret keys in sensor networks is predeployed keying. The degree of key sharing between the nodes in the system is a very important design consideration for security protocols that is based on symmetric keys. There can be two extremes for data encryption and authentication at one side there could be a network-wide keys and at the other extreme there could be a key sharing approach in which all the secure communication is based on keys that are shared pairwise between two nodes. The former case has the lowest storage costs and is efficient in-terms of energy because there is no need for communication between nodes for establishing additional keys. The security disadvantage associated to this scheme is that the compromise of a single node reveals the global key and hence the whole network becomes at risk. The latter case is ideal in-term of security because the compromise of node does not reveal any keys that are used by other nodes in the network.

The unique issue that needs to be considered in sensor network before selecting a key sharing approach is its impact on the effectiveness of in-network processing. The proposed algorithm support in-network processing and also provide security properties similar to those provided by pairwise key sharing schemes. Similar to leap+ and other protocol the proposed solution is also based on the observation that different types of messages exchanged between sensor nodes have different security requirements which leads us to the conclusion that a single keying mechanism is not suitable for meeting these different security requirements. Like leap+ the proposed algorithm also support the establishment of four different types of keys

- **Individual key:** Shared with the base station
- **Pairwise key:** Shared with another sensor node
- **Cluster key:** With multiple neighboring nodes it is shared
- **Global key:** Shared by all nodes in the network

3.1 Assumption In-Term of Network and Security

The following important assumption has been made while studying and designing the protocol

- A static sensor network where nodes are not mobile
- The base station work as a controller
- The power supplied to the base station is supplied with long-lasting power
- All nodes are equal in computational and communication capabilities
- Every node has enough space to store hundred of bytes of keying materials
- Nodes installation can be done both either through aerial scattering physical installation.
- In advance The immediate neighboring nodes of any sensor node are not known
- All the information a node holds becomes known to the attacker If it is compromised
- Attacks of the physical layer and media access control layer are not considered

3.2 Basic Overview of the Algorithm

Sensor nodes exchange different kinds of packets based on different criteria as an example:

- Control Packet Vs Data Packets
- Broadcast Packets Vs Unicast Packets
- Queries or Commands Vs Sensor Readings and so on

The security requirements for each type of packets are different. Authentication is required for all types of packets whereas confidentiality is required for some types of packets. As it has been explained that a single keying mechanism is not appropriate for all the secure communications needed in sensor networks. So the algorithm support four types of keys for each sensor node:

3.2.1 Individual Key

To have secure communication between the node and the base station this key is used. Every node in the network have its own individual key. Individual key is also important in the sense that it can be used to compute the message authentication code if the message is to be verified by the base station. This can be also used to send alert to the base station if there is any abnormal behavior observed. Base station can use individual key to encrypt any sensitive information such as keying material or special instructions to individual node. It is important to mention that individual key is preloaded into the network before deployment. The individual key is generated as follows:

$$IKu = fKm\,(u) \tag{1}$$

Where *IKu* is the individual key, *f* is a pseudo-random function, *Km* is the master key, and *u* is any node for which we want to find individual key.

It is is also possible to save storage needed to keep all the individual keys the controller might only keep its master key. It compute the individual key on the fly when needed. The individual key has been find by encrypting the node id with the global key (master key) by using Advance encryption standard algorithm.

3.2.2 Establishing Pairwise Keys

The most important step is to have pairwise key between nodes. It is very important from the security point of view as if the key is compromised its effect is localized. The pairwise key is shared between one-hop neighbors. A sensor node communicate with its immediate neighbor through pairwise key. The important assumption made to establish pair wise keys are:

- Node doesn't know its neighbor pairwise key before deployment. Pairwise key is created after deployment
- The nodes of the network are stationary nodes
- A node that is added to the network will discover most of its neighbors at the time of deployment.

The pairwise is generated by following these steps:

1- Key Predistribution

An initial key generated by the controller is loaded to each node. Each node then derive its master key as:

$$Ku = fKin\,(u) \tag{2}$$

Where *Ku* is the master key generated by the node *u*, *f* is a pseudo-random function, *Kin* is the initial key, and *u* is any node for which we want to derive the master key.

2- Neighbor Discovery

Every node try to find its neighbor by broadcasting a HELLO message. This Hello message contains id of the node. Also a timer is started which fires after time Tmin. This node then wait for any node *v* which respond tho this HELLO with an ack message having the id of the node *v*. Ack from neighbor is authenticated using master key *kv*. The master key is derived as:

$$Kv = fKin\,(v) \tag{3}$$

Where *Kv* is the master key of node *v*, *f* is a pseudo-random function, *kin* is the initial key, and *v* is any node that want to find its master key.

3- Pairwise key Establishment

Any two nodes between the network let say node *u* and *v* compute the pairwise key as:

$$Kuv = fKv\,(u) \tag{4}$$

Where *kuv* is the pairwise key between node *u* and *v*, *f* is a pseudo-random function, *kv* is the master key of node, and *u* is the node id of any node *u*.

4- Key Erasure

When the timer expires after *Tmin*, node *u* erases *Kin* and all the master keys of its neighbor which was computed during the neighbor discovery phase.

3.2.2.1 Analysis in Term of Performance

The pairwise key establishment scheme implemented has the following computational overhead. The two nodes that want to establish the pairwise key has to verify a message authentication code from every neighbors and also evaluate a pseudo-random function to create pairwise key. The ack message has two fields one for a node Id and one for a MAC. Hello message only includes node id. The storage space required is only for one key *Kin*. Hence it can be easily concluded that the computational, communication and storage overhead for establishing pairwise key for our scheme is small.

3.2.2.2 Analysis in Term of Security

This scheme is very efficient in-term of security thanks to the pairwise key established between two nodes that even if a key is compromised its effect is localized and the whole network is saved from being compromised.

3.2.3 Establishing Cluster Keys

Cluster key is established between a node and all its neighbors. Using cluster key a node encrypt broadcast message. To establish the cluster key any node u generate a random key and then encrypt this random key with the pairwise key already generated. Then the generated cluster key is transmitted to each neighbors.

3.2.4 Global Key Establishment

A key shared by the base station and every node is the global key. It is important and is used when the base station (controller) want to generate a confidential message.

4 Our Contribution

The critical assumption that leap+ has considered is that within *Tmin* a node cannot be compromised. This idea seems practical but only in an extreme ideal condition. There is possibility that *Tmin* would in reality be greater than the one assumed. As an example if nodes are dropped and scattered from airplanes, the scattered nodes may arrive in different parts of the network at different times even if dropped simultaneously and hence will need some time to setup the network and exchange pairwise key. Taking the advantage of this an adversary may observe a node and obtains the key and if the global key is compromised the whole network becomes at risk. Since this is a very serious threat to security different algorithms has been studied and proposed a model that detect the compromised node and take necessary action to delete the compromised node from the network [7] [8] [9] [10] [11]. Some of the algorithms available in literature to detect compromise nodes are:

- Detecting Compromised Nodes in Wireless Sensor Network by Rick Mckenzie, Min song, Mary Mathews, sachin shetty,
- A Framework for identifying Compromised Nodes in sensor Network by Qing Zhang, Ting Yu, Peng Ning
- Malicious Node Detection in wireless sensor Networks using by Idris M.Atakli, Hongbing Hu, Yu chen
- Sensor Node Compromise Detection, The location Perspective by Hui song and liang xie

A time called Tp has been chosen and run it iteratively after a certain period of time to check if there is any node compromised. It can be run directly after pairwise keys are exchanged and the nodes start communication so that to be sure that none of the node is compromised and the network has successfully exchanged all the keys securely. The steps to do the tasks would be as follows:

Step 1: Periodic check for Node detection

An iterative-periodic *PERIODIC-CHECK (Tp)* routine has been run on every node to check if it is under attack or not. Duration of "*Tp*" is the tradeoff of the "Complexity" and "Attacker Threat". In the former case *Tp* could be increased to so as to minimize over all network complexity in terms of packet exchange. In the latter case, Attacker Threat, *Tp* could be minimized to have more frequent checks on Node compensation.

Now suppose a node is compromised exactly after the CHECK period just finished, that is "*Tp+t*" In this case rather than to wait for the next period (t=2Tp suppose), the node itself send a Help broadcast message. The base station receive the help broadcast message and take the necessary action explained in the next step.

Step 2: Suspension of the compromised node

Upon Reception of the HELP from the node under attack, the Base Station broadcasts an ALERT message. The alert message contains id of the sender of the

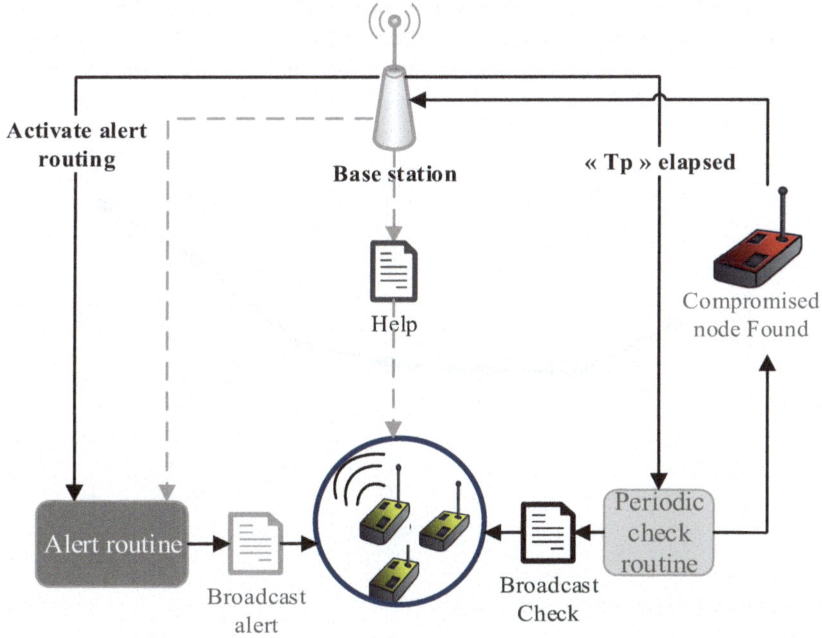

Fig. 1 Proposed Model schematic Diagram

HELP (i.e the node under attack, which has been named as "in_danger_id"). Nodes receiving ALERT message, inspects for the "in_danger_id" and then compares with the the list of ids present in its NEIGHBOR LIST. If a match is there, node deletes the pair-wise key already established with that node. If no-match is found, node just keeps this id for a specified amount of time, and whenever initiates PAIR-WISE key exchange process, compares this key in order not to establish any pair-wise key with a node that is infected. In fig .1, we present the proposed scheme model.

5 Performance and Security Analysis

TOSSIM is used to simulate the Entire TinyOS applications. The goal is to achieve accurate and scalable simulation of TinyOS application. TOSSIM is a TinyOS library which work by replacing components with simulation implementations [12] [13].

5.1 Performance Evaluation

5.1.1 Pairwise Key Generation Time Analysis

As the time available for the generation of pairwise key is short therefore the time for the successful generation of the pairwise key with different nodes model has been analyzed. The main interest was to check if the density of network has any effect in the successful generation of pairwise key with in the available time. Interestingly it has been noticed that the algorithm successfully generate pairwise

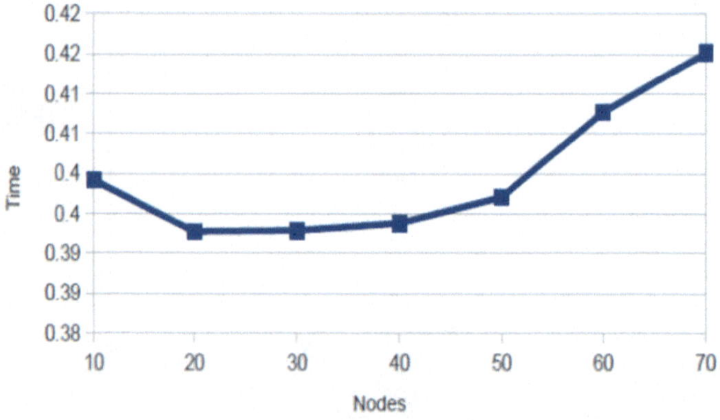

Fig. 2 Pairwise key generation time analysis

key on different node model within the available time. It has been also noticed that each pair of neighbor successfully generate the same pairwise key that is uniquely identified to the specific pair of nodes. The experiment has been performed with a difference of two, five and ten nodes repeatedly ten times each. Almost the same result has been noticed for each experimental model with successful generation of the pairwise key.

- Performed with a difference of 10 nodes ten times repeatedly

5.1.2 Individual Key Generation Time Analysis

The most important task was to check the successful completion of pairwise key generation. Apart from that the time for generation of individual keys for different node models has also been analyzed. It has been confirmed that each pair of node successfully generate a unique individual key.

- Performed with a sample of ten Nodes

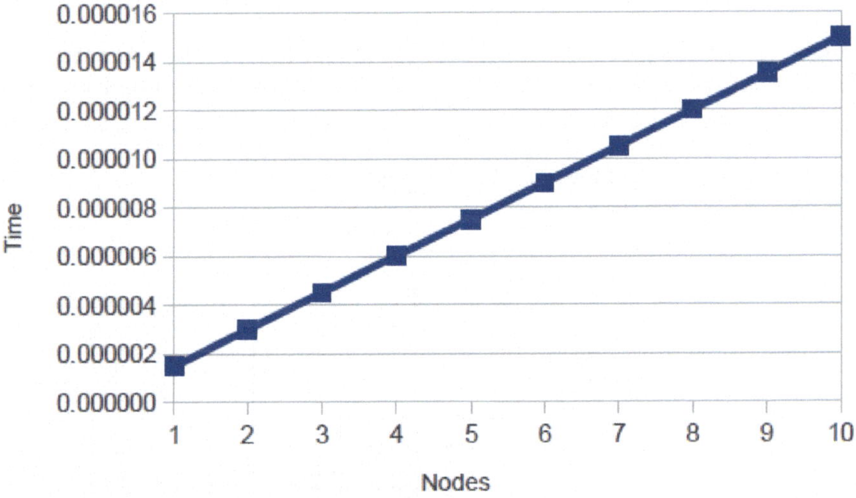

Fig. 3 Individual key generation Time analysis

5.1.3 Energy Efficiency

Energy efficiency is an essential metric and a critical criterion of choice to adopt or not a cryptography solution in WSNs. As shown in figure 4, By adopting a symmetric system, our model like LEAP is more suitable and among rapid in terms of calculation and energy saving, and also scalable to larger network sizes.

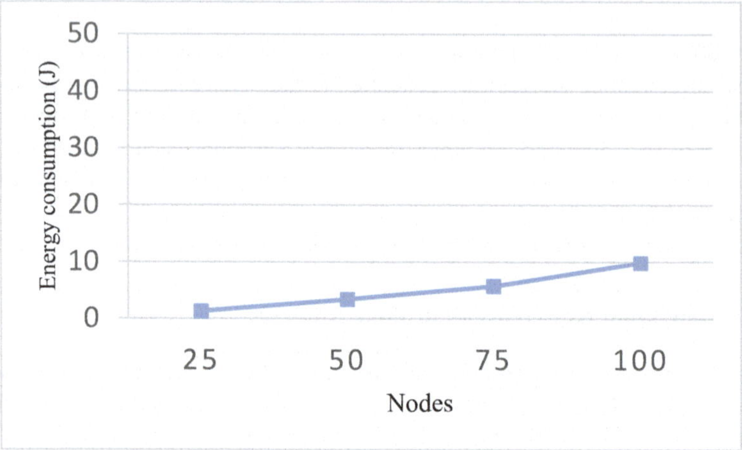

Fig. 4 Energy utilized

5.2 Comparative Analysis

In table 1, we compare our model with LEAP and SPINS. This comparison is based on several criteria (scalability, connectivity, resistance against attacks and

Table 1 Comparison with LEAP and SPINS schemes

Scheme	Criteria of comparison							
	Scalability	Connectivity	Resistance against attacks			Resource cost used		
			Information collection	Communication perturbation	Data aggregation and resource exhausted	Memory (key store)	Calculation and energy consumption	Renewal and revocation
Our scheme	Good	★★★		★★☆		●●○	●○○	●●●
SPINS scheme	Limited	★★☆		★☆☆		●●●	●●○	●●●
LEAP scheme	Good	★★★		★★☆		●●●	●○○	●●●

resource cost used). Note that the storage in memory assessed in the table takes into account only the size of the keys stored in the nodes and not the size of the code algorithms and cryptographic primitives. The stars of the table shall designate a quality. We have established three, two or one black star in the column "Connectivity" to note that the diagrams have respectively a high, medium or low connectivity. Now that we have established three, two or one star in the column "Resistance against the attacks" to note that the diagrams have respectively a high, medium or low resistance against attacks. On the other hand, the circle shall designate a default. We have established three, two and one black circle in the column "Cost of resources used" to note that the diagrams are costing respectively a high, medium or low use of resources.

5.3 Security Analysis

The comparison made in this part concluded that our method has a higher level of security against attacks compared to those described in our literature review. The table shows that our proposal scheme has a low key storage cost, while it has a cost equivalent to LEAP in execution time and energy consumption. Our data encryption is symmetrical, the time consumed in cost and energy is equivalent to other symmetrical methods. While our model and LEAP scheme provides better connectivity between network nodes. Note that the symmetric diagrams are costly in operations (if they exist) of renewal and revocation of keys since they use secret keys in order to exchange other secret keys. The problem is simpler in the asymmetric diagrams since the public keys do not need to be secret. Overall, we conclude that the protocols proposed in this study are scalable and efficient enough in storage, communication, and computation.

6 Conclusion

Sensor networks are vulnerable against external and internal attacks due to their unique characteristics like they can be physically access, they are limited with computation, communication and memory capabilities. Classic security mechanism has to be avoided in WSNs due to limited constraints. Thus it has been concluded to use symmetric shared key schemes. One of the important observation for any type of key-management scheme is that a single keying mechanism is not suitable for meeting different security requirements. During this work different state of the art algorithms has been studied in depth for the proposal of a key-management scheme based on symmetric shared keys. Our model base shows a good performance in comparison with other symmetric schemes as LEAP and SPINS.

The rapid growth of WSNs in different areas especially critical system is attracting a large number of researchers to work on its security.

References

1. Huang, A.: Security Primitives for Ultra-Low Power Sensor Nodes in Wireless Sensor Networks, Faculty of Engineering, the Built Environment and Information Technology University of Pretoria (2005)
2. Gaubatz, G., Kaps, J., Sunar, B.: Public Key Cryptography in Sensor Networks—Revisited, Department of Electrical & Computer Engineering Worcester Polytechnic Institute, 100 Institute Road, Worcester, MA 01609, U.S.A.
3. Perrig, A., Szewczyk, R., Tygar, J.D., Wen, V., Culler, D.E.: SPINS: security protocols for sensor networks **8**(5), 521–534, September 2002
4. Deng, J., Han, R., Mishra, S.: INSENS: intrusion-tolerant routing for wireless sensor networks. Computer Communications **29**(2), 216–230 (2006)
5. Zhu, S., Setia, S., Jajodia, S.: LEAP+: efficient security mechanisms for large-scale distributed sensor networks. ACM Trans. Sen. Netw. **2**(4), 500–528 (2006)
6. Lim, C.H.: LEAP++: a robust key establishment scheme for wireless sensor networks. IEEE (2008). doi:10.1109/ICDCS.Workshops.2008.93
7. Wen, H., Luo, J., Zhou, L.: Lightweight and effective detection scheme for node clone attack in wireless sensor networks. IET Wirel. Sens. Syst. **1**(3), 137–143 (2011). doi:10.1049/iet-wss.2010
8. Choi, H., Zhu, S., La Porta, T.F.: SET: detecting node clones in Sensor Networks, Department of Computer Science and Engineering, Pennsylvania State University
9. Zhang, Q., Yu, T., Ning, P.: A Framework for Identifying Compromised Nodes in Sensor Networks, North Carolina State University
10. Wang, Y.-T., Bagrodia, R.: ComSen: a detection system for identifying compromised nodes in wireless sensor networks. In: SECURWARE 2012: The Sixth International Conference on Emerging Security Information, Systems and Technologies
11. Atakli, I.M., Hu, H., Chen, Y., Ku, W.-S., Su, Z.: Malicious node detection in wireless sensor networks using weighted trust evaluation. In: Simulation of Systems Security (SSSS 2008), Ottawa, Canada, April 14–17, 2008
12. Levis, P., Lee, N., Welsh, M., Culler, D.: TOSSIM: accurate and scalable simulation of entire tinyos applications. In: SenSys 2003, Los Angeles, California, USA, November 5–7, 2003
13. Levis, P., Lee, N.: TOSSIM: a simulator for TinyOS networks, September 17, 2003

Part VI
Advances in Image Processing Applications for Pervasive systems

Computer Aid Diagnostic in Mammogram Image Using SUSAN Algorithm and Hierarchical Watershed Transform

Chaimae Anibou, Mohammed Nabil Saidi and Driss Aboutajdine

Abstract This work is directed toward a conception of a computer aid diagnosis (CAD) system to detect suspicious area in digital mammogram and classify them into normal and abnormal. Original image is preprocessed to separate the breast region from it's background with pectoral muscle suppression to reduce false positive rate.

The suspicious regions are extracted using a modified SUSAN algorithm, followed by a function that extract dense regions, then an hierarchical watershed transforms applied to detect edges of suspicious regions.

For detected edges Fourier Descriptors are computed and stored as feature vector. A support vector machine is used to classify suspicious regions into normal or abnormal. The proposed system is tested on Mini-Mias database.

Keywords Mammogram image · Preprocessing · Segmentation · Modified · Hierarchical watershed transform · Fourier descriptor · Support Vector Machine (SVM)

1 Introduction

Breast cancer has become one of the significant and frequent forms of cancer for women all over the world, it's the first cancer among women and the third one of all registered cancer cases. Most studies conducted have proven that an early diagnosis can increase the survival rate, thus the use of computer systems as a detection tool could be viewed as essential to helping with this disease.

C. Anibou(✉) · D. Aboutajdine
University Mohamed V-Agdal, Rabat, Morocco
e-mail: anibou.chaimae@gmail.com, aboutaj@fsr.ac.ma

M.N. Saidi
INSEA, Rabat, Morocco
e-mail: msaidi@insea.ac.ma

E. Sabir et al. (eds.), *The International Symposium on Ubiquitous Networking*,
Lecture Notes in Electrical Engineering 366,
DOI: 10.1007/978-981-287-990-5_28

A mammogram is an x-ray image of the breast tissue which allows better visualization of internal structure of the breast. Computer-aided techniques for detecting, classifying, and annotating diagnostic features on the mammogram images is an efficient tool for an early detection of breast cancer. A fast and an efficient computer-aided diagnosis (CAD) system to detect the presence of micro calcifications and masses from the breast mammogram images is the need of the hour. It would be of great help to humanity especially for people living in those places where human expertise is not so easily available.

All of the CAD systems require, as a first stage, the segmentation of each mammogram into its representative anatomical regions, i.e., the breast border and pectoral muscle after an image enhancement operations to improve the appearance of images, to eliminate noise and error. Different studies were conducted to extract breast region and remove a pectoral muscle [5][2] [12] [11][6], because it allows the search for abnormalities to be limited to the region of the breast tissue without undue influence from the background of the mammogram. The presence of pectoral muscle in mammograms biases detection procedures as the pectoral muscle appears approximately at same density as the dense tissues inside breast region and cause problems by affecting the results of different image analysis methods, which recommends removing the pectoral muscle during mammogram image pre-processing.

Much works has been done to detect masses in the mammography images and they have used different methods for this purpose. [13] proposed a CAD system to detect abnormalities or suspicious areas in digital mammograms and classify them as malignant or non malignant using Zernike Moments and SVM classifier, [7] identify the breast cancer using texture and classify them as normal or abnormal, [9] proposes a detection method for abnormal mammograms by extracting a few of discriminative features, first-order statistical intensities and gradients and [10] uses a form of template matching at multiple scales to locate pixels in the image, which may be part of a mass for identification of abnormal masses in digital mammography images and [4] detect breast masses in mammogram images using cellular neural networks to segment the regions that might contain masses and Support vector machines is used to classify the candidate regions as masses or non-masses.

This work proposes a CAD system to classify mammogram images into normal and abnormal. Preprocessing steps are described in the Preprocessing section. Region of interest (ROI) extraction technique using modified Susan algorithm ,dense region extraction and edge detection by hierarchical watershed transform are described in the Segmentation section. The "Feature Extraction and Classification" section explains Fourier Descriptor moments used as features along with support vector machine (SVM) used for mammogram classification. Experimental results are summarized in the "Experimental results section". Finally, the "Conclusion" section concludes the work.

2 Proposed Methods

In this study a computer aid diagnosis (CAD) system is presented to detect suspicious area in digital mammogram and classify them into normal and abnormal. In this system, digital mammography images contains labels, these may produce unnecessary disturbances during mass detection process. Hence, it should be removed, also the pectoral muscle represents a predominant density region. Hence it will severely affect the result of image processing. For better detection accuracy pectoral region should be also removed from mammogram image. After a preprocessing step we extract suspicious area using a modified Susan algorithm instead of extracting edges, then we browse filtered image by a sliding window that extract dense regions in the images to reduce the false positive. Hierarchical watershed transform is used to detect edges of the suspicious area. Fourier's Descriptors are computed and stored as feature vector. A support vector machine is used to classify suspicious regions into normal or abnormal.(cf. Fig. 1)

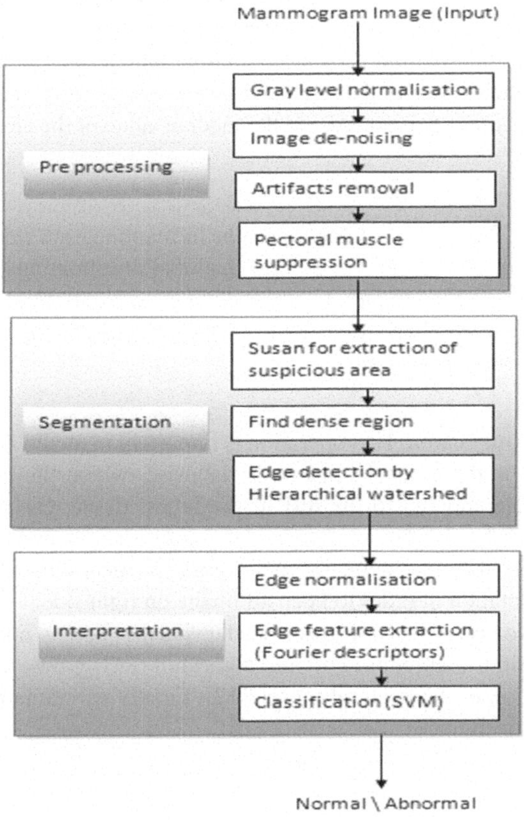

Fig. 1 The proposed CAD system

2.1 Preprocessing

The objective of preprocessing is to improve the quality of image in order to limit the search for abnormalities by removing artefacts and unwanted parts without having undue influence on the background of mammograms. It aims to enlarge the intensity difference between objects and background to produce reliable representation of breast tissue structures. This section deals with pre-processing and enhancement activities such as removal of film artefacts and labels, filtering the image, normalization and removal of pectoral muscle region.

Normalisation. Because this algorithm should operate similarly on different mammograms, it is recommended to normalize images before any processing. Normalisation is a process that changes the range of pixel intensity values. All images are changed to range 0 to 255.

Noise Reduction. In order to filter out the noise present in image, it's recommended to use a low pass filter which smoothes and often removes the sharp edges. A special type of low pass filter that we used is the Median filter which takes an area of image 3 x 3, observes all pixel values in that area and puts it into the array called element array. Then, the element array is sorted and the median value of the element array is found out.

Artefacts Removal. Removing artefacts present in mammogram image is an essential step before any processing. We first use a global thresholding of the filtered image then we use an opening binary area of the obtained binary image to keep the region whose area is greater than a threshold, finally we convolve the filtered image with the image of opening binary area.

Pectoral Muscle Suppression. This operation is important in medio lateral oblique view (MLO) which is the case of mini Mias database, where the pectoral muscle, slightly brighter compared to the rest of the breast tissue, can appear in the mammogram.

The first step is a skin line detection of original image followed by a breast localisation and orientation in order to keep all image on right side.

Once the orientation is known, a seed is placed inside the pectoral muscle (the first not black pixel). A statistical region growing algorithm (RG) grows from this seed to fill the whole region of the pectoral muscle[12]. Finally we refine the boundary by a morphological operator.

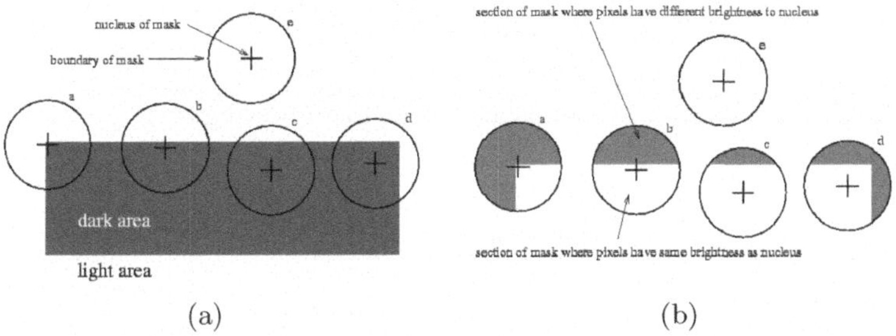

Fig. 2 Susan principle

2.2 *Segmentation Process*

The segmentation stage aims to identify the breast regions with greater possibility of being masses. In this work, segmentation is performed through SUSAN algorithm followed by dense region extraction , then an hierarchical watershed transform is applied to detect edges.

Extraction of Suspicious Regions Using Susan Algorithm. We first introduce the SUSAN(Smallest Univalue Segment Assimilating Nucleus) principle, which employs intensity information of images directly and efficiently for edges and corner detection[14]. Implementation of SUSAN edge detector is performed by sliding a circular mask of radius R from the top left corner to the bottom right corner of an image. When a circle mask is moving in an image, if the intensity difference between an inner pixel and the nucleus is smaller than a given threshold, we consider that the pixel has the same (or similar)intensity as the nucleus. For convenience, the set of such pixels which satisfy the above condition is called USAN [8]. (cf. Fig. 2) This is equivalent to the following:

$$S(r_0) = \Sigma_{r \in N(r_0)} C(r_0, r) \tag{1}$$

where $C(r_0, r)$ is a judgement function of the pixels defined by:

$$C(r_0, r) = \begin{cases} 1 & if \quad |f(r_0) - f(r)| \leq T \\ 0 & if \quad |f(r_0) - f(r)| > T \end{cases} \tag{2}$$

where r_0 and r represent, respectively, the positions of the nucleus and of any other point within the mask; $f(r_0)$ and $f(r)$ denote the corresponding intensities; T is the threshold of brightness difference. Each point in the input image is used as the nucleus of the circular mask, and the associated USAN is found. In fact, equation 2 is not stable in practice, and an improved comparison function 3 is more often used because of its efficiency.

$$C(r_0, r) = \exp(-(\frac{f(r) - f(r_0)}{T}))^6 \quad (3)$$

The SUSAN edge detector is then created using the following rule:

$$R(r_0) = \begin{cases} G - S(r_0) & if \quad S(r_0) < G \\ 0 & otherwise \end{cases} \quad (4)$$

where G is set equal to $\frac{3Smax}{4}$ / $and Smax$ is the maximum value of S.[8]

In this study, the modified Susan is used to detect regions instead of detecting edges with a pre processing step to reduce the search area in the mammographic image knowing that masses are concentrated in high dense regions.

Algorithm

1. Place a circular mask with radius 4 on each pixel of image
2. if $f(r_0) \leq$ threshold than $S(r_0) = 0$
   else :
3. Calculate number of pixels which have gray level near to the nucleus

$$sum(r_0) = \exp(-(\frac{f(r) - f(r_0)}{T}))^6$$

4. Calculate the central pixel output

$$S(r_0) = \begin{cases} sum(r_0) - G & if \quad G < sum(r_0) \\ 0 & otherwise \end{cases}$$

The threshold is the mean of non zero pixels of the pre processed image and T is fixed to 4.

Find Dense Region. We browse the resulted image from Susan algorithm by a sliding window of $10 * 10$ pixels and we calculate the min of each bloc and we affected to the central pixel to keep just the region with high density. An morphological operator is applied to make region more visible.

Edge Detection Using Hierarchical Watershed Transform. Image segmentation based on watershed transform [16] has proved to be an efficient method provided but it has a drawback which is the over segmentation process provided if the watershed transform is applied directly to the image to be segmented. Among the solution to overcome this problem of over segmentation is an approach based on hierarchical segmentation of the image aiming at merging the catchment basins of the watershed image belonging to almost homogeneous regions.[3] An hierarchical watershed transform is applied to the image of dense region to detect edges.

2.3 Interpretation Stage

Edge Feature Extraction. In this stage, the objective is to extract descriptive measurements from the regions of interest segmented in the previous stage, in order to generate feature vectors to be used in the classification stage. For that, shape feature is used. The shape of the regions of interest used in this work is described through Fourier Descriptor [17] which represent the shape in frequency domain.

Before applying Fourier transform on the shape signature, shape is first sampled to fixed number of points. In general, the number of data points of the object can have different sizes. For matching purposes, the shape boundary or the shape signature of objects and models must be sampled to have the same number of data points. In order to facilitate the use of the fast Fourier transform (FFT), the number of sampled points is chosen to be N=30 points.

The Fourier Descriptor of the shape is given by:

$$DF_n = \frac{1}{N}\Sigma_{t=0}^{N-1} r(t)\exp\left(\frac{-i2\pi nt}{N}\right), n = 0, 1, ...N-1, \tag{5}$$

where N=number of normalised points and $r(t)$ is the centroid distance function (shape signature), defined by:

$$r(t) = \sqrt{(x(t) - x_c)^2 + (y(t) - y_c)^2} \tag{6}$$

which represent the distance of the boundary points from the centroid of the shape (x_c, y_c) defined by:

$$x_c = \frac{1}{L}\Sigma_{t=0}^{L-1} x(t),\ y_c = \frac{1}{L}\sigma_{t=0}^{L-1} y(t) \tag{7}$$

which is the average of the boundary coordinates.

Classification. Classification refers to as assigning a physical object or incident into one of a set of predefined categories. In the classification stage, the dataset is divided into two sets: training and test. The sample is separated in two subsets: normal and abnormal from which one subset is chosen for training and the remaining ones are used for test. In this work,the support vector machine(SVM) was used with Sigmoid kernel. It was initiated by [15], is primarily a linear classification approach to two classes. It try to separate individuals from two classes (+1 and -1) seeking the optimal hyperplane that separates the two sets, guaranteeing a large margin between the two classes.

3 Experimental Results

To evaluate the mass detection methodology proposed in this work, several tests were performed. This section presents and discusses the results obtained by the various approaches used. The proposed system is tested over 235 image of Mini-Mias database[1]. It is a reduced version of the original MIAS Database (digitized at 50 micron pixel edge) that has been reduced to 200 micron pixel edge and clipped or padded so that every image is 1024x1024 pixels. This database consists of 332 mammograms of right and left breast, from 161 patients, where 53 were diagnosed as being malignant, 69 benign and 206 normal with ages ranging from 50 to 65.

3.1 Preprocessing

Th original images of Mini-Mias database has been preprocessed as seen in section 2.1. Figure 3 show some examples of preprocessing step, but this methods still have a drawback which is an over-segmentation of the breast in cases with dense tissue, where the contrast between the muscle and the tissue is fuzzy as shown in figure 3(i).

3.2 Segmentation

The processed images are segmented through SUSAN algorithm followed by function for extracting of dense region, then an hierarchical watershed transform is applied to detect edges as described in section 2.2. Figure 4 show experimental results obtained after segmentation process on some images of Mini-Mias database.

3.3 Interpretation

Training and Test Data Selection. The dataset is randomly divided into two sets: training and test using sampling-test method (cf. Fig. 5) from which 188 images are used for training and 47 images for test. This process is repeated 30 times.

The Performance of the proposed CAD system is evaluated in terms of Accuracy. A comparison with other methods to detect masses in mammograms, is performed in Table 1, in order to put the quality of our work in perspective.

---

[1] http://peipa.essex.ac.uk/info/mias.html

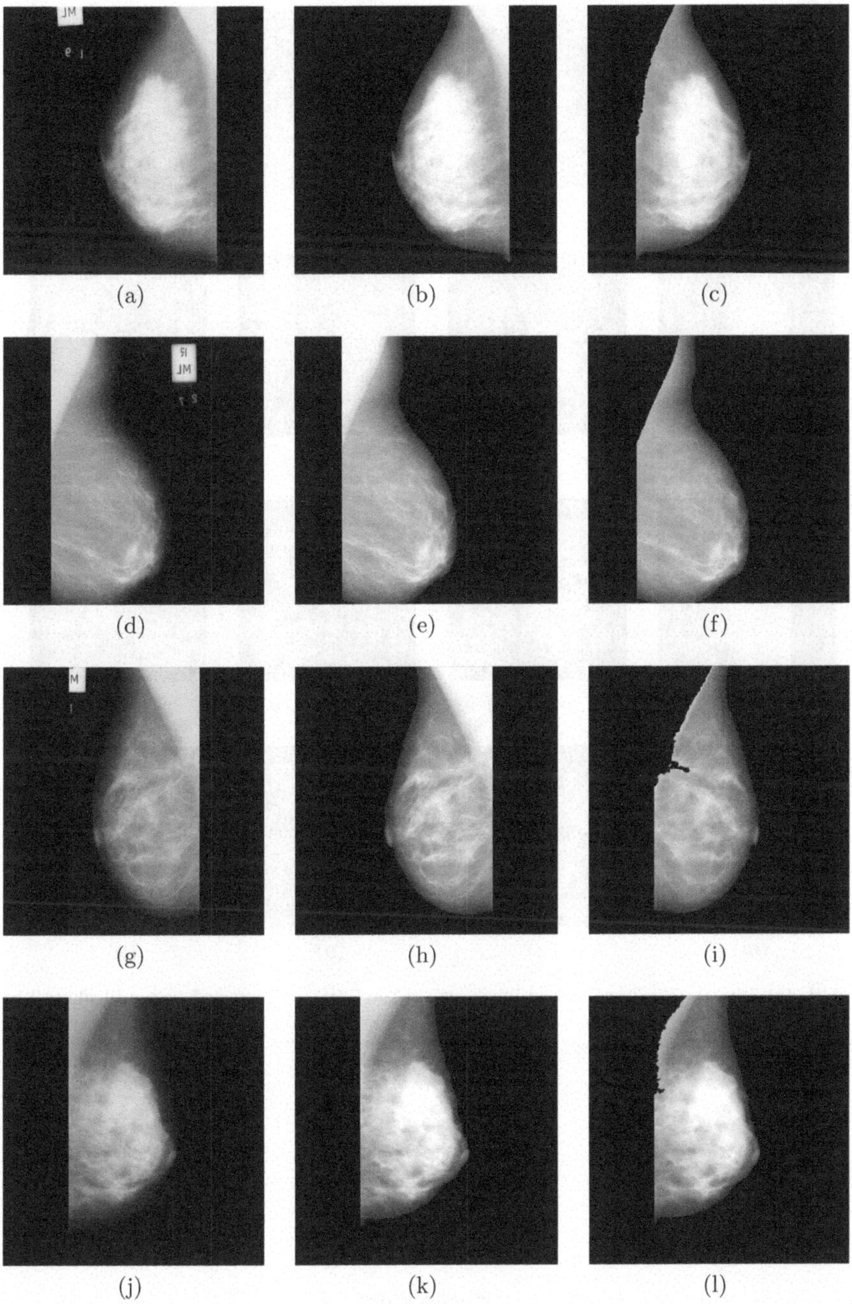

Fig. 3 (a), (d), (g), (j) Original image and (b), (e), (h), (k) enhanced and background removed image and (a), (f), (i), (l) pectoral muscle removed image of mdb03, mdb096, mdb045, mdb226 respectively

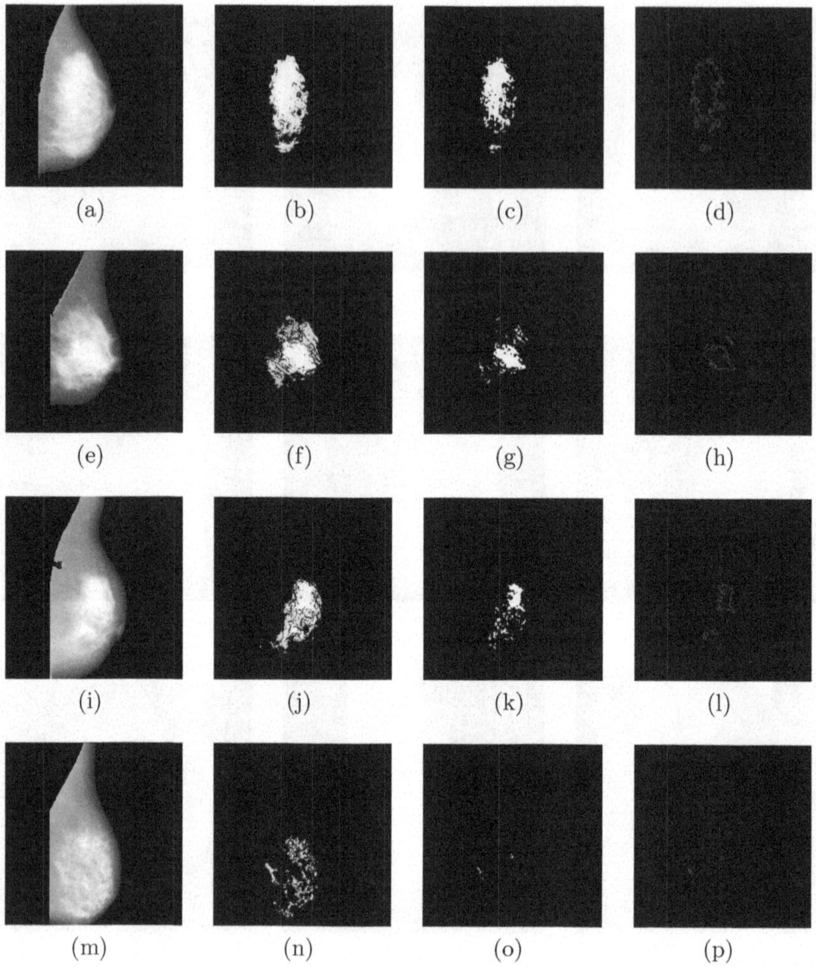

Fig. 4 (a), (e), (i), (m) Preprocessed image and (b), (f), (j), (n) Extracted region of interest by modified SUSAN algorithm and (c), (g), (k), (o) Dense region extraction of mdb03, mdb001, mdb162, mdb099 respectively

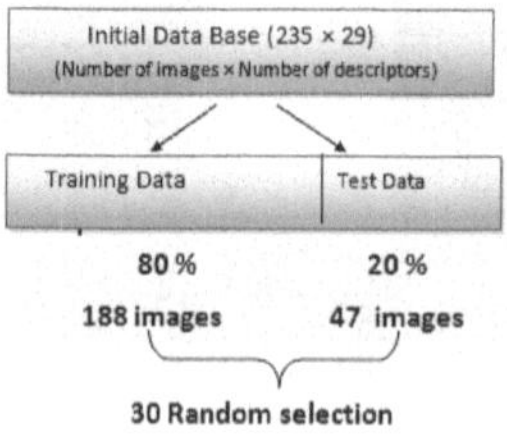

Fig. 5 Training and Test Data selection

Table 1 Comparison with other techniques in term of accuracy

Methods	Segmentation algorithm	Features	Classifier	Dataset	Accuracy(%)
[1]	Small patches of 128 x 128 pixels	2DPCA	SVM	IRMA reference DDSM	80.07
[4]	CNN	shape + Texture	SVM	DDSM	84.72
Proposed method	Modified Susan and Watershed Transform	Fourier (centroid distance)	SVM(Sigmoid Kernel)	Mini-Mias	78.77

4 Conclusion

In this paper, we have proposed a CAD system to detect suspicious area in digital mammogram and classify them into normal and abnormal. Original image is pre-processed to separate the breast region from it's background with pectoral muscle suppression to reduce false positive rate. Then the suspicious regions are extracted using a modified SUSAN algorithm, followed by a function that extract dense regions, then an hierarchical watershed transforms applied to detect edges of suspicious regions.

For detected edges Fourier Descriptors are computed and stored as feature vector. A support vector machine is used to classify suspicious regions into normal or abnormal. The proposed system is tested on Mini-Mias database.

As a future work, we want to validate our approach on other mammographic database and combine descriptors to improve classification rate, also we intent to make classification more specific, it means classify abnormal masses into benign or malign.

References

1. Computer-aided diagnostics of screening mammography using content-based image retrieval (2012)
2. Alam, N., Mohammed, J.I.: Pectoral muscle elimination on mammogram using k-means clustering approach. International Journal of Computer Vision and Signal Processing (2014)
3. Beucher, S.: Watershed, hierarchical segmentation and waterfall algorithm. In: Computational Imaging and Vision. Springer, Netherlands (1994)
4. Sampaio, W.B., Diniz, E.M., Silva, A.C., De Paiva, A.C., Gattass, M.: Detection of masses in mammogram images using CNN, geostatistic functions and SVM. Computers in Biology and Medicine (2011)
5. Boss, R.S.C., Thangavel, K., Daniel, D.A.P.: Automatic mammogram image breast region extraction and removal of pectoral muscle. International Journal of Scientific & Engineering Research (2013)
6. Camilus, K.S., Govindan, V.K., Sathidevi, P.S.: Pectoral muscle identification in mammograms. Journal of Appied Cinical Medical Physics (2011)

7. Fatehia, B.G., Mawia, A.H.: Classification of breast tissue as normal or abnormal based on texture analysis of digital mammogram. Journal of Medical Imaging and Health Informatics (2014)
8. Gao, C., Zhu, H., Guo, Y.: Analysis and improvement of susan algorithm. Signal Processing (2012)
9. Jen, C., Yu, S.: Automatic detection of abnormal mammograms in mammographic images. Expert Systems with Applications (2015)
10. Bandyopadhyay, S.K.: Detection of abnormal masses in mammogram images. International Journal of Computer Science and Information Technologies (2010)
11. Mirzaalian, H., Ahmadzadeh, M.R., Sadri, S., Jafari, M.: Pre-processing algorithms on digital mammograms. In: Conference on Machine Vision Applications (2007)
12. Raba, D., Oliver, A., Mart, J., Peracaula, M., Espunya, J.: Breast segmentation with pectoral muscle suppression on digital mammograms. Springer, Heidelberg (2005)
13. Sharma, S., Khanna, P.: Computer-aided diagnosis of malignant mammograms using Zernike moments and SVM. Journal of Digit Imaging (2015)
14. Smith, S., Brady, J.: Susan-a new approach to low level image processing. International Journal of Computer Vision (1997)
15. Vapnik, V.: Statistical Learning Theory. Wiley (1998)
16. Vincent, L., Soille, P.: Watersheds in digital spaces: an efficient algorithm based on immersion simulations. IEEE Transactions on Pattern Analysis and Machine Inteligence (1991)
17. Zhang, D., Lu, G.: A comparative study of fourier descriptors for shape representation and retrieval. In: 5th Asian Conference on Computer Vision (ACCV), pp. 646–651. Springer (2002)

Improving the Performance of CBIR with Genetic Approach and Feedback

Youssef Bourass, Abdelkhalak Bahri and Hamid Zouaki

Abstract Today, mass production multimedia is observed with high variability of form and content, which complicates the management of databases. The use the research techniques based on content has become increasingly necessary rather than the metadata, such as keywords, tags, or descriptions associated with the image. To dynamically associate the most appropriate search technic to each type of image, an intelligent system is required. However, it is very difficult to determine the adequate descriptor and distance for the analysis of a given image, the system quickly becomes unstable. In this paper we develop an application for the implementation and test of the most classic color and texture descriptors, in order to combine them using Entropy Impurity and Mutation approach. Our objective is to increase system performance and stability.

Our application is based on a web interface, able to perform an experimental comparison of several methods used in image retrieval, in terms of accuracy and relevance of texture and color descriptors. Distances, between different descriptors are also calculated for four references of multimedia databases.

1 Introduction

With the explosive growth in image records and the rapid increase of computer power, retrieving images from a large-scale image database becomes one of the most active research fields [1]. We consider the problem of content-based image retrieval, for applications such as object recognition or similar image retrieval. This problem requires producing a description of the image, typically a fixed-size vector. This description must be discriminative, but sufficiently invariant to handle the transformations the image may have undergone (cropping, different viewpoints,

Y. Bourass(✉) · A. Bahri · H. Zouaki
Department of Mathematics and Computer Science, Faculty of Science,
Modélisation Mathématiques et Informatique Décisionnelle, El Jadida, Morocco
e-mail: Bourass.youssef@gmail.com, {Bahri_abdelkhalak,Hamid_zouaki}@yahoo.fr

E. Sabir et al. (eds.), *The International Symposium on Ubiquitous Networking*,
Lecture Notes in Electrical Engineering 366,
DOI: 10.1007/978-981-287-990-5_29

rotation, etc.). Most of the Content-Based Image Retrieval (CBIR) system uses the low-level features such as color, texture and shape to extract the features from the images. However, most features correspond to particular properties of an image (e.g., color histograms describe only the color distribution of images and global texture features and Tamura describe only texture properties), for most scenarios a combination of features is typically the most successful approach. Consider this problem, we will show an automatic technique for combining color and texture descriptors, using Entropy Impurity.

In CBIR systems, the searching process works as follows. The user queries the system, usually, by using a query image. Its properties are encoded into feature vectors and then compared against the feature vectors of the database images that had been previously extracted. The comparisons are made by computing distance values and those values are used to rank the database images according to their similarities to the query image. The most similar images are finally shown to the user. The image descriptor is involved in this process in the extraction of images properties and in the distances computations. However, the paramount challenge in CBIR is the so-called semantic gap between the low-level visual features and the high-level semantic concepts. To bridge the semantic gap, relevance feedback (RF) methods were proposed to learn the user's intentions. The conventional process of RF is as follows: (1) from the retrieved images, the user labels a number of relevant samples as positive feedbacks, and a number of irrelevant samples as negative feedbacks; (2) the CBIR system then refines its retrieval procedure based on these labeled feedback samples to improve retrieval performance.

In this paper we develop an application for the implementation and test the most classic color and texture descriptors, in order to combine them by using Entropy Impurity approach. The system can use machine learning techniques based feedback to refine its similarity measure and thus to improve the results. The evaluation (+ or -) is recovered in the form of mutation, this latter will help us to give a weight to each descriptor in the combination generated by Entropy Impurity.

Our main objective is to increase system performance and stability after each iteration.

Another important aspect of our study is related to scalability and diversity. Descriptors are analyzed in a Web environment with four databases containing more than 400 thousand images with a very heterogeneous content. The experimental analysis considers efficiency and effectiveness aspects. Web environment takes into account how much the descriptors agree with the human perception of semantic similarity, by asking a pool of users to annotate the relevance of answers for each query.

2 Proposed Methods Using Mutation Feedback and Entropy Impurity

One way to improve the effectiveness of the retrieval system is to try to select the best suited feature vector for a query image q. In a general image retrieval scenario, we have observed that for different query different feature vectors have the best effectiveness. Hence, given a set of FVs, we would like to select the best one for performing the similarity search for q. Our hypothesis is that a good feature vector is expected to have a certain level of coherence in the answer set, i.e., we expect to retrieve similar images at the first positions of the ranking list.

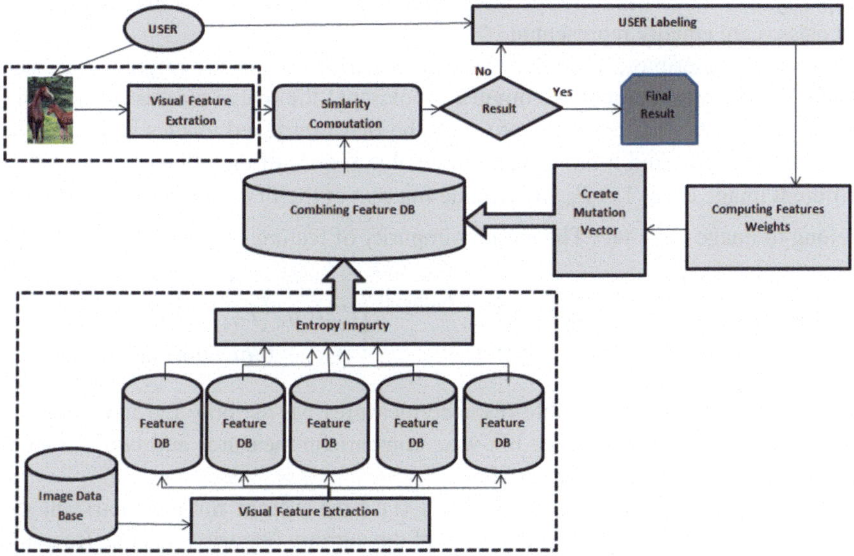

Fig. 1 Image retrieval system flow chart

2.1 Combination of Descriptors Using Entropy Impurity

The large-scale, heterogeneous nature of the Web can benefit from the use of descriptors in combination. Although the combination of descriptors is a complex topic, with many competing techniques, and thus outside the scope of this work, we perform a general analysis of the complementarity of the best descriptors, which should be taken into account when selecting them for combined use.

We concentrate our evaluation on color and texture descriptors. We decide not to include shape descriptors in our study, because almost all of them are segmentation dependent. As is known, segmentation is still a hard and extremely application dependent task. Therefore, shape descriptors are still not mature for a heterogeneous environment like the Web.

To improve the images search in term of precision, the metric function used to perform the similarity query (the so-called query distance) corresponds to a dynamic combination of metrics, thus the metric function may change on each performed query. The weight of each metric is obtained by implementing a query processor based on the entropy impurity method [17] to compute the dynamic weights for each descriptor vector. This method uses a reference dataset that is classified in object classes (in our case, we used the classified subset of the images database).

The entropy impurity is a well-known measure used in the context of decision tree induction, where it measures the "impurity" of a node N of the tree w.r.t. the elements assigned to N. If these elements all have the same class label, then the impurity is 0, otherwise it is a positive value that increases up to a maximum when all classes are equally represented.

For each descriptor, a similarity query is performed on the reference dataset. Then, the entropy impurity is computed looking at the image classes of the first *t* retrieved objects: It is equal to zero if all the first *t* retrieved images belong to the same image class, and it has a maximum value if each of the *t* object belongs to a different image class. Let P_{w_j} denote the fraction of the first *t* retrieved images that belong to image class w_j . The entropy impurity of feature vector *i*

$$Entropy\ Impurety(i) = \sum_{j=1}^{No.of\ Classes} \begin{cases} P_{w_j} log_2\left(P_{w_j}\right) & if\ \ P_{w_j} > 0 \\ & 0\ otherwise \end{cases}$$

Instead of listing all possible distance measures we list only the most popular and successful ones here: three bin-wise comparison measures and two distances which take into account different bins.

The weight value for feature vector i (i.e., the weight for the metric in the combination) is computed as the inverse of the entropy impurity plus one (to avoid dividing by zero), i.e.,

$$w_i = \frac{1}{1 + entropy\ Impurty(i)}$$

(We used $t = 3$ for our experiments.)

Let$\mathbb{M} =< \delta_i >$ be a vector of metric functions, and let $\mathbb{W} =< w_i >$ be a vector of weights, with $|\mathbb{M}| = |\mathbb{W}| = m$ and $\forall i\ w_i\ \in [0, 1]$.

The linear multi-metric (or linear combined metric function) is defined as:

$$\Delta_w(O_1, O_2) = \sum_{i=1}^{m} w_i . \delta_i\,(O_1, O_2)$$

2.2 FeedBack

Similarity-based Relevance Feedback algorithm is maybe the most popular query reformulation strategy in Information. At each step, the algorithm uses the feedback from the user to automatically refine the request by adding to the original query a vector based on the feedback information such that the relevant data instances are promoted, whereas the non-relevant data instances get lower ranks. The optimal query is defined as the query that maximizes the average query-data similarity for the relevant data instances, and at the same time minimizes the average query-data similarity for the non-relevant data instances. The new query is considered a better approximation to the user request than the initial query, and therefore, it should produce better retrieval results. Then, the algorithm ranks the data set by using a similarity measure, and returns to the user a short list of top ranked data instances (the top most 100), considered to be the most relevant to the given query, for further feedback information. The process continues iteratively until the user is satisfied with the search results or decides to quit the search.

Generally, the feature selection based Relevance Feedback methods can adjust weights associated with various dimensions of the feature space to enhance the importance of those dimensions that help in retrieving the relevant images and to reduce the importance of those dimensions that hinder the retrieval performance.

In this paper we will further exploit neighborhood relations to weight feature subsets according to their relevance to user's needs. The basic idea behind the proposed weighting mechanism is that the feedback from the user implicitly defines which images should be considered as neighbors of each other (i.e., the relevant images), and which images should not (i.e., non-relevant images should not be in the neighborhood of relevant images)

2.3 DNA Mutation

Mutations are random, but their frequency can be increased by mutagens, sometimes called agents or mutagenic factors.

Although mutations that cause change in protein sequences can be harmful to an organism; on occasions, the effect may be positive in a given environment. In this case, the mutation may enable the mutant organism to withstand particular environmental stresses better than wild-type organisms, or reproduce more quickly. In these cases a mutation will tend to become more common in a population through natural selection.

Evolutionary biologists have theorized that higher mutation rates are beneficial in some situations, because they allow organisms to evolve and therefore adapt more quickly to their environments.

In our approach, eliminating images that do not represent a satisfactory degree of similarity, is done through the feedback user evaluation. their rating of these create a mutation in the weights of descriptors used and which keeps those who give good semantic results.

In the GP approach, the individuals represent programs that undergo evolution. The fitness evaluation consists of executing these programs, and measuring their degrees of evolution. Genetic programming, then, involves an evolution-directed search in the space of possible computer programs that best solve a given problem.

ALGORITHM

- User indication of query image q
- Show the initial set of images
- While the user is not satisfied do
    - User indications of the relevant images
    - Generate an initial population of individuals
    - for N generations do
        - Calculate the fitness of each individual
        - Select the individuals to genetic operations
        - Apply reproduction
        - Apply crossover
        - Apply mutation
    - end for
    - Rank the database images
    - Show the L most similar images
- end while

3 Feature Descriptors

Both the effectiveness and the efficiency of content-based image retrieval systems are very dependent on the image descriptors that are being used. The image descriptor is responsible for characterizing the image properties and to compute their similarities. In other words, the image descriptor makes it possible to rank images according to their visual properties.

For research purposes, we have implemented 6 different feature vectors from the literature. However, as we learned from our experiments, the effectiveness of a given feature vector cannot be assessed for the general case, as it depends on the specific type of image that one wants to search. For example, we have observed that the best effectiveness for "plants images" is achieved using the color histogram, but the best effectiveness for "medical images" is achieved using the SURF feature.

In this section we give an overview of the features tested:

Table 1 List of descriptors tested

Feature name	Type
Color Histogram	Color representation
Bag-of-Colors	Color representation
CEDD	Color representation
Global texture	Texture representation
Tamura feature	Texture representation
FCTH	Color and texture representation
SURF	Shape representation

Color Descriptors

One of the most important visual properties identified by human vision is color, making it one of the most used properties in CBIR systems.

Color Histogram: are among the most basic approaches and widely used in image retrieval [4,5,6,7,8]. To show performance improvements in image retrieval systems, systems using only color histograms are often used as a baseline. The color space is partitioned and for each partition the pixels with a color within its range are counted, resulting in a representation of the relative frequencies of the occurring colors.

Bag-of-Colors: The spatial organization of colors is an important aspect of the structure of an image and definitively one of the first to be perceived. It is an intermediate feature between low-level content such as color histograms and image semantics. BOC save the spatial information for each visual word if is present in the image [9].

Texture Descriptors

Texture is an important property for the characterization and recognition of images. This fact is observed by the great amount of research involving texture analysis of images [10, 11, 12]. As shown in [13]: ‘‘its structure is simply attributed to the repetitive patterns in which elements or primitives are arranged according to a placement rule’’; ‘‘an image texture is described by the number and types of its primitives and the spatial organization or layout of its primitives’’.

Global Texture: In [14] a texture feature consisting of several parts is described: Fractal dimension measures the roughness of a surface. The fractal dimension is calculated using the reticular cell counting method [15]. Coarseness characterizes the grain size of an image. It is calculated depending on the variance of the image. Entropy of pixel values is used as a measure of disorderedness in an image. The spatial gray-level difference statistics describe the brightness relationship of pixels within neighborhoods.

Tamura Feature: Tamura feature are designed in accordance with psychological studies on the human perception of texture. In [16] the authors propose six texture features corresponding to human visual perception: coarseness, contrast, directionality, line-likeness, regularity, and roughness. From experiments testing the significance of these features with respect to human perception, it was concluded that the first three features are very important.

Color and Edge Directivity Descriptor (CEDD) and Fuzzy Color and Texture Histogram (FCTH)

The structure of these descriptors consists of n texture areas. In particular, each texture area is separated into 24 sub regions, with each sub region describing a color.

CEDD and FCTH use the same color information, as it results from 2 fuzzy systems that map the colors of the image in a 24-color custom palette.

To extract texture information, CEDD uses a fuzzy version of the five digital filters proposed by the MPEG-7 EHD, forming 6 texture areas. In contrast, FCTH uses the high frequency bands of the Haar wavelet Transform in a fuzzy system, to form 8 texture areas. When an image block (rectangular part of the image) interacts with the system that extracts CEDD and FCTH, this section of the image simultaneously goes across 2 units. The first unit, the color unit, classifies the image block into one of the 24 shades used by the system. Let the classification be in the color m,m E [0,23]. The second unit, the texture unit, classifies this section of the image in the texture area a,a E[0,n]. The image block is classified in the bin a X 24 +m. The process is repeated for all the image blocks of the image. On the completion of the process, the histogram is normalized within the interval [0:1] and quantized for binary representation in a three bits per bin quantization.

Shape Representation (Speeded Up Robust Features "SURF")

In recent years, Speeded-Up Robust Features (SURF) [19] has emerged as a popular choice for interest point detection and region description. Building upon previous work (e.g. SIFT [1]). It is widely used in most of the computer vision applications. The SURF has been proven to achieve high repeatability and distinctiveness. It uses a Hessian matrix-based measure for the detection of interest points and a distribution of Haar wavelet responses within the interest point neighborhood as descriptor.

An image is analyzed at several scales, so interest points can be extracted from both global and local image details. In addition to that, the dominant orientation of each of the interest points is determined to support rotation-invariant matching.

4 Experimentations

4.1 Databases

We performed a number of range queries using four real word databases:

Table 2 Image databases

Database	**Images**	**Queries**
Microsoft Research Cambridge Object Recognition Image	4320	4320
Wang	1000	1000
Oxford plant	8189	8189
University of Washington	1109	1109

Wang[1]

The WANG database is a subset of 1,000 images of the Corel stock photo database which have been manually selected and which form 10 classes of 100 images each. The 10 classes are used for relevance estimation: given a query image, it is assumed that the user is searching for images from the same class and therefore the remaining 99 images from the same class are considered relevant and the images from all other classes are considered irrelevant.

Microsoft Research Cambridge Object Recognition Image Database

The MSRC[2] database was published by the Machine Learning and Perception Group from Microsoft Research, Cambridge, UK and is available online39. It consists of 4,320 images from 33 classes such as aeroplanes, bicycles/general, bicycles/sideview, sheep/general, sheep/single and is generally considered a difficult task [18].

Experiments on the MSRC database are carried out in a leaving-one-out manner. That is, each image is used as a query to retrieve relevant images (i.e. images from the same class) from the remainder of the database.

Oxford Plants[3]

Collection 8189 pictures divided on two data sets of flowers images gathering from various websites. The first dataset is a smaller one consisting of 17 different flower categories, and the second dataset is much larger, consisting of 102 different categories of flowers common to the UK.

University of Washington

The database created at the University of Washington (UW) consists of a roughly categorized collection of 1,109 images. These images are partly annotated using keywords. The remaining images were annotated by our group to allow the annotation to be used for relevance estimation.

4.2 *Results*

Extensive experimentation has been carried out to assess the relative performance of the algorithms proposed and also to fully characterize their behavior from different viewpoints.

To properly evaluate the different algorithms, the process of user interaction to retrieve images has been simulated in the following way: at each iteration, the

---

1 http://wang.ist.psu.edu/docs/related/

2 http://research.microsoft.com/en-us/downloads/b94de342-60dc-45d0-830b-9f6eff91b301/default.aspx

3 http://www.robots.ox.ac.uk/~vgg/data/flowers/

system is a ware of both positive and negative images that are used to produce a ranking on the remaining images in the database. The first k images in this ranking are then disclosed (shown to the user) along with their class labels (user judgments) in order to incorporate them to the system either as positive or negative information.

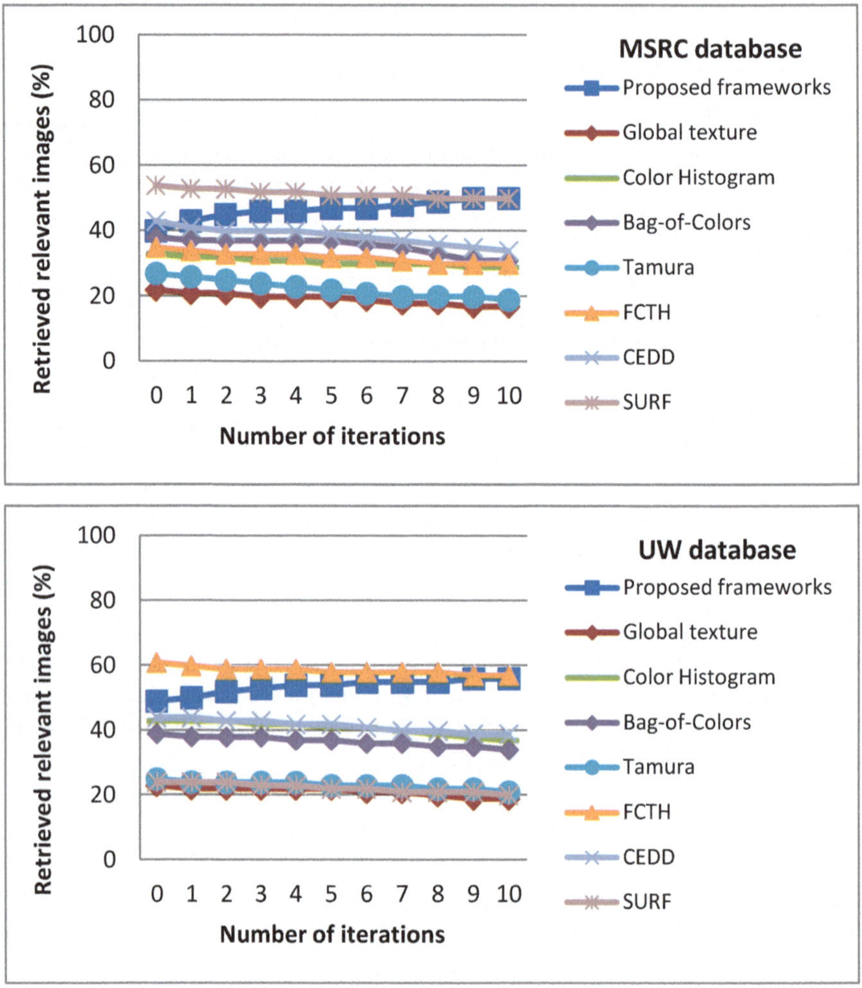

Fig. 2 Comparison between the proposed frameworks and descriptor collection, considering 25 images displayed per iteration

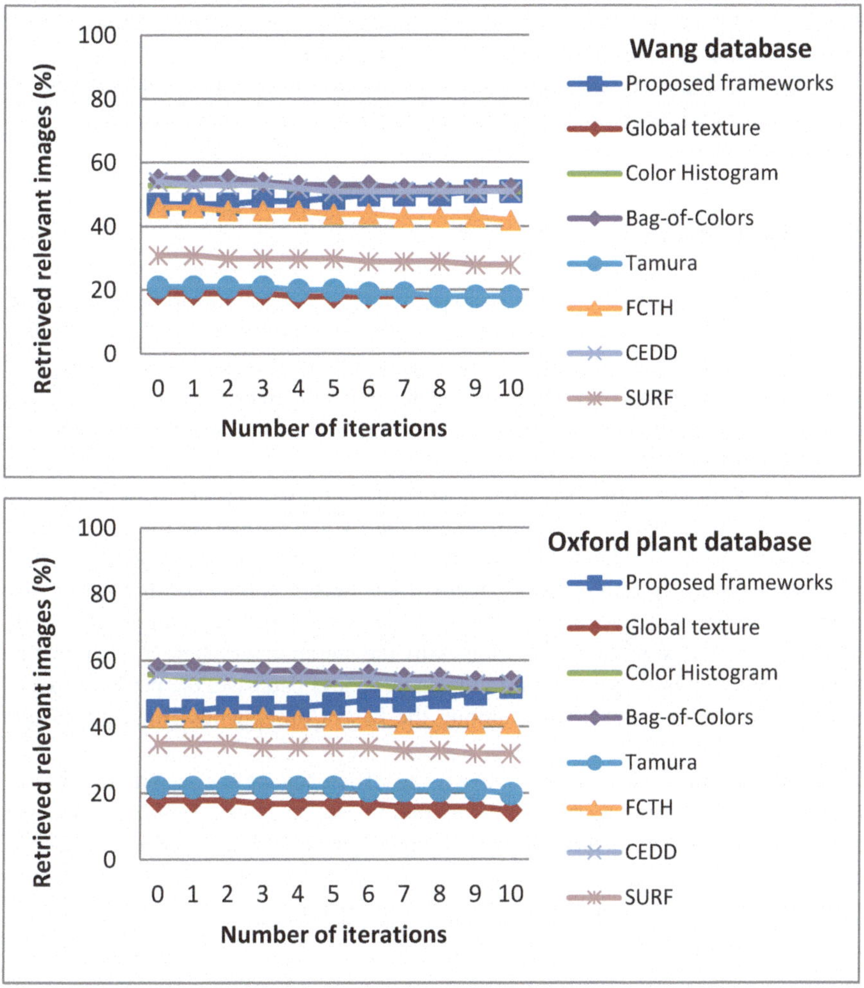

Fig. 2 (*Continued*)

The previous figures gives an overview of the performance of proposed framework; its relevance is still close to that of the best descriptor, despite the change of distance or the database, which provides better system stability. So we can see the goal of our experience, despite changing the type of descriptor, distance and database, the performance of the combination remains always among the best.

Considering the descriptors used as baselines for comparisons, there is a clear predominance in the use of descriptors based on color representation. The great majority of images in web are colored; therefore, in most cases, color descriptors give very good results. But even in the case where the performance of the color descriptor is less than the other descriptors (Results on the Microsoft Research

Cambridge Object Recognition and UW Image Database), combinations correct the search, and switches to the most efficient descriptor.

In the end, the proposed experiments also shows that the research obtained by the use of combination method (Entropy Impurity) and mutation approach with Feedback lead to desired results and a significant gain of stability.

5 Conclusion

After performing a large number of experiments, we find descriptors which give better results than others for each type of image. Such as radiological medical images, content based image retrieval conduct with local feature descriptors, gives best result than others. So, the solution is to find a hybrid method which can combine a multiple descriptors, and she can adjust the weight of each one according to type of images.

In this work we have proposed a potential line of improvement of current retrieval methods, by hybridizing methods that which allows the improvement of research by fostering the weight of the best descriptor. In particular, a simple weighted ranking entropy impurity method has been combined with a parametric dynamic weight selection strategy based mutation and feedback that progressively increase the precision of exploration with the iteration number. We show that despite changing the type of descriptor, distance and database, the performance of the combination remains always among the best. Consequently, increase system performance and stability.

In future work, we plan to improve the efficiency of the search in term of accuracy by combining them with the textual vocabulary, for this purpose, we will merge the techniques presented in this paper and textual annotation, we expect that, by combining all these technique in one index structure, we will be able to further improve the efficiency and stability of CBIR.

References

1. Lowe, D.: Distinctive image features from scale invariant key points. IJCV **60**(2), 91–110 (2004)
2. Sivic, J., Zisserman, A.: A text retrieval approach to object matching in videos
3. Mikolajczyk, K., Schmid, C.: A Performance Evaluation of Local Descriptors. IEEE Transactions on Pattern Analysis and Machine Intelligence **27**(10), 1615–1630 (2005a). In: Proc. ICCV (2003), vol. 2, pp. 1470–1477
4. Smeulders, A.W.M., Worring, M., Santini, S., Gupta, A., Jain, R.: Content-Based Image Retrieval at the End of the Early Years. IEEE Transactions on Pattern Analysis and Machine Intelligence **22**(12), 1349–1380 (2000)
5. Faloutsos, C., Barber, R., Flickner, M., Hafner, J., Niblack, W., Petkovic, D., et al.: Efficient and Effective Querying by Image Content. Journal of Intelligent Information Systems **3**(3/4), 231–262 (1994)

6. Puzicha, J., Rubner, Y., Tomasi, C., Buhmann, J.: Empirical evaluation of dissimilarity measures for color and texture. In: International Conference on Computer Vision, Corfu, Greece, vol. 2, pp. 1165–1173 (1999)
7. Deselaers, T.: Features for Image Retrieval. Master's thesis. Human Language Technology and Pattern Recognition Group, RWTH Aachen University, Aachen, Germany (2003)
8. Swain, M.J., Ballard, D.H.: Color Indexing. International Journal of Computer Vision **7**(1), 11–32 (1991)
9. Bahri, A., Zouaki, H.: Improving the Bag of Colors, And Adapting LSH for Dynamic Similarity Search In Multi-Metric Spaces. International Journal of Multimedia (2014). ISSN: 2249-2801 & E-ISSN: 2249-281X
10. Xu, F., Zhang, Y.: Evaluation and comparison of texture descriptors proposed in MPEG-7. Journal of Visual Communication and Image Representation **17**(4), 701–716 (2006)
11. Tuceryan, M., Jain, A.K.: Texture Analysis, pp. 235–276 (1993)
12. Zegarra, J.A.M., Leite, N.J., da Torres, R.S.: Wavelet-based feature extraction for fingerprint image retrieval. Journal of Computational and Applied Mathematics **227**, 294–307 (2009)
13. Haberäcker, P.: Praxis der Digitalen Bildverarbeitung und Mustererkennung. Carl Hanser Verlag, München (1995)
14. Tamura, H., Mori, S., Yamawaki, T.: Textural Features Corresponding to Visual Perception. IEEE Transaction on Systems, Man, and Cybernetics **8**(6), 460–472 (1978)
15. Bustos, B., Keim, D., Saupe, D., Schreck, T., Vranić, D.: Using entropy impurity for improved 3D object similarity search. In: Proc. IEEE International Conference on Multimedia and Expo (ICME 2004), pp. 1303–1306. IEEE (2004)
16. Winn, J., Criminisi, A., Minka, T.: Object categorization by learned universal visual dictionary. In: International Conference on Computer Vision, Beijing, China, vol. 2, pp. 1800–1807, October 2005
17. Bay, H., Tuytelaars, T., van Gool, L.: SURF: speeded up robust features. In: European Conference on Computer Vision (2006)

Multi-scale Image Co-segmentation

Rachida Es-salhi, Imane Daoudi, Jonathan Weber, Hamid El Ouardi, Saida Tallal and Hicham Medromi

Abstract This paper focuses on producing accurate segmentation of a set of images at different scales. In the process of image co-segmentation, we turn our attention to the task of computing dense correspondences between a set of images. These correspondences are calculated in a dense grid of pixels, where each pixel is represented by an invariant descriptor computed at a unique, manually selected scale, this scale selection limits the efficiency of image co-segmentation methods when the common foregrounds appear at different scales. In this work, we use scale propagation to compute dense correspondences between images by assuming that if two images are being matched, scales should be assigned by considering feature point detections common to both images. We present both quantitative and qualitative tests, demonstrating significant improvements to segment images with large scale variation.

Keywords Image co-segmentation · Scale selection · Dense correspondence · Image matching

1 Introduction

Co-segmentation aims at extracting common objects from a set of images. The advantage over the traditional segmentation techniques is the fact that object of interest is being segmented using several related images. Moreover, it is computationally desirable to segment similar objects "jointly" from various images rather than segment the salient object separately from each image. This is suitable for large scale

R. Es-salhi(✉) · I. Daoudi · H. El Ouardi · S. Tallal · H. Medromi
Systems Architecture Research Group, ENSEM, Hassan II University, Casablanca, Morocco
e-mail: {rachida.es-salhi,i.daoudi,h.elouardi}@ensem.ac.ma

I. Daoudi
Greentic, Hassan II University, Casablanca, Morocco

J. Weber
LORIA, Lorraine University, Nancy, France
e-mail: jonathan.weber@loria.fr

E. Sabir et al. (eds.), *The International Symposium on Ubiquitous Networking*,
Lecture Notes in Electrical Engineering 366,
DOI: 10.1007/978-981-287-990-5_30

image dataset segmentation and many others practical applications such as medical imaging, data-driven image synthesis as well as for drone-based applications. In the past few years, numerous co-segmentation models have been proposed. However, most existing co-segmentation methods are still limited to directly handle arbitrary images, they were shown to work well on popular co-segmentation datasets (namely the Microsoft Research Cambridge database- MARC and iCoseg) where the target objects are sufficiently clear and appear nearly at the same visual scale.

In this paper, we propose a multi- scale co-segmentation method which perform on images that depict a large variation of scale. We work on the object discovery and segmentation in internet images algorithm [1] and we use propagate scale method [2] to estimate dense correspondences between images. The experimental results demonstrate the improvement over the original work [1].

2 Related Work

Co-segmentation. The concept of image co-segmentation was first introduced by Rother et al. [3] in the context of simultaneously segmenting a person or object of interest from an image pair. The earlier work on co-segmentation considered only two classes; foreground/ Background and required that the two images have similar foreground and different background. Later, several authors have extended their work setting to deal with a large group of images containing different instances of the same class. Many of these methods formulated co-segmentation as a optimisation problem, in the initial work [3] Rother et al. used the Markov Random field (MRF) to extract the objects from image pairs through adding the constraint of foreground similarity to traditional MRF-based methods. Following this work Many other methods were proposed [4, 5, 6] Mukherjee et al. [4] exploited the pseudo-Boolean method to optimize the energy function. Hochbaum and Singh [5] rewarded foreground similarity instead of penalizing foreground difference, which simplified the energy function optimisation. Chang et al. [6] proposed a novel global energy term which considered both foreground similarity and background consistency, and used the graph-cut algorithm to solve the optimisation.

A part MRF-based methods many other co-segmentation methods were proposed. Joulin et al. [7] presented a discriminative clustering framework which combined existing tools for bottom-up image segmentation such as normalised cuts, with kernel methods commonly used in object recognition. To exploit knowledge about image more directly, Batra et al. [8] proposed an interactive method, which segments common objects through human interaction guided by an automatic recommendation system. Chai et al. [9] introduced a bi-level algorithm, which did not assume the similarity of either global geometric shape or foreground color distributions throughout the image group. Joulin et al. [10] exploited spectral clustering and discriminative clustering methods for co-segmentation of multiple classes and a significantly larger number of images. Kim et al. [11] created an efficient spectral segmentation framework that was robust on single images as well as on large-scale heterogeneous datasets.

A scale invariant method was proposed by Mukherjee et al.[12] which required that the rank of the matrix corresponding to foreground regions should be one. Vicente et al. [13] proposed a method for object co-segmentation, which borrowed recent ideas from object recognition: the use of powerful features extracted from a pool of candidate object-like segmentations. In [14] Kim et al. proposed a distributed co-segmentation approach via sub-modular optimisation on anisotropic diffusion for a highly variable large-scale image collection. Meng et al. [15] designed a digraph to represent the local region similarities according to the feature distance and the saliency map, and thus formulate the co-segmentation problem as a shortest path problem. Rubio et al. [16] proposed a multiple-scale multiple-image generative model, which jointly estimated the foreground and background appearance distributions from several images.

Image Alignement. Image matching is defined as the establishment of correspondences between two or more images [17]. Dense alignement which means matching all pixels of one image to pixels in another image has been the subject of many computer vision application such as stereoscopic vision [18], deep estimation [19, 20, 21] and image segmentation [1]. However, in all cases cited above the same scale was assumed for the images involved, either by performing global matching of the images [22] or in the best cases considering the assumption that in large image collection there exists at least one which present the same objects in the same scale [1]. Objects appear in different scales in different images, thus matching pixels across scale require robust, per-pixel representation. To deal with this issue many solutions have been proposed in the literature: Ignore scale differences by using Dense SIFT descriptor (DSIFT) [23] which is calculated in dense grid of pixels (one descriptor per pixel), in this setting the scale is fixed beforehand so we use single, user-selected scale for all pixels, this scale selection greatly limits its applicability. Multi-Scale representation: In scale invariant descriptor (SID) [24] image intensities around each pixel were transformed to log-polar coordinate systems. Doing so, converted scale and rotation to translation and translation invariance was then introduced by applying Fourier Transform Modulus (FTM), thus obtaining the Scale Invariant Descriptors (SID). Though SID descriptors were shown to be scale and rotation invariant, even on a dense grid, their use of image intensities directly implies that they are not well suited for matching images of different scenes [25]. Also in Scale-Less SIFT (SLS) representation [25], the authors produce at each pixel multiple SIFT descriptors, at multiple scales. Both the SID and SLS descriptors work match better than the original Dense SIFT in terms of accuracy and they were shown to be highly robust to scale changes, but the cost of this was a quadratic inflation in the descriptor size which is shown in Table 1, therefore making them difficult to apply in practice.

In the field of image co-segmentation establishing reliable correspondences between pixels in different images enable us to learn a common appearance model. The method described here is multi-scale co-segmentation, which aims to produce accurate segmentation from group of images depicting object at different sizes. We focused particularly on the work of Rubinstein et al [1] and we turn our attention to the task of computing dense correspondences between images. In the work of

Table 1 Comparison of different descriptor dimensions [21]

Descriptor	Dim
Dense SIFT	128D
SID	3,328D
SLY	8,256D

Rubinstein et al, SIFT-flow [26] was used to perform pixel correspondence, this method achieve high accuracy to align images with different scene. Although an important step, SIFT-flow relies on dense SIFT descriptor [23], and therefore assume that visual information in images appears at the same scale. Here, we propose a new alternative to align images, in order to improve the co-segmentation results of [1].

3 The Proposed Method

In this paper, we propose a multi-scale image co-segmentation algorithm to deal with local scale changes in the image group. The idea is to exploit the dataset structure and similarity between image regions to establish scale-invariant correspondences between pixels in different images. Thus, the proposed method can achieves accurate segmentation even if the images present large scale differences.

To align two images we select the scale at each pixel, which allow to us assigning scale value to all pixels in the image. We propose to use propagate scale algorithm [2]. The idea is to exploit a few detected pixels, by multi-scale descriptor as SIFT, their scale values are used to estimate the scale assignments for the remaining image pixels. Thus a scale map $S_I(p)$ of the image I is produced, whose elements are the scale value σ_p at each pixel $p(x, y)$.

Formally, the global cost for a scale assignment is defined as follow:

$$C(S_I) = \sum_p \left(S_I(p) - \sum_{q \in N(p)} \omega_{pq} S_I(q) \right)^2 \tag{1}$$

Here, scales are assigned to all image pixels, by minimizing (1), using only the known scales of pixels selected by a multi-scale feature detector- SIFT. This function shows the requirement that the scale assigned to pixel p should be as similar as possible to a weighted average of the scales of its similar pixels, denoted by $q \in N(p)$. The weight ω_{pq} refer to an affinity function, which expresses the degree to which the scale of one pixel is assumed to influence another.

Assuming that neighboring pixels with similar intensities, should be assigned with similar scales. We define this affinity function as follow:

$$\omega_{pq} = 1 + \frac{1}{\sigma_p}((I(p) - \mu_p)(I(q) - \mu_p)) \tag{2}$$

Where $I(p)$ is the intensity value at a pixel p and σ_p, μ_p, are the mean and variance of the intensities in the neighborhood of pixel p.

In this paper we consider matching-aware propagate scale which means that when two images are being matched, scales should be assigned by considering feature point detections common to both images. the process is as follow: 1) we compute correspondences between the scale invariant descriptors (SIFT), extracted at sparse locations. 2) We take the 20% of the correspondences with the best closest to second-closest SIFT match ratio [27], and only their scales are considred to seed scale propagation in each image.

Formally lets w_{ij} the flow field (warp) between image i and image j, and given the binary masks b_i, b_j. The energy function of SIFT-flow used in [1] becomes:

$$\begin{aligned} F(w_{ij}, b_i, b_j) = &\sum_p b_i(p)(b_j(p + w_{ij}))\left\| f(p, S_i(p)) - f(p + w_{ij}(p), S_j(p + w_{ij}))\right\|_1 \\ &+ (1 - b_j(p + w_{ij}(p)))C_0 + \sum_{q \in N} \alpha \left\| w_{ij}(p) - w_{ij}(q)\right\|_2 \end{aligned} \tag{3}$$

The function f in (3) represents the SIFT feature transform, where S_i and S_j are the scale maps of image i and j, used for computing the descriptors. The difference between this objective function and that used in [1] is that it encourages matching foregrounds pixels in image i with foreground pixels in image j even if they appear in different sizes since the SIFT descriptors are calculated at dense scale selection given by the scale map.

4 Experimental Results

We tested the proposed method on one of the standard co-segmentation datasets; iCoseg [8], which contain 30 classes with varying number of images per class, also we report results for various datasets containing images that exhibit large variation of scale to show the efficiency of the proposed algorithm. We present both qualitative and quantitative results, as well as a comparaison with the original work [1]. Quantitative evaluation is performed against the hand-annotated ground-truth mask collected using the Labeled annotation toolbox [28]. The comparison between the algorithm's output and the ground truth is performed on two evaluation metrics, commonly used for evaluation in image segmentation, i.e., the precision which is the ratio of correctly labelled pixels, and the Jaccard index (J) which is defined as the ratio of the area of the intersection region between the segment and the ground-truth to the area of their union region ($J(segment, ground - truth) = \frac{segment \bigcap ground-truth}{segment \bigcup ground-truth}$).

The Fig. 1 shows per-class accuracy for the iCoseg dataset, compared with the original work of Rubinstein et al [1]. These results demonstrate that for the classes

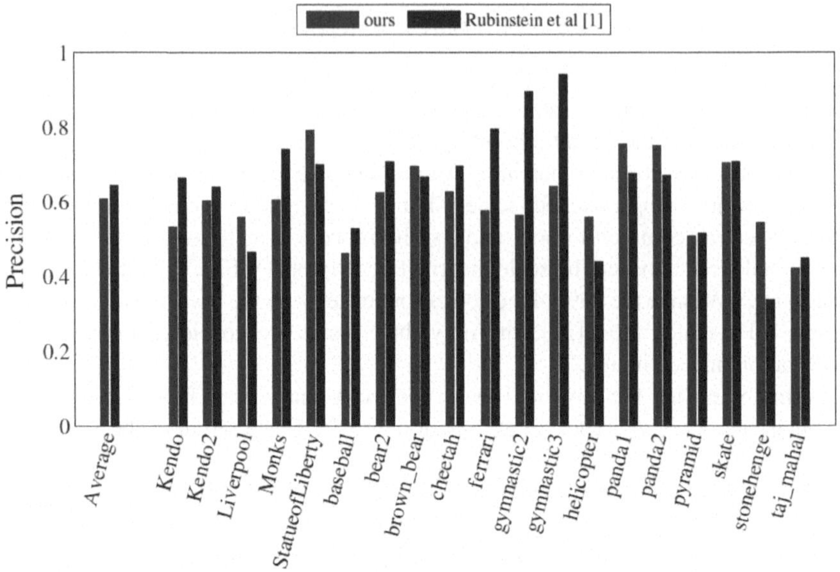

Fig. 1 Per-class precision on iCoseg dataset. the plot shows the precision computed per class and compared with the original work of Rubinstein et al. [1]

($Statueofliberty$, $panda1$, $penda2$, $Liverpool...etc$) containing images that depict the object of interest in different sizes, our algorithm perform better than the original work. Fig. 2 displays some qualitative results on statue of liberty class. The statue appears at different visual scales in different regions of each image. We obtain perfect segmentations of the statue (Fig. 3, bottom row), while, DRIFT used in [1] performs worst (Fig. 3, top row).

Figure 3, 4 and 5 display qualitative results on $taj-mahel$, $flowers$, $stop$ dataset respectively, we clearly see that multi-scale co-segmentation algorithm (bottom row) performs better than the original algorithm (top row). It should be also noted that the object discovery algorithm [1] fails to discover the object of interest as shown at Fig. 3 (first column), Fig. 4 (third column) and Fig. 5 (fourth column).

Table 2 presents the Jaccard index's values on several classes. Given a group of images, the average Jaccard index's values of all images are used as the objective result of the image group. A good co-segmentation result corresponds to a large Jaccard index's value. Otherwise, small value will be obtained for bad co-segmentation results. We can see that the proposed method achieves larger Jaccard index's values compared with the original work.

The figure Fig. 6 shows the per-dataset precision in comparaison with the object discovery algorithm [1]. The performance on all datasets is better than the original algorithm. Particulary on stop dataset the accuracy is considerably better, since images in this dataset depict a large scale variation as shown in Fig. 5 Admittedly using propagate scale to estimate reliable correspondences between images improve the segmentation results even though the common object appears at different scales.

Fig. 2 Comparaison of results on statue of liberty class. The first row present the results obtained with object discovery and segmentation [1], ours results are displayed in the second row. For each image we show a pair of the original (left) and the segmentation result (right).

Fig. 3 Qualitative results on taj mahel dataset.

Fig. 4 Qualitative results on flowers dataset.

It is essential to mention that the object discovery and segmentation algorithm [1] outperform our proposed algorithm in several object classes like $Kendo$, $gymnastic2$, $gymnastic3$, $Monks$ (as shown in Fig. 1) due to the assumption that these image collections are large enough such that at least one image will portray the same information in the same scales.

Fig. 5 Qualitative results on stop dataset.

Table 2 The Jaccard index results on various classes

Method	Multi-scale coseg	Object Discovery
Flowers	56.66	45.04
Taj-mahel	24.02	16.22
Stop	46.83	40.1
Statue of Liberty	67.2	57.07

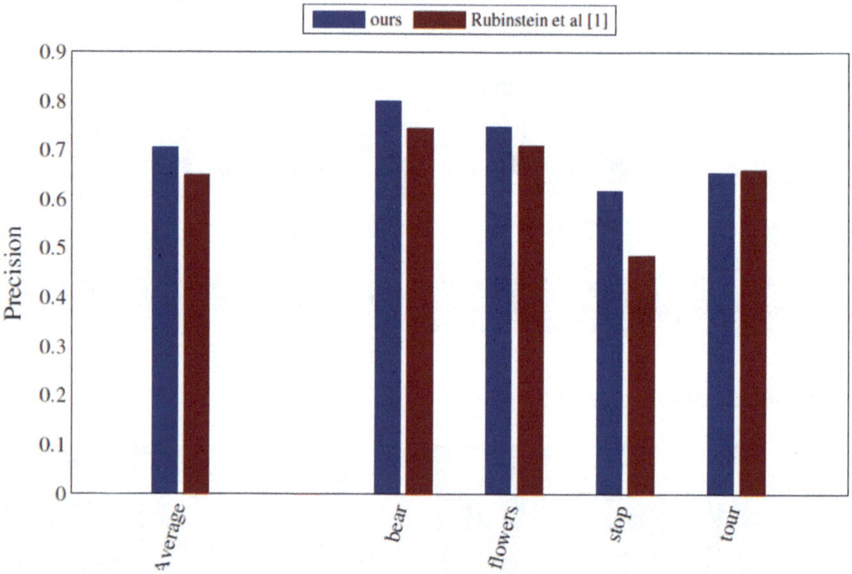

Fig. 6 Per-dataset precision and comparaison with the object discovery algorithm [1]

5 Conclusion

In this paper, we proposed a multi-scale co-segmentation method. The propagate scale algorithm was used to estimate dense correspondences in the image collection, the algorithm therefore discovers and segments the object of interest although it appear in different visual scale and at different regions of each image. The experimental results demonstrate the efficiency of the proposed method.

References

1. Rubinstein, M., Joulin, A., Kopf, J., Liu, C.: Unsupervised joint object discovery and segmentation in internet images. In: IEEE Conference on Computer Vision and Pattern Recognition (CVPR), pp. 1939–1946. IEEE (2013)
2. Tau, M., Hassner, T.: Dense correspondences across scenes and scales. arXiv preprint arXiv:1406.6323 (2014)
3. Rother, C., Minka, T., Blake, A., Kolmogorov, V.: Cosegmentation of image pairs by histogram matching - incorporating a global constraint into MRFs. In: 2006 IEEE Computer Society Conference on Computer Vision and Pattern Recognition, vol. 1, pp. 993–1000, June 2006
4. Mukherjee, L., Singh, V., Dyer, C.R.: Half-integrality based algorithms for cosegmentation of images. In: IEEE Conference on Computer Vision and Pattern Recognition, CVPR, pp. 2028–2035. IEEE (2009)
5. Hochbaum, D.S., Singh, V.: An efficient algorithm for co-segmentation. In: IEEE 12th International Conference on Computer Vision, pp. 269–276. IEEE (2009)
6. Chang, K.Y., Liu, T.L., Lai, S.H.: From co-saliency to co-segmentation: an efficient and fully unsupervised energy minimization model. In: IEEE Conference on Computer Vision and Pattern Recognition (CVPR), pp. 2129–2136. IEEE (2011)
7. Joulin, A., Bach, F., Ponce, J.: Discriminative clustering for image co-segmentation. In: IEEE Conference on Computer Vision and Pattern Recognition (CVPR), pp. 1943–1950. IEEE (2010)
8. Batra, D., Kowdle, A., Parikh, D., Luo, J., Chen, T.: iCoseg: interactive co-segmentation with intelligent scribble guidance. In: IEEE Conference on Computer Vision and Pattern Recognition (CVPR), pp. 3169–3176. IEEE (2010)
9. Chai, Y., Lempitsky, V., Zisserman, A.: BICoS: a bi-level co-segmentation method for image classification (2011)
10. Joulin, A., Bach, F., Ponce, J.: Multi-class cosegmentation. In: IEEE Conference on Computer Vision and Pattern Recognition (CVPR), pp. 542–549. IEEE (2012)
11. Kim, E., Li, H., Huang, X.: A hierarchical image clustering cosegmentation framework. In: IEEE Conference on Computer Vision and Pattern Recognition (CVPR), pp. 686–693. IEEE (2012)
12. Mukherjee, L., Singh, V., Peng, J.: Scale invariant cosegmentation for image groups. In: IEEE Conference on Computer Vision and Pattern Recognition (CVPR), pp. 1881–1888. IEEE (2011)
13. Vicente, S., Rother, C., Kolmogorov, V.: Object cosegmentation. In: IEEE Conference on Computer Vision and Pattern Recognition (CVPR), pp. 2217–2224. IEEE (2011)
14. Kim, G., Xing, E.P., Fei-Fei, L., Kanade, T.: Distributed cosegmentation via submodular optimization on anisotropic diffusion. In: IEEE International Conference on Computer Vision (ICCV), pp. 169–176. IEEE (2011)
15. Meng, F., Li, H., Liu, G., Ngan, K.N.: Object co-segmentation based on shortest path algorithm and saliency model. IEEE Transactions on Multimedia **14**(5), 1429–1441 (2012)
16. Rubio, J.C., Serrat, J., López, A., Paragios, N.: Unsupervised co-segmentation through region matching. In: IEEE Conference on Computer Vision and Pattern Recognition (CVPR), pp. 749–756. IEEE (2012)
17. Schenk, T.: Digital photogrammetry-volume i background, fundamentals, automatic orientation procedures. Terra Science, USA (1999)
18. Hartley, R., Zisserman, A.: Multiple view geometry in computer vision. Robotica **23**(2), 271–271 (2005)
19. Hassner, T., Basri, R.: Example based 3D reconstruction from single 2D images. In: CVPRW 2006 Conference on Computer Vision and Pattern Recognition Workshop, pp. 15–15. IEEE (2006)
20. Karsch, K., Liu, C., Kang, S.B.: Depth extraction from video using non-parametric sampling. In: Computer Vision–ECCV 2012, pp. 775–788. Springer (2012)
21. Hassner, T., Basri, R.: Single view depth estimation from examples. arXiv preprint arXiv:1304.3915 (2013)

22. Hassner, T.: Viewing real-world faces in 3D. In: IEEE International Conference on Computer Vision (ICCV), pp. 3607–3614. IEEE (2013)
23. Vedaldi, A., Fulkerson, B.: VLFeat: An open and portable library of computer vision algorithms. In: Proceedings of the International Conference on Multimedia, pp. 1469–1472. ACM (2010)
24. Kokkinos, I., Yuille, A.: Scale invariance without scale selection. In: IEEE Conference on Computer Vision and Pattern Recognition, CVPR 2008, pp. 1–8. IEEE (2008)
25. Hassner, T., Mayzels, V., Zelnik-Manor, L.: On sifts and their scales. In: IEEE Conference on Computer Vision and Pattern Recognition (CVPR), pp. 1522–1528. IEEE (2012)
26. Liu, C., Yuen, J., Torralba, A., Sivic, J., Freeman, W.T.: Sift flow: dense correspondence across different scenes. In: Computer Vision–ECCV 2008, pp. 28–42. Springer (2008)
27. Lowe, D.G.: Distinctive image features from scale-invariant keypoints. International Journal of Computer Vision **60**(2), 91–110 (2004)
28. Torralba, A., Russell, B.C., Yuen, J.: Labelme: Online image annotation and applications. Proceedings of the IEEE **98**(8), 1467–1484 (2010)

A Novel Approach for Computing the Coefficient of ART Descriptor Using Polar Coordinates for Gray-Level and Binary Images

Abderrahim Khatabi, Amal Tmiri and Ahmed Serhir

Abstract ART (Angular Radial Transform) are widely used in many applications of domains of image processing and pattern recognition, this last is a region based shape descriptor adopted in MPEG-7; among the property, that descriptor is invariant to rotation. The direct computation of ART coefficients characterized by the basis function, which is defined in polar coordinates over a unit disk and digital images are usually defined in cartesian coordinates, which produces two types of errors namely geometric error and integral approximation error and invariant to rotation of ART is not verified. In this paper, we propose an algorithm to compute the coefficients in polar coordinates systems to eliminate these errors, which have been produced by the conventional method. We use cubic interpolation for generation of the pixel to system Polar image, the results show that the proposed approach improves the polar rotation invariance. In order to apply this approach at the gray level and the binary images, which we can compare the time of execution of this last.

Keywords ART descriptors · Shape · Binary images · Gray level images · Extraction

1 Introduction

Digital images are one of the effective media to present a variety of information in graphic form, due to the increase of huge amount of images are generated every

A. Khatabi · A. Tmiri
Department of Computer Science, Chouaib Doukkali University, El Jadida, Morocco
e-mail: khatabiabdo5@gmail.com, B_tmiri@yahoo.fr

A. Serhir(✉)
Department of Mathematics, Chouaib Doukkali University, El Jadida, Morocco
e-mail: A.serhir@gmail.com

E. Sabir et al. (eds.), *The International Symposium on Ubiquitous Networking*,
Lecture Notes in Electrical Engineering 366,
DOI: 10.1007/978-981-287-990-5_31

day in the diagnosis medical, military and registration, geographical images, etc... Therefore, the search for images by visual content has become an active area of research. Several techniques are developed to search for images accurately and efficiently, based on specific image features such as color, texture, shape.

The shape became more popular, where the representation of the shape is significant distribution image retrieval [1]. System is considered an important visual element in the recognition shape Retrieval [2]. According to Kim [3], a good representation of the form must be compact and retains the essential characteristics, and to extract this feature we need a shape descriptor. Angular radial transform (ART), is the region based shape descriptor for MPEG-7[4], which is a shape descriptor by region approach characterizes the distribution of pixels in a 2D object or region. It is based on the study of frontiers and internal pixels of the region to describe. It applies to a large number of objects, such as complex objects consisting of multiple discrete regions or simple objects with or without hole.

ART has been widely utilized in applications of domains of image processing, such as shape retrieval [5], video security systems, logo recognition system [6], and image watermarking [7].

In order to reduce its computational complexity, fast algorithms [8, 9], have been developed to make it most suited for several real time applications where rotation invariant global shape descriptors are required. Furthermore, the ART coefficients defined for binary images and is represented by 140 bits.

In recent years, active efforts have been made in the shape recognition based on the ART descriptor; however, the ART has properties like invariance to image rotation is not ideal by computation in Cartesian coordinates. This computation produces two types of geometric approximation error and integral approximation error are inherent and unavoidable. In this article, we try to improve the invariance of ART rotation by computing coefficients more accurately.

We show that if coefficients ART are calculated in the polar coordinate system eliminated the geometric and integral error. We present a detailed description of an algorithm for calculating the coefficients ART in the polar coordinate system and to obtain the values of image of the Pixel in the polar coordinates ous let us calculate the convolution of the function of image of intensity by the cubic interpolation. The method increases the exactitude of ART and improves consequently invariance with rotation and the rebuilding of image execution. The rest of the paper is organized as follows: In Section 2, an overview of ART is given. The convotional method is described in Section 3. Section 4 aims to give a comparison between the computational time and some experimental results. Conclusion and concluding remarks are presented in Section 5.

2 Angular Radial Transform (ART)

ART is a complex orthogonal unitary transform defined on a unit disk based on complex orthogonal sinusoidal basis functions in polar co-ordinates [10, 11, 12]. The ART coefficients, of order n and m, are defined by:

$$F_{nm} = \int_0^{2\pi} \int_0^1 V_{nm}^*(r,\theta), f(r,\theta) r dr d\theta \tag{1}$$

Where $f(r,\theta)$ is an image intensity function in polar co-ordinates and $V_{nm}^*(r,\theta)$ is a basis function, which is complex conjugate of $V_{nm}(r,\theta)$ defined in polar coordinates over a unit disk. These are expressed in a separable form of both radial and angular parts as follows:

$$V_{nm}(r,\theta) = A_m(\theta) R_n(r) \tag{2}$$

The kernel functions $V_{nm}(r,\theta)$ are separable into radial function $R_n(r)$ and angular function $A_m(\theta)$.

The indices n and m are nonnegative integers. The real valued radial polynomial and the angular basis function are defines as:

$$R_n(r) = \begin{cases} 1 & n = 0 \\ 2\cos(\pi n \rho) & n \neq 0 \end{cases}$$

And $$A_m(\theta) = \frac{1}{2\pi}\exp(jm\theta) \qquad \textit{Where} \quad j = \sqrt{-1}$$

The important characteristics of ART is the rotational invariance, the magnitude values of ART are unaffected and remain identical for image functions before and after rotation. The original image is represented by the intensity image function $f(r,\theta)$ having ART is rotated counterclockwise by angle α; the transformed image function is $g(r,\theta) = f(r,\theta - \alpha)$. The ART coefficients of original and rotated images are F_{nm} and $F_{nm}^{rot} = \exp(-jm\theta) F_{nm}$, the magnitude values are identical, where $\|\exp(-jm\theta)\| = 1$.

The default ART has 35 coefficients (n=3, m=12). The real parts of the ART basis functions are shown on Fig.

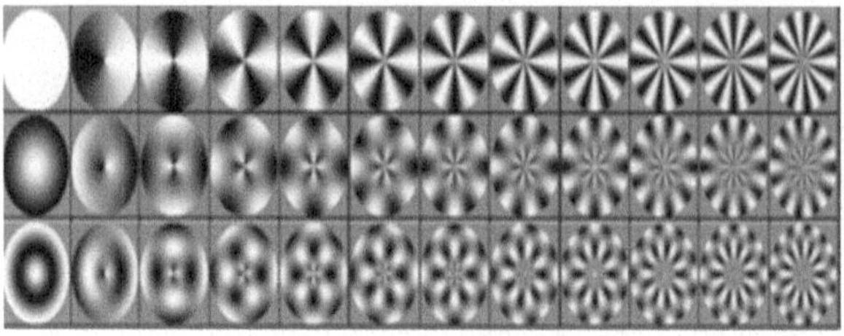

(a)

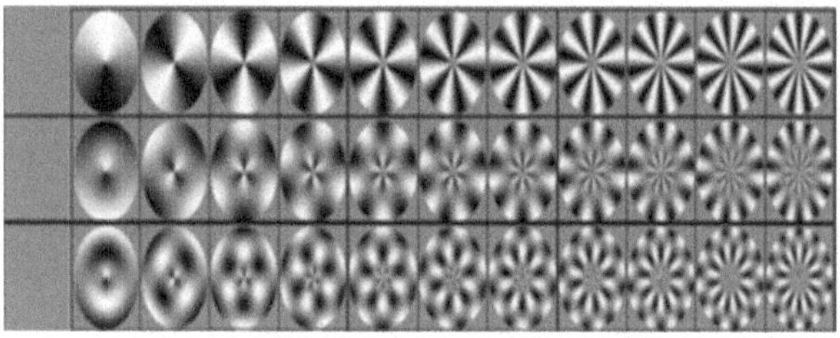

(b)

Fig. 1 Real (a) and imaginary (b) parts of ART basis functions

The difference operation between two ART coefficient vectors use formed as:

$$Dist_{q,s}^{m,n} = \sum_{i=1}^{m\times n} \left| x_i^{m,n} - y_i^{m,n} \right|$$

Then the similarity

$$Sim_{ART} = \frac{1}{Dist_{q,s}^{m\times n}}$$

2.1 Conventional Method

ART can be defined as a linear combination of bases functions, where the ART basis functions are defined in polar coordinates over a unit disk and the image intensity function is always defined in Cartesian coordinates (x,y). Therefore, the conventional method for computing ART coefficient based on zero order approximation (ZOA). This approximation produces two types of errors namely geometric and numerical errors, those who degraded the quality of the computed descriptors. The first error exists owing to mapping [13] of the image covered by square pixels to a unit disk, whereas the latter results from the calculation of integrals of the ART bases functions are replaced by summations, consequently this method is not applicable in cases on real-time implementation of large databases. The approximate ART coefficients are:

$$F_{nm} = \sum_{i=1}^{N} \sum_{j=1}^{N} R_{nm}(r_{ij}) \exp(-\hat{\imath}\,\theta_{ij}) f(i,j) \qquad (3)$$

Where $r_{ij} = \sqrt{(q_1 i + q_2)^2 + (q_1 j + q_2)^2}$ And $\theta_{ij} = \tan^{-}\left(\frac{q_1 j + q_2}{q_1 i + q_2}\right)$

We are two approaches; first all pixels outside the unit disk are ignored, which result in a loss of some image information in second approach, the whole square image is mapped inside the unit disk, where the center of the image is assumed to be the coordinate's origin where we can use I this paper. The transformed coordinates are defined:

$q_1 = \frac{\sqrt{2}}{N}$ And $q_2 = -\frac{1}{\sqrt{2}}$ With i, j=1, 2… N and $0 \le r_{ij} \le 1$

A digital image of size is $N \times N$ an array of pixels with N rows and N columns. Centers of these pixels are the points (x_i, y_j), the coefficients ART for digital images are computed on a unit disk, the coordinate (x_i, y_j) a unit disk are given by:

$x_i = \frac{2i + 1 - N}{H}$ $y_j = \frac{2j + 1 - N}{H}$ i, j=0, 1, 2… N-

$$H = \begin{cases} N, & for\ inner\ disk \\ \sqrt{2}N, & for\ outer\ circular\ disk \end{cases}$$

In this paper, we use the outer circular disk and for ART coefficients we use n<3 and m<12, which gives a feature vector containing 36 moments.

The zeroth order approximation is generally used to find transform ART coefficients after converting the polar into its Cartesian equivalent.

$$F_{nm} = \sum_{i=0}^{N-1}\sum_{j=0}^{N-1} f(x_i, y_j) V^*(x_i, y_j) \Delta x_i \Delta y_j \tag{4}$$

$$\Delta x_i = \frac{2}{H}, \quad \Delta y_j = \frac{2}{H}$$

$$F_{nm} = \frac{1}{H^2} \sum_{i=0}^{N-1}\sum_{j=0}^{N-1} f(x_i, y_j) V^*(x_i, y_j) \tag{5}$$

Where $V^*(x_i, y_j)$ is the complex conjugate of $V(x_i, y_j)$ which can be obtained by replacing $r = \sqrt{x^2 + y^2}$ and $\theta = \tan^{-1}(y/x)$

2.2 Coefficients of ART in Polar Coordinates SYSTEM

The 2D ART about n and of the repetition m of an image function $f(r,\theta)$ in polar coordinates over a unit disc are defined by:

$$F_{nm} = \frac{1}{2\pi}\sum_i\sum_j f(r_i,\theta_{i,j})H(r_i,\theta_{i,j}) \quad (6) \qquad \text{Where} \qquad H_{nm}(r_i,\theta_{i,j}) = I_n(r_i)I_m(\theta_{ij})$$

Real-valued radial write

$$I_n(r_i) = \int_{u_i}^{u_{i+1}} R_n(r)rdr \qquad (7)$$

And angular

$$I_m(\theta_{ij}) = \int_{u_i}^{u_{i+1}} \exp(-\hat{\imath}m\theta)d\theta \qquad (8)$$

Mapped Cartesian image pixels to polar image pixels we use the higher and lower limits are defined using the following vectors:

$$\begin{aligned} U_{i+1} &= (2i+1)/N; \qquad & V_{i,j+1} &= \theta_{i,j} + \pi/(8i+4) \\ U_i &= 2i/N; \qquad & V_{i,j} &= \theta_{i,j} - \pi/(8i+4) \end{aligned} \qquad (9)$$

One fear of writing In and Im:

$$I_m(\theta_{ij}) = \int_{V_{i,j}}^{V_{i,j+1}} \exp(-\hat{\imath}m\theta)d\theta \Rightarrow \mathrm{I}_m(\theta_{i,j}) = \frac{\hat{\imath}}{m}(\exp(-\hat{\imath}mV_{i,j+1}) - \exp(-\hat{\imath}mV_{i,j}))$$

$$I_n(r_i) = \int_{u_i}^{u_{i+1}} 2\cos(n\pi r)rdr \Rightarrow I_n(r_i) = \begin{cases} (u^2_{i+1} - u^2_i)/2 & n=0 \\ \left(\frac{2r}{u_{i+1}.\pi}\sin(u_{i+1}\pi r) + \frac{2}{(u_i.\pi)^2}\cos(u_i\pi r)\right) & n\neq 0 \end{cases}$$

The unit disc is uniformly divided along the radial direction into U and V sections.

We still have to integrate the image at the same coordinates; we will study it in the next step.

$$F_{nm} = \frac{1}{2\pi}\sum_i^U\sum_j^V f(r_i,\theta_{i,j})H(r_i,\theta_{i,j})$$

2.2.1 Image in Polar Coordinates

Digital images are usually defined in Cartesian coordinates, what is not compatible with the basic function, for that one will use the convolution of interpolation cubic to solve this problem, There is a certain number of existing techniques of interpolation [14-15] that we can employ to determine the Pixel polar, such as the close method nearest, the bilinear method and the method bicubic. In our case we try to adopt the interpolation bicubic [16] to produce polar image. The value of a polar Pixel is determined by the 16 Pixel Cartesian neighbors. The function of the core 1-D is a cubic groove:

$$h(x)=\begin{cases}1-\frac{5}{2}|x|^2+\frac{3}{2}|x|^3 & |x|\le 1\\ 2-4|x|+\frac{5}{2}|x|^2-\frac{1}{2}|x|^3 & 1\prec|x|\le 2\\ 0 & otherwise\end{cases} \tag{10}$$

The image value can be estimated via the 2-D convolution between the image function $f(x_i, y_j)$ and the kernel h(x) h(y),

$$f(r,\theta)=\sum_{i=0}^{U}\sum_{j=0}^{V} f(x_i, y_j)h(r\cos\theta-x_i)h(r\sin\theta-y_j) \tag{11}$$

Combining (7), (8), (10) and (11), we obtain the coefficients ART in polar coordinates by without introducing any geometric error or integral error.

$$F_{nm}=\frac{1}{2\pi}\sum_{i=0}^{U}\sum_{j=0}^{V} f(x_i, y_j)h(r\cos\theta-x_i)h(r\sin\theta-y_j)H(r_i,\theta_{i,j})$$

2.2.2 Algorithm of ART

Algorithm to compute coefficients of ART in polar coordinates of and Extraction

Base computation:

```
1: for n = 1 to N do
2:     for m = 1 to M do
3:         ART (n, m) ← functionBase (n, m) ×image(r, theta)
           Image(r, theta) ← kernel function(x, y) ×image(x, y)
4:     end for
5: end for
```

Coeffecients extraction:

```
1: for n = to N do
2:   for m = 1 to M do
3:     for x = 1 to size of image do
4:         for y = 1 to size of image do
5:         ARTcoeffs (n, m) ← ARTcoeffs (n, m) + functionbases(n,m)×img(x, y)
6:         end for
7:       end for
8:     end for
9: end for
```

This algorithm helps explain how the coefficients of ART was well calculated using interpolation cubic, from where:
FunctionBase (n, m): $V_{nm}(r,\theta)$
Kernel function(x, y): The function of the core 1-D is a cubic.
ARTcoeffs (n, m): Coefficients of ART.

3 Simulation Results

In this section, we test the validity of the proposed method could be proved in our work we use MPEG-7 for binary and Flavia dataset gray level images to test the algorithm for computing coefficients of and execution time of extraction. The efficiency according to the proposed method. Finally, a comparison between the computation times for extracting the characteristics of shape of the object in an image. All the experiments in this section are implemented on PC with Dual Core 2.40 GHz and 2GB RAM and the development is made by using Matlab.

In this work we leave the MPEG 7 binary database as it is and we convert each RGB image of Flavia into grayscale. All images are scaled in 101x101 resolution. After converting the image to grayscale we find the connected components of image and use the centroid property to find the center of mass of the region, so we can move the image to the center. This is necessary for the ART coefficients calculation to extract and calculate features of two images of MPEG-7 and Flavia. In our work we test thé validity of our algorithm and computational time of two database MPEG 7 and flavia.

3.1 Results and Analysis

The image of figure Used for our experiment binary and gray level image of size 101 x 101 pixels.

(a) (b) (c)

Fig. 2 a) RGB image, b) Grayscale image, c) Binary image

The defaults ART coefficients for n=3 and m=12 which gives a feature vector containing 36 moments.

3.2 *Computational Time*

The execution time (in seconds elapsed CPU) is the amount of time during which ART coefficients of order n and m extract objects in an image.

Feature Extraction (ART)	Feature extraction time (in s)
for single **Binary image**	0.179809 s
for single **Gray level image**	0.348863 s
for full database MPEG7(**Binary image)**	368.316 s
for full database Flavia **Gray level image)**	556.229 s

4 Figures

CPU elapsed times of ART coefficients computation for a digital gray scale image and binary image of size 101×101 pixel are displayed in graph. We noticed that the gray image needs the execution time for the computation ART coefficients of that binary image.

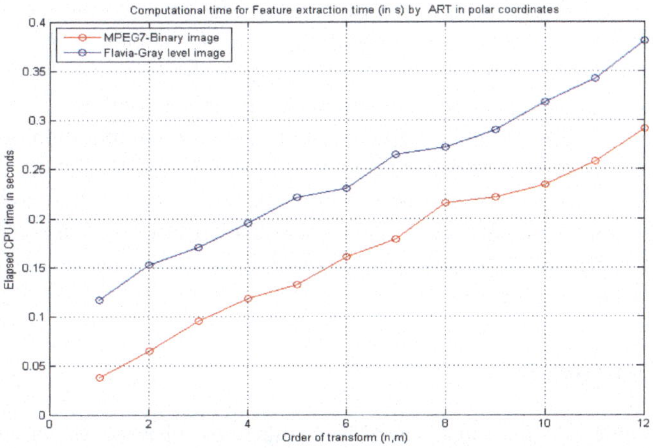

Fig. 3 Computational time for feature extraction time (in s) by ART in polar coordinates.

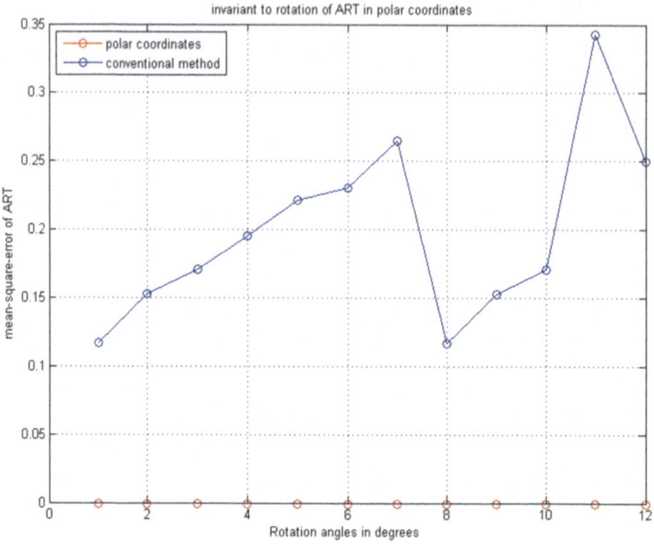

Fig. 4 Invariant to rotation of ART in polar coordinates.

Among the most important properties of ART is the magnitude invariance to image rotation. If ART is rotated counterclockwise by angle α; the transformed image function is $g(r,\theta)=f(r,\theta-\alpha)$. The ART coefficients of original and rotated images are F_{nm} and $F_{nm}^{rot}=e^{-jn\theta}.F_{nm}$, the magnitude values are identical. However, for a digital image, this property has to be compromised. The average absolute difference between the magnitudes of the coefficients ART of the images in Fig. 4 and their 20°-rotated counter parts. It can be seen that the magnitudes of ART deviate considerably from invariance under image rotation.

This non-ideal rotational invariance stems from the computation of coefficients ART when this computation has been made by conventional method.

According the results of Fig.4, where the red curve is the result from the Cartesian approach, while the other is the result from the polar approach. It is evident that the polar method greatly outperforms the Cartesian method in terms of rotational invariance.

5 Conclusion

Shape is one of the most valuable features to identify or describe objects represented in images. The results show that our algorithm has been improved the rotational invariance of ART through accurate computation in polar coordinate system and destroyed the errors, which have been produced by the conventional method. According the graph, we find that our approach does not consume the execution time as the conventional method.

References

1. Rui, Y., Huang, T.S: Image retrieval: Past, Present, and Future. Journal of Visual Communication and Image Representation (1997)
2. Walia, E., Vesal, S., Pal, A.: An Effective and Fast Hybrid Framework for Color Image Retrieval. Sensing and Imaging **15**(1), 1–23 (2014)
3. Hwang, S., Kim, W.: Fast and efficient method for computing art. IEEE Trans. Image Process. **15**(1), 112–117 (2006)
4. Bober, M.: MPEG-7 visual shape descriptors. IEEE Trans. Circuits Syst. Video Technol. **11**(6), 716–719 (2001)
5. Choi, M.-S., Kim, W.-Y.: The description and retrieval of a sequence of moving objects using a shape variation map. Pattern Recognition Letters **25**, 1369–1375 (2004)
6. Omar, M.W., Omar, K., Nasrudin, M.F.: Logo recognition system using angular radial transform descriptors. J. Comput. Sci. **7**, 1416–1422 (2011)
7. Lee, S.H., Sharma, S., Sang, L., Park, J.-I., Park, Y.G.: An intelligent video secu-rity system using object tracking and shape recognition. In: ACIVS. LNCS, vol. 6915, pp. 471–482. Springer-Verlag, Heidelberg (2011)
8. Mokhtarian, F., Abbasi, S., Kittler, J.: Efficient and robust retrieval by shape content through curvature scale space. In: Proc. International Workshop on Image Databases and Multimedia Search, Amsterdam, The Netherlands, pp. 35–42 (1996)
9. Kim, H.K., Kim, J.D.: Region-based shape descriptor invariant to rotation, scale and translation. Signal Processing: Image Communication **16**, 87–93 (2000)
10. Kim, W.-Y., Kim, Y.-S.: A new region-based shape descriptor. ISO/IEC MPEG99/M5472, TR15-01, Maui, Hawaii, December 1999
11. Broumandnia, A., Shanbehzadeh, J.: Fast Zernike wavelet moments for Farsi character recognition. Image and Vision Computing **25**, 717–726 (2007)
12. Amanatiadis, A., Kaburlasos, V.G., Gasteratos, A., Papadakis, S.E.: Evaluation of shape descriptors for shape-based image retrieval. Image Processing **5**, 493–499 (2011)
13. Hosny, K.M.: Fast Computation of Accurate Zernike moments. Journal of Real-Time Image Processing **3**(2), 97–107 (2008)
14. Unser, M., Thevenaz, P., Yaroslavsky, L.: Convolution-based interpolation for fast, high-quality rotation of images. IEEE Transactions on Image Processing **4**(10), 1371–1381 (1995)
15. Meijering, E., Unser, M.: A note on cubic convolution interpolation. IEEE Transactions on Image Processing **12** (2003). (LIB-ARTICLE-2003-008)
16. Wang, X., Ding, Y., Liu, M.Y., Yan, X.L.: Efficient implementation of a cubic-convolution based image scaling engine. Journal of Zhejiang University SCIENCE C **12**(9), 743–753 (2011)

A New Image Interpolation Using Laplacian Operator

Said Ousguine, Fedwa Essannouni, Leila Essannouni, Mohammed Abbad and Driss Aboutajdine

Abstract In this paper, a novel method for image interpolation is proposed. This method is based on the application of the Laplacian operator for the purpose of detecting the edge-directions and then interpolating the missing pixels using the cubic convolution. We start applying a down-sampled by a factor of two to the gray high-resolution image in order to obtain a low-resolution image. Then, the preprocessed image is reconstructed by using the proposed interpolation method. The proposed method is implemented and tested over several gray images, and also compared to many interpolation methods in the state-of-the-art. The simulation results are shown to be superior compared to the other interpolation methods in both of objective measurement in terms of PSNR, SSIM and FSIM, and visual quality of image results.

Keywords Image interpolation · Laplacian operator · Image reconstruction · Super-resolution

1 Introduction

Image interpolation refers to reconstructing a high-resolution image (HR) from its low-resolution (LR) counterpart; it is an essential task in super-resolution algorithms during all tasks that are performed on a high-resolution image such as registration, down-sampling, denoising and enhancement. The concept behind image interpolation is the estimation of the underlying representation of the data point or the missing pixels in the area of the image processing to obtain a high-resolution image with better resolution and best visual quality, also to get accurate results out of an interpolation with a trade off between the interpolated image result and computational cost complexity. The interpolation can be used in image resizing, upscaling, enlargement and

S. Ousguine · F. Essannouni · L. Essannouni · M. Abbad · D. Aboutajdine(✉)
LRIT Research Laboratory (associated unit to CNRST, URAC no 29),
Faculty of Sciences, Mohammed V University, Rabat, Morocco
e-mail: lamak07@hotmail.fr

E. Sabir et al. (eds.), *The International Symposium on Ubiquitous Networking*,
Lecture Notes in Electrical Engineering 366,
DOI: 10.1007/978-981-287-990-5_32

image enhancement. In addition, there are numerous applications which require applying image interpolation such as remote sensing, surveillance, medical imaging, computer vision, high definition television and consumer electronics.

Interpolation methods can be classified into two main categories: the first one is the non-adaptive interpolation, which process on the whole image in uniform way based on mathematical techniques of interpolation like nearest neighbor, bilinear, bi-cubic, b-spline and Sinc interpolation. These classical methods use the adjacent pixels, but suffer from a various problems such as blur or jaggy, and may cause discontinuity or zigzag effects around edge area. The second class of interpolation methods is the adaptive interpolation, this class adjusts their operation according to the content of the local structure of image, such as edge information and variation intensity among image pixels. The important groups between the adaptive algorithms are edge-directed interpolation methods which have as a common objective: the analysis of an image using information about localisation and orientation of extracted edges, to ensure the success of interpolation process along these edges.

Numerous interpolation methods have been proposed in the literature either in adaptive class or non-adaptive class or by combining the two classes [1-19]. G. ramponi [1] proposes to use the warped distance to the interpolated pixel instead of a regular one, thus modifying the one-dimensional kernel of a separable interpolation filter. An amelioration of this method was made by Huang and Lee [2] which consists in modifying the two-dimension interpolation Kernel by the characteristics of local gradient. A method of edge detection was introduced by Jensen and Anastassion [3]; they apply a projection onto orthonormal basis to detect the edge direction, and modify the interpolation process to avoid interpolating across the edges to improve the visual perception of the interpolated image. An iterative edge-directed scheme was developed in [16]. The authors use an estimate of the edge mapping of the high-resolution image to guide a bilinear interpolation such that interpolation across edges are avoided. A New Edge-Directed Interpolation (NEDI) interpolation was presented by Li and Orchard in [4]. The authors used the local covariance of low-resolution images to estimate the covariance of high-resolution image. Asuni and Giachetti [13] propose Improved New Edge-Directed Interpolation (INEDI) to improve the algorithm of Li [4] by adopting circular windows and adaptive selecting the size of windows. However, in the region of fast luminance change, the windows are too small to get a stable solution. Li and Zhang [6] propose an Edge-guided algorithm via directional filtering and data fusion, they have estimated the missing pixels which can be interpolated, by two sets of observations which are defined in two orthogonal directions, these latter are fused through linear minimum mean-square-error (LMMSE). The authors in [9] propose a regularity-preserving interpolation method which consists in extrapolating a new wavelets sub-bands based on the decay of the finest scale. In Markov Random Field Model-Based Edge-Directed Image Interpolation (MRF-EDI) interpolation [18], the authors use the Markov rand field model in edge-directed interpolation method, by searching the minimal energy state of a two-dimensional random field, so the energy function in the MRF model that is along the edge direction is strongly suppressed to achieve smoothness, while the energy that is across the edge is much less suppressed to preserve the sharpness of the edge. An another adaptive interpolation method was proposed by K. adamczyk and A.walczak [19]. The authors use localization of extracting edges by using of the 2D Anistrisotropic wavelets, to interpolate new pixels of an image depending on localization with respect to the extracted edges. Muresan and Park [5] extended this strategy which based on the influence of a full cone sharp edge in the wavelet scale spaces, for estimation the coefficient of scale by an optimal recovery theory. An interpolation scheme was proposed by Giachetti and Asuni [8], its principal is to interpolate locally the missing pixels in the two diagonal directions, while the second order image derivative is lower, these interpolated pixels value are modified using an iterative refinement to minimizing the differences in second order image derivative. Cha and Kim [7] describe an interpolation method by using a bilinear interpolation and correcting the errors by adapting the interpolation error theorem in an edge-adaptive way. The authors in [11] develop an interpolation approach based on edge-oriented algorithm, they classify the image into two partitions, the homogenous zones which its missing pixels are interpolated by a bilinear interpolation, and the edge area, which its missing pixels are interpolated using all neighboring pixels, this neighboring pixels contains the original pixels and the interpolated pixels in the homogenous area. Zhou and Shen [10] propose an image zooming using cubic convolution interpolation with detecting the edge-direction, this method is based on the detection of edge-direction of the missing pixels by computing the gradient of horizontal, vertical, 45° diagonal and 135° diagonal directions, then they interpolated the missing pixels using a cubic convolution interpolation.

In this paper, we propose a new method of image interpolation using the Laplacian operator; which the missing pixels are interpolated along the detected edge direction, this latter is estimated by the Laplacian of Gaussian in the horizontal and vertical directions, a cubic convolution interpolation is used to interpolate the missing pixels along the strong edge. The paper is organized as follow: in section 2 describes the proposed interpolation method. Simulation and results are shown in Section 3. Finally, the conclusion is drawn in section 4.

2 The Proposed Method

The Laplacian filters are derivative filters used to detect areas of abrupt change (edges) in images. Since derivative filters are very sensitive to noise, it is common to smooth the image (e.g., using a Gaussian filter) before applying the Laplacian. This two-step process is calling the Laplacian of Gaussian (LoG) operation. The Laplacian of a function $f(x, y)$ is defined as follows:

$$LP(x, y) = \nabla^2 f(x, y) = \delta^2 f(x, y)/\delta x^2 + \delta^2 f(x, y)/\delta y^2. \quad (1)$$

To include a smoothing Gaussian filter, we combine the Laplacian and Gaussian functions to obtain a single equation:

$$LOG(x, y) = -1/\pi\sigma^4[1 - (x^2 + y^2/2\sigma^2)]\exp^{-((x^2+y^2/2\sigma^2))}, \tag{2}$$

where σ is the standard deviation and the gaussian function is centered ($\mu = 0$).

The LoG operator takes the second derivative of the image. Where the image is basically regular, the LoG will give zero. Wherever a change occurs, the LoG will give a positive response on the darker side and a negative response on the lighter side. At a sharp edge between two regions, the response will be:

- zero at a long distance
- positive just to one side of edge
- negative just to the other side of edge
- zero at some point in between on the edge itself

An LR image I_l can be considered to be a directly down sampled version of the HR image corresponding to I_h, where the downsampling factor is 2. The high resolution image I_h is restored by copying the LR image I_l pixels into an enlarged grid and then filling with the missing pixels.

Here we propose an estimation method of the strong edge for a missing pixel location. The proposed method can be described as follows:

Step 1: Compute the Laplacian of Gaussian of the low resolution image I_l :

$$[LP_x; LP_y] = LOG(I_l)$$

Step 2: For every pixel to be estimated (i, j), the laplacian in the 45° and 135° diagonal directions are computed in the 7x7 neighbour of this pixel, at the same way in the second step, we estimate the strength of the edges on the horizontal and vertical directions in the 5x5 neighbour. They are four immediate diagonal pixels to be known, the orientation of the Laplacian of Gaussian at the central location (i, j) can be determined using the given equations:

$$d_1(i, j) = \sum_{k=3,\pm1} \sum_{l=3,\pm1} |LOG(i + k; j - l) - LOG(i + k - 2; j - l + 2)| \tag{3}$$

$$d_2(i, j) = \sum_{k=3,\pm1} \sum_{l=3,\pm1} |LOG(i + k; j + l) - LOG(i + k - 2; j + l - 2)| \tag{4}$$

Using $d_1(i, j)$ and $d_2(i, j)$, the edge direction for a pixel location (i,j), can be estimated as follows:

$$\begin{cases} \text{if } d_1(i, j) > d_2(i, j) \\ \quad \text{the Laplacian is greater in the horizontal or 45° direction} \\ \text{else} \\ \quad \text{the Laplacian is greater in the vertical or 135° direction} \\ \text{end} \end{cases}$$

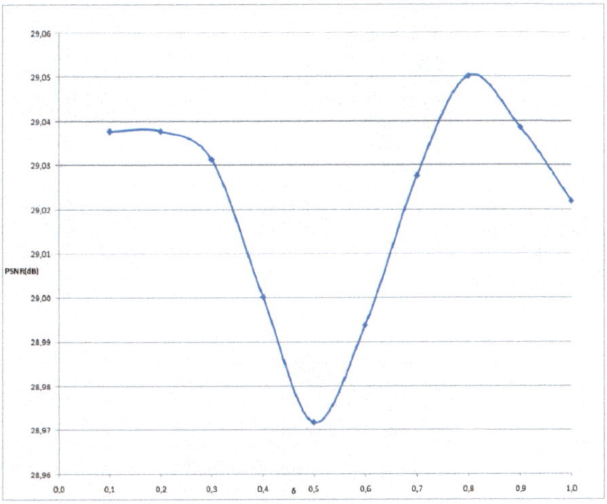

Fig. 1 Selection of the standard deviation σ.

Step 3: Compute the intensity value at the missing pixel position using the estimated edge direction. Indeed, the intensity pixel value can be computed using the cubic convolution by interpolating the four neighbour pixels along the strong edge.

The results of the proposed method depend on the choice of the standard deviation σ. The parameter cannot be directly determined for a given LR image because they depend on the types of the images. We derive σ through training. Twenty-four 512×768 Kodak colour images [23] were used as training samples, which were first converted into grey images and then downsampled the grey images to obtain their LR counterparts. The HR images were reconstructed from the LR ones using the proposed method with the different σ. The parameter σ was separately set from 0.1 to 1 by step length 0.1. The average peak signal-to-noise ratio (PSNR) curves is shown for different cases in Fig. 1. It can be seen that the higher PSNR values can be achieved when the greater σ is used. On the other hand, we can see from this figure that the PSNR achieves its maximum values when $\sigma = 0.8$.

3 Simulation and Results

In order to demonstrate the performance of the proposed method, several simulations are made to explore the effectiveness of the proposed method, this latter has been implemented in Matlab and tested on several images, and compared with the state of art of interpolation algorithms in terms of both quantitative metrics and subjective high quality, cubic convolution interpolation (CC) [12], directional filtering and data fusion (DFDF) [6], new edge-directed interpolation (NEDI) [4], improvement new edge-directed interpolation (iNEDI) [13] and fast artifacts-free Image interpolation (ICBI) [8].

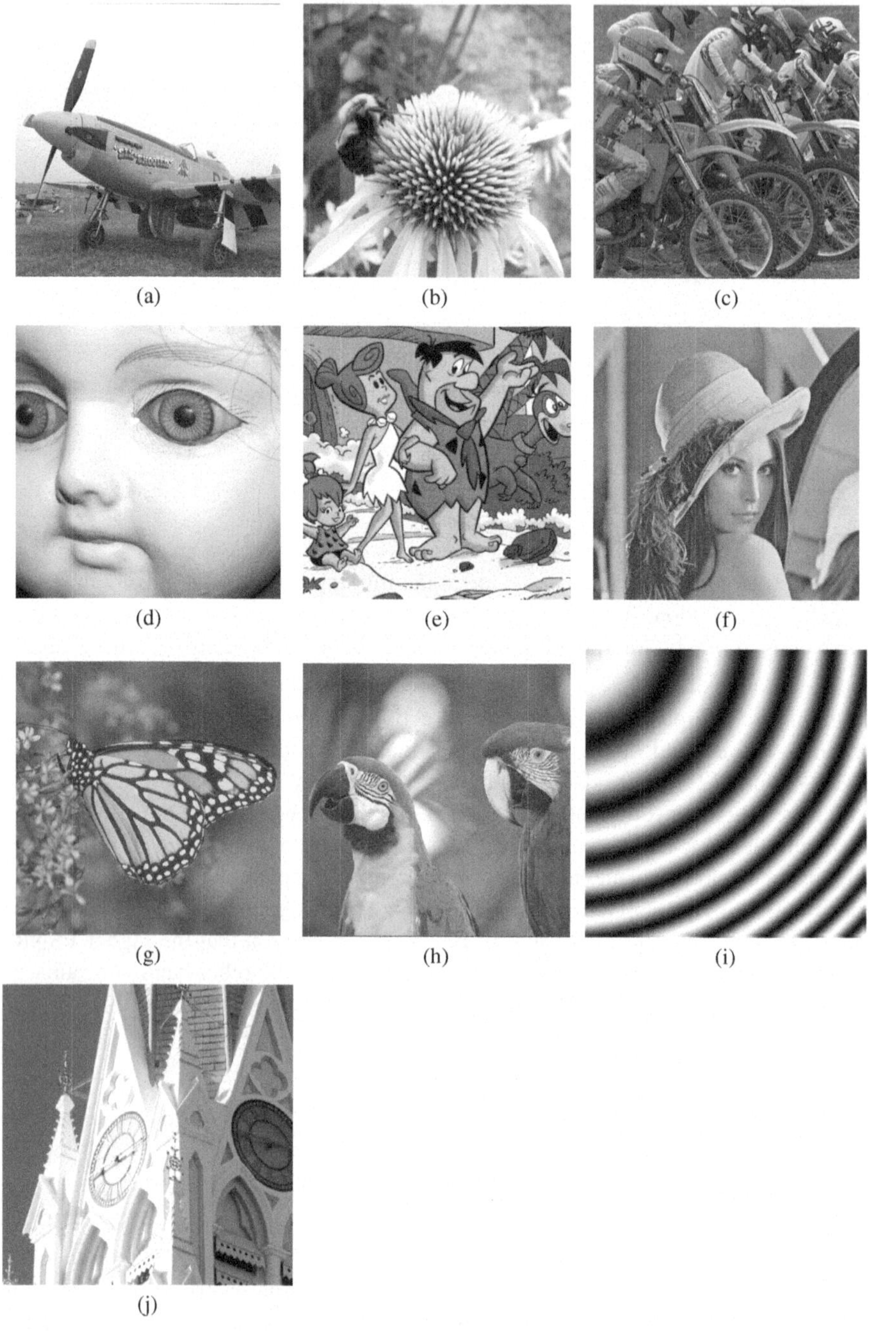

Fig. 2 Set of images test (a) Airplane; (b) Bee; (c) Bike; (d) Face; (e) Flinstones; (f) Lena; (g) Monarch; (h) Parrot ; (i) Rings; (j)Watch

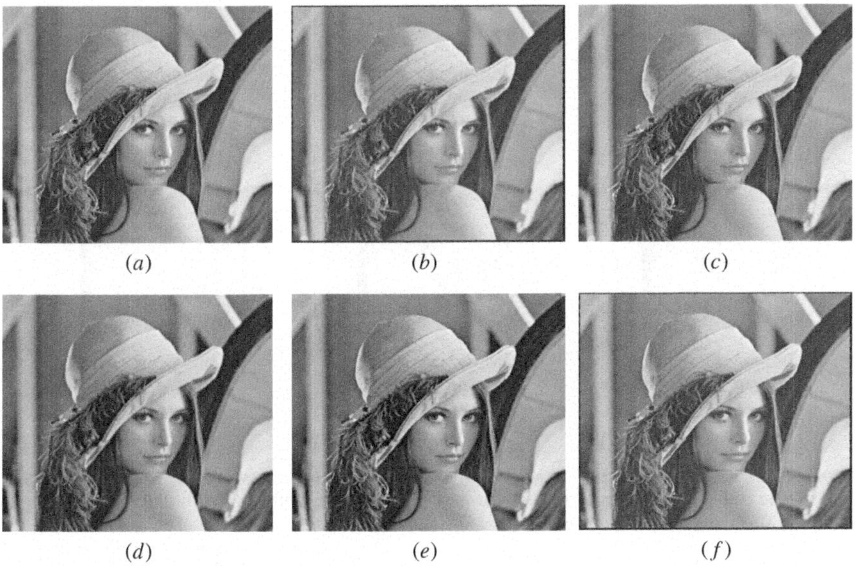

Fig. 3 The interpolated images results of 'Face' (a) Cubic convolution; (b) DFDF; (c) ICBI; (d) NEDI; (e) INEDI; (f) Proposed

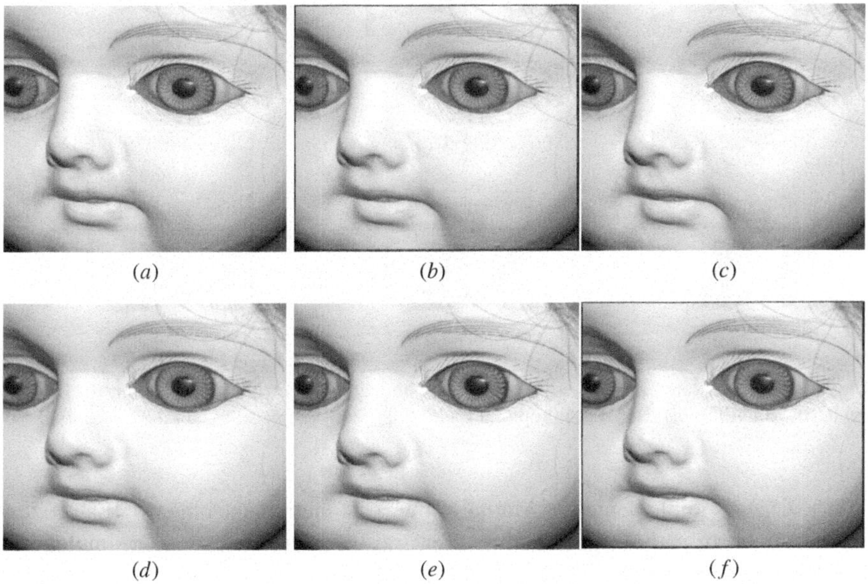

Fig. 4 The interpolated images results of 'Face' (a) Cubic convolution; (b) DFDF; (c) ICBI; (d) NEDI; (e) INEDI; (f) Proposed

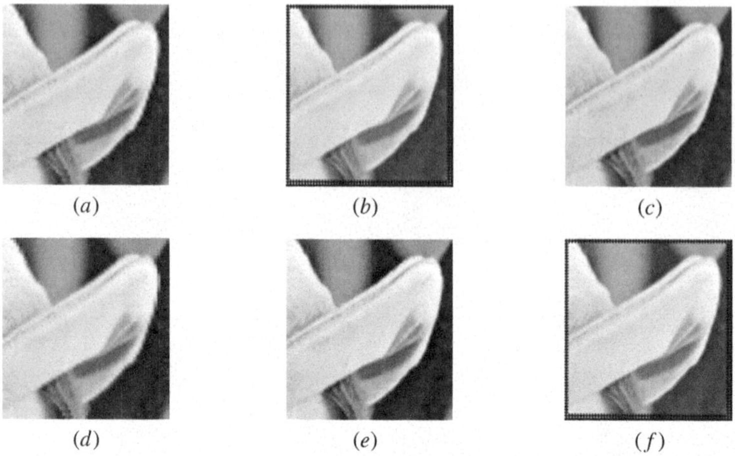

Fig. 5 The interpolated images results the part of 'Lena' (a) Cubic convolution; (b) DFDF; (c) ICBI; (d) NEDI; (e) INEDI; (f) Proposed

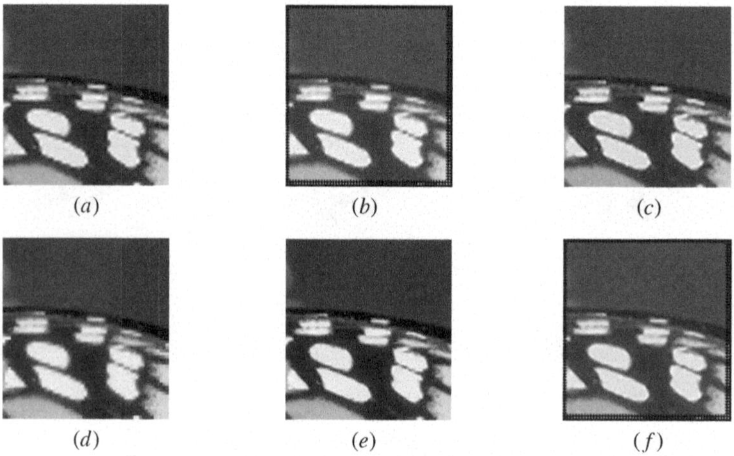

Fig. 6 The interpolated images results the part of 'monarch' ((a) Cubic convolution; (b) DFDF; (c) ICBI; (d) NEDI; (e) INEDI; (f) Proposed

The CC was implemented by MatLab "interp2" function, the other MatLab codes were available from the original authors. Ten gray test images are used for simulation; Fig.2 displays the ten gray images used for test.

We started from an original high-resolution gray image; this latter was down-sampled by a factor of two to obtain the low resolution image. The LR image was reconstructed by a set of interpolation methods (for the proposed method, the standard deviation σ has been set to 0.8). For evaluation of the proposed method in comparing with the other interpolation methods, several experimental factors are

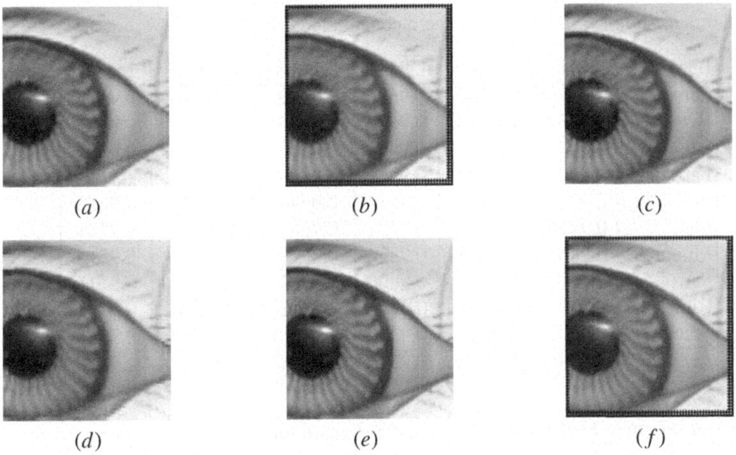

Fig. 7 The interpolated images results the part of 'Face' (a) Cubic convolution; (b) DFDF; (c) ICBI; (d) NEDI; (e) INEDI; (f) Proposed

Table 1 Comparison of PSNR (db) results of the reconstructed images

Image	CC [12]	DFDF [6]	ICBI [8]	NEDI [4]	INEDI [13]	Proposed
Airplane	29,77	29,65	30,20	29,44	30,01	**30,32**
Bee	34,38	35,05	34,31	32,11	33,32	**35,14**
Bike	25,56	25,18	25,60	24,90	25,65	**26,09**
Face	40,53	40,84	39,79	38,41	39,19	**41,27**
Flinstones	26,96	27,00	27,03	25,28	27,03	**27,62**
Lena	33,81	33,88	33,77	32,79	33,20	**34,13**
Monarch	30,16	30,85	30,71	28,92	**31,26**	31,19
Parrot	33,32	33,10	33,32	32,20	33,49	**34,03**
Rings	45,61	42,18	46,53	30,11	28,66	**54,11**
Watch	31,91	31,80	31,68	30,70	31,84	**32,66**
Average	33,20	32,95	33,29	30,49	31,36	**34,66**

set-up to examine the quality of interpolated images; the first factor is the subjective visual quality of the output images, for this, the image results for each interpolation methods for the tested images 'Lena' and 'Face' are shown in Figs. 3–7 respectively.

The second factor is the statistic parameter which is the quantitative measure: the Peak signal to noise ratio (PSNR) was measured for each of the output images, to determine the difference between original image and the reconstructed image. The structural similarity (SSIM) measures the similarity between two images [20] and the Feature Similarity Index (FSIM) measures the feature similarity index between two images [21]. Tables 1, 2 and 3 show the resulting PSNR, SSIM and FSIM scores of the different methods.

The above proposed method was compared to the several interpolation methods. In the part of the subjective appearance of the resulted image, it is clear that the proposed method produces a sharp high-resolution images and improves their

Table 2 Comparison of SSIM results of the reconstructed images

Image	CC [12]	DFDF [6]	ICBI [8]	NEDI [4]	INEDI [13]	Proposed
Airplane	0,9680	0,9654	0,9690	0,9659	0,9668	**0,9683**
Bee	0,9934	0,9940	0,9926	0,9883	0,9917	**0,9941**
Bike	0,9492	0,9446	0,9464	0,9356	0,9470	**0,9524**
Face	0,9926	0,9927	0,9915	0,9889	0,9916	**0,9928**
Flinstones	0,9633	0,9620	0,9632	0,9540	0,9637	**0,9654**
Lena	0,9720	0,9700	0,9721	0,9688	0,9692	**0,9741**
Monarch	0,9858	0,9862	0,9862	0,9820	0,9869	**0,9874**
Parrot	0,9812	0,9799	0,9806	0,9782	0,9807	**0,9811**
Rings	0,9999	0,9995	0,9996	0,9892	0,9825	**0,9999**
Watch	0,9809	0,9813	0,9790	0,9728	0,9824	**0,9846**
Average	0,9786	0,9776	0,9780	0,9724	0,9763	**0,9800**

Table 3 Comparison of FSIM results of the reconstructed images

Image	CC [12]	DFDF [6]	ICBI [8]	NEDI [4]	INEDI [13]	Proposed
Airplane	0,9717	0,9701	0,9805	0,9767	0,9695	**0,9812**
Bee	0,9865	0,9869	0,9948	0,9906	0,9841	**0,9959**
Bike	0,9740	0,9718	0,9729	0,9648	0,9701	**0,9774**
Face	0,9871	0,9870	0,9954	0,9923	0,9854	**0,9968**
Flinstones	0,9668	0,9663	0,9751	0,9629	0,9658	**0,9781**
Lena	0,9787	0,9772	0,9868	0,9831	0,9750	**0,9876**
Monarch	0,9786	0,9801	0,9866	0,9794	0,9796	**0,9885**
Parrot	0,9823	0,9814	0,9860	0,9879	0,9807	**0,9910**
Rings	0,9999	0,9989	0,9998	0,9964	0,9840	**0,9999**
Watch	0,9774	0,9782	0,9839	0,9771	0,9774	**0,9885**
Average	0,9803	0,9798	0,9862	0,9811	0,9772	**0,9910**

resolution, which then demonstrates the outperformance to the images produced by the other interpolation methods. In the other hand, the highest PSNR value of each row is shown in bold. As can be observed from table 1, the proposed method gives a higher PSNR results, on an average a PSNR improvement of 1.37 dB is achieved compared to the second best method for all the testing images. The better PSNR improvement was 5.93 dB over the second best method for the Rings image.

4 Conclusion

A novel method of image interpolation is presented. This method consists in interpolating the missing pixel along the detected edge direction by using the well-known Laplacian operator and cubic convolution interpolation. The experimental results shows that the proposed method achieve a visible improvement compared to the images generated by the other interpolation methods. In the other hand, considerable objective results are observed with the proposed method, and comparable to the state-of-art interpolation method results, with similar PSNR, SSIM and FSIM scores. Therefore, the proposed method can preserve sharp edge and can achieve the reduction of the noise, blurring and aliasing occurred by the traditional interpolation methods.

References

1. Ramponi, G.: Warped distance for space-variant linear image interpolation. IEEE Trans. Image Process. **62**(5), 629–639 (1999)
2. Hwang, J.W., Lee, H.S.: Adaptive image interpolation based on local gradient features. IEEE Signal Process. Lett. **11**(3), 359–362 (2004)
3. Jensen, K., Anastassiou, D.: Subpixel edge localization and the interpolation of still images. IEEE Trans. Image Process. **4**(3), 285–295 (1995)
4. Li, X., Orchard, M.T.: New edge-directed interpolation. IEEE Trans. Image Process. **10**(10), 1521–1527 (2001)
5. Muresan, D., Parks, T.W.: Prediction of image detail. In: Process. Int. Conf. Image 2000, vol. 2 (2000)
6. Zhang, L., Wu, X.: An edge-guided image interpolation algorithm via directional filtering and data fusion. IEEE Trans. Image Process. **15**(8), 2226–2238 (2006)
7. Cha, Y., Kim, S.: The error-amended sharp edge (EASE) scheme for image zooming. IEEE Trans. Image Process. **16**(6), 1496–1505 (2007)
8. Giachetti, A., Asuni, N.: Fast artifacts-free image interpolation. In: Br. Mach. Vis. Conf. Leeds UK, pp. 123–132 (2008)
9. Carey, W.K., Hemami, S.S.: Regularity-preserving image interpolation. IEEE Trans. Image Process. **8**(9), 1293–1297 (1999)
10. Dengwen, Z., Xiaoliu, S., Weiming, D.: Image zooming using directional cubic convolution interpolation. IET Image Process. **6**(6), 627–634 (2012)
11. Chen, M.J., Lee, W.L.: A fast edge-oriented algorithm for image interpolation. Image Vis. Comput. **23**(9), 791–798 (2005)
12. Keys, R.: Cubic convolution interpolation for digital image processing. IEEE Trans. Acoust. Speech Signal Process. **29**(6), 1153–1160 (1981)
13. Asuni, N., Giachetti, A.: Accuracy improvements and artifacts removal in edge based image interpolation. In: Proc. 3rd Int. Conf. Comput. Vis. Theory Appl. VISAPP 2008 (2008)
14. Dube, S., Hong, L.: An adaptive algorithm for image resolution enhancement. In: Thirty-Fourth Asilomar Conf. Signals Syst. Comput. 2000 Conf. Rec., vol. 2 (2000)
15. Kimmel, R.: Demosaicing: image reconstruction from ccd samples. IEEE Trans. Image Processing **23**(9), 1221–1228 (1999)
16. Allebach, J., Wong, P.W.: Edge-directed interpolation. In: Proc. Int. Conf. Image Process., vol. 3, pp. 707–710 (1996)
17. Cok, D.R., Lee, W.L.: Signal processing method and apparatus for producing interpolated chrominance values in a sampled color image signal. US Pat. No 4642678 (1987)
18. Li, M., Nguyen, T.: Markov random field model-based edge-directed image interpolation. In: International Conf. Image Proc., pp. 93–96 (2007)
19. Adamczyk, K., Walczak, A.: Application of 2d anisotropic wavelet edge extractors for image interpolation. In: Human-Computer Systems Interaction, pp. 205–222. Springer-Verlag, Heidelberg (2012)
20. Wang, Z., Bovik, A.C., Sheikh, H.R., Simoncelli, E.P.: Image quality assessment: From error visibility to structural similarity. IEEE Transactions on Image Processing **13**(4), 600–612 (2004)
21. Zhang, L., Zhang, L., Mou, X., Zhang, D.: A Feature Similarity Index for Image Quality Assessment. IEEE Transactions on Image Processing **20**(8), 2378–2386 (2011)
22. Ousguine, S., Essannouni, F., Essannouni, L., Aboutajdine, D.: High Resolution Image Reconstruction Using the Phase Correlation. J. Inf. Organ. **2**(3), 128–134 (2012)
23. Dubois, E.: http://www.site.uottawa.ca/edubois/demosaicking

3D Objects Comparison Using New Approach Based Similarity Index

Khalid Aznag, Nouhoun Kane, Ahmed El Oirrak and Essaid El Bachari

Abstract In this work, a new method is presented for the representation of 3D objects with binary matrix. This method is based on two steps: normalization and quantization. This representation allows us to compare 3D objects by computing the similarity between them.

Keywords 3D objects · Binary matrix · Similarity index

1 Introduction

The size of multimedia data (text, images, audio, video, 3D objects, etc.) used on the Internet and in computer systems has become enormous, especially with the rapid advancement of technology acquisition and storage of objects 3D. Also the digital databases of 3D objects which are used in various domains (e-commerce, games, medicine, etc) become large. Therefore, an efficient method that allows users to find similar 3D objects in a 3D model is needed. Many content based search systems and search engines for 3D models are available on the web [1, 2, 3, 4]. Several approaches to extract the similarity between the 3D objects are described in the literature. Hekzko et al. [5] proposed the images based descriptor which extracts feature vectors from several images obtained by orthogonal projections of the object. Chen et al. [6] and Ansary et al. [7]use the view based approach in which a number of views of the models are used in order to generate a shape descriptor. This approach is based on the idea that if two shapes are similar they should look similar from all viewing angles. Generally, the 2D shape descriptors are used in the view based case. As a non-feature vector approach. Hilaga et al. [8] proposed the method using Reeb graphs based on geodesic distances between points on the mesh, which provides a

K. Aznag(✉) · N. Kane · A. El Oirrak · E. El Bachari
Faculty of Sciences Semlalia, Computer Science Department, LISI, Marrakech, Morocco
e-mail: khalid.aznag@ced.uca.ma, {n.kane,oirrak,elbachari}@uca.ma

E. Sabir et al. (eds.), *The International Symposium on Ubiquitous Networking*,
Lecture Notes in Electrical Engineering 366,
DOI: 10.1007/978-981-287-990-5_33

rich representation of particular forms capable of integrating the object topology. The approach based on the skeleton is used by Sundar et al. [9] present the object by its skeleton, calculated by applying the thinning algorithm on the voxelization of a solid object. Vranic et al. [10] proposed the basic approach of rays, which extract the extended center of mass of the object's surface. Feature vectors constructed using this method are presented in the frequency domain using the spherical harmonics [11].

In this work we present a new representation for 3D objects; it is the binary matrix representation of objects. To apply our approach we need two important steps : firstly, the phase of normalization which allows to have coordinates in [0, 1]. Secondly, the phase concerning the quantization which attributes every interval a label. We applied this representation for 2D objects[12], we have shown the advantages of this representation in terms of comparison of 2D objects (binary image, 2D curve) and especially noisy the curves.

In this work the voxelization is automatic and is simple to reach but in [13] they use two approaches to set the box size. The first approach examines the histograms of the extremities along x, y and z axes of the pose normalized objects in the database and selects the box size such that the majority of the objects will remain entirely in the box. The second approach calculates the cropped proportion over all the objects in the database in terms of surface area for varying box sizes and then chooses the minimum size that keeps the lost surface proportion below a threshold. In [14] they use the analytical algorithm presented in [15] for calculating the surface areas S_{ijk} that intersect the voxel V_{ijk}. Each voxel V_{ijk} stores the real value S_{ijk}/S , where S is the total surface area of the object witch is equal to, $\sum\sum\sum S_{ijk}$, then use an octree structure in order to store the voxels, this avoid explicit storage of non-occupied part of the voxel grid.

The rest of this paper is organized as follows: section 2 describe our new approach. Section 3 describe the similarity index between 3D objects. The conclusion is stated in Section 4.

2 Application to 3D Objects

In this section our objective is to determine the 3D binary matrix of the 3D objects. To determine the 3d binary matrix of an object we need two steps: normalization and quantization.

2.1 Normalization

The normalization is a technique which allows to have coordinates in normalized interval [0, 1].

$$x_n = \frac{|x - min(X)|}{|max(X) - min(X)|} \tag{1}$$

for each x∈X with x_n ∈[0,1]

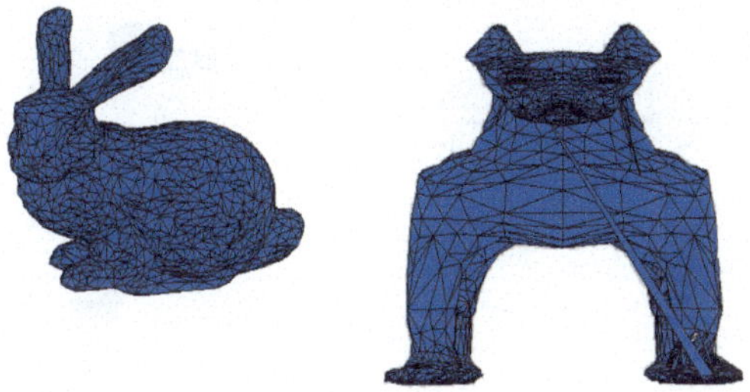

1.a 1.b

Fig. 1 1.a) Bunny object 1.b) Dog object

x_n denote normalized coordinate.

$$y_n = \frac{|y - min(Y)|}{|max(Y) - min(Y)|} \tag{2}$$

for each y∈Y with $y_n \in [0,1]$
y_n denote normalized coordinate.

$$z_n = \frac{|z - min(Z)|}{|max(Z) - min(Z)|} \tag{3}$$

for each z∈Z with $z_n \in [0,1]$
z_n denote normalized coordinate.

2.2 *Quantization*

2.2.1 Vector Quantization

Quantization technique is one of the most popular compression technique [16]. In our case we make the following substitutions:

$$\begin{cases} values\, in[i/10, (i+1)/10]\, are\, replaced\, by \\ the\, label\, (i+1)\, for\, each\, i \in [0, .., 9] \end{cases} \tag{4}$$

Equation (4) allows quantization until level 10 for high level. Another solution K-means [17] algorithm is used to obtain label for coordinates x,y and z.

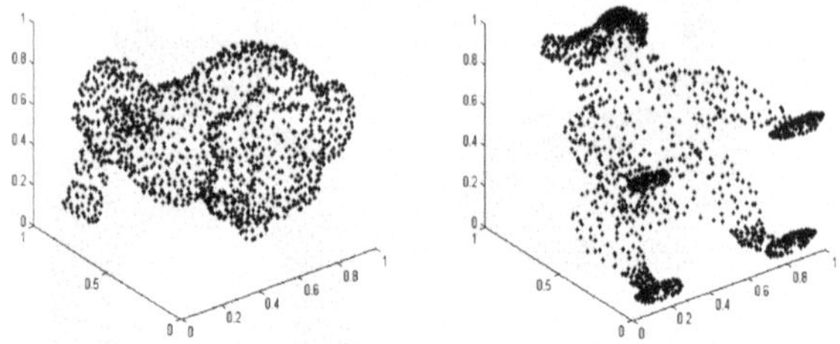

2.a 2.b

Fig. 2 2.a) Normalisation for bunny object 1.a 2.b) Normalisation for object dog 1.b

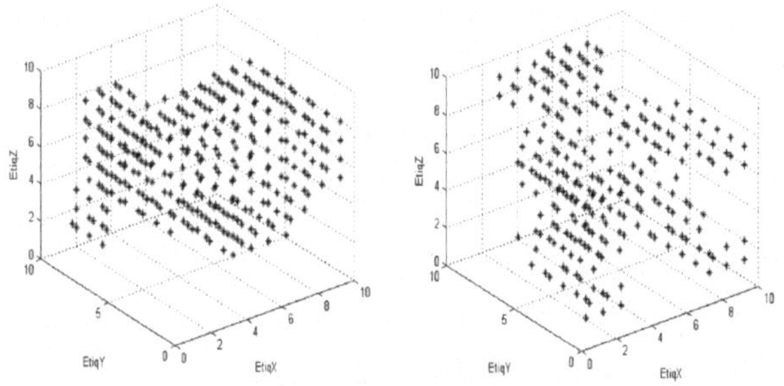

3.a 3.b

Fig. 3 3.a) Binary matrix for object 1.a 3.b) Binary matrix for object 1.b

2.3 Calculate the Binary Matrix

After the step of normalization which allows to have coordinates in [0, 1]. The quantization allows to assign a label ∈ [1, .., 10] to each interval, all the elements of binary matrix are deduced after browse of each of the three vectors (X_n, Y_n, Z_n).

Let **Voxel** denote the binary matrix then

Voxel(i, j, k)= 1 $(1 \leq i \leq 10, 1 \leq j \leq 10 \ and \ 1 \leq k \leq 10)$.

if there exist $X_p, Y_q \ and \ Z_r$ such as:

$[X_p = i, Y_q = j \ and \ Z_r = k]$

3 Similarity Index Betwen Two 3D Objects

3.1 Jaccard Coefficient

The Jaccard Index (JI), also known as the Jaccard Similarity Coefficient[18], is a measurement used to identify the degree of similarity and diversity of two data windows.

Let us consider two data windows, X_i and X_j. The coefficient measures the degree of overlap between two windows by computing the ratio of the number of shared attributes between X_i and X_j. For simplicity, let us consider two sets A and B in place of data windows. The region of intersection $(A \cap B)$ and union $(A \cup B)$ between these two sets can be measured according to set theory. Thus, the Jaccard index is calculated as follows:

$$Jaccard(A, B) = \frac{(A \cap B)}{(A \cup B)} \tag{5}$$

To identify the similarity between two data windows, X_i and X_j the Intersection between these windows is obtained by calculating the cardinality of the same value attributes in the windows as depicted in Eq. 6.

$$\overrightarrow{X_i} \cap \overrightarrow{X_j} = \sum_{k=1}^{n} X_{ki} \cap X_{kj}, X_{ki} \in \overrightarrow{X_i} \; and \; X_{ki} \in \overrightarrow{X_{kj}} \tag{6}$$

where,

$$X_{ki} \cap X_{kj} = \begin{cases} 1, & if \; X_{ki} = X_{kj}; \\ 0, & otherwise. \end{cases}$$

In standard set theory, the union of two sets is the set of all distinct elements in the sets. While calculating the Jaccard similarity measurement, the Union between two data windows $\overrightarrow{X}_i$ and $\overrightarrow{X}_j$ is obtained by using Eq. 7.

$$\overrightarrow{X_i} \cup \overrightarrow{X_j} = \sum_{k=1}^{n} X_{ki} \cup X_{kj}, \; X_{ki} \in \overrightarrow{X_i}; \; X_{kj} \in \overrightarrow{X_j} \; and \; X_{ki} \cup X_{kj} = 1 \tag{7}$$

$$= Dimension \; of \; the \; data \; window \tag{8}$$

Thus, the Jaccard similarity measurement or Jaccard Index, Jaccard($\overrightarrow{X_i}$, $\overrightarrow{X_j}$) between the two windows $\overrightarrow{X_i}$ and $\overrightarrow{X_j}$ is :

$$Jaccard(\overrightarrow{X_i}, \overrightarrow{X_j}) = \frac{cardinality \; of \; Intersection}{cardinality \; of \; Union} \tag{9}$$

3.2 Results and Discusion

In this work we present a new representation for 3D objects. We represented the 3D objects as mesh. We have a database of 3D objects as files (.off). So we develop a program in java that extracts the facets and points (X, Y and Z) for each object. So once we have for each object (facets, Points) we apply our approach that can represent each object as 3D binary matrix.

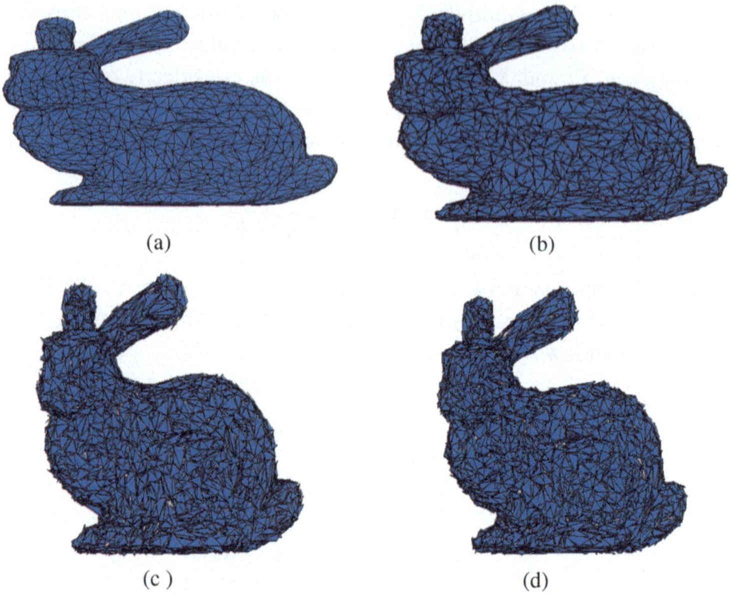

Fig. 4 (a):original bunny model ,(b),(c),(d): noise model bunny

Table 1 *Jaccard Similarity between Binary matrix 3D of differents objects*

	(a)	(b)	(c)	(d)
(a)	1	0.40	0.50	0.33
(b)	0.40	1	0.66	0.44
(c)	0.50	0.66	1	0.66
(d)	0.33	0.44	0.66	1

In table 1 we present similarity index between differents model.

4 Conclusion

In this work, we present a new method for the representation of 3D objects. Our approach is based on two steps; the first concerning the normalization and the secondly step is the quantization which allows to attribute to every interval a label. In futur work instead of using k-means separately on X, Y and Z we will use it for the tuple (X, Y, Z).

References

1. Dejan Vranic's 3D Search Engine. http://merkur01.inf.uni-konstanz.de/CCCC/
2. http://3d.csie.ntu.edu.tw/
3. http://shape.cs.princeton.edu/search.html
4. http://www-rech.enic.fr/3dretrieval/
5. Heczko, M., Keim, A.D., Saupe, D., Vranic, D.V.: Verfahren zur ahnlichkeitssuche auf 3d objekten: (methods for similarity search on 3d databases). Datenbank-Spektrum **2**, 54–63 (2002)
6. Chen, D.Y., Ouhyoung, M., Tian, X.P., Shen, Y.T.: On visual similarity based 3D model retrieval, pp. 223–232. Eurographics, Granada (2003)
7. Filali Ansary, T., Daoudi, M., Vandeborre, J.P.: A Bayesian 3d search engine using adaptive views clustering. IEEE Transaction On Multimedia **9**, 78–88 (2007)
8. Hilaga, M., Shinagawa, Y., Kohmura, T., Kunii, T.L.: Topology matching for fully automatic similarity estimation of 3D shapes. In: ACM SIGGRAPH 28th Annual Conference on Computer Graphics and Interactive Techniques, Los Angeles, pp. 203–212 (2001)
9. Sundar, H., Silver, D., Gagvani, N., Dickinson, S.J.: Skeleton based shape matching and retrieval, Shape Modeling International (SMI03), pp. 130–142. IEEE Computer Society, Washington, DC (2003)
10. Vranic, D.V., Saupe, D.: 3d model retrieval with spherical harmonics and moments, DAGM 2001. In: Radig, B., Florczyk, S. (eds.), Munich, Germany, pp. 392–397, September 2001
11. Lmaati, E.A., El Oirrak, A., Kaddioui, M.N., Ouahman, A.A., Sadgal, M.: 3D Model Retrieval Based on 3D Discrete Cosine Transform. Int. Arab J. Inf. Technol. **7**(3), 264–270 (2010)
12. Aznag, K., Kane, N., El Oirrak, A., Aboutajdine, D.: A new representation for 2D object: binary matrix. In: ISIVC(International Symposium on signal, Image, Video and Communications) (2014)
13. Dutaac, H., Sankur, B., Yemez, Y.: Subspace methods for retrieval of general 3D models. Computer Vision and Image Understanding **114**(8), 865–886 (2010)
14. Lmaati, E.A., El Oirrak, A., Kaddioui, M.N., Ouahman, A.A., Sadgal, M.: 3D Model Retrieval Based on 3D Discrete Cosine Transform. Int. Arab J. Inf. Technol. **7**(3), 264–270 (2010)
15. Kaufman, A., Shimon, E.: 3D scan-conversion algorithms for voxel-based graphics. In: Proceedings of the 1986 workshop on Interactive 3D graphics, pp. 45–75. ACM, January 1987
16. Wang, W.J., Huang, C.T., Liu, C.M., Su, P.C., Wang, S.J.: Data embedding for vector quantization image processing on the basis of adjoining statecodebook mapping. Information Sciences **246**, 69–82 (2013)
17. Andrada, M.F., Vega-Hissi, E.G., Estrada, M.R., Martinez, J.C.G.: Application of k-means clustering, linear discriminant analysis and multivariate linear regression for the development of a predictive QSAR model on 5-lipoxygenase inhibitors. Chemometrics and Intelligent Laboratory Systems **143**, 122–129 (2015)
18. Rahman, M., Hassan, M.R., Buyya, R.: Jaccard Index based Availability Prediction in Enterprise Grids. Procedia Computer Science **1**, 2701–2710 (2012)

Part VII
Advances in Electrical Engineering and Embedded Systems

Multilevel MPSoC Performance Evaluation, ISS Model with Timing and Priority Management

Abdelhakim Alali, Ismail Assayad and Mohamed Sadik

Abstract To deploy the enormous hardware resources available in Multi-Processor Systems-on-Chip (MPSoC) efficiently, rapidly and accurately, methods of Design Space Exploration (DSE) are needed to evaluate several design choices. In this article, we provide a framework that makes fast simulation and performance evaluation of MPSoC early in flow of design, therefore reducing the time of design. In this platform and within the Transaction Level Modeling (TLM) approach, we present a new definition of ISS level by introducing two complementary modeling ISST and ISSPT sublevels. This later, that we illustrate an arbiter modeling approach that allows a high performance MPSoC communication. A round-robin method is chosen for algorithm arbiter modeling because it is simple, minimizes the communication latency and has an accepted speed-up. Two applications are tested and used to validate our platform: Game of life and JPEG Encoder.

The performance of the proposed approach has been analyzed in MPSoC platform based on multi-MicroBlaze. Results of simulation show with ISSPT sublevels gives a high simulation speedup factor of up to 32 with a negligible performance estimation error margin.

Keywords Embedded multiprocessor systems · Estimation of performance · MPSoC · TLM · SystemC · ISS · CABA · Priority management

1 Introduction

The literature shows that much of the design time is spent in the performance evaluation. In addition, the iterations in the design flow become prohibitive for complex systems. Therefore, achievement of high performance MPSoCs is a

A. Alali(✉) · I. Assayad · M. Sadik
Ecole Nationale Supérieure d'Electricité et de Mécanique, RTSE Team,
HASSAN II University, Route D'El Jadida, Casablanca, Morocco
e-mail: hakim.alali@gmail.com, {i.assayad,m.sadik}@ensem.ac.ma

E. Sabir et al. (eds.), *The International Symposium on Ubiquitous Networking*,
Lecture Notes in Electrical Engineering 366,
DOI: 10.1007/978-981-287-990-5_34

challenge. The solution is strongly linked to the availability of fast and accurate methods for the design and performance evaluation [1]. A modeling approach to reduce the time of design and validation time for MPSoCs is to use the Transaction Level Modeling models (TLM) [2]. So with TLM we can validate the behavior for both the hardware and the software components of MPSoC platform as well as the interaction between them.

Besides, TLM cosimulation also allows the performance evaluation of the whole system at the earlier stages of the design flow before making a prototype, which is faster than HDL register-transfer level (RTL) simulation [3] [4].

For this work open-source ISS from SocLib [6] are used and components modeled with SystemC language Version 2.2.0 [5] and TLM methodology.

In this paper, we adopt a strategy for estimating the performance at two levels Cycle Accurate Bit Accurate (CABA) and ISS. We also introduce a new sub-level of the ISS level called ISSPT, so the ISS level is now composed of two sub-levels, ISST which is the original ISS level and ISSPT which is the new one.

Our objectives in this publication are:

- To present a tool for fast MPSoC design performance exploration at different-level of speed and accuracy;
- To show that even though it may require slightly more modeling effort, the ISSPT models offer better alternative than currently used approaches. ISSPT result in simulations with higher accuracy and acceptable additional modeling effort whereas traditional approaches suffer either from a significant loss in simulation accuracy (ISST), or from low simulation speed (CABA) and/or from high modeling effort cost (CABA).

The rest of this paper is organized as follows: an overview of related related work on existing simulation platform with TLM for MPSoC is provided in section 2. Section 3 describes the architecture of the multi-MicroBlaze system. Section 4 presents the simulation platform and ISSPT molding with round-robin approach. Section 5 describes the Timing estimation in ISSPT. Section 6 presents examples of software tested in MPSoC platform. Finally Section 7 describes the results of the applications running on the platform.

2 Related Work

A lot of works on design exploration and performance evaluation for MPSoC embedded systems have been conducted. As a result of these researches, many of exploration environments are proposed, such as SESAM/Par4All [7], STARSoC [8] and SimSoC [9]. The work presented in this paper can be seen as complementary to these environments.

Since the first appearance of TLM in 2000 [10], an increasing number of research projects have examined the problem of its definition, which led to several frameworks [11] [12] [13] and a multitude versions latest is TLM 2.0.1. All these studies have two factors in common:

1. TLM is featured on several levels;
2. The aspects of communication and computing platforms are separated.

Viaud and al. [14] were the first who proposed have an efficient TLM with timing modeling and simulation environment based on parallel discrete event principles. They obtained a long runtime simulation factor but they did not measure this runtime on real applications. Their model is also different from ours. Firstly, with our approach we can be applied for hierarchical or distributed MPSoC design, and secondly, it is open-source.

Stattelmann [15] and Boukhechem [8] proposed a new technique for HW/SW co-simulation for heterogeneous MPSoC platforms in timing model PVT, we have all advantages of PVT that we refined in order to add it as a priority management. Also we integrated computation and communication simulation.

3 Architecture

The basic architecture of the platform implemented in VHDL and generated from Xilinx Platform Studio, consists of: 1, 2 or 3 MicroBlazes each one connected with a private memory of 64 KB BRAM via the LMB bus processors. Processors are also connected to the OPB bus [17] and an SRAM memory of 32MB [16], an interrupt handler, VGA controller, timer and GPIO. A high-level view of the architecture of multi-core MicroBlaze is illustrated in Fig.1.

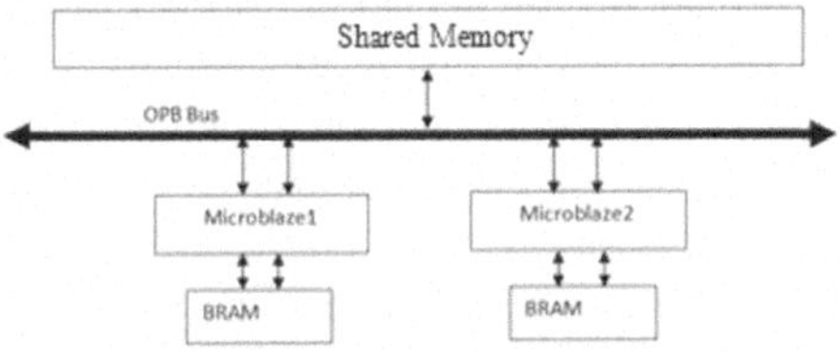

Fig. 1 Multi-MicroBlaze basic architecture

In the simulation environment, the complete design is a SystemC implementation of a multi-MicroBlaze MPSoC architecture including 2 MicroBlazes, an interrupt handler, VGA controller, timer, GPIO and SRAM, as follows:

In our case study, in ISS (Instruction Set Simulator) simulation and by using inter-process communication, we connect two ISSs (two processors) with SystemC communication models. Therefore, it is easy to add or to remove a processor from the MPSoC design. The interconnection is based on OPB bus [17] described in TLM SystemC. Communication model uses communication mechanism for the shared memory, the bus arbitration mechanism is managed by bus arbiter which provides by the round-robin arbitration policy (described in Network interconnection model section).

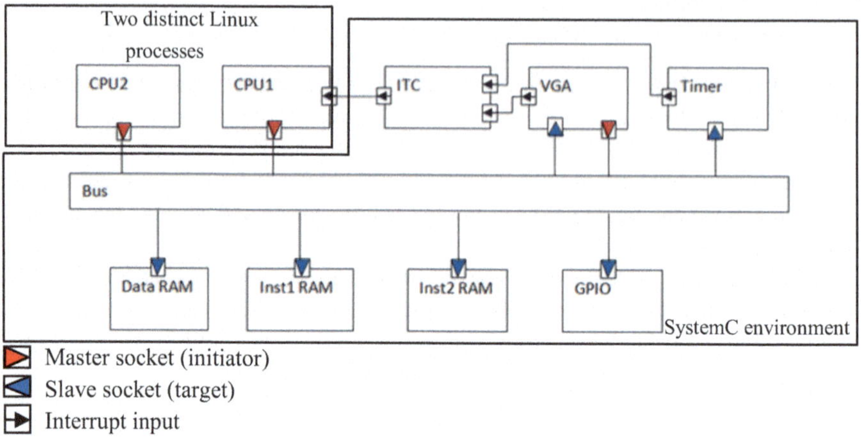

Fig. 2 Platform architecture with 2 processors

4 SystemC Simulation Platform

4.1 Processor Model and Simulation

With the TLM approach, the behavior of a processor has two major descriptions ISS and CABA (Cycle Accurate/Bit Accurate). In the ISS, the processor description is modeled with a specific instruction level simulator. Instructions are executed sequentially without reference to the micro-architecture of the component. CABA models the behavior of the system at each cycle similar to the RTL level, Indeed, the CABA level modeling is based on the theory of "finite state machine (FSM) interconnected synchronous" (Synchronous Communicating Finite State Machines) [19] [20] [21].

The estimated performance in new ISSPT (Instruction Set Simulation with timing and priority management) level returns to evaluate performance of two parts calculation and communication time.

For calculation time, to assess the time of each task we used the simulator MicroBlaze processor ISS level but adding time. For this we mainly identified the number and type of instructions executed as relevant activities in the processor component.

Timing execution instructions of MicroBlaze processor is estimated from the technical documentation provided by Reference Guide of MicroBlaze [16].

Below is an example of our thread implementation that its functionality of the calculation part (processor) described in ISSPT level. For the communication time is detailed in Bus and Network interconnection model section.

```
void MicroBlazeIss::step(void) {
/* decode of instruction outstanding */
IDecode(m_ir, &ins_opcode, &ins_rd, &ins_ra, &ins_rb,
&ins_imm);
switch (ins_opcode) {
//execution of instruction
case OP_ADD:
next_pc = r_npc + 4;
Wait(ADD_delay,sc_core::SC_NS)
break;
.........
//load of data
case OP_LW:
.... .
  LOAD(READ_WORD, addr, time);
  next_pc = r_npc + 4;
Wait(Transaction_delay,sc_core::SC_NS)
  break;
............ .
    }
}
```

[Example a thread implementation of MicroBlaze ISS]

4.2 Memory Model

The memory module that we designed is a passive "slave" component composed of two concurrent parts, one for instructions and one for data. A memory transaction includes two methods: read and write. This structure allows us to accelerate the simlation. These two methods are called and executed directly in the thread initiator connected to the memory component.

In our environment, the target port is connected directly to the bus. Data part memory is shared between the processors. Access time and cycle time parameters are added to the component description to estimate performance.

4.3 Bus and Network Interconnection Model

Our architecture platform is designed around the OPB bus (On-Chip Peripheral Bus) whose architecture was developed by IBM [17]. The bus supports various features depending on the desired bus operations: single cycle read/write, multiple masters, block transfer. In our work we use OPB Bus connected Xilinx MicroBlaze processor.

In this paper, we have limited develop an integrated Bus crossbar, which is based on two main features routing and arbitration see Fig. 3 and 4. The router is a generic component that directs a request from an initiator to the target in question, using a routing table specified. When a new transaction get from initiator, the router reads the corresponding addresses and selects an output port, this process is illustrated in Fig. 3.

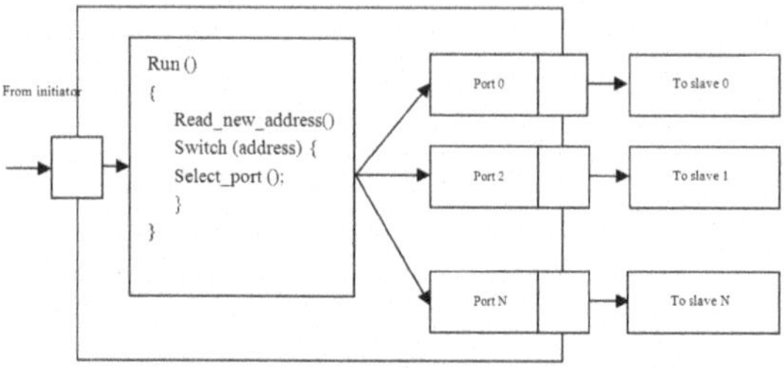

Fig. 3 Router component

To manage conflicts between multiple simultaneous requests to a target, we developed an active component called "arbiter" to schedule the access to shared resources. When initiator needs access to a shared target, via the interconnection network, it sends a request using the corresponding communication channel and waits for the response. At the arbiter, one thread reads queries present in the FIFO of each communication channel and selects the priority request based on the arbitration round-robin strategy. After processing by the target, the arbiter transmits the response in the corresponding FIFO of the communication channel.

The initiator retrieves the response and completes the transaction. This communication management blocks a router during the processing of the request by the target, which can be a drawback. However, it has the advantage, simplification of the protocol and reduced the number of ports. These two factors can accelerate the simulation. Our arbiter plays a second role very important, is used for the estimation of delays in the interconnection network.

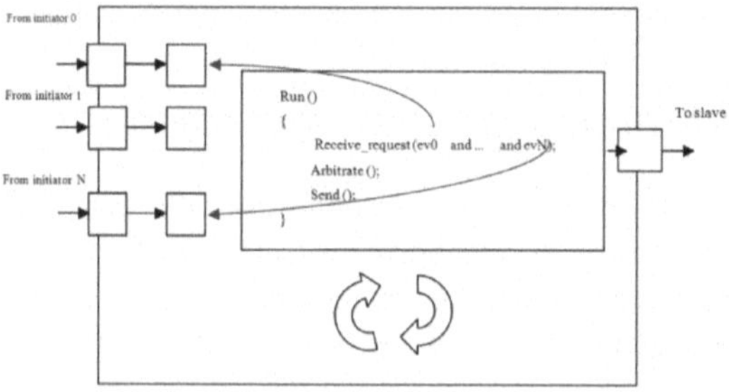

Fig. 4 Arbiter component

Fig. 5 shows the implementation of a crossbar from the two modules "router" and "arbiter". This architecture is relatively simple, but sufficient to achieve our objective to observe the restraints and retrieve information about latencies. Several interconnection topologies can be designed as multi-stage network.

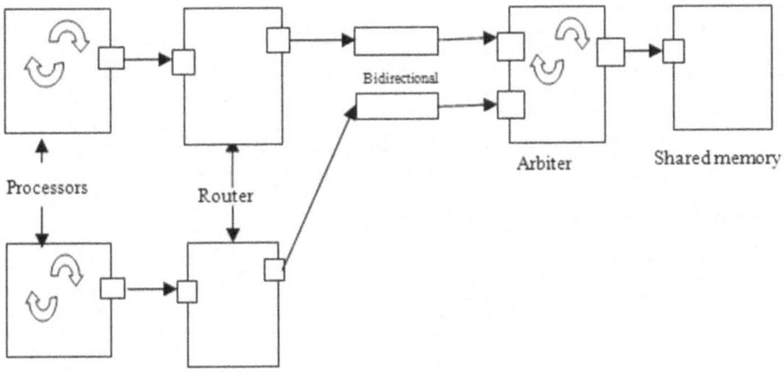

Fig. 5 Crossbar implementation

5 Timing Estimation in ISSPT

The moment when a processor performs its corresponding memory access can affect the access time of the other processor in a collision. Fig. 6 shows an example of contention detection error in the interconnection network due to non - compliance with the deadlines events (Packet setup, Routing and Arbitration). When the transmitted packet from processor 1 arrives at the router (R&A in Fig. 6), there is no possibility of detecting the occupation of the router by the processing of the packet coming from processor 0. In effect, events in the ISST sublevel are instantly executed (zero delay). This abstraction changes the behavior of the party in the communication system, which reduces the precision of performance estimation. To solve this problem, we have improved the ISS sublevel by introducing synchronization instructions. They take into account the time of component activities, delays in forwarding packets and finally the communication protocol. These are the characteristics of the ISSPT level.

To compare the estimation error between ISSPT level and CABA level in our platform, we had to implement the specifications of the OPB protocol. To emulate the same behavior of the OPB protocol in ISSPT sublevel and observe the behavior of components, the wait () statements have been added in the description of the components and before the transmission of orders and answer queries. The wait () statements added require arguments expressed in units of time such as nano seconds (ns) or number of cycles. In our experiments, these arguments are measured from CABA platform. Table 1 shows time made in the ISSPT sublevel.

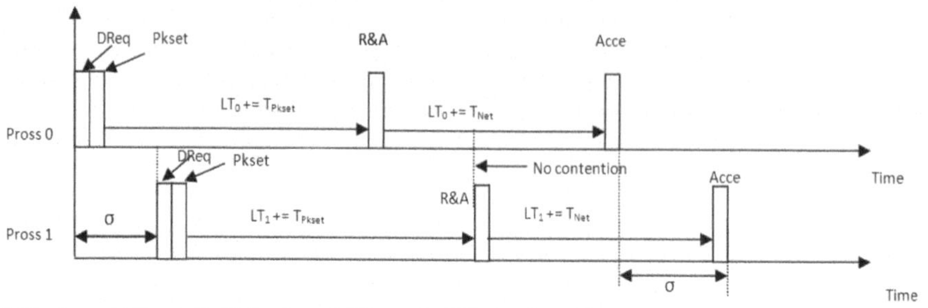

LT$_0$: Local Timer 0; **LT**$_1$: Local Timer 1; **DReq**: Data Request
Pkset: Packet Setup; **R&A**: Routing and Arbitration; **Ace**: Memory Access
T$_{Pkset}$: Packet Setup time; **T**$_{Net}$: Network time

Fig. 6 Timing estimation in ISSPT sublevel

Table 1 Time activities used in the experiments

Activities	Time (cycles)
Preparing an OPB command request	4
Preparing an OPB response request	5
Execution of an instruction	1
read memory access	2
write memory access	2
VGA	360000

6 Software Integration

The application layer has two softwares and was tested in the platform:

- The game of life is an infinite two-dimensional orthogonal grid of square cells, each of which is in one of two possible states, alive or dead. Every cell interacts with its eight neighbors, which are the cells that are horizontally, vertically, or diagonally adjacent [18],
- JPEG Encoder is a minimalistic JPEG encoder written in C. It is both "portable" (tested on x86 and MicroBlaze) and "lightweight" (around 600 LOC). Application allows us to write JPEG compressed images from input image data on memory. It works in "grayscale only" (monochrome JPEG file): there is no support for color so far,

– It produces baseline, DCT-based (SOF0), JFIF 1.01 (APP0) JPEG-s,
– It supports "8x8 blocks only",
– It includes default quantization and Huffman tables that are not customizable at runtime.

Generally, for each application, it executed by 1, 2 or 3 processors, it is stored in their local memory and they executed in parallel and synchronized by the same clock system.

7 Results and Discussion

We present results that we carried out to validate our platform and evaluate its performances. We also compare performances among the different abstraction levels.

The same environment was employed for simulation on different abstraction levels. The results of simulation were gotten by running the platform on a core 2 duo with a RAM memory size of 1GB, based on Linux Fedora 8 Core 3.1. ISS was built by the GNU cross-compiler (GCC version 3.4.6).

7.1 Simulation Results in CABA, ISST and ISSPT

Fig. 7 and Fig. 8 show Speedup and precision simulation results for CABA, ISST and ISSPT with Game of Life, and the same simulation shown in Fig. 9 and 10 with JPEG Encoder. Speedup corresponds to the simulation or execution time of software at the different simulation abstraction levels.

t2 = 'end time', t1 = 'start time'.

Speed-up formula: $$\frac{(t2-t1)x}{(t2-t1)bit} \quad (1)$$

Precision formula: $$\Delta_x - \Delta_{bit} \quad (2)$$

With 'x' = 'CABA',' ISST' or 'ISSPT'.

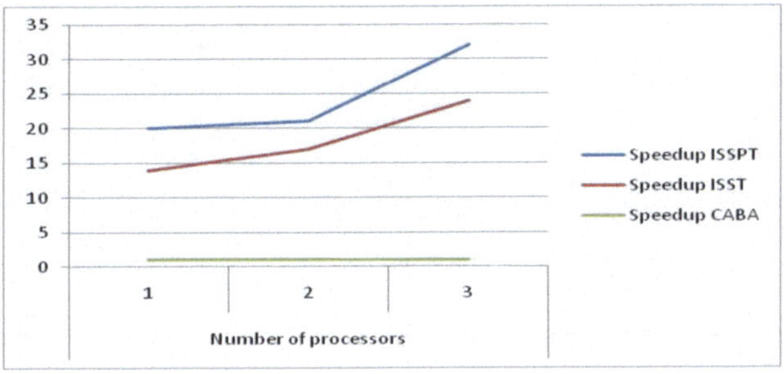

Fig. 7 Speedup simulation results for CABA, ISST and ISSPT, software used: Game of life

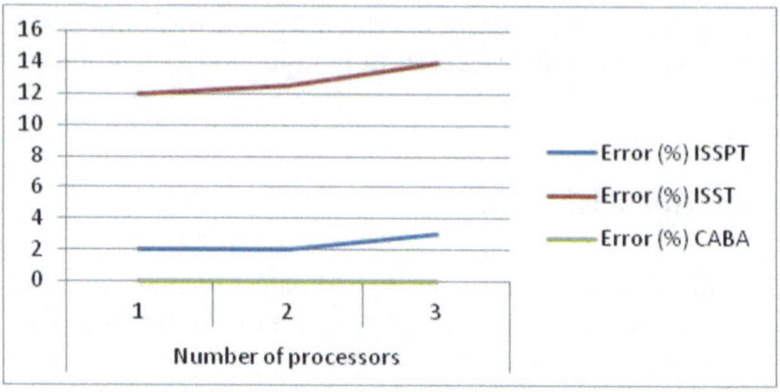

Fig. 8 Precision simulation results for CABA, ISST and ISSPT, software used: Game of life

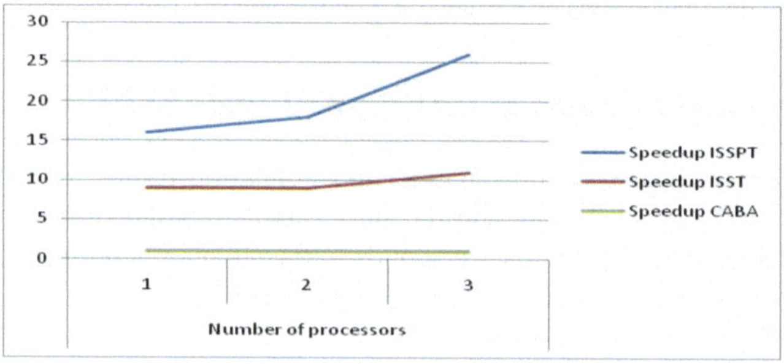

Fig. 9 Speedup simulation results for CABA, ISST and ISSPT, software used: JPEG Encoder

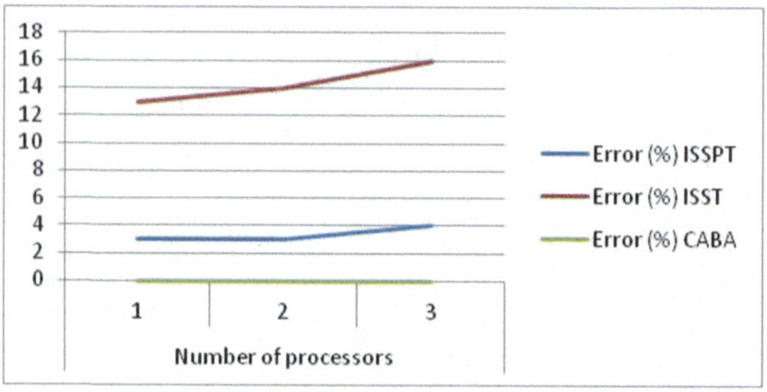

Fig. 10 Precision simulation results for CABA, ISST and ISSPT, software used: JPEG Encoder

Models were simulated: ISSPT, ISST and CABA, we tested CABA model by its comparing a synthesizable RTL model (using VHDL generated from Xilinx Platform Studio and simulated with ModelSim) [19]. The first two models are used to validate the software and the system architecture (they include the ISSs, bus model at transaction level, and models of other components, all blocks use SystemC models).

We validated the simulation by the same conditions, after running the environment we obtained these experimental results:

- The CABA has an important precision [19]
- ISSPT model is about 20 times faster than the CABA model.
- ISSPT model is about 2 times faster than the ISST model.
- The model of ISSPT has a good accuracy without significantly penalizing the simulation time.
- The addition of new processors in the system increases the acceleration factor, which is explained by the amplification of the communication between the processors and shared memory modules.
- The nature of the software running on the platform impacts performance in differences levels.

A precise analysis of the trace produced by the SystemC simulator shows that 80% of the simulation time is made for the execution of the function of the bus while the simulation time of the calculation part is low which reflects our choice to treat the case of ISSPT.

7.2 Modeling Effort

So far we have shown the usefulness of our approach in terms of acceleration of the simulation and in terms of performance estimation. However, this approach has proven effective also in terms of modeling effort. It allows designers the development and validation of MPSoC systems in less time. Table 2 presents the modeling effort expressed in terms of lines of code (LOC) needed to design an MPSoC system in the CABA and ISSPT levels. According to the results, the modeling effort with ISSPT is reduced of a factor of 59%. The use of a multi-level simulation strategy (with objectives) quickly allows focusing on a subset of MPSoC systems without having to increase the modeling efforts for each level of abstraction.

Table 2 Comparing the modeling effort

Abstract level		CABA	ISSPT
Modeling effort(LOC)	Processor	1578	1259
	Bus	399	170
	Memory	312	133
	VGA	650	167
	Timer	340	231
	Total	3279	1960
Reduction (%)			59%

8 Conclusion

The use of open source ISSs models of MicroBlaze wrapped under SystemC by using UNIX inter-process communication that we have presented and validated our methodology for MPSoC co-simulation at a high level of abstraction (SystemC-TLM) within a single simulation environment. Comparing three different abstraction levels, namely, ISSPT which implement ISS (Instruction Set Simulator) with priority and timing management, ISST level which implement ISS with timing and finally, CABA Cycle Accurate Bit Accurate.

The experimental results show that the use the ISSPT approach with SystemC-TLM reduces the design validation time and permit developing models rapidly with an acceptable precision.

This motivates our choice for SystemC and TLM as a system design methodology, dedicated to architecture exploration in our project which is the main contribution of this work.

As perspective, we think to develop models for estimating the energy consumption at different levels.

References

1. Benini, L., et al.: MPARM: Exploring the Multi-Processor SoC Design Space with SystemC. J. of VLSI Signal Processing (2005)
2. Ghenassia, F.: TLM with SystemC Concepts and Applications for Embedded Systems. Springer (2005)
3. Huang, C.-Y., et al.: SoC HW/SW verification and validation. In: 2011 16th Asia and South Pacific Design Automation Conference (ASP-DAC) (2011)
4. Hines, K., Borriello, G.: Dynamic communication models in embedded system co-simulation. In: Proceedings of the 34th Design Automation Conference (DAC 1997), Anaheim, Calif, USA, pp. 395–400 (1997)
5. Systemc: http://www.systemc.org/
6. http://www.soclib.fr/trac/dev/wiki/Component
7. Ventroux, N., et al.: SESAM/Par4All: a tool for joint exploration of MPSoC architectures and dynamic dataflow code generation. In: Proceedings of the 2012 Workshop on Rapid Simulation and Performance Evaluation: Methods and Tools. ACM (2012)
8. Boukhechem, S., Bourennane, E.B.: TLM platform based on systemC for STARSoC design space exploration. In: AHS 2008. NASA/ESA Conference on Adaptive Hardware and Systems, Noordwijk (2008)
9. Helmstetter, C., Joloboff, V.: SimSoC: a SystemC TLM integrated ISS for full system simu-lation. In: APCCAS 2008, Macao, China (2008)
10. Gajski, D., et al.: SpecC: Specification Language and Methodology. Kluwer (2000)
11. Donlin, A.: Transaction level: flows and use models. In: CODES+ISSS 2004, Stockholm, Sweden (2004)
12. Cai, L., et al.: Transaction level modeling: an overview. In: CODES+ISSS 2003, New York, USA (2003)
13. Benini, L., et al.: SystemC cosimulation and emulation of multiprocessor SoC designs. IEEE Computer **36**(4) (2003)

14. Viaud, E., Pecheux, F., Greiner, A.: An efficient TLM/T modeling and simulation environment based on parallel discrete event principles. In: DATE 2006, Munich, Germany (2006)
15. Stattelmann, S., Bringmann, O., Rosenstiel, W.: Fast and accurate source-level simulation of software timing considering complex code optimizations. In: Proceedings of the 48th Design Automation Conference. ACM (2011)
16. MicroBlaze Processor v5.4. Reference Guide, UG081 (v5.4) (2014)
17. On-Chip Peripheral Bus Architecture Specifications V2.1
18. http://en.wikipedia.org/wiki/Conway%27s_Game_of_Life
19. Alali, A., Assayad, I., Sadik, M.: Modeling and simulation of multiprocessor systems MPSoC by SystemC/TLM2. International Journal of Computer Science Issues (IJCSI) **11**(3) (2014)
20. Sgroi, M., Lavagno, L., Sangiovanni-Vincentelli, A.: Formal Models for Embedded System Design. IEEE Design and Test of Computers **17**(2), 14–27 (2000)
21. Lavagno, L., Sangiovanni-Vincentelli, A., Sentovich, E.: System-level synthesis, chapter Models of computation for embedded system design, pp. 45–102. Kluwer Academic Publishers (1999)

Hybrid Electrical Architecture for Vertical Takeoff and Landing Unmmaned Aerial Vehicule

Souad Berradi, Fouad Moutaouakkil and Hicham Medromi

Abstract Recently, Unmanned Aerial Vehicles (UAV) become a significant research area due to their multiple domains of applications such as: search and rescue operations, aerial surveying of crops, inspecting power lines and pipelines, delivering medical supplies to remote or otherwise inaccessible regions. However the UAV performances (such as autonomy, endurance, maximum flight altitude, maximum takeoff weight, maximum speed etc) depend mainly on its energy storage system (batteries, fuel cells, ultra capacitors). The more the drone's complexity grows the more energy it consumes. In this paper we will discuss the different electrical architecture used in UAV and we'll propose a hybrid solution that optimizes the energy consumption for multi-rotor.

Keywords Unmanned Aerial Vehicles · Multi-rotor · Electrical architecture · Photovoltaic panel · Energy storage devices · Electrical consumption · Optimization

1 Introduction

Unmanned Aerial Vehicules (UAV) is an aircraft that can fly autonomously along a pre-defined flight path without pilot on board [1, 2].

Missions for such an aircraft would mainly be focused around surveying and surveillance applications:

- ✓ Surveying applications include aerial photography and remote sensing.
- ✓ Surveillance applications entail **civilian reconnaissance** such as monitoring of pipelines or power lines, search and rescue operations, aerial surveying of crops or **military reconnaissance** as the border patrols.[2]

S. Berradi(✉) · F. Moutaouakkil · H. Medromi
EAS Research Group, ENSEM, Hassan II University of Casablanca, Casablanca, Morocco
e-mail: berradi.souad@gmail.com, fmoutaouakkil@hotmail.com, hmedromi@yahoo.fr

E. Sabir et al. (eds.), *The International Symposium on Ubiquitous Networking*,
Lecture Notes in Electrical Engineering 366,
DOI: 10.1007/978-981-287-990-5_35

Historically, the aircraft propulsion system was based on internal combustion engine which use fossil fuel. However the depletion of fossil fuel resources and the pollution of the environment are the major problems that made researcher look for an alternative clean energy which are less harmful to the environment, more efficient and more durable than fossil fuel. [3]

Alternative energy includes: fuel cells, solar cells, winds etc. In order to maximize the efficiency of this energy, it will be coupled with an energy storage system such as battery, fuel cell, ultra capacitor.

Recently, the electrical propulsion system has received more attention than the conventional combustion propulsion system because of:

- The gain of the weight: too much fuel will increase the weight of the airplane and then minimize the endurance.
- It's a clean energy so there is no dioxide emission and no noise.

The power consumed by the UAV propulsion system is changing from aircraft taxying, take off, cruising to landing. Most air vehicles require about twice as much power for take off and climbing than for cruising.[4]

In this paper, a survey of UAV's electrical architecture was discussed and a proposed architecture for multi-rotor was presented.

This paper is organized as follows: In section 1 the technologies that is used in most electrical architecture was presented. The UAV's architecture and proposed architecture were presented and discussed in Section 2. Finally, the last section gives some concluding remarks and a vision of ours future work.

2 Storage Device and Photovoltaic Panels

The efficiency of any UAV depends on the energy storage system used and the technologies that generate power or store it. The image in figure compare the energy density of the most used energy storage system.

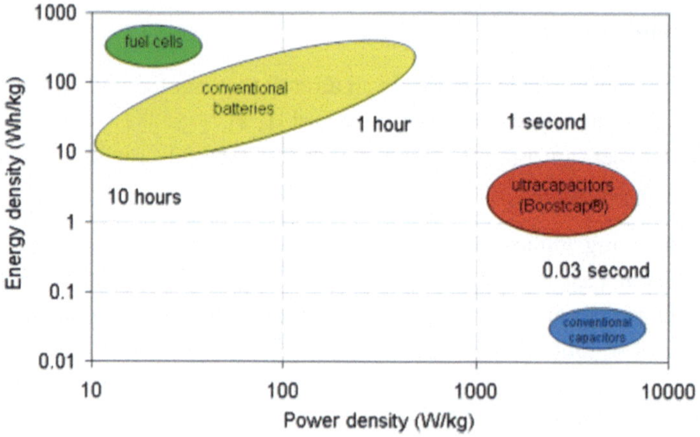

Fig. 1 Ragone Chart

2.1 *Fuel Cell*

Fuel Cell is a device that generates electricity through a chemical reaction between oxygen and Hydrogen.[4]

Conventional fuel cell includes the fuel cell stacks, energy storage component, reactants tank, and auxiliary components. It can be categorized into non-regenerative and regenerative types. The regenerative fuel cell includes electrolyser stack in it.

It consists of an anode, an anode catalyst layer which speeds the reaction of the anodes, an electrolyte that allows charges to move between the two sides of the fuel cell, a cathode and a cathode catalyst layer. In order to achieve higher voltage pack, multiple Fuel cells is arranged in series or parallel in a stack.[5]

Advantages:

- high energy density;
- high conversion efficiency of fuel to electricity;
- it's a green energy because hydrogen is non-polluting ;
- fuel durability and reliability.[4]

Drawbacks:

- It's expensive due to the expensive cost of the Hydrogen (difficult and complex production's process and storage);
- The response time is longer compared to other storage devices. That why they require additional storage system such as batteries and ultra-capacitor.[6]

2.2 *Lithium Battery*

Device that convert chemical reaction in the electrodes into electrical energy.

Advantages:

- High power density which limited to short period;
- High power density;
- Good performances at high temperature;
- Low memory effect;
- High specific power 300W/kg;
- High specific energy 100Wh/kg;
- Long battery life o 1000 cycles. [4, 6]

Drawbacks:

- Longer charging time which can take from 1 hour to several hours for full charge;

- Battery dynamics that change during time so the output voltage is not constant;
- Operation outside a specified range of temperature, cell voltage can damage the battery or even cause an explosion;
- Batteries have a limited life cycle that depends on: the operating temperature, the depth of discharge, the number of discharge cycles. [4-8]

2.3 Ultra-Capacitor

Ultra-capacitor is high-capacity capacitors that store electrical energy. It contains: An electrical double layer which is made of a nano-porous material to improve storage density and a separator that separates and holds the electrical charges.

Advantages:

- Longer lifecycle (500,000 cycles),
- High power density;
- Very high rate of charge/discharge;
- Low internal resistance (minimum heat loss and good reversibility);
- Efficiency cycle of approximately 90% due to low internal resistance.[5]

Drawback:

Ultra-capacitor have a low energy density (The amount of energy stored per unit weight of the ultra-capacitor is between 3 and 5 W h kg).[9]

The chart bellows present a comparison between each storage device:

Storage device	Ultra-capacitor	Rechargeable battery	Fuel cell
Charging duration	1s<t <30s	1hour< t <5 hours	-
Discharging duration	1s<t <30s	0.3hour<t<3 hours	As long as the fuel cell is supplied with hydrogen and oxygen
Efficiency	between 85% and 98%	between 70% and 80%	Between 40 % and 60%
Power density (W/kg)	10^4	$< 10^3$	
Energy density (Wh/kg)	between 1 and 10	between 10 and 100	Up to 450
Number of cycle	10^6	10^3	-

From this chart, each storage device has an advantages and drawbacks that made it the best candidate for a specific application.

2.4 Photovoltaic Panel

A photovoltaic Panel (PV) is used to convert the sunlight's energy into electrical energy.

The total amount of power obtainable is depending on:

- Day of the year: it determines the amount of sunlight that is shone on the earth's surface. The intensity of the sunlight in summer is more than in winter;
- Time of day: the intensity of the irradiance is at it maximum in noon;
- Irradiance: the energy density falling on the earth over some period of time for example daily or hourly irradiance;
- Efficiency of the Photo voltaic cell: defined as the ratio of input irradiance to output electrical power of the solar cell.
- Surface area of the Photo voltaic cell: the surface area of the solar cell which is exposed to the sunlight;
- Geographic position: is defined as the latitude, longitude and altitude.[10]

The main constraint of the PV in UAV's platform is ensuring enough energy to propel the aircraft during the day and the night flight. So it's must be used with an energy storage device.

3 Electrical UAV's Platform and the Proposed Architecture

3.1 Meyer's Architecture

The architecture proposed in [2] is designed for Low Altitude Long Endurance Solar powered UAV and can fly 1400Km in just 48 hours.

The considerations addressed to design a long endurance solar UAV include:

- The design of the UAV wing for long endurance flights (the design of the airframe, wing design, aerofoil design, fuselage and tail design, etc.)
- The determination of the required initial energy for all UAV's component.
- The determination of the available solar power

The electric architecture, shown in fig. 2, consists of solar panel, two electric Direct Current (DC) motors, six lithium polymer batteries used as energy storage device, battery management circuits and a power management computer.

Solar Panel: Solar energy is the main source of electrical power for this UAV which can be collected by the photovoltaic cells mounted on the wing surface of the UAV.

Battery Management Systems: Battery management system manage and protect a lithium polymer batteries from self-discharge, operating outside its Safe Operating Area, monitoring its state, reporting its parameters data to the power management system.

Isolation Switches: Isolation switches are used to isolate the batteries from the electric power busses of the UAV.

Energy Storage Devices: Lithium polymer batteries are used to store the excess electrical energy collected by the photovoltaic cells.

Electric Motors: In this architecture Brushless DC motors was chosen because DC brushless motor offers high torque at high efficiency of around 88%.

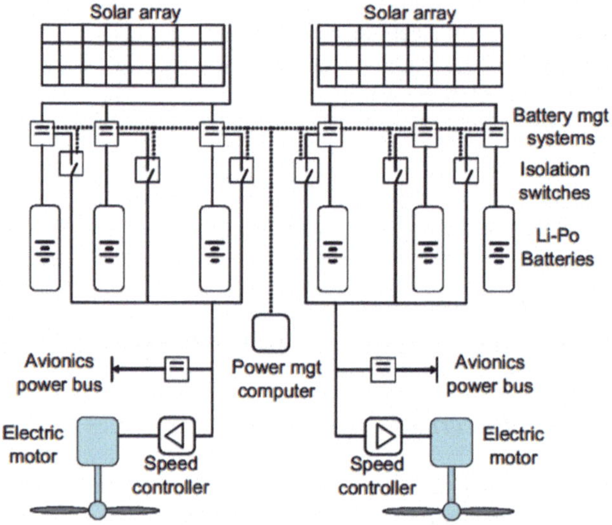

Fig. 2 Meyer's architecture

The design of this structure depends strongly on the weather conditions presented in 2.4 and also the forme and the surface of the aircraft's wing.

To meet the energy requirement for night flight, the size of the battery must be very large, thus its will add additional weight to the airplane, consequently it's will minimize the endurance of the UAV.

3.2 Albaker's Architecture

The architecture presented in [11] contains PV Panel, Maximum Power Point Tracker (MPPT), power management system, battery modules, motors and DC to DC converter.

Since the photovoltaic energy dependent on many factor listed in 2.4 and in order to provide the maximum power to UAV's motors and UAV's avionics system, the MPPT is necessary to overcome the uncertainty of solar power. [12]

This architecture resemble to the first architecture listed in this paper. The PV is the main source energy that converts sunlight's irradiance into electrical energy. The battery modules are used as an emergency device if additional power needed for the load. The Maximum Power Tracking used enhances the efficiency of the PV.

This architecture is more efficient than the first one due to the use of MPPT system that enhances the efficiency of the PV Panels. But it's still depends on the type of the algorithm chosen.

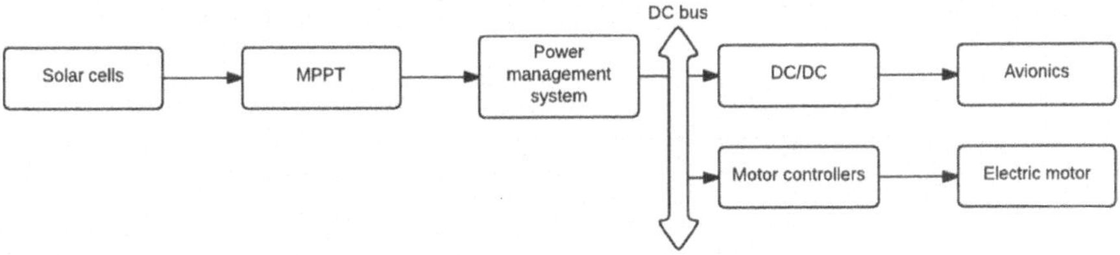

Fig. 3 albaker's architecture

3.3 Hao's Architecture

The architecture presented in [1]is designed for Long powered solar Endurance UAV. It combines between PV Panel, rechargeable fuel cell and rechargeable battery.

Hybrid PV panel/fuel cell power system can provide enough energy for UAV during the whole day. The energy extracted from the PV panel will be used to power the UAV's equipment and the extra energy will be used to electrolyse the water to produce the hydrogen (which will be used by the fuel cell in future).

During the flight, the hybrid PV panel/fuel cell power system provides a constant power and it's not able to answer to peak power demand (climbing, take off, electrical load...) because of the slow response of the fuel cell. To meet the peak power requirement during the flight, a rechargeable battery was added to provide the difference between the burst load power and the output power of the PV panel and/or fuel cell and it can be also be recharged by the PV panel. The hybrid system designed in this paper can provide 500W average power with the peak ability of 1500W, and 800W take off power

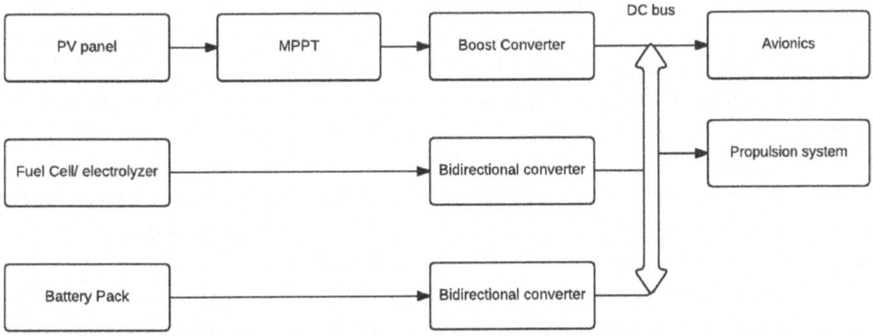

Fig. 4 Hao's architecture

Hao's architecture is the best architecture that can be used for glider.
This chart presents a comparison between the three architectures:

Architecture	Structure	Drawbacks	Advantages
Meyer's architecture	PV and Li-po battery	Depend strongly on the weather conditions and also the forme and the surface of the aircraft's wing. the size of the battery must be very to meet the energy requirement during night flight	Long endurance 48hours Used for glider
Albaker's architecture	PV and MPPT module	Depend on the type of the algorithms choosing of the MPPT	Long endurance Used for glider
Hao's architecture	PV panels	-	Long endurance 500W average power with the peak ability of 1500W, and 800W take off power Used for glider

All these architecture are recommended for glider and not recommended for Vertical Takeoff and Landing (VTOL). The small surface that has the VTOL make the use of the PV panel not advantageous.

3.4 Proposed Architecture for Vertical Take off and Landing

Vertical Take-Off and Landing (VTOL) Unmanned Aerial Vehicles (UAVs) is an aircraft that take off, and land vertically such as multi-rotor aircrafts. The advantage of multi-rotor aircraft is the simplicity of rotor mechanism required for flight control. The flight is achieved by varying the speed of each rotor to change the thrust and torque produced. They are 3 types of multi-rotors: quadcopter, hexacopter and octocopter that refer to 4, 6, 8 rotor.[13]

The electrical architecture for this rotorcraft usually uses rechargeable battery modules that power the propulsion system and the UAV's avionics. The use of the PV panel is not advantageous because of the small surface that has the VTOL.

The electrical architecture that we propose is hybrid architecture that enhance more the efficiency of the rotorcraft by using two complementary energy storage system such as the lithium polymer and the supercapacitor.

As shown in fig. 1, the battery have a high energy density (i.e., more energy is stored per weight than other technologies) in comparison to ultra-capacitor, while the ultra-capacitor have a high power density in comparison to battery. Since the batteries and ultra-capacitors work as complementary to each other, the efficiency of the UAV will grows.

The proposed models contains:

Permanent source: we use lithium-polymer battery as the primary energy source because of its high energy density.

Emergency source: we use the ultra-capacitor as an emergency storage device for peak power demand (take-off…etc) and its protect also the battery from high peak currents and extend the battery's lifetime.

Battery management system: Circuit that manage the use of energy sources Propulsion system: contains ESC and brushless motor

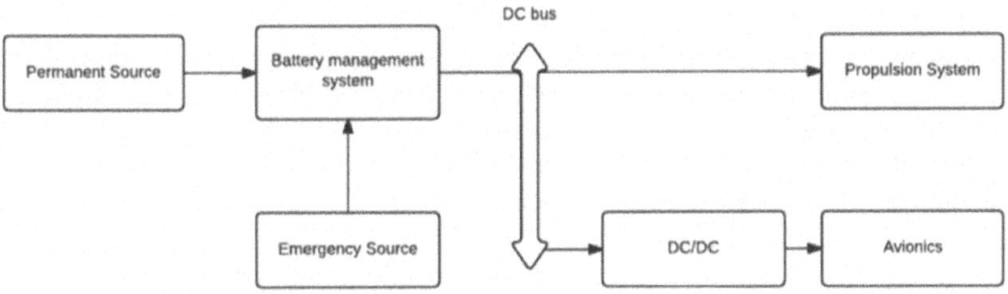

Fig. 5 proposed architecture

4 Conclusion and Future Works

To power the UAV's avionics system and propulsion system, a hybrid system that combine between renewable energy and a storage device is usually used. The different architectures presented in this paper use PV panels as main source energy and a storage device to ensure enough power for night flight. In a multi-rotors aircraft, the use of PV is not recommended due to the small surface.

Since the lithium polymer battery provides high energy density and ultra-capacitor provides high power density, we propose to use both of them to ensure the energy required for multi-rotor aircraft. Peak power demand will be provided by the Ultra-capacitor, while the average power demand will be provided by the battery.

In the future we will detail more this architecture and simulate it using Matlab Software.

References

1. Hao, C., Khaligh, A.: Hybrid energy storage system for unmanned aerial vehicle (UAV). In: IECON 2010 - 36th Annual Conference on IEEE Industrial Electronics Society (2010)
2. Meyer, J., et al.: Design considerations for a low altitude long endurance solar powered unmanned aerial vehicle. In: AFRICON 2007 (2007)

3. Bohwa, L., et al.: Active power management system for an unmanned aerial vehicle powered by solar cells, a fuel cell, and batteries. IEEE Transactions on Aerospace and Electronic Systems **50**(4), 3167–3177 (2014)
4. Karunarathne, L., Economou, J.T., Knowles, K.: Fuzzy logic control strategy for fuel cell/battery aerospace propulsion system. In: Vehicle Power and Propulsion Conference, VPPC 2008. IEEE (2008)
5. Hannan, M.A., Azidin, F.A., Mohamed, A.: Hybrid electric vehicles and their challenges: A review. Renewable and Sustainable Energy Reviews **29**, 135–150 (2014)
6. Khaligh, A., Zhihao, L.: Battery, Ultracapacitor, Fuel Cell, and Hybrid Energy Storage Systems for Electric, Hybrid Electric, Fuel Cell, and Plug-In Hybrid Electric Vehicles: State of the Art. IEEE Transactions on Vehicular Technology, **59**(6), 2806–2814 (2010)
7. Steinhorst, S., et al.: Smart cells for embedded battery management. In: 2014 IEEE International Conference on Cyber-Physical Systems, Networks, and Applications (CPSNA) (2014)
8. Podhradsky, M., et al.: Battery model-based thrust controller for a small, low cost multirotor Unmanned Aerial Vehicles. In: 2013 International Conference on Unmanned Aircraft Systems (ICUAS) (2013)
9. Karangia, R., et al.: Battery-supercapacitor hybrid energy storage system used in Electric Vehicle. In: 2013 International Conference on Energy Efficient Technologies for Sustainability (ICEETS) (2013)
10. Wilkins, G., Fourie, D., Meyer, J.: Critical design parameters for a low altitude long endurance solar powered UAV. In: AFRICON 2009 (2009)
11. Albaker, B.M.: Preliminary architectonic design for a smart solar-powered UAV. In: 2013 IEEE Conference on Clean Energy and Technology (CEAT) (2013)
12. Yu-En, W., Chih-Lung, S., Chia-Yu, W.: Research and improvement of maximum power point tracking for photovoltaic systems. In: International Conference on Power Electronics and Drive Systems, PEDS 2009 (2009)
13. Che, et al.: A multifunctional HIL testbed for multirotor VTOL UAV actuator. In: 2010 IEEE/ASME International Conference on Mechatronics and Embedded Systems and Applications (MESA) (2010)

SMART: Implementing a New Flight System Based on Multi-agent and Embedded on a Real-Time Platform

Firdaous Marzouk, Mohamed Ennaji and Hicham Medromi

Abstract Current work in the field related to the drones mainly aimed at improving their autonomy related to the environment. The purpose of this paper focuses on the areas of improvement of autonomy, flexibility and adaptability of these aircraft during a flight. The developing of flying machines characterized by a pseudo-autonomy, better flexibility and adaptability to situations are mainly blocked by order constraints equipment, such as for architecture requires more complexity. To meet this need autonomous control architecture has been developed; it supports various aspects of the development process of the agent, from the design of the agent architecture, to the implementation on the hardware. An original architecture has been developed that allows the real-time control and manage different decisions by evaluating the path to follow and the speed of the UAV (unmanned aerial vehicle). It is deployed on an embedded system platform that provides good computing power for this kind of tasks in addition to managing communication with the ground operator.

Keywords Control · UAV · Drone · Autonomy · Multi-agents · SMA · Embedded system · Intelligent systems · Autopilot

1 Introduction

UAVs, or drones, are a relatively new research area; we see more and more amateurs and academic projects. Several research projects model airplanes are more efficiently implemented and provide basic performance at a reasonable price, but few projects offer full autonomous control of the drone. Piloting drones requires sophisticated training, so to control these devices we must limit human instruc-

F. Marzouk(✉) · M. Ennaji · H. Medromi
EAS Team, LISER Laboratory, ENSEM, Hassan II University, Casablanca, Morocco
e-mail: {marzouk.f,ennaji.moh}@gmail.com, hmedromi@yahoo.fr

E. Sabir et al. (eds.), *The International Symposium on Ubiquitous Networking*,
Lecture Notes in Electrical Engineering 366,
DOI: 10.1007/978-981-287-990-5_36

tions and replace them with software instructions. The complexity of driving in an unfamiliar environment requires the UAV to have more autonomy in both decision and equipment. This autonomy is particularly necessary when the drone must move and interact with an unknown environment. It must include a set of features and equipment, enabling it to respond to all these instructions. We were thus led to work on the implementation of two types of architecture: structural architecture and operational architecture. In this paper, we first propose a physical architecture (distributed architecture) which allow the drone to meet the needs in real time and with more performance, then we propose a multi-agent architecture which contains several agents that interact among themselves to increase the autonomy and the intelligence of the drone, and finally we present the practical results relating to the study done before.

2 Overall Description of the System

The mechanics of the drone is designed to maintain weight as well as ensure permanent protection of the device and its environment. Figure 1 illustrates the mechanical elements the UAV.

Note that this is a VTOL (vertical take-off and Landing) made up of six propulsion systems, fixed pitch, coupled with brushless electric motors attached to the ends of a rigid body in the form of stars, the drone has therefore six propellers. These have 14 inches in diameter. They are made of carbon fiber composite to ensure maximum rigidity.

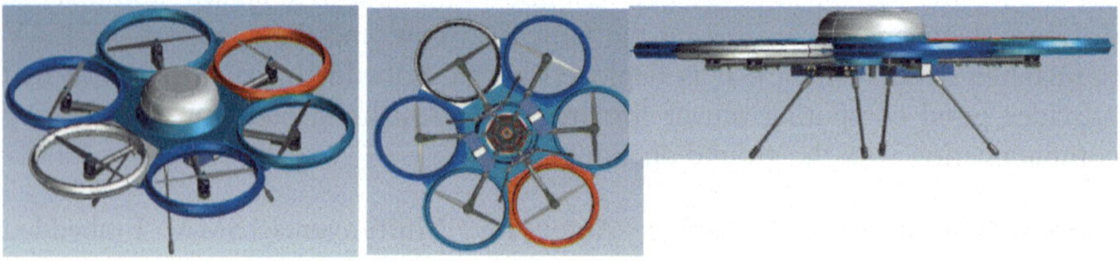

Fig. 1 Mechanical Model

3 System Design

3.1 Distributed Architecture

The embedded system platform is based on an embedded architecture (μC, avionics equipment). The architecture of the drone is mainly based on a multi-agent system; each agent provides a well-defined function from the basic functions such as takeoff and landing, to the most complex functions such as ensuring stability and the robustness of the control via the autopilot. These agents also have to get the interfacing functions (communication, command and control). Our system defined by its ability to react constantly to the demands of the environment by complying with a number of temporal constraints. In real time, our system must acquire and process data and events, characterize the temporal evolution of this environment and make appropriate decisions and turn them into actions.

To meet these needs we chose a parallel architecture: MIMD (multiple instructions, multiple data).

In computing, MIMD (Fig. 2): (Multiple Instruction Multiple Data) is a technique employed to achieve parallelism. Vehicles using MIMD have a number of processors that function asynchronously and independently. At any time, different processors may be executing different instructions on different pieces of data.

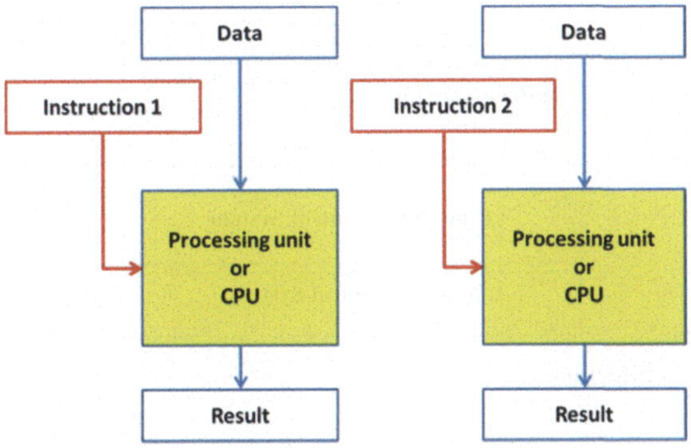

Fig. 2 Multiple Instruction Multiple Data

3.2 Hardware Design

The physical architecture (fig. 3) of the system is made up of different systems interacting with each other to better carry out the different missions requested to the drone:

- **GCS** (ground control station) is essential for monitoring data and sending missions.
- **Communication system** with GCS ensures a permanent connection with the UAV.
- **Video Metric System** handles all that is capturing photos and videos of the outside and send them to the GCS via the communication system.
- **Avionic instrumentation system** integrates wholes drone sensors, and allows the acquisition of information in the outside environment.
- **Decision making System** is implemented to well lead the control of the drone and receives commands and flight plans.
- **Propulsion Control System** acts directly on the motors.
- **Energy Control System** displays energy consumption battery and photovoltaic plate.

The Decision Making System, the Communication system, the Propulsion and Energy Control Systems are deployed in each microcontroller, whereas the Avionic Instrumentation System is deployed in another type of microcontroller.

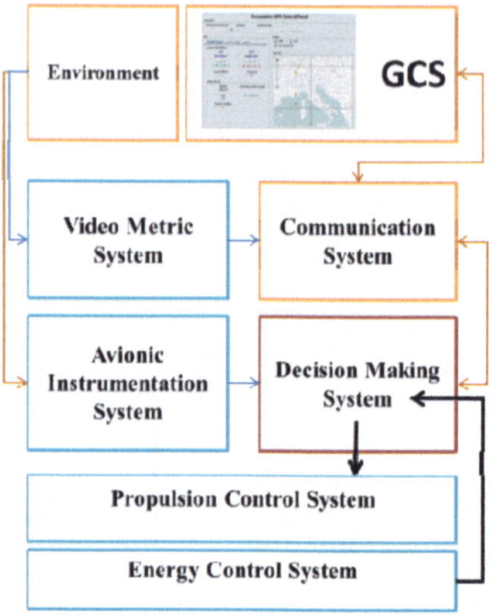

Fig. 3 Hardware architecture

This system needs a certain degree of intelligence; in fact we chose to adopt the paradigm of multi-agent system.

3.3 Multi-agent Architecture

Multi-agent systems have so many applications in the field of artificial intelligence; they reduce the complexity of solving a problem by dividing the necessary knowledge into subsets, combining independent intelligent agent to each of these subassemblies and coordinating the activity of these agents. Our architecture system called (SMART: System Multi-agent Real Time) consists of several agents deployed in different systems.

An agent is a software entity that functions continuously and autonomously in a particular environment, often inhabited by other agents and processes.

In multi-agent systems, the global behavior arises from the interaction between the constituent agents. Agents interact (cooperate, coordinate or negotiate) with the others, either to achieve a common goal, or because it is necessary for them to achieve their own goals. Fig. 4.

We describe the different systems of the drone and the agents interacting in each system in the following figure (Fig. 4).

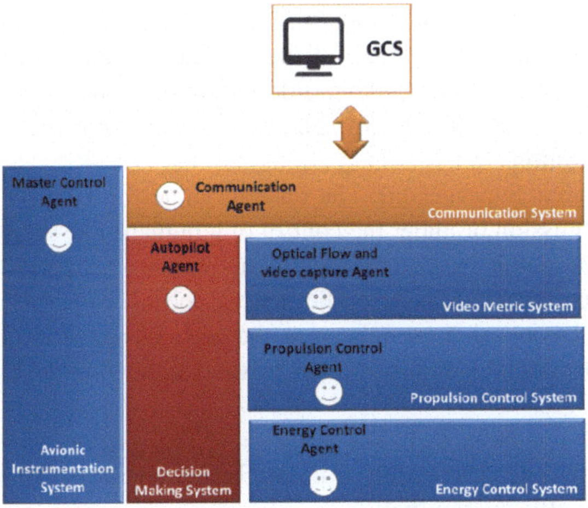

Fig. 4 SMART System

The proposed Multi-Agent architecture is designed to increase the autonomy of the drone and assure the communication in real time with the overall system of the drone.

4 Strategy and Intelligence

4.1 Autonomy of Decision

The autonomy of a vehicle is asked when this vehicle can be controlled at all times by an operator. It is therefore to provide the gear decision functions so it can "get by" particularly to events or contingencies that may occur during the mission. The architecture embedded on an unmanned vehicle that will carry out missions in a dynamic environment must implement a minimum of decision-making autonomy for the vehicle to achieve the objectives of the mission while ensuring its survival and adapting to hazards. Architecture is thus the link between the hardware components of the vehicle (sensors, payload, and actuators) and software components required for the decision-making autonomy (perception, situation monitoring, decision / action planning).

The US Office of Naval Research (ONR) provides a classification into 6 levels. This classification relates to the position of the center of decision. It goes from the operator to the vehicle, through several levels of shared decision making.

- Level 0 (Remotely Operated Vehicle): All system activity is the result of input / human commands. The system does not make decisions.

• Level 1 (Vehicle humanly attended): The system can act in parallel with operator's input, it enhances the success of the operator on the activity performs but does not have the ability to act without orders from operator.

• Level 2 (Delegation of control): The system can control its activities on the basis of the delegation.

• Level 3 (Human Supervision): The system can perform a wide range of activities with the permission or under the control of a human. The system returns sufficient information about its internal operations and its behavior for a human to supervise and redirect if necessary. The system doesn't have the capacity to take initiatives on behaviors that are related to the tasks assigned initially. This is the first step of "operational autonomy", which was validated in the first prototype in 2011.

• Level 4 (shared Initiative): The human and the system can both take action based on sensor's data. The system can coordinate their behavior with that of humans both (explicitly and implicitly).

• Level 5 (full autonomy): The system needs no human intervention to carry out its mission and adapts to its environment, autonomy is not only to follow a plan, but also to be able to reschedule online their mission after the occurrence of events that degrade or invalidate the plan.

Our UAV will have a variety of uses including monitoring and control of waterways, pollution detection, monitoring of high-voltage lines, homeland security missions (traffic surveillance, protection of reserved or prohibited areas, natural disaster assessment. ..), inspection of structures such as bridges, viaducts, dams, agricultural spraying, monitoring of cultivated areas, communications relay. All this requires a relatively high degree of autonomy; it must be able to successfully accomplish missions with minimal intervention of the driver.

4.2 Mission Planning

The decision-making autonomy defines the ability of the UAV to react in quasi-autonomously and deal with all possible events. Thus, the UAV must be in harmony with the environment in which it operates and the various embedded devices should operate to improve the decision-making autonomy.

To increase the autonomy of the drone and order in real time, it is proposed to incorporate a multi-agent architecture, each agent is deployed in a microcontroller, Figure fig. 4, represents the different agents and different systems of the drone and their interaction with the ground station.

Each agent has several functions that run in real time; the agents communicate to better carry out flight missions requested.

1. Communication Agent: Reactive

• **Role**	• Ensure continuous connectivity
• **Functions**	• Switching • Communication BLT • Communication Radio • Communication WIFI • Communication 3G • Debugging • Download, Upload

2. Optical Flow and video capture Agent: Reactive

• **Role**	• Capture video and imaging flow • Treat Visio Metric Information
• **Functions**	• Capture Camera H • Capture Camera V • Capture Optical flow

3. Master Control Agent

• **Role**	• Ensure the acquisition of permanent data • Allow the perception of the external environment
• **Functions**	• GPS acquisition • CAP acquisition • Height acquisition (sonar, altimeter) • Temperature acquisition • Horizontal speed acquisition • Vertical speed acquisition • Acceleration acquisition • Angular position acquisition • Moments acquisition • Recording checkpoints (flight recorder)

4. Propulsion Control Agent: Reactive

• **Role**	• Ensure the permanent control of propulsion to the order
• **Functions**	• Speed control • Balancing propulsion

5. Energy Control Agent : Reactive

• **Role**	• Monitoring of energy level
• **Functions**	• Displays level of energy

6. Autopilot Agent : Cognitive

• **Role**	• Ensure the execution of the flight plan • Ensure the return of the drone to the starting point in the event of signal loss • Ensure landing in case of insufficient power • Sending magnetic disturbance • Waypoint navigation
• **Functions**	• Take off • Landing • Cruising • Automatic Return • Forced landing • Predicting the position • Characterization of the dynamic behavior • Path planning • Collision avoidance

4.3 Summary

Our architecture is deployed in a platform based on different microcontrollers and multi-agent systems to increase the autonomy and the real time of the acquisition data and control of the UAV and its stability during flight no vacancy.

5 Experimental Results

To test our architecture, we use:
APM 2.5+ for Avionic Instrumentation System:

- MPU-6000,
- Six-Axis (Gyro + Accelerometer) MEMS Motion Tracking
- HMC5883L-TR, 3-Axis Digital Compass.
- MS5611
- MEAS High Resolution Altimeter

Arduino MEGA ADK for Decision Making System:

- Microcontroller ATmega2560,
- Flash Memory 256 KB of which 8 KB used by bootloader
- SRAM 8 KB
- EEPROM 4 KB
- Clock Speed 16 MHz
- USB Host Chip MAX3421E)

Arduino mini pro for Engine Control System, Energy Control System and Communication System:

- Microcontroller ATmega168,
- Flash Memory 16 KB (of which 2 KB used by bootloader) ,
- SRAM 1 KB, EEPROM 512 bytes,
- Clock Speed 8 MHz (3.3V model) or 16 MHz (5V model))

Raspberry Pi computer, (ARM1176JZF-S CPU 512 and MB SDRAM) is used for Video Metric System.

We tested our platform during a full flight (takeoff, followed waypoints and landing). Figure 4 below show the screenshots of the experimental test.

Fig. 5 The proposed solution embedded in our UAV

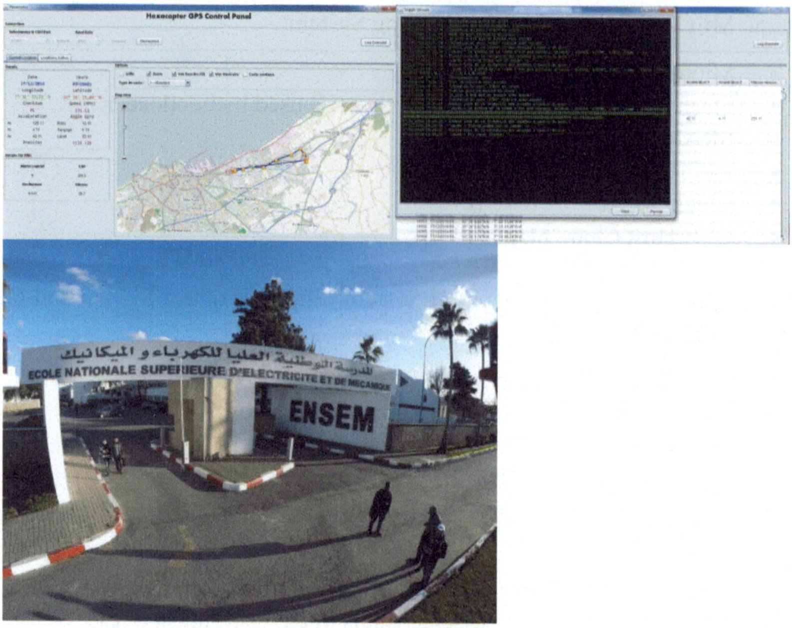

Fig. 6 Screenshots of a full flight

6 Conclusion

The main objective of the previous work was mainly to develop and design a flexible flying autonomous vehicle, integrating Multi-Agent Systems, on embedded platform and controlled in real-time.

This work has enabled the development of a collaborative platform mobilizing jointly various disciplines. This vehicle mobilized the knowledge and the expertise developed by different teams of researchers of the Laboratory Information System and Renewable Energies ENSEM.

This craft can perform different types of near Mission (tactical) from the recognition to the payloads delivery. But functional properties represent a major limitation compared to the functional altitude of the drone and also its range of action, so that is why the drone is classified in the category of tactical UAVs.

It is for this reason that work is being carried to optimize these features and extend its areas of action for altitude and range. With the aim to operate the UAV in, at first, average altitude UAV MALE (Medium Altitude Long Endurance) then, to extend it to operate in mode HALE (High Altitude, Long Endurance).

References

1. Chen, B., Zhang, A., Cao, L.: Autonomous intelligent decision-making system based on Bayesian SOM neural network for robot soccer. Neurocomputing **128**, 447–458 (2014)
2. Li, X.: A Software Scheme for UAV's Safe Landing Area Discovery. AASRI Procedia **4**, 230–235 (2013)
3. Toratani, D.: Research and development of double tetrahedron hexa-rotorcraft (Dot-HR). In: Proc. 28th Int. Congr. Aeronaut. Sci., pp. 1–8 (2012)
4. Rullán-Lara, J.L., Salazar, S., Lozano, R.: Real-time localization of an UAV using Kalman filter and a Wireless Sensor Network. J. Intell. Robot. Syst. Theory Appl. **65**, 283–293 (2012)
5. Leong, B.T.M., Low, S.M., Ooi, M.P.-L.: Low-Cost Microcontroller-based Hover Control Design of a Quadcopter. Procedia Eng. **41**(Iris), 458–464 (2012)
6. Selecký, M., Meiser, T.: Integration of autonomous UAVs into multi-agent simulation. Acta Polytech. **52**, 93–99 (2012)
7. Chrobocinski, P., Makri, E., Zotos, N., Stergiopoulos, C., Bogdos, G.: DARIUS project: deployable SAR integrated chain with unmanned systems. In: 2012 Int. Conf. Telecommun. Multimedia, TEMU 2012, pp. 220–226 (2012)
8. Tahar, K.N., Ahmad, A., Abdul, W., Wan, A., Akib, M.: Unmanned aerial vehicle technology for low cost. In: 11th South East Asian Surv. Congr. (11th SEASC 2011) (2011)
9. Bristeau, P.-J., Callou, F., Vissière, D., Petit, N.: The navigation and control technology inside the AR. drone micro UAV. In: Proc. 18th IFAC World Congr. 2011, vol. 18, pp. 1477–1484 (2011)
10. Vokřínek, J., Novák, P., Komenda, A.: Ground tactical mission support by multi-agent control of UAV operations. In: Lect. Notes Comput. Sci., (LNAI, LNBI), vol. 6867, pp. 225–234 (2011)
11. Bhattacharya, S., Likhachev, M., Kumar, V.: Multi-agent path planning with multiple tasks and distance constraints. In: Proc. - IEEE Int. Conf. Robot. Autom., pp. 953–959 (2010)
12. Bertuccelli, L.F., Choi, H., Cho, P., How, J.P.: Real-time multi-UAV task assignment in dynamic and uncertain environments. Construction, 1–16 (2009)
13. Albusac, J., Castro-Schez, J.J., Lopez-Lopez, L.M., Vallejo, D., Jimenez-Linares, L.: A supervised learning approach to automate the acquisition of knowledge in surveillance systems. Signal Processing **89**(12), 2400–2414 (2009)
14. Jang, M.W., Agha, G.: Scalable agent distribution mechanisms for large-scale UAV simulations. In: 2005 Int. Conf. Integr. Knowl. Intensive Multi-Agent Syst. KIMAS 2005 Model. Explor. Eng., vol. 2005, pp. 85–90 (2005)

Analytical Study of Nonlinear Controls Applied to Wind Energy Conversion Systems Using a DFIG

Abdelmajid Berdai, Moussa Reddak, Abdelaziz Belfqih, Boukherouaa Jamal, Faissal El Mariami and Abdelhamid Hmidat

Abstract Variable-speed wind energy conversion systems (WECSs) have become increasingly important in recent years as they enable obtainment of maximal output power in the fields of the low and average wind speeds. In this study, we focus on WECS using doubly-fed induction generators (DFIGs) and present models for the turbine and doubly-fed induction generator. The main objective is to study two types of nonlinear controls; sliding mode and backstepping used for control of the grid-side converter will be also studied and analyzed. Matlab / Simulink simulation results are included.

1 Introduction

Increases in energy demand are posing an ever greater challenge across the world. Rapid growth of industrial activity in developed countries and in investments by companies in countries that ensure lower production costs has led to spiraling increases in energy demand worldwide –increases expressed in rising prices for oil, which remains the leading energy source. World oil reserves continue to diminish and in future years there will not be enough to meet demand. The earth's climate is deteriorating and natural sources of water are becoming scarcer. Nuclear energy is not available to everybody for political, technological or financial reasons; its installation is expensive and it can be dangerous ecologically. Use of such conventional sources is therefore either limited or not encouraged for environmental reasons. And so the world is turning to renewable sources – sun, wind, underwater currents etc. – to produce electricity [1].

A. Berdai(✉) · M. Reddak · A. Belfqih · B. Jamal · F. El Mariami · A. Hmidat
ESE Research Group, Department of Electrical Engineering, National School of Electricity and Mechanics ENSEM, University Hassan II, Casablanca, Morocco
e-mail: {a.berdai,moussa.reddak}@gmail.com, a-belfqih@hotmail.com,
{j.boukherouaa,f_elmariami,a.hmidat}@yahoo.fr

E. Sabir et al. (eds.), *The International Symposium on Ubiquitous Networking*,
Lecture Notes in Electrical Engineering 366,
DOI: 10.1007/978-981-287-990-5_37

At present, variable-speed wind systems based on the doubly fed induction generator (DFIG) are the systems most commonly used in land wind farms. Their implementation in renewable energy conversion systems has increased dramatically due to their many advantages, which include low cost, small size, and the ability to produce maximum power under various wind and rotational speeds. The DFIG enables functioning on a range of speeds of ± 30 % around the simultaneous speed, so ensuring reduced sizing of the static converters, as these latter are connected between the rotor winding [2].

Growing interest is being shown in this machine, largely due to the degree of freedom it provides, owing to its rotor's accessibility and the consequent possibility of supplying it via a converter on the stator side and rotor side alike, as well as extending the speed range (sub-synchronous, synchronous and above-synchronous) [3].

2 Modeling of DFIG System

2.1 Model of the Wind Turbine

If the kinetic energy of a mass of air moving at speed v could be fully recovered using a device with surface A, located perpendicular to wind speed direction, instantaneous power would be [4]:

$$P_e = \frac{1}{2}\rho A v^3. \tag{1}$$

In reality, however, a wind turbine extracts a power P lower than available power Pe, as a result of the non-null speed of air masses behind the aero-engine. Wind energy conversion aptitude is given by the power coefficient Cp, defined by the relationship:

$$C_p = \frac{P}{P_e}\,; C_p < 1. \tag{2}$$

$$\lambda = \frac{R\Omega_t}{v}. \tag{3}$$

With $A = \pi R^2$

$$P_t = \frac{\rho \pi R^2 v^3}{2} C_p. \tag{4}$$

2.2 *Model of DFIG*

In complex notation, the DFIG equations are derived from the Park model expressed in a reference frame d-q rotating at synchronous speed ωs. The electrical energy conversion system is described by the induction machine equations shown in [5]:

$$\begin{cases} V_{ds} = R_s I_{ds} + \frac{d\varphi_{ds}}{dt} - \omega_s \varphi_{qs} \\ V_{qs} = R_s I_{qs} + \frac{d\varphi_{qs}}{dt} + \omega_s \varphi_{qs} \\ V_{dr} = R_r I_{dr} + \frac{d\varphi_{dr}}{dt} - (\omega_s - \omega_r)\varphi_{qr} \cdot \\ V_{qr} = R_r I_{qr} + \frac{d\varphi_{qr}}{dt} + (\omega_s - \omega_r)\varphi_{rs} \\ C_{em} = P(\varphi_{sd} i_{sq} + \varphi_{sq} i_{sd}) \end{cases} \tag{5}$$

The expressions of stator and rotor fluxes in the dq-axis system:

$$\begin{cases} \varphi_{ds} = L_s I_{ds} + M I_{dr} \\ \varphi_{qs} = L_s I_{qs} + M I_{qr} \\ \varphi_{dr} = L_s I_{dr} + M I_{ds} \\ \varphi_{qr} = L_s I_{qr} + M I_{qs} \end{cases} \tag{6}$$

Active and reactive powers exchanged between stator and grid may be expressed as follows:

$$\begin{cases} P_s = V_{ds} I_{ds} + V_{qs} I_{qs} \\ Q_s = V_{qs} I_{ds} - V_{ds} I_{qs} \end{cases} \tag{7}$$

We use the simplification of oriented flux equations:

$$\begin{cases} \varphi_{ds} = L_s I_{ds} + M I_{dr} = \varphi_s \\ \varphi_{qs} = L_s I_{qs} + M I_{qr} = 0 \end{cases} \tag{8}$$

3 Architecture of Control System

The architecture of the control system is presented in Figure 1. It is based on the three-phase model of the channel of electro-mechanical wind-system conversion.

As shown in Figure 1, three detailed controls are then necessary to insure functioning of the wind turbine:

- Control of extraction of maximum wind power via a control known as "MPPT" (Maximum Power Point Tracking).
- Control of the machine-side converter (RSC) by controlling the electromagnetic torque and the DFIG's reactive stator power.
- Control of the GSC by controlling the continuous bus voltage and the active and reactive powers exchanged with the grid.

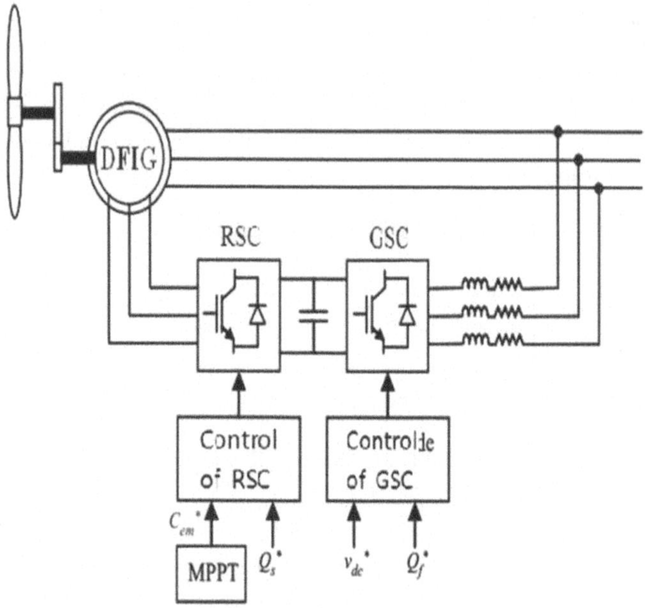

Fig. 1 Architecture of the wind energy control system

3.1 MPPT Control Strategy

Computation of Reference Torque [6]
In this case, system control has to impose a reference torque so as to enable the DFIG to turn at an adjustable speed to ensure an optimal functioning point in terms of maximum power extraction. In this context, the ratio of the speed of wind turbine λ must be maintained at its optimal value ($\lambda = \lambda_{opt}$) on a set range of wind speeds, so maintaining the power coefficient at its maximal value ($C_p = C_{p\_max}$). In this case, the aerodynamic torque may be expressed as:

$$C_{aero} = \frac{1}{2\Omega_t} C_{p\_max} \rho S v^3 \tag{9}$$

$$C_{g\_ref} = \frac{1}{G} C_{aero} \tag{10}$$

Assuming that the angle of orientation of blades ß is constant, wind speed can be estimated as follows:

$$v = \frac{R\hat{\Omega}_t}{\lambda_{opt}} \tag{11}$$

The electromagnetic reference torque may therefore be expressed as follows:

$$C_{em}^{*} = \frac{1}{2\Omega_t G} C_{p\_max} \rho S v^3 \tag{12}$$

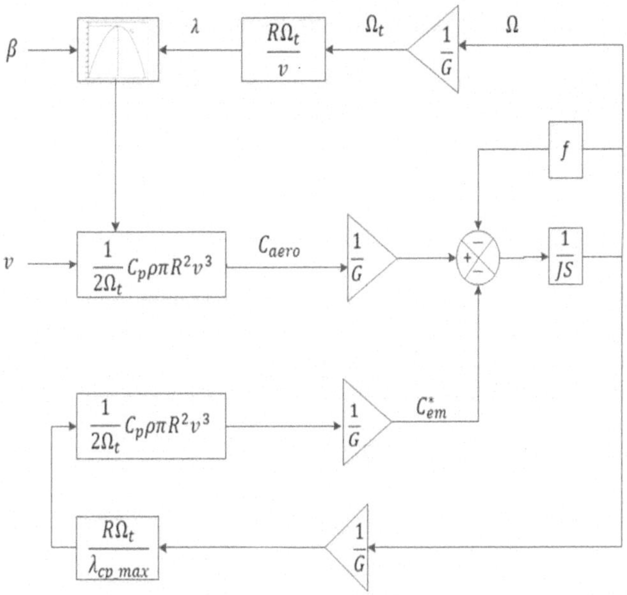

Fig. 2 MPPT control strategy

3.2 *Rotor-Side Converter Control*

Electromagnetic torque and stator reactive power control will be obtained by controlling the DFIG dq rotor currents.

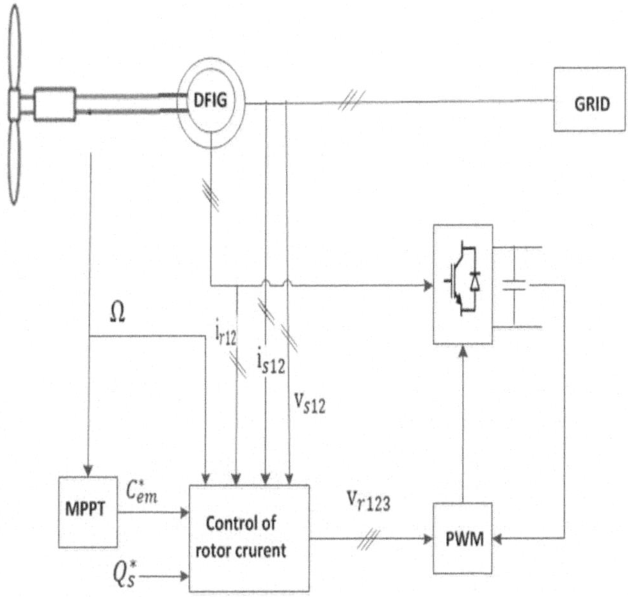

Fig. 3 Rotor-side converter control

3.3 Generation of Rotor Current References

Knowledge of the stator flux is required for generation of the dq axes reference rotor currents (see Figure 3). In our study, the electricity grid is assumed to be stable and the selected dq reference point is connected to the rotating stator field: thus, d axis stator flow may be estimated in open-loop from measurements of stator and d-axis rotor currents:

$$\varphi_{sd-est} = L_s . i_{sd} + M . i_{rd} \tag{13}$$

By estimating the stator flux, we generate the dq axis rotor reference currents. As the electromagnetic torque is proportional to q axis rotor current (according to equation 5), we can establish a relationship between current irq* and electromagnetic torque Cem* from the control MPPT block, by:

$$i_{rq}^* = -\frac{L_s}{p.M.\varphi_{sd-est}} C_{em}^* \tag{14}$$

Expression of current ird* according to reactive power Qs* is established by:

$$i_{rd}^* = \frac{\varphi_{sd-est}}{M} - \frac{L_s}{M.v_{sq}} Q_s^* \tag{15}$$

3.4 Grid-Side Converter Control [7][8]

The grid-side converter (GSC) is connected between the continuous bus and the grid via an RL filter. This converter has two roles: to maintain the bus' constant

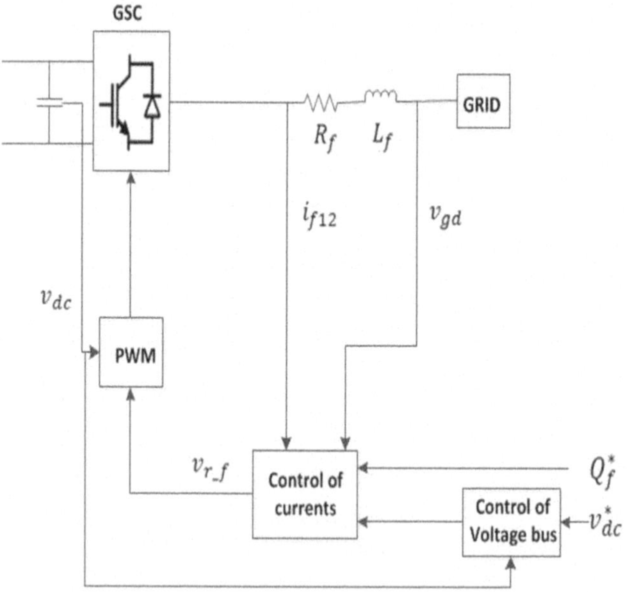

Fig. 4 .Grid-side converter control principle

continuous voltage whatever the amplitude and direction of the MADA rotor power flow and to maintain a power factor equal to 1 at the point of connection with the grid. Figure 4 describes the GSC control, which has a dual function: - Control of currents circulating in filter RL, Control of continuous bus' voltage.

3.5 Control of Filter Currents

The dynamic model of the grid-side inverter side in the reference frame rotating synchronously with the grid voltage vector is:

$$\begin{cases} L_f \frac{di_{d-f}}{dt} = v_{rd} - R_f i_{d-f} + \omega L_f i_{q-f} \\ L_f \frac{di_{q-f}}{dt} = v_{rq} - R_f i_{q-f} - \omega L_f i_{d-f} + Vgd \end{cases} \tag{16}$$

$$\begin{cases} P_f = \frac{3}{2}(V_{gd} i_{d-f} + V_{gq} i_{q-f}) \\ Q_f = \frac{3}{2}(V_{gd} i_{d-f} - V_{gq} i_{q-f}) \end{cases} \tag{17}$$

Where vrd and vrq are the converter d-axis and q-axis voltage components; Vgd is the grid voltage component in the d-axis; idf and iqf are the grid's d-axis current and q-axis current; Lf is the filter inductance; Rf is the filter resistance, and x is the network angular frequency.

$$\begin{cases} P_f = \frac{3}{2} V_{gd} i_{d-f} \\ Q_f = -\frac{3}{2} V_{gd} i_{q-f} \end{cases} \tag{18}$$

By ignoring losses in resistance Rf of the RL filter and taking account of the orientation of the dq reference voltage linked to d-axis (vgq = 0), the equations become:

$$\begin{cases} i_{dr-f} = \frac{2P_f^*}{3V_{gd}} \\ i_{qr-f} = -\frac{2Q_f^*}{3V_{gd}} \end{cases} \tag{19}$$

Accordingly, active and reactive power control can be achieved by controlling direct and quadrature current components respectively. Moreover, the basic aim of the grid-side converter control is to produce no reactive power and to transfer all active power produced by the DFIG to the grid.

Where $id_{\rm rf}$ and $iq_{\rm rf}$ are the desired values of d-axis current and q-axis current respectively. In addition, q-axis current$iq_{\rm rf}$, is directly given from the outside of the controller and it sets to zero to achieve unity power factor control.

4 Examples of Nonlinear Control

4.1 Sliding Mode Control

In order to regulate the current components iq and id to their references, the sliding surfaces must be defined as follows:

$$\begin{cases} S_{d-f} = i_{dr-f} - i_{d-f} \\ S_{q-f} = i_{qr-f} - i_{q-f} \end{cases} \tag{20}$$

It follows that:

$$\begin{cases} \frac{dS_{d-f}}{dt} = \frac{di_{dr-f}}{dt} - \frac{di_{d-f}}{dt} = \frac{di_{dr-f}}{dt} - \frac{1}{L_f}(v_{rd} - R_f i_{d-f} + \omega L_f i_{q-f}) \\ \frac{dS_{q-f}}{dt} = \frac{di_{qr-f}}{dt} - \frac{di_{q-f}}{dt} = \frac{di_{qr-f}}{dt} - \frac{1}{L_f}(v_{rq} - R_f i_{q-f} - \omega L_f i_{d-f} - Vgd) \end{cases} \tag{21}$$

When the sliding mode occurs on the sliding surfaces:

$$\begin{cases} S_{d-f} = \frac{dS_{d-f}}{dt} = 0 \\ S_{q-f} = \frac{dS_{q-f}}{dt} = 0 \end{cases} \tag{22}$$

Combining (21) and (22), controls of d axis and q axis voltages are defined by:

$$\begin{cases} v_{rd}^{eq} = L_f \frac{di_{d-f}}{dt} + R_f i_{d-f} - \omega L_f i_{q-f} \\ v_{rq}^{eq} = L_f \frac{di_{q-f}}{dt} + R_f i_{q-f} + \omega L_f i_{d-f} + Vgd \end{cases} \tag{23}$$

Expression of the sliding terms is given by:

$$\begin{cases} V_{dr-n} = L_f K_{d-f} sgn(S_{d-f}) \\ V_{qr-n} = L_f K_{q-f} sgn(S_{d-f}) \end{cases} \tag{24}$$

As a result, the control voltages of q axis and d axis are defined by:

$$V_{r-f} = V^{eq} + V_{r-n}$$

$$\begin{cases} v_{dr-f} = L_f \frac{di_{d-f}}{dt} + R_f i_{d-f} - \omega L_f i_{q-f} + L_f K_d sng(S_{d-f}) \\ v_{qr-f} = L_f \frac{di_{q-f}}{dt} + R_f i_{q-f} + \omega L_f i_{d-f} + Vgd + L_f K_q sng(S_{q-f}) \end{cases} \tag{25}$$

4.2 Backstepping Control

The basic idea behind Backstepping design is the use of so-called "virtual control" to systematically decompose a complex nonlinear control design problem into simpler, smaller ones. In general terms, Backstepping design is divided into a series of steps, each essentially dealing with an easier, single-input-single-output design problem, and each providing a reference for the next design step. Overall stability and performance are achieved by a Lyapunov function for the whole system [8].

In the first step, the system must follow a given trajectory for each output variable. To do so, two functions are defined, where i_{d-f} and i_{q-f} are the dq-axis current references respectively. The rotor and stator currents tracking error are defined by [9]:

$$\begin{cases} e_1 = i_{d-f} - i_{dr-f} \\ e_2 = i_{q-f} - i_{qr-f} \end{cases} \tag{26}$$

The derivative of Eq. (13) gives:

$$\begin{cases} \dot{e_1} = \dot{i_{d-f}} - \dot{i_{dr-f}} \\ \dot{e_2} = \dot{i_{q-f}} - \dot{i_{qr-f}} \end{cases} \tag{27}$$

By replacing the equation, we obtain the following expressions;

$$\begin{cases} \dot{e_1} = \dot{i_{d-f}} - \frac{1}{L_f}(v_{rd} - R_f i_{d-f} + \omega L_f i_{q-f}) \\ \dot{e_2} = \dot{i_{q-f}} - \frac{1}{L_f}(v_{rq} - R_f i_{q-f} - \omega L_f i_{d-f} - Vgd) \end{cases} \tag{28}$$

To improve control performances in terms of stability, we use the following Lyaponov functon:

$$V = \frac{1}{2}(e_1^2 + e_2^2) \tag{29}$$

Using eq (28), the derivative of eq. (29) is written as follows:

$$\dot{V} = \dot{e_1}e_1 + \dot{e_2}e_2$$

$$e_1\left(\frac{1}{L_f}(v_{dr-f} - R_f i_{d-f} + \omega L_f i_{q-f}) + K_1 e_1\right) + e_2.(\frac{1}{L_f}(v_{qr-f} - R_f i_{q-f} - \omega L_f i_{d-f} - Vgd + K_2 e_2) - K_1 e_1^2 - K_2 e_2^2$$

The derivative of the complete Lyapunov function eq. (26) can be negative definite if the quantities between parentheses in eq. (26), are chosen equal to zero.

$$\begin{cases} \frac{1}{L_f}(v_{dr-f} - R_f i_{d-f} + \omega L_f i_{q-f}) + K_1 e_1 = 0 \\ \frac{1}{L_f}(v_{qr-f} - R_f i_{q-f} - \omega L_f i_{d-f} - Vgd + K_2 e_2) = 0 \end{cases}$$

$$\dot{V} = -K_1 e_1^2 - K_2 e_2^2 \leq 0$$

Rotor voltages may then be deduced as follows:

$$\begin{cases} v_{dr-f} = R_f i_{d-f} - \omega L_f i_{q-f} - K_1 e_1 \\ v_{qr-f} = R_f i_{q-f} + \omega L_f i_{d-f} + Vgd - K_2 e_2 \end{cases}$$

5 Simulation Results

5.1 Sliding Mode Control

As we mentioned in Part III, control of active power is achieved by controlling direct component Id of the rotor current, and control of reactive power is achieved by controlling the component in quadrature Iq.

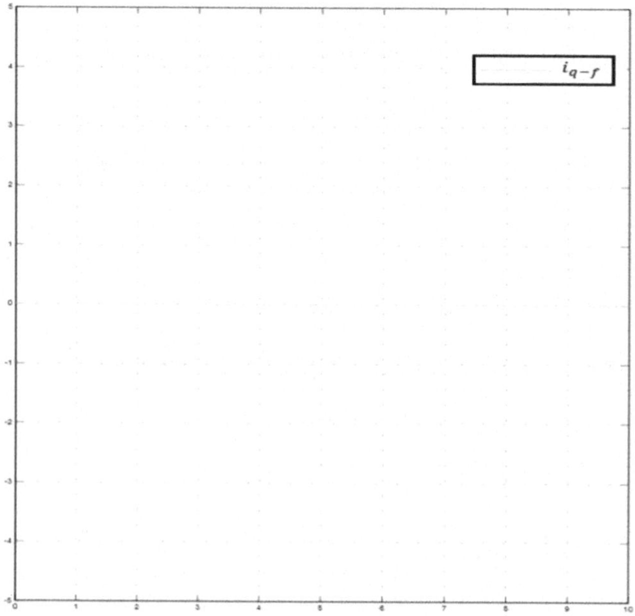

Fig. 5 Iq current response

The Figure 5 presents the temporal evolution of the reactive power. The wind system works in Unitarian power factor because the reactive power of reference Qf* is imposed equal to zero.

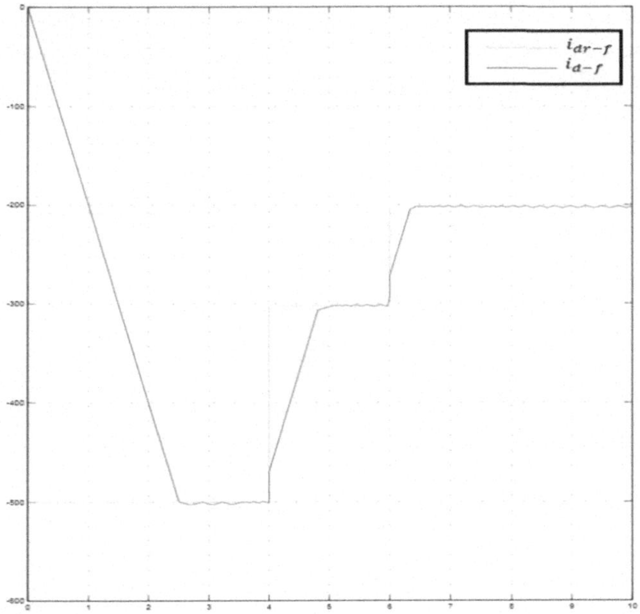

Fig. 6 Control of Id component

5.2 Backstepping Control

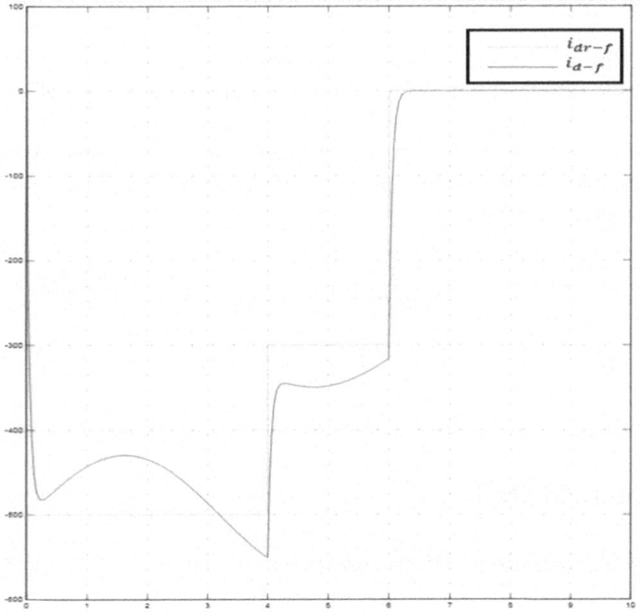

Fig. 7 Control of active power by Id component current

The Figures 6 and 7 shows the regulation of the active power that controlled by the component Id. The power imposed varies in three levels. Which mean that the active power is dependent at the request of the consumers.

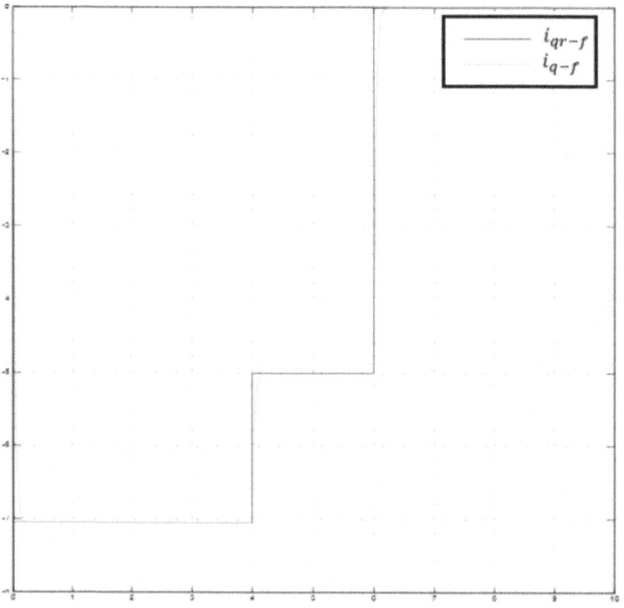

Fig. 8 Control of reactive power by Iq component current

Between times t = 4 and t = 6, we have changed the reactive power reference, which means that to the reference changes on the rotor current ird (Figure) 8). We notice that the time of response of the system is satisfactory and that the reactive power supplied by the wind system is regulated well in its reference value.

The parameters of the grid and the controllers (Backstepping and sliding mode) are given by the following table.

Parameter of the grid	Parameter of backstepping control	Parameter of sliding mode control
L_f =0.22	K_1 =20	K_1 =200
R_f =1.3	K_2 =30	K_1 =6

6 Analytical Study

6.1 The 'Chattering' Phenomenon

The control technique described in the sliding mode control part brings about unwanted behavior in the closed-loop system. However, ideally, it requires an infinite switching of actuators. Such oscillation close to the surface is known as 'chattering'. Figure (10) shows the effect of chattering in convergence of the system. Chattering is not desirable as it induces high frequency dynamics in the system, increasing energy consumption which can damage the actuators [10].

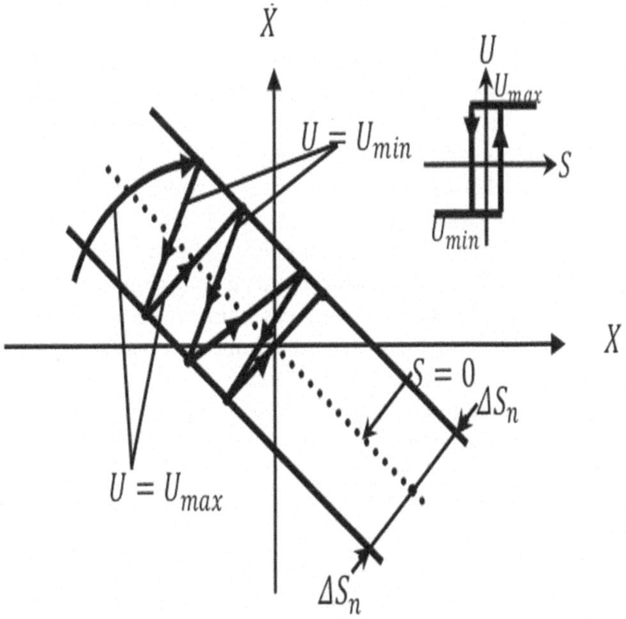

Fig. 9 Demonstration of the 'chattering' phenomenon

Chattering is the main disadvantage of sliding mode control. Numerous studies have been undertaken with the aim of reducing or eliminating the problem – for example, proposing solutions by limitation of the sliding conditions, by using an observer, etc. In this section, we shall describe techniques for limitation of the sliding condition, as they are the most often used for applications in real time. Such techniques are based on definition of a zone around surface S within which a sliding condition less strict than the sign condition is applied. Thus, the term sgn (S) in the sliding part of the control is often replaced by a term with lighter variation, such as integral control or control with a threshold [10].

6.2 *Advantage of Backstepping Control*

One major advantage of Backstepping control is that nonlinearities can be treated in several ways, with useful nonlinearities contributing to stabilization retained in the Lyapunov function and non-useful nonlinearities replaced by linear control, requiring a control effort and, better yet, optimal resulting control laws guaranteeing a degree of robustness [11].

7 Conclusion

This work describes various control strategies applied to wind energy conversion systems (WECSs) based on DFIGs. We have designed and compared two control strategies for the grid-side converter (GSC): nonlinear backstepping control and sliding mode control.

To the diversity of methods for analysis of nonlinear systems, there is a corresponding diversity of design methods. Depending on the importance of the nonlinear phenomenon present in the system, these methods provide necessary and adequate conditions for good functioning.

Through response characteristics, we observe good performances even in the presence of instruction variations.

Pursuit of power is assured as is decoupling, stability and convergence towards balance. Furthermore, such regulation presents a very simple and robust algorithm control, which has the advantage of being easily integrated into computer control.

As presented, the backstepping method assumes availability of measurement of the state of the whole process. When a part of such state is not measurable, backstepping cannot be applied. A solution then consists of replace the state by an estimation made by an observer.

References

1. Lazhar, R.: Modélisation et Commande de la Génératrice à Induction Double Alimentée Couplée sur un Réseau Electrique (2012)
2. Gaillard, A.: Docteur de l'Université Henri Poincaré, Nancy-I
3. Poitiers, F.: Etude et commande de generatrices asynchrones pour l'utilisation de l'energie eolienne-Machine asynchrone à cage autonome-Machine asynchrone à double alimentation reliée au réseau, Université de Nantes (2003)
4. Multon, B., Roboam, X., et al.: Aérogénérateurs électriques (2006)
5. Ghennam, T., Berkouk, E.M., François, B., et al.: Modeling and Control of a Doubly Fed Induction Generator (DFIG) Based Wind Conversion System (2009)
6. Cherif, B.: Simulation de la commande vectorielle par régulateurs à mode glissant d'une chaîne éolienne à base d'une machine asynchrone à double alimentation (2014)
7. Errami, Y., Ouassaid, M., Maaroufi, M., et al.: A performance comparison of a nonlinear and a linear control for grid connected PMSG wind energy conversion system. Int. J. Electr. Power Energy Syst. **68**, 180–194 (2015)
8. Guessas, M.L.: Backstepping Backstepping adaptatif pour le contrôle la poursuite et la synchronisation des systèmes dynamiques non linéaires chaotiques, Université de Biskra (2012)
9. Khedher, A., Khemiri, N., Mimouni, M.F., et al.: Wind Energy Conversion System Using DFIG Controlled by Backstepping and Sliding Mode Strategies. Int. J. Renew. Energy Res. IJRER **2**(3), 421–430 (2012)
10. El Mahdi, A.S.A.: Commande en vitesse par mode glissant d'une Machine Asynchrone à Double Alimentation
11. Benaskeur, A.R.: Aspects de l'application du backstepping adaptatif à la commande décentralisée des systèmes non linéaires (2000)

Part VIII
Advances in IT Governance and Information Systems

Statistical Learning Based Framework for Random Networks Knowledge Extraction Applied in Smart Cities

Smail Tigani, Mohammed Ouzzif and Rachid Saadane

Abstract Smart Cities are a future reality that emerged recently. They became a wide research field around the world. These cities will combine the power of ubiquitous communication networks and wireless sensors with the efficient management systems to solve daily challenges and create exciting services. In this work, we involve the power of artificial intelligence to solve one of the serious challenges in big cities. This concerns the traffic management and prediction. This work proposes a statistical model serving the analysis of a random graph that represents, in reality, roads on map. Using those models and collected data from sensors or human agents, we can extract useful hidden knowledge for the best decision making. To prove the reliability of the approach, a Monte Carlo simulation algorithm is designed and results confirms the added-value of the approach.

Keywords Learning theory · Bayesian modeling · Random graph · Smart Cities · Traffic prediction

1 Introduction

Easy transportation and decreasing urban populations stress are considered one of main Smart Cities challenges. The new vision to solve this problem is the deployment of smart IT systems that allows population to have necessary information to decide

S. Tigani(✉)
Computer Science Department, ENSEM, Casablanca, Morocco
e-mail: sma.tigani@gmail.com

M. Ouzzif
Computer Science Department, EST, Casablanca, Morocco
e-mail: ouzzif@gmail.com

R. Saadane
Electrical Department, EHTP, Casablanca, Morocco
e-mail: rachid.saadane@gmail.com

E. Sabir et al. (eds.), *The International Symposium on Ubiquitous Networking*,
Lecture Notes in Electrical Engineering 366,
DOI: 10.1007/978-981-287-990-5_38

the adequate road in adequate time. Modern information technologies such as Cloud computing, mobility and big data contributes to deploy necessary infrastructure and services for Smart Cities goals [1].

We focus, in this work, on the transport problem in big cities. That comes to represent the map (of some city) with a mathematical network (a graph). Mathematical Networks data and statistical methods have emerged in recent years and seem to appear in many research areas. Introduction to this large literature may be done from the surveys in Stephen E. Fienberg [2] and [3]. Some network structures are static and others are dynamically evolving. Networks are usually represented in terms of graphs with the nodes representing entities, for example, people, and the links or edges representing ties or relationships. Edges may be directed or undirected depending on the application. Complex networks are involved in many fields: epidemic modeling introduced in work of F. Darabi Sahneh and al in [4] and also the internet traffic classification based on Bayes theory discussed by F. Darabi Sahneh and al in [5]. Learning machines applied on complex network allows the prediction of the weight of some link based on historical data [6], [7]. That comes to predict all the state of the graph.

The content of this paper is organized as follows: the section 2 introduces the mathematical definitions and models serving the graph knowledge extraction. Theoretical aspects proposed in it are used the the third section as bases of practical aspects. The database relational model design is given in the third section with all necessary functions and stored procedures to extract discussed knowledge. In the end, we present concluding remarks followed by perspectives and projections to next works.

2 Mathematical Modelling

In this section we present three main mathematical definitions that are the bases of proposed framework. We have **Weight Average Per Link** which is the first definition, the second is **Best Links Set Given the Instant**, and then **Graph State Given the Instant**.

2.1 Preliminaries and Notations

The main goal is the definition of a random graph that changes, or perhaps not, weights at any instant. Let $\mathcal{G}$ be a graph defined by a set of nodes denoted $\mathbb{X} = \{x_1, ..., x_p\}$ and an other set of links $\mathbb{A} = \{(x, y) \in \mathbb{X}^2\}$. We suppose, for the application discussed on this work, that we have one weight observation for each hour and for each link. During a day, we will collect 24 values for each link of the graph. Let call the hours interval $[\![0, 23]\!]$.

The observed value for each link is in fact a physical measure of some phenomena. It can be traffic in a road or in a computer network or telephonical network or also the transported size of goods in some logistic network... The weight of a link represents this physical measure, let call it ω. This logical sequence takes three parameters and gives the weight as an percentage (value in [0, 100]). Those parameters are: the link (element of $\mathbb{A}$), the hour (element of $[\![0, 23]\!]$) and the index of the observation (an integer $\mathbb{N}^*$). Formally:

$$\omega : \mathbb{A} \times [\![0, 23]\!] \times [\![1, Q]\!] \longrightarrow [0, 100] \tag{1}$$

During a process of data collection, we need to have a large data set on which learning algorithms will run. Let Q be the number of collected data and $\mathcal{D}^Q = \{(i, n, l, \omega_n^i(l)\}$ the set containing the Q observations with: $i \in [\![0, Q]\!]$ represents the observation index and $n \in [\![0, 23]\!]$ is the observation hour of the observed link $l \in \mathbb{A}$. The ith weight of the link l at time n is denoted $\omega_n^i(l) \in [0, 100]$.

2.2 *Weight Average Per Link*

Based on main definitions in the previous subsection, we define rules to extract some hidden useful information. The set $\mathcal{D}^Q$ contains all data for all instants. However, before running learning algorithms we should pinpoint adequate data in order to avoid noisy data. By noisy data we mean, all data that can mislead algorithms and then give wrong results. Let give an example by considering traffic cars problem. It is very normal that the traffic measure in some periods in the day is very high: at the morning, mi-day and the end of the day but very calm in other parts of the day. To let the algorithm find the reality of a given road from 08:00 to 08:59 for example, we have to run it on all historical observations between 08:00 and 09:00 on the same road. This is why we define the quantity $\overline{\omega}_{n_0}(l_0)$ as the average of weights if we fix the hour at n_0 and the link at l_0. We consider $W_{N,L}$ a random variable following some unknown probability law $\mathbb{P}$. This variable may takes many random values $\omega_n^i(l)$ when the index i is changing. Formally:

$$\overline{\omega}_{n_0}(l_0) = \mathbb{E}[W_{N,L} | N = n_0, L = l_0] \tag{2}$$

The expectation in 2 can be computed practically using the empirical conditional density. The probability of the event $(W_{N,L} = \omega_n^i(l))) \cup (n = n_0) \cup (l = l_0)$ is equivalent of $(W_{N,L} = \omega_{n_0}^i(l_0))$. The last probability is, in an other way, the number of observations of $\overline{\omega}_{n_0}(l_0)$ divided by the observation number Q. The expression 2 becomes 3:

$$\overline{\omega}_{n_0}(l_0) \triangleq \sum_{i=1}^{Q} \omega_{n_0}^i(l_0).\frac{\mathbb{P}(W_{N,L} = \omega_{n_0}^i(l_0))}{\mathbb{P}(n = n_0, l = l_0)} \tag{3}$$

2.3 Best Links Set Given the Instant

The set of best links when we fix the instant (the hour in this case) is a set containing all links of the graph $\mathcal{G}$ optimizing the weight average at the fixed instant. The optimization here can be minimization or maximization according to what the weight value represents. Let illustrate this, if the weight represents a cost of something, so the best link is, naturally, the link minimizing the weight average. However, if it represents a gain then the best link should be the ones maximizing the weight average.

Definition 1. Let call $\Omega_{n_0}^+$ the set of best links at the instant n_0. We suppose that weight represents some costs, that comes to find all links minimizing the weight average. Formally we write:

$$\Omega_{n_0}^+ \triangleq \{l \in \mathbb{A}, l = \arg\min_{x \in \mathbb{A}} \overline{\omega}_{n_0}(x)\} \tag{4}$$

Following the same method, we can define the set of worst links denoted $\Omega_{n_0}^-$. This set contains non tolerable links at some instant n_0, those maximizing weight average (for weight representing costs).

Question 1. How the best links set can serves in real modern life? Let suppose the problem of traffic prediction in big cities. A driver that is looking for a free road on a map must have access to centralized database containing all collected data. The access can be found by a mobile application that consuming a web service implementing the definition 4 to extract necessary information from the database. By this, the driver makes right decision and chooses the quick and safe road.

2.4 Best Instants Set Given the Link

The set of best instants, when we fix the link l_0, contains all instants in the interval $[\![0, 23]\!]$ on which the average of weights of the fixed link. A best instant is the one that minimizes the average of weights for the fixed link.

Definition 2. Let call $\Gamma_{l_0}^+$ the set of best instants for the link l_0. We suppose that weight represents some costs like in the set of best links, that comes to find all instants minimizing the weight average. Mathematically:

$$\Gamma_{l_0}^+ \triangleq \{n \in [\![0, 23]\!], n = \arg\min_{x \in [\![0,23]\!]} \overline{\omega}_x(l_0)\} \tag{5}$$

Following the same method, we can define the set of worst instants denoted $\Gamma_{l_0}^-$. This set contains all instants on which the links weight average is not tolerable: those maximizing weight average.

Question 2. How the best instants set can serves in traffic problem? The driver life becomes easier if he can enjoy a panoramic view of the traffic charge on all the map (materialized by the graph) at a given instant. The model 5 allows the extraction, based on the history, of worst and best instants given a link.

2.5 *Graph's Instantaneous State*

One of main notion in the current work is the graph state.that comes to replace the real valued weights with discreet valued ones. This transformation is done by a conditional binary discrimination, defined formally the application Θ. Binary valued weights in real life represent the fact that a link's weight is tolerable (1) or not (0).That pushes to think the possibility to introduce a fixed constant ω_0 representing the tolerable average. Formally:

$$\begin{cases} \Theta : [\![0, 23]\!] \times \mathbb{A} \longrightarrow \{0, 1\} \\ \Theta_n(l) = \mathbf{1}_{\{\overline{\omega}_n(l) \leq \omega_0\}} \end{cases}$$

Definition 3. The real valued graph $\mathcal{G}(\mathbb{X}, \mathbb{A})$ becomes a binary valued one when we add on it the component Θ defined by the equation 2.5. We talk at this level about the state of the graph $\mathcal{G}$ at the instant n and we formalize with:

$$\mathcal{G}_n^{\Theta}(\mathbb{X}, \mathbb{A}), n \in [\![0, 23]\!], \forall l \in \mathbb{A} \tag{6}$$

Let give an example in order to clarify the proposed generic model. Let $\mathbb{X} = \{x_1, x_2, x_3\}$ and $\mathbb{A} = \{(x, y) \in \mathbb{X}^2\}$. We suppose that tolerable average is $\omega_0 = 80\%$. Let $\omega_i^n(l)$ be the ith phenomenon measure for a given link l at the instant n. The table 1 contains supposed measures from $08:00$ to $08:59$ $(n = 8)$.

Based on data on the table 1 we compute the weights average for each link. The link (x_1, x_2), at $n_0 = 8$, has 89.00% as weight average. However (x_1, x_3) takes the small weight average 22.50%.

Table 1 Illustration: Supposed Measures at $n = 8$.

Observation Id (i)	Link (l)	Measure ($\omega_i^n(l)$)
1	(x_1, x_2)	90.00%
2	(x_1, x_3)	05.00%
3	(x_1, x_2)	88.00%
5	(x_1, x_3)	20.00%

Question 3. How can we interpret those results? The weight average from $08:00$ to $08:59$ ($n_0 = 8$) of the link (x_1, x_2) is 89.00% greater than tolerable value $\omega_0 = 80\%$. That classifies the link (x_1, x_2) as intolerable at this moment. **In real life, the driver must avoid this link at this moment**.

3 Simulation and Discussion

All proposed models in this work are based on the weight average $\overline{\omega}_n(l)$ given some link l and instant n. It is crucial to think to validate proposed models by simulation and also to introduce the theorem on which the models are based. This section focuses on the simulation strategy explanation and also the proposed key performance indicators to validate the work.

3.1 Simulation Strategy

The graph structure must be defined before running any simulation algorithm on it. It is easy de define this structure for some special cases, but very hard on the general case because the simulation results could not be generalized for any other graph. Let suppose a graph what ever it is, the motion from a first node (let call it N_s to say the start node) to an other node (so called N_e to say the ending node) must be done by visiting an intermediary node (let call it N_i to say a node having the index i). We have defined by this, a subgraph witch must be included in any complex other graph: it is the one composed by a starting node N_s and an ending node N_e and a set of intermediary nodes $\{N_1, ..., N_k\}$. The figure 1 shows proposed subgraph that is used in the simulation:

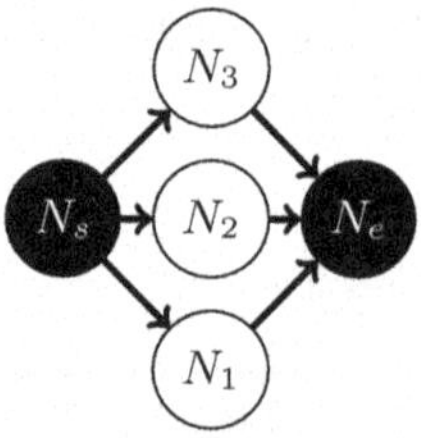

Fig. 1 Proposed elementary graph for the simulation

In all possible graphs, we can find a sub-graph having the proposed forme in 1 except the number of intermediary nodes that can be different. There is a special case that worth to be discussed: it is the case on which we have only a starting node N_s connected directly with the ending one N_e without any intermediary nodes.

In this special case there is no need to use this framework because of the absence of a dilemma: the presence of multiple intermediary nodes make the decision making harder. However, the direct link makes the decision evident because we have just one choice. That is the reason why this case is ignored.

The aim of the simulation is to compute key performance indicators in order to compare the proposed approach performance and some other classical approach. That is the reason to define a simulation scenario. We suppose that we make two trips starting from the starting node N_S of the elementary graph 1 to reach N_e of the same graph. The first trip must reach the destination N_e by passing from an intermediary node that must be **chosen randomly**. Let suppose a uniform distribution on the intermediary nodes set $\mathbb{X} \setminus \{N_s, Ne\})$. The choice of uniform distribution is done to give the same chance to all nodes, other distributions can be used also. The second trip must also reach the destination N_e from N_s but this time the choice of the road must be done intelligently based on the proposed learning process in this work.

In this simulation, the previous scenario is done hourly during one day but after a learning period of 1000 days. For each hour, we compute the cost for each trip. Let define N_{rc} as the randomly chosen node for the classical trip and N_{ic} the intelligently chosen node for the trip driven by proposed approach. For each hour h, the simulation program computes the cost for the random trip so called $C_{rc}(h)$ and the one for the intelligently driven trip denoted $C_{ic}(h)$. By cost we mean, the sum of weights of the two links of the trip (N_s, N_{rc}) and (N_{rc}, N_e) for random trip and (N_s, N_{ic}) and (N_{ic}, N_e) for the intelligent one. For each hour h we define also the gain KPI that represents the cost economy that we can have by adopting the intelligent approach. This gain is the difference of tow costs given formally by $C_{rc}(h) - C_{ic}(h)$. Let add that this quantity can be lost but in some very few cases because this is a stochastic process. For that reason we have defined the cumulative cost in order to see the final result. Is is formally given by the equation 7:

$$G(h) \triangleq \sum_{k=0}^{h} [C_{rc}(k) - C_{ic}(k)] \tag{7}$$

Let summarize the simulation steps as the following:

1. *Choose Randomly a Node:* $N_{rc} \leftarrow \mathcal{U}(\mathbb{X} \setminus \{N_s, Ne\})$.
2. *Compute the Cost:* $C_{rc}(h) \leftarrow \omega(N_s, N_{rc}) + \omega(N_{rc}, N_e)$.
3. *Choose Intelligently a Node:* $N_{ic} \leftarrow \arg\min_{l \in \mathbb{X} \setminus \{N_s, Ne\}} \overline{\omega}_h(l)$.
4. *Compute the Cost:* $C_{ic}(h) \leftarrow \omega(N_s, N_{ic}) + \omega(N_{ic}, N_e)$.
5. *Compute Cumulative Gain:* $G(h) \leftarrow G(h) + [C_{rc}(h) - C_{ic}(h)]$.

3.2 Simulation Algorithm

In order to compute previous KPIs, the simulation program have to generate a random graph for each hour of each day of the simulation. For that reason, this section presents an algorithm that generates a random graph (graph's weights matrix). This operation is done by generating random values for each link as a weight. The gaussian low is adopted in this generation with an expectation $\mu(h)$ and variance $\sigma(h)$ that depends on the observation's hour. The choice of parameterized expectation and variance is used to control the traffic - represented by weights - for some hours of the day. That makes the simulations more realistic. See the algorithm 1:

Algorithm 1. Random Graph Generator Algorithm

Input: h : Observation hour.
Output: (ω_{ij}) : Weights Matrix

for $i = 1$ **to** $Card(\mathbb{X})$ **do**
  for $j = 1$ **to** $Card(\mathbb{X})$ **do**
    Next node choice
    $\omega_{ij} \leftarrow X \sim \mathcal{N}(\mu(h), \sigma^2(h))$

return (ω_{ij})

Monte Carlo methods are a broad class of computational algorithms that rely on repeated random sampling to obtain numerical results as in [10]. They are often used in physical and mathematical problems and are mainly used in three distinct problem classes: optimization, numerical integration and estimation... Those simulations are very useful when it is difficult or maybe impossible to have a deterministic model to compute desired parameters. The main idea is to write the parameter to compute as an expectation expression that we can approximate with an average [11]. The algorithm 2 is designed to implement Monte Carlo principles in order to compute introduces KPIs in the previous sub-section:

Algorithm 2. Monte Carlo Algorithm for Graph's Knowledge Extraction

Output: $(\mathbf{C}_{rc}, \mathbf{C}_{ic}, \mathbf{G})$

for $i = 1$ **to** n **do**
  for $h = 0$ **to** 23 **do**
    Next node choice
    $N_{rc} \leftarrow \mathcal{U}(\mathbb{X} \setminus \{N_s, Ne\})$
    $N_{ic} \leftarrow \arg\min_{l \in \mathbb{X} \setminus \{N_s, Ne\}} \overline{\omega}_h(l)$
    Daily cost for two approaches
    $C_{rc}(h) \leftarrow \omega(N_s, N_{rc}) + \omega(N_{rc}, N_e)$
    $C_{ic}(h) \leftarrow \omega(N_s, N_{ic}) + \omega(N_{ic}, N_e)$
    Cumulative gain
    $G(i) \leftarrow G(h) + [N_{rc}(h) - N_{ic}(h)]$

return $(\mathbf{C}_{rc}, \mathbf{C}_{ic}, \mathbf{G})$

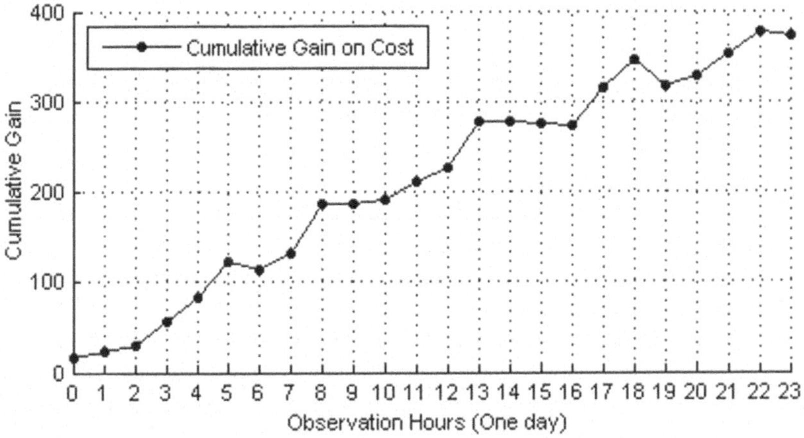

Fig. 2 Cumulative gain evolution during the 24 hours of the simulations day

Table 2 Intelligent and random trip cost comparative study.

Hour (h)	Cost ($C_{rc}(h)$)	Cost ($C_{ic}(h)$)	Satisfaction
00:00 - 00:59	156.00	140.00	✓
01:00 - 01:59	171.00	163.00	✓
02:00 - 02:59	148.00	142.00	✓
03:00 - 03:59	164.00	138.00	✓
04:00 - 04:59	164.00	138.00	✓
05:00 - 05:59	179.00	138.00	✓
06:00 - 06:59	147.00	157.00	✗
07:00 - 07:59	167.00	149.00	✓
08:00 - 08:59	194.00	139.00	✓
09:00 - 09:59	141.00	141.00	✓
10:00 - 10:59	147.00	142.00	✓
11:00 - 11:59	165.00	146.00	✓
12:00 - 12:59	154.00	138.00	✓
13:00 - 13:59	189.00	138.00	✓
14:00 - 14:59	144.00	144.00	✓
15:00 - 15:59	161.00	163.00	✗
16:00 - 16:59	140.00	142.00	✗
17:00 - 17:59	187.00	144.00	✓
18:00 - 18:59	167.00	138.00	✓
19:00 - 19:59	144.00	171.00	✗
20:00 - 20:59	148.00	138.00	✓
21:00 - 21:59	171.00	146.00	✓
22:00 - 22:59	176.00	152.00	✓
23:00 - 23:59	162.00	167.00	✗

3.3 Simulation Results

The aim of this section is the proof, by simulation, that the intelligently driven trip gives low cost than the random one. The figure 2 shows obtained results after computing the gain KPI defined in 7 and this shows that proposed approach is considerably more economical. See the figure:

The table 2 contains the observation hour and the cost values for the two trips. The satisfaction column shows that the intelligent trip is more economical or not. When the intelligent trip cost is less than the other we symbolize by a the symbol ✓ and the sign ✗ in the other case.

During 24 hours of the simulation, the number of satisfied trips is 19 in 24 trips in total. That gives a satisfaction average of 79.19% for just one day. The learning process for many other days will contribute to increase the satisfaction percentage.

4 Conclusion

In this paper we have proposed mathematical definitions for a complex networks knowledge extraction. That finds large domains of applications such as Smart Cities field. The work focuses on the application of learning machines to extract relevant information from a graph representing traffic roads in a big city. We believe that proposed framework has the potentials to build smart IT systems for new generation of cities. Many optimizations and enrichments are suggested in next works such as the use of cartographical coordinates to localize the node and some spatial statistics for more amazing features.

Acknowledgment We are grateful to anonymous reviewers for their constructive comments which improved significantly the quality of the article.

References

1. Khan, Z., Anjum, A., Kiani, S.L.: Cloud based big data analytics for smart future cities. In: Proceeding of the IEEE/ACM 6th International Conference on Utility and Cloud Computing, Dresden, Germany, December 2013
2. Fienberg, S.E.: Introduction to papers on the modeling and analysis of network data. The Annals of Applied Statistics **4**(1), 1–4 (2010)
3. Fienberg, S.E.: Introduction to papers on the modeling and analysis of network data II. The Annals of Applied Statistics **4**(2), 533–534 (2010)
4. Sahneh, F.D., Scoglio, C., Van Mieghem, P.: Generalized Epidemic Mean-Field Model for Spreading Processes Over Multilayer Complex Networks. IEEE/ACM Transactions on Networking **21**(5), 1609–1620 (2013)
5. Zhang, J., Chen, C., Xiang, Y., Zhou, W., Xiang, Y.: Internet Traffic Classification by Aggregating Correlated Naive Bayes Predictions. IEEE Transactions on Information Forensics and Security **8**(1), 5–15 (2013)

6. Lv, Y., Duan, Y., Kang, W., Li, Z., Wang, F.-Y.: Traffic Flow Prediction With Big Data: A Deep Learning Approach. IEEE Transactions on Intelligent Transportation Systems **16**(2), 865–873 (2015)
7. Liebig, T., Piatkowski, N., Bockermann, C., Morik, K.: Route Planning with Real-Time Traffic Predictions. The Annals of Applied Statistics **4**(2), 533–534 (2010)
8. Dempster, A.P., Laird, N.M., Rubin, D.B.: Maximum likelihood from incomplete data via the EM algorithm. Journal of the Royal Statistical Society B **39**(1), 1–38 (1977)
9. Vapnik, V.N.: An Overview of Statistical Learning Theory. IEEE Transactions on Neural Networks **10**(5), 988–999 (1999)
10. Leite da Silva, A.M., de Resende, L.C., da Fonseca Manso, L.A., Miranda, V.: Composite Reliability Assessment Based on Monte Carlo Simulation and Artificial Neural Networks. IEEE Transactions on Power Systems **22**(3), 1202–1209 (2007)
11. Véliz, F.F.C., Borges, C.L.T., Rei, A.M.: A Comparison of Load Models for Composite Reliability Evaluation by Nonsequential Monte Carlo Simulation. IEEE Transactions on Power Systems **25**(2), 649–656 (2010)

Just in Time: A Social Computing Approach for Finding Reliable Answers in Large Public Spaces

Nasim Mahmud

Abstract While people are away from their regular social terrain, they are usually exposed to newer situations, where they often need to seek for information or help. Nowadays people can find information from the Internet by using smartphones even when they are traveling. However, for many real–life questions, the Internet is not a suitable source of a 'reliable answer,' especially when the information–need or the question is too context–sensitive. Furthermore, it is also difficult to compose a context–sensitive real–life question effectively to find suitable answers. Therefore, along with other reasons, such as, individual's ability or experience, people seek for help or assistance from other people. And most of the cases, they need personalized support which is tailored for a particular context.

With the recent growth of computer mediated online social networks, people can relatively easily ask their social peers for help. However, these networks are *not yet suitable* for composing questions with rich–media (e.g., with audio) and with contextual information. Aspects of a question (e.g., timeliness, demand for details) become much clearer when the context in which a question being asked is exchanged. In order to address this, a *Social Computing* system called *Just in Time* has been developed which is a context and social aware 'question–and–answer' system. It utilizes the users' context and social network to formulate a question. It helps the users to get answers from trustable social peers. A formative evaluation was conducted with a small number of users that used the system for two days. Some interesting side effects were observed, such as, users started using the system as a context–aware 'instant messaging' system. It showed that there is a clear benefit in sharing context in order to get relevant answers. And the result was inspiring for further development of social computing research works.

Keywords Social computing · Mobile social network · Context–aware · Q&A

N. Mahmud(✉)
Dynamics Lab, UCD Geary Institute, University College Dublin (UCD), Dublin, Ireland
e-mail: nasim.mahmud@ucd.ie
http://www.dl.ucd.ie, http://www.mahmud.be

E. Sabir et al. (eds.), *The International Symposium on Ubiquitous Networking*,
Lecture Notes in Electrical Engineering 366,
DOI: 10.1007/978-981-287-990-5_39

1 Introduction

When we face a newer situation or we are to solve any problem, usually we search for relevant and reliable information on how to tackle the problem. For tackling a problem there are different sources of information that people can utilize. People utilize libraries, journals, browse the Internet and so forth to fulfill their needs for information. People also turn to their social peers such as friends, families or colleagues when they need information, support or guidance to solve a problem or to a make decision. The process of finding required information and being able to identify the right piece of information that a person can rely on requires time and effort. In many situations both in work and in the context of a social life, it is often expedient to consult a knowledgeable person over searching for the information alone [16]. If we have a specific question from a field in which a friend or a colleague is an expert, it is more efficient to consult her or him directly [6, 15, 16].

In modern days, more people move from one place to another for various reasons (e.g., education, work, social.) This mobility causes them to seek for more and more context sensitive information. At the same time, the 'timeliness' factor becomes more crucial for travelers in terms of receiving an answer in time. This mobility of the people also leads to information relevance not only being determined by the *content* but also by *extrinsic properties,* related to users' current context and urgency.

Popular search engines are constantly updating their search–techniques to cope with the abundance of information as well as to retrieve the correct and most relevant result. Mainstream search engines (e.g., Bing and Google) have applied *social search* techniques to retrieve and rank search results using social network information. The web search provides way to find web pages that contains the information that one is looking for. However, there are other approaches followed by search engines that attempt to find 'people' who actually have an answer (such as Aardvark[1]). Most of these approaches assume that the users are searching the Web on a *stationary computing device.* The most modern search techniques often take the users' location information as a part of the context for filtering and improving the search. And those approaches do not take into account the user's real–life current *context,* leaving a very small information–filtering, and context–aware help and support for nomadic users. With the recent growth of computer mediated online social networks, people can easily 'repurpose' the online social networking sites to ask their social peers for help, in addition to their regular social and playful use of the system. However, that lacks the ability for sharing the context and rich media (e.g., audio.)

In this paper, a context and social aware 'question and answer system,' called *Just in Time* has been introduced. It is a social computing system that addresses a very basic problem rooted into our everyday social life, the lack of 'common understanding of the context of a query' being shared among the people who are communicating from spatially different places. The system uses mainly audio–visual messages for placing a question. It also enables live text–chat facility and utilizes the users' context and social network to formulate a question. All these together help the user to get

[1] Aardvark, http://vark.com/ - Acquired by Google

an answer from a reliable social peer. The prototype system has been developed for Android based mobile phones as it utilizes number of available sensors (e.g., GPS) that are easily accessible on Android devices. The system receives users' availability status from users' calendar.

We describe our design of the system and grounding reasons behind our approach that helped us develop a useful and necessary system. In this extended abstract we present the primary qualitative results of a formative user test of two users (and an actor) that used the system for two days.

2 The Intended User Experience

This scenario is adapted from a pilot study; it helps elucidate the primary capabilities of the *Just in Time* system.

> *Noah is a PhD student. He is working in his research laboratory on a usual office day. He is looking through an electronic microscope lenses to a 'cultivated rat brain' that he has been working on for about two years. Today he has found something exciting, but he was not totally sure about his findings. So, he wanted to verify what he has found right now is really what he was looking for.*
> *First Noah wanted to search the Internet again for more recent information from existing literature. Soon he realized that his query is so problem specific that it is not easy to find proper and reliable information on the Internet. He felt the urge to discuss it with someone reliable and knowledgeable. He felt that it is best to ask professor John Doe from the same lab. May be John is available, however, he was not at his desk. Most probably a senior researcher can also help Noah, but Noah does not know who will be interested and interruptible at this moment for such a query.*
> *Noah turned to his Just in Time system for asking his question. Using the system he took a picture of the 'slide' from the microscope monitor, it was not OK on the first attempt; he took a second picture and then asked his question. He decided to disclose his location detail and marked the question as 'urgent.' He has also set the mode of communication as 'voice call' that means while replying the this request, a help–provider can initiate a voice call. Then he sent his query to a 'selected group' of researchers who are working on closely related topic. His professor was available and was notified by his Just in Time system. Moments later, John received an answer from his professor who was taking coffee in the coffee corner before going to give his lecture.*

This scenario shows five of the basic capabilities that has been built into the *Just in Time* system. The *Just in Time* system can be used to:

- Compose a question with a picture of the subject of interest and details of the question as audio
- Select a group of users from the social network of the user to send the question
- Adjust the location detail according to the user's preference
- Select the urgency of the question and
- Select the preferred mode of communication such as reply in text, voice message or phone call

3 The *Just in Time* Design and Implementation

The *Just in Time* system was designed with the intention to make the whole system completely mobile. That means, considering the possibility that while both the help–seeker and the help–provider are on the move, the system must be equally useful. There are question and answer systems that do not take the users' mobility into account, such as commercial online question and answer sites, for example 'Yahoo! Answers' and 'Google Answers' (closed since December 2006). Some non–commercial systems such as VizWiz [5] considers mobility partially into account. It assumes the help seeker is mobile whereas the help–providers provide their responses using standard computer.

Online social network of a user is often more diverged than the traditional social network. There is additional challenges, especially while dealing with trust in online social interaction. Social trust is a fuzzy concept with its different unmeasurable components [11]. People have different views on their friends or peers on online social network. Often the interaction pattern and the boundary between personal and public on online social network interaction is complicated [4, 8]. In order to alleviate this problem, while supporting the users to post a query easily, the *Just in Time* system suggests a number of group of users for posting the query, however, it allows the user override the decision if required (See figure 1(c).)

The system is designed to help find answers from within the social network of the user. This is how it ensures that the answer or the recommendation is coming from a reliable source, as opposed to commercially crowdsourced system. The system keeps track of the users' availability and interruptibility status utilizing the calender and available to–do list. So that it can properly notify a user about an incoming request for help. The *Just in Time* system is designed to help the users to automatically dispatch context information such as current location, current heading, users' social network information utilizing the Friend of a Friend (FOAF)[2] profile, urgency and so on, so that the question being asked is context rich to the answer providers. Context plays a crucial role to convey the question being asked in a meaningful way.

[2] Friend of a Friend Project. http://www.foaf-project.org/

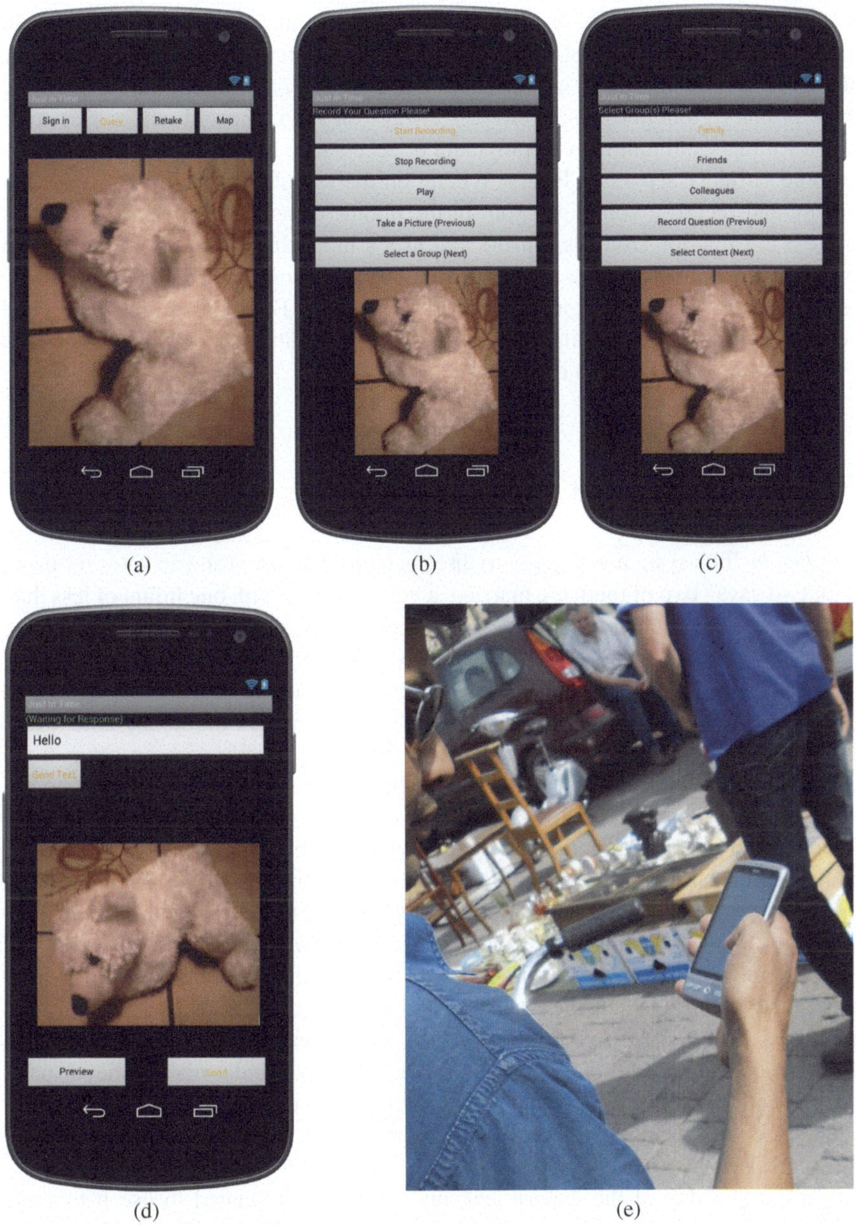

Fig. 1 The steps involved in using the *Just in Time* system. 1(a) The user takes a picture using the camera. 1(b) Shows the voice recording screen. The user can ask a voice–question along with the picture. 1(c) Shows the group selection screen. A group is consists of users of a particular category, e.g., close–friends. 1(d) Shows that the user has just sent a question together with the context of the question and rich–media (e.g., audio, image) and then waiting for a reply and 1(e) Shows a user is composing a question using the system.

In order to ask for help, or provide help to others, any system should deal with three basic aspects. It should *a)* run always, *b)* it should preserve users' privacy and *c)* it should be viable to use [17]. The *Just in Time* system conforms these three primary guidelines, as it provides always-on service, avoid real or perceived privacy threats and minimize deployment or acceptance barriers. The system works in three virtually separated layers. Where, on the top of the system the user takes picture and record a voice question. In the middle–layer of the system, it collects context information from its sensors and user's input. In the back–end the system communicates and negotiates to find an appropriate person or a group of persons who can help within the mobile social network of the user. For the communication and negotiations purposes, it utilizes a mobile 'social computing' supporting framework which was developed based on an earlier work [14].

4 Users and the User Evaluation

The *Just in Time* system was given to three persons, two users and an actor for using it for two days. Two of them are married, a young couple with one infant of less than two years of age. The young couple often chose to refrain from going out together. Instead, when necessary, one of the parents stay home to take care of the kid while the other goes out to carryout regular tasks such as buying grocery. Occasionally, one of them takes the infant out while the partner handles the house hold work. The third user, who was a family friend of the young couple, was an actor in the experiment. The participants have received thirty minutes of training on how to use the system and they were allowed some time to play with it, so that they become familiar with the basic functionality of the system. Then each of them was asked to compose and ask a question using the system to ensure that they are familiar with the basic functionalists of the system. Initially it was planned to give the *Just in Time* system to them for using it for a day for the formative evaluation. It was a usual Friday. After using for a day, we interviewed them for thirty minutes. However, after using the system for a day, the two users (the young couple) found it interesting and expressed their wish to use it for the next day. On the next day one of them had a plan to visit a traditional European Saturday market. The particular market they usually visit, is not only a place for many to sell their 'not so used any more' products but also a place for socializing. The young couple were fond of visiting that Saturday market and they try to visit almost every week whenever the weather permits.

On the first day of the experiment, one of the users wanted to use it to share his office environment with his friend (the actor) and his wife. Instead of asking a question, which was basically expected within frame of the experiment, surprisingly, he chose to share his 'experience' from his office environment. He shared a picture of his office wall that was full of many pictures of his dear ones, together with his voice message, he was joking, "Whose office is this?" That day, the user also wanted to share some pictures of 'slides' from a microscope in his laboratory with his wife, even though his wife was not an expert on it. In the interview he has explained the reason behind sharing pictures from his research work. He argued that he likes to share interesting moments of his life with his dear ones. Its fun and his wife also likes it.

On the second day, one of the users went to the Saturday market, and there he found a red toy–car and a teddy bear. He was hesitant whether he should buy either one or both of them for his son. He decided to ask his wife for her opinion. He turned to his *Just in Time* system and took a picture of the toy and asked his question (See figure 1). Both users found the system interesting because it gave them a 'feeling connected' sense by sharing spontaneous moments without interfering other's work.

5 Related Work

Asking questions and seeking for help from others is one of the most common and well established ways of solving problems in a social environment. When looking for expert knowledge, people usually seek help from their personal social network [7, 13]. Systems that support people to share their knowledge or expertise have been a research topic within Computer Supported Collaborative Work (CSCW) in recent years. So far this field of research has been largely studied within organizational settings (e.g., employees help seeking behavior) or online communities [1] and these also have been explored in a series of CSCW studies [3]. Systems that utilize social networks have also been studied [10, 12]. These systems attempt to leverage social networks within an organization or community to help find and locate suitable persons who can help. However, while many systems have been built, there is still a lack of exploration in context–aware help system that takes both users' context and social whereabouts into account.

Community based question answering site has recently become popular, where a member can post a question on online and another member may reply to the question. In this kind of question answering site, usually the users receive reward points for answering questions. Popular sites, such as 'Yahoo! Answers' offers free service whereas 'Google Answers' used to impose fees for asking questions. Use of similar question answering site is one of the popular alternatives for searching answers for those subjective questions that usually search engines (such as Bing, Google, Yahoo!) cannot help with. Search engines can not help to find useful solution when the problem is hard to define.

Answer Garden [2] is one of the earliest cooperative tool, providing an expert locating system that is designed to help improve 'organizational memory' by providing database for commonly and frequently asked questions. Whereas, new questions are automatically routed to appropriate experts. Answer Garden is not the first one of its kind, but it showed how relatively simple combination of well–known concepts can provide a platform for a new kind of cooperative work application in form of a question answering system. Answer Garden system is designed for stationary computer users and does not utilize any kind of context information.

CityFlocks [6] is a newer and context–aware mobile system that enables nomadic visitors or new residents in a city to acquire knowledge about the city from the local residents. It also allows the users to share their experiences with the local residents and other users by digitally annotating, commenting and rating any artifacts in the city. It specifically aims to lower the existing access barrier for information. However the system does not consider the contexts of the information seeker and information provider.

VizWiz [5] is one of the most recent question answering system. It allows a blind person to recruit remote sighted workers (i.e., crowdsource) to help him with visual problems. The sighted persons who are recruited by the blind person for answering a question are provided by workers on Amazon Mechanical Turk[3]. It is a picture based mobile question and answering system for blind persons that allows posing a question to a potential answer provider together with a picture of an object of interest. The answer provider remains physically apart from the scene of the question.

VizWiz system is a partially mobile system, meaning that half of the system rely on traditional stationary computer systems. The recruited persons who are going to provide answer of any question asked by the user, typically are replying remotely from a stationary computer. VizWiz does not take into account the context of the user who is asking the question, nor the context of the sighted person helping or answering the question. Moreover there is no social relation between the person asking the question and the person replying, making the answer an issue of *reliability and comfort.*

6 Discussion

The scenario presented in this paper, reveals a work environment where a colleague was benefited from a trustable acquaintance. The user test reveals several other situations that are not work–environment rather in social settings. This way, we present the usefulness of a context and social aware multi–modal communication system namely, *Just in Time* system. It is useful both in work and in social settings from small scale e.g., inside the university campus to a large scale public spaces, e.g., in a city environment.

Research in the social search space has focused on the informational benefits of collaboration with others during web and workplace information seeking. However, social interactions, especially during complex tasks, or decision making process can have cognitive benefits as well [9]. This research also shows that the users uses the context and social aware communication systems to harness help in a ‘help–me–best–help–me–fast’ situation. They also used it to exchange context and question whenever they want to take a collective social decision such as, whether or not buying a particular toy for the kid. In this experiment, the users were expected to use the *Just in Time* system to seek help when needed. However, some interesting side effects were observed, the users simply and conveniently started using it to share

[3] Amazon Mechanical Turk, http://www.mturk.com/

their whereabouts in rich media, which is not necessarily a query or help–seeking situation. In the interview they expressed that they did it, because they wanted to share, and it was convenient to do so with the *Just in Time* system.

The web is not situation–aware, and real life scenario is yet too dynamic to be incorporated and made available for being searched through online search. Moreover, the structure of the web is as such that it does not take the context of use into account. That is why a peer to peer interaction is necessary for many reasons. Not only for certainty, comfort and immediate feedback but a peer to peer interaction is sometimes necessary for people with different range of abilities. This is another reason why we often prefer to ask another human being over web based information. And here we successfully strive the effectiveness of our proposed social computing system.

Acknowledgments The author wants to thank the participants and the colleagues from the Dynamics Lab, UCD Geary Institute, University College Dublin (UCD), Ireland as well as the colleagues from the HCI Lab, Expertise Center for Digital Media (EDM), Hasselt University, Belgium.

References

1. Ackerman, M.S.: Augmenting organizational memory: a field study of answer garden. ACM Trans. Inf. Syst. **16**, 203–224 (1998)
2. Ackerman, M.S., Malone, T.W.: Answer garden: a tool for growing organizational memory. In: Proceedings of the ACM SIGOIS and IEEE CS TC-OA Conference on Office Information Systems, COCS 1990, pp. 31–39. ACM, New York (1990)
3. Ackerman, M.S., Wulf, V., Pipek, V.: Sharing Expertise: Beyond Knowledge Management. MIT Press, Cambridge (2002)
4. Baym, N.K., Boyd, D.: Socially mediated publicness: an introduction. Journal of Broadcasting & Electronic Media **56**(3), 320–329 (2012)
5. Bigham, J.P., Jayant, C., Ji, H., Little, G., Miller, A., Miller, R.C., Tatarowicz, A., White, B., White, S., Yeh, T.: Vizwiz: nearly real-time answers to visual questions. In: Proceedings of the 2010 International Cross Disciplinary Conference on Web Accessibility (W4A), W4A 2010, pp. 24:1–24:2. ACM, New York (2010)
6. Bilandzic, M., Foth, M., De Luca, A.: Cityflocks: designing social navigation for urban mobile information systems. In: Proceedings of the 7th ACM Conference on Designing Interactive Systems, DIS 2008, pp. 174–183. ACM, New York (2008)
7. Bonnie, S.W., Nardi, A., Schwarz, H.: It's not what you know it's who you know. First Monday **5**(5), May 1, 2000. http://firstmonday.org/htbin/cgiwrap/bin/ojs/index.php/fm/article/view/741/650 (accessed: April 7, 2011)
8. Boyd, D.: Social network sites as networked publics: affordances, dynamics, and implications. In: Networked Self: Identity, Community, and Culture on Social Network Sites, pp. 39–58 (2010)
9. Evans, B.M., Kairam, S., Pirolli, P.: Do your friends make you smarter?: An analysis of social strategies in online information seeking. Information Processing & Management **46**(6), 679–692 (2010)
10. Foner, L.N.: Yenta: a multi-agent, referral-based matchmaking system. In: Proceedings of the First International Conference on Autonomous Agents (Agents 1997), pp. 301–307. ACM Press (1997)
11. Golbeck, J.: Computing with social trust. Springer Science & Business Media (2008)
12. Kautz, H., Selman, B., Shah, M.: The hidden web. AI Magazine **18**, 27–36 (1997)

13. Mahmud, N.: Exploiting Context-Awareness and Social Interaction to Provide Help in Large-Scale Environments. PhD thesis, Computer Science, April 2012
14. Mahmud, N., Luyten, K., Coninx, K.: Context aware help and guidance for large-scale public spaces. In: Proceedings of the 2009 Fourth International Workshop on Semantic Media Adaptation and Personalization, SMAP 2009, pp. 105–110. IEEE Computer Society, Washington, DC (2009)
15. McDonald, D.W., Ackerman, M.S.: Just talk to me: a field study of expertise location. In: Proceedings of the 1998 ACM Conference on Computer Supported Cooperative Work, CSCW 1998, pp. 315–324. ACM, New York (1998)
16. McDonald, D.W., Ackerman, M.S.: Expertise recommender: a flexible recommendation system and architecture. In: Proceedings of the 2000 ACM Conference on Computer Supported Cooperative Work, CSCW 2000, pp. 231–240. ACM, New York (2000)
17. Smith, I., Consolvo, S., Lamarca, A., Hightower, J., Scott, J., Sohn, T., Hughes, J., Iachello, G., Abowd, G.D.: Social disclosure of place: from location technology to communication practices. In: Pervasive Computing, pp. 134–151. Springer (2005)

A Decision Approach to Select the Best Framework to Treat an IT Problem by Using Multi-Agent System and Expert Systems

A. Chakir, M. Chergui, S. Elhasnaou, H. Medromi and A. Sayouti

Abstract This work is registered in two disciplinary axes that are the decision making system, and the practices of the IT GRC. Many organizations deployed integrated the practices of the IT GRC, the problem that arises it is how to choose the good one practices to satisfy a precise need. Our work is motivated by the need to make decisions by understanding and by incorporating perceptions, decisions and actions to make the best choice. The objective of the research is to build a decision-making model to satisfy a precise need IT. The proposed approach bases on three main stages to set up a decision-making model. The model takes in entrance the strategic needs, the first stage consists in reducing the size of the problem by dividing it into many problems, by basing itself on the mapping between all the reference tables and methods of the GRC and also this stage is going to allow us to assure the sequencing of these under problems according to the variables of the environment as for example the type of the organization. In the second stage, it is a question of formalizing every under problems according to the criteria stored in the datawarehouse to generate the best choice of the good practice by using methods of aggregation multi criterion to satisfy the need IT. The third stage consists in estimating the satisfaction IT and helps to make decisions at the level of every chosen reference table.

Keywords Expert system · IT governance · Decision method · Data warehouse

1 Introduction

Three axes of the IT, Governance, Risk, and Compliance (GRC), assures the alignment of the objectives of company with regard to the needs for stakeholder,

A. Chakir(✉) · M. Chergui · S. Elhasnaou · H. Medromi · A. Sayouti
EAS Team, LISER Laboratory, ENSEM, Casablanca, Morocco
e-mail: {aziza1chakir,chergui.meriyem,elhasnaoui.soukaina,sayouti}@gmail.com, h.medromi@ensem.ac.ma

E. Sabir et al. (eds.), *The International Symposium on Ubiquitous Networking*,
Lecture Notes in Electrical Engineering 366,
DOI: 10.1007/978-981-287-990-5_40

conditions and options putting the management. This is shown by the strategic progress of a given organization and also by the taken decisions [1].

The variety of methodologies, standards, and best practice of the IT GRC [2] puts the persons in charge of Information system in front of a problem of joint of these reference tables to reach precise goals, especially as there are practices of the IT GRC which aim to be global and which handle all the fields of activity of the IT without being detailed enough and those who handle a domain in particular in detail without a global view. The diversity of the practices of the IT GRC puts difficulties for companies to make the adequate choice of the practices of the IT GRC. It's necessary to implement a decision-making model which assures the selection of the best solution of the GRC.

The request expressed by nature strongly qualitative, subjective and rich in thought of the users will be translated into quantitative and formal data to be exploited by the decision-making system with the supreme purpose to improve the learning and the communication by endowing our system by a system multi-agent and an expert system.

To exploit well, and to increase the utility of the GRC, especially for the not experts, the decision-making system is going to allow the end users to see the relevance of the chosen reference table.

2 State of the Art

2.1 GRC IT

The approach Governance, Risk and Compliance (GRC) allows to master an organization with the minimum of effort by staying in compliance with its internal politics and with the external regulations by assuring one strategic alignment and an effective improvement of its processes and its projects [3][10].

Three aspects GRC allows the companies to check well their activities. What translates by:

- A good control of their processes and their projects
- An effective governance of their organization
- A good management of risks associated to their activities

The internal control and the external statutory requirements of an organization, recently, are the object of an increasing attention where from the necessity of integrating three aspects IT (GRC). The following plan shows the collaboration of three pillars of the GRC within an organization to assure that it respects its objectives.[9]

3 Decision-making Systems

The decision-making systems give to the decision-makers, in a transparent way, the possibility of reflecting about their practices, of binding explicitly actions and effects and of seeing the reasoning behind applied practices. This improves the perspectives for the learning, the negotiation, the implementation of new politics and to proceed has collective actions [16].

3.1 Data Warehouse

The data warehouse is a collection of sent, integrated, not volatile subject of data and historized, organized for the support of a process of decision-making [11] [12] [17].

The data warehouse it is only simple copied by the data of production, making a reference to the definition given by Ralph Kimball ''A datawarehouse is a copy of transaction data specifically structured for query and analysis.'' [14][15]

3.2 The Stages of a Methodology of Decision-making Support

The Formulation Multi criterion of a Problem of Decision

The formulation of the problem is a very important stage because she involves of identifying the problem in a explicit way, there partitioning the problem in many together restricts that possible to choose the best reference table (problem of choice) [20] and to tidy up the actions to be made according to the type of activity of an organization, the most important for the least important. According to the terminology introduced by Vansnick (1990) [20] the formulation multi criterion of a problem of decision can be defined as the model "A, A/F, E" where:

- A is the set of the potential actions. This group can be explicitly defined (finished), the constraints being implicit, or implicitly (generally infinity), the constraints being explicit. In this second case, we resort to the mathematical programming with multiple objectives (PMOM) and we often indicate all the acceptable actions by the symbol X;
- A/F is the set finished by the attributes or the criteria, generally conflicting, from which the actions will be estimated;
- and E is all the benchmarking of the actions according to each of the attributes or the criteria, that is all the vectors of performances, a vector by action.

Generally, this formulation allows to simplify the problem but she does not allow to handle the problem of decision, what requires the call to the methods multi criteria to release the preferences of an organization.

Types of the Problem in Multi Criterion to the Decision

There are four types of problem in help multi criterion to the decision problems of choices, problems of sorting, problems of arrangement and problems of description. [20][21]

- Problems of choice P.α

This type of problem consists in selecting a subset so restricted as possible of better solutions by eliminating the other solutions by basing itself on procedure of selection of the operational research.

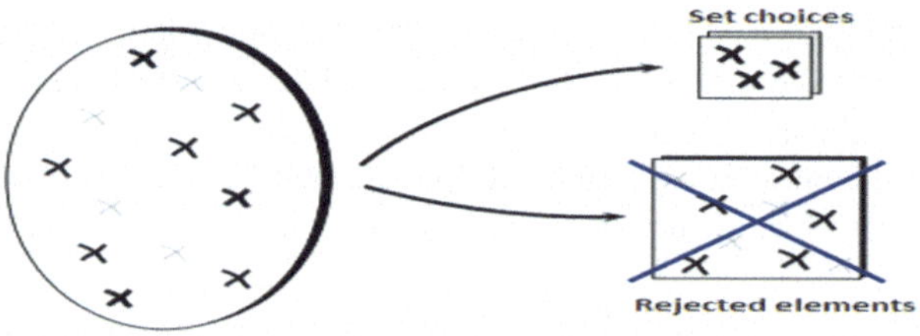

Fig. 1 Problems of choice

- Problems of sorting P. ß

This type of problem consists in allocating the actions to predefined categories by comparing the actions of the set with the reference actions, by basing itself on the procedures of affectation in categories.

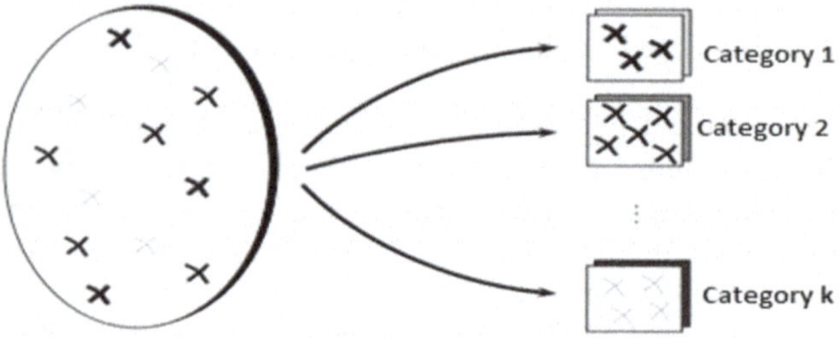

Fig. 2 Problems of sorting

- Problems of arrangement P.γ

In this type of problem, we order all the actions of the best the least good, it is a kind of affectation in classes of equivalence led by this preorder, orderly categories.

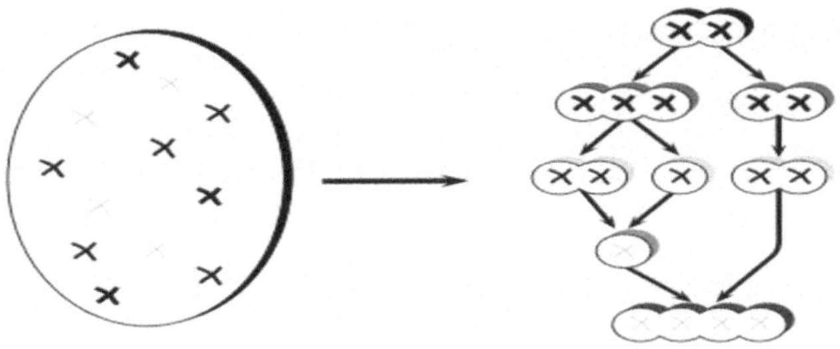

Fig. 3 Problems of arrangement

- Problems of description P.δ

This type of problem consists in determining all the potential actions by taking into account a set of parameters such as threshold of indifference and rather, level of pursuit … Generally this type of problem became an adequate type when the decision-maker does not manage to define the problem, or to express the type of result whom he would like to obtain. This type of problem is implemented by cognitive procedures.

Comparing with our problem which consists in turning the best reference table or the best reference tables, we opted in the procedures of selection.

Methods Multi Criteria

There are several types of methods multi criteria, the most known type it is the mathematical programming with multiple objectives PMOM. Most of its methods base themselves on the previous formalization especially if all the actions, defined by a set of the explicit constraints, is implicit to release or the best actions (in our case or the best reference) [19].

To answer the problem of the GRC IT and propose the best reference table or the best reference tables, it is necessary to apply the procedure or the adequate procedures.

- Example of method:[22]

ELECTRA I (Problems of choice P.α):This method has for objective to split a set of the actions, the reference tables, which contains the best alternatives among which the one is that the decision-maker will choose.

4 Expert Systems

It is the system which is capable of reproducing a reasoning by trying to analyze a problem as a human expert in a precise domain would make it. For example the failure detection, the medical diagnosis.

An expert system consists by:

- A knowledge base:
  - Base of Rules: model the knowledge of the considered domain.
  - Base of Facts: contains the information concerning the handled problem.
- An interference engine:
  - Argue from the information contained in the base of facts and in the base of rules
  - Capable of making deductions or inferences
- A user interface (and an interface with the expert)
  - Possibility of evolution of the expert system during the execution

5 Multi Agent System

- What is an agent?

An agent is an autonomous, real or abstract entity, which is capable of acting on herself and on its environment, which, in a universe multi-agents, can communicate with other agents, and whose behavior is the consequence of its observations, its knowledge and the interactions with the other agents [7] [8]. Experts multi-agent systems have classified agents into three major categories according to essential criteria that is the representation of its environment, and are therefore: Reagent agents, Cognitive agents and Hybrid agents.

- What is a multi agent system

A system multi-agents is a compound distributed system of a set of agents.

A SMA is so characterized:

- Every agent has information or capacities of resolution of limited problems (so, every agent has a partial point of view);
- There is no global control of the system multi-agents;
- The data are decentralized;
- The calculation is asynchronous.

- Why Multi-agent system?

The SMA presents a common point by report the governance IT which is the management by process, indeed this way of managing prepares perfectly with the governance IT which is nothing else than of a set of process which interact between them for a better management of information technologies.[11]

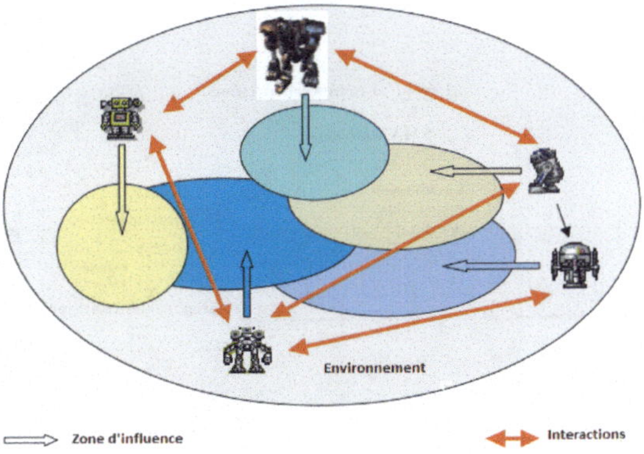

Fig. 4 Multi agent system

6 The Proposed Approach

6.1 Decision-making Model

The proposed approach consists in setting up a decision-making model which is going to assure the selection of the best reference table or the best reference tables according to the need or the strategic needs asked by an organization.

The decision-making model has two levels, the first level assures the choice of the best reference table or the best reference tables according to need IT and the second level is going to allow us to handle any kind of decision after the treatment of the need IT by basing itself on the chosen reference table, and it is going to allow us to assure the satisfaction of the applicant by basing itself on performance indicators communicated after every made treatment.

The following plan illustrates the proposed decision-making model:

- Level 1:

The first level sets up two layers, every layer has a precise feature has to assure.

The first layer "SMA sequencing" arranges two under layer "Categorization decision 1.1" and "Categorization decision 1.2".

The first one under layer "Categorization decision 1.1" has for objective to repeat the strategic needs according to a matrix of priority, which translates the mapping of the objectives IT expressed by Cobit [4] [6] and the other reference tables (ITIL[5], ISO 270001, ISO 270002, ISO 270005, PMBOK, CMMI) and the methods of the GRC (MEHARI, EBIOS…) And which arranges as information the classification of the reference tables of less detailed in the most detailed, this under layer allows us to make a joint between the matrix of the strategic needs IT and the matrix of the priority to produce a reduced matrix which will be handled by the second under layer of the decision-making model.

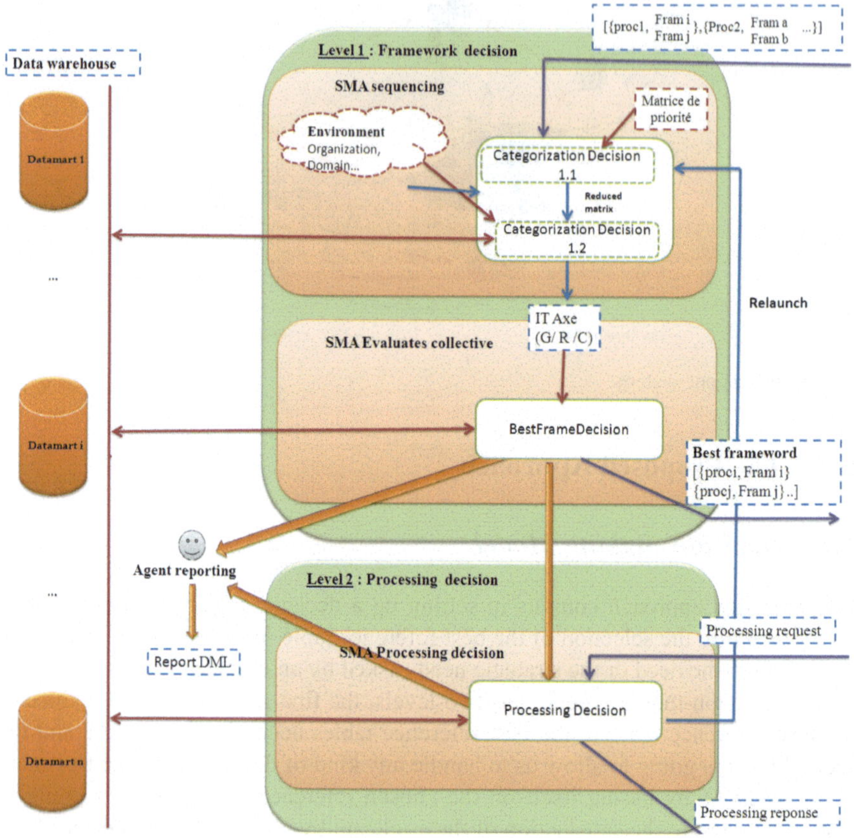

Fig. 5 Decision-making Model

The following plan illustrates the size of the matrix of the priority

	Reference and methods of the GRC IT				
Objectives IT	ITIL	CMMI	ISO 27005	ISO 27002	ISO 27001
Define a strategic IT plan	All, 1	All,2	Null	Null	Null
Make sure of the conformity with the obligations extern	Null	Null	All, 1	All,2	Null
Assure the safety of the System	Network management, 1	Null	Null	All, 1	All, 2

Fig. 6 Matrix of the priority

The second under layer "Categorization decision 1.2" takes in entrance the matrix produced by the first one under layer, the type of activity of an organization and the data stored in one dated warehouse to attribute an order number to needs IT to assure the sequencing of their execution by basing itself on the algorithms of the sequencing of the operational research.

The second layer "SMA Evaluates collective" takes in entrance the matrix produced by the first layer and treats every objective IT as one under problem. This layer formalizes every under problem by taking in entrance the set up reference tables, their versions, the certification or not employers the organization and it also takes the dimensions and the indicators stored in data warehouse IT as criteria to exploit any kind of information, to generate the best choice of the good practice IT GRC, by using methods of aggregation multi criterion to satisfy the need IT expressed in entrance and by setting up an expert system.

The following plan illustrates performance indicators proposed by date warehouse IT:

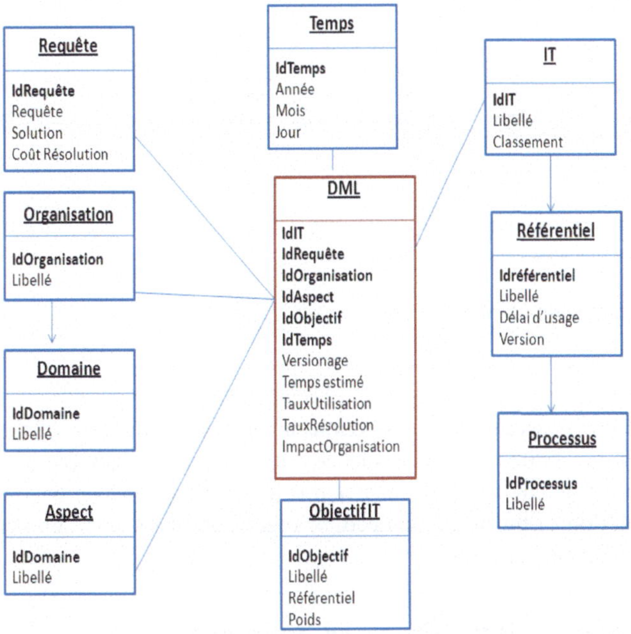

Fig. 7 The star schema data warehouse

The following plan illustrates the functioning of our expert system which is going to assure the collective expertise of the decision-making model.

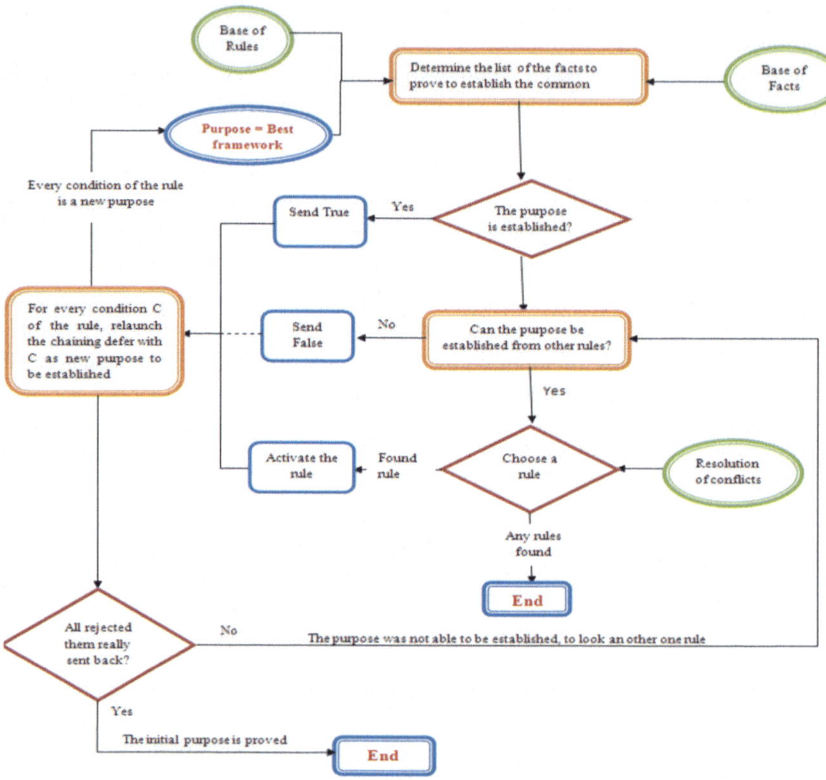

Fig. 8 The functioning of our expert system

- Level 2:

The second level assures the satisfaction or the not satisfaction of the choice by basing itself for example on the success rate, if the success rate is upper to a threshold thus it is OK for the choice and if it is not the case we have to see again the reformulation of the strategic need, or regenerate the second choice by basing itself on of other one criteria.

6.2 *Simulation of the Proposed Approach*

The following plan shows a simulation of our decision-making model which is going to take in entrance the following objectives IT:

- Objective 1: define a strategic IT plan
- Objective 2: make sure of the conformity with the obligations extern
- Objective 3: assure the safety of the System

The first objective can be handled by the reference ITIL which is a part of the axis governance (G) or by the reference CMMI which is a part of the axis governance (G).

The second objective can be handled by the reference ISO27002 which is a part of the axis risk (R) or by the reference ISO27005 which is a part of the axis Compliance (C).

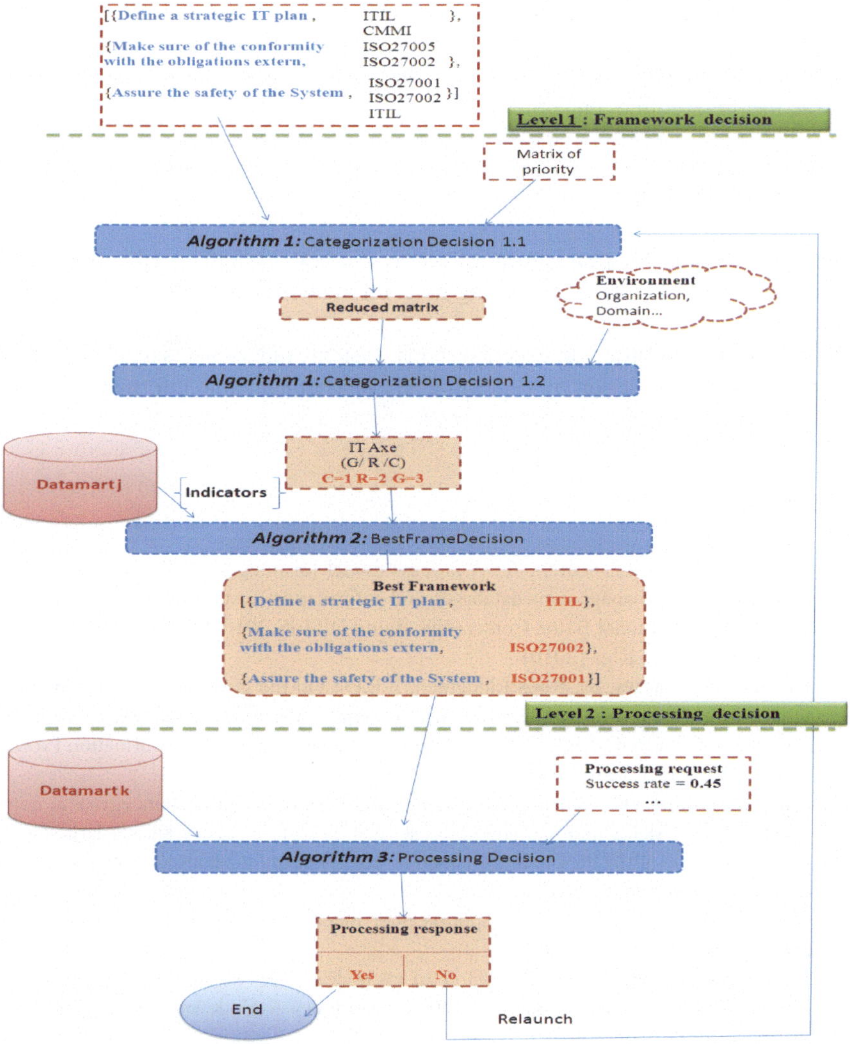

Fig. 9 Simulation of decision making system

The third objective can be handled by the reference ISO27001qui been a part of the axis risk (R) or by the reference ISO27002 which is a part of the axis (R) or by the reference ITIL which is a part of the axis governance (G).

The first level produced got out of it the best reference by objective IT, for the first objective gives as result ITIL and for the second objective gives as result ISO27002 and for the third objective give ISO 27001.

And the second level assures the satisfaction or the not satisfaction of the choice by basing itself for example on the success rate, if the success rate is upper to a threshold thus it is OK for the choice and if it is not the case we have to see again the reformulation of the strategic need, or regenerate the second choice by basing itself on of other one criteria.

7 Conclusion

This paper handles the problems of selection of a better reference table or the best reference tables of IT GRC by basing itself on methods for decision-making support multi criteria.

Our approach is based on expert systems, systems multi-agents and methods of aggregation multi criterion by exploiting any kind of available information by the organization to satisfy their strategic need.

References

1. Racz, N., Weippl, E., Seufert, A.: A process model for integrated IT governance, risk, and compliance management databases and information systems. In: Proceedings of the Ninth International Baltic Conference, Baltic DB&IS 2010, pp. 155–170. University of Latvia Press, Riga (2010)
2. Kooper, M.N., Maes, R., Roos Lindgreen, E.E.O.: On the governance of information: introducing a new concept of governance to support the management of information. International Journal of Information Management: The Journal for Information Professionals **31**(3), 195–200 (2011)
3. Racz, N., Panitz, J.C., Amberg, M., Weippl, E., Seufert, A.: Governance, risk & compliance (GRC) status quo and software use: results from a survey among large enterprises. In: ACIS 2010 Proceedings, paper 21 (2010). http://aisel.aisnet.org/acis2010/21 (retrieved 13 December 2010)
4. Stachtchenko, P.: COBIT 5, ses apports pour management et la gouvernance du SI, Janvier 25, 2013
5. Delbrayelle, Introduction à ITIL V3 et au cycle de vie des services, juillet 2011. ISO office, Information technology— Security techniques— Code of practice for information security management (2005)
6. ITGI and OGC, Aligning CobiT® 4.1, ITIL® V3 and ISO/IEC 27002 for Business Benefit (2008)
7. Ferber, J.: Les systèmes multi-agents, vers une intelligence collective. InterEditions, 63–144 (1995)

8. Shoham, Y.: Agent-oriented programming. Artificiel Intelligence. Stanford, USA, February 1992
9. Racz, N., Weippl, E., Bonazzi, R.: IT governance, risk & compliance (GRC) status quo and integration. an explorative industry case study. In: Proceedings of the 1st International Workshop on IT GRC, ITGRC 2011. IEEE, Washington (2011)
10. Chakir, A., Medromi, H., Sayouti, A.: La gouvernance du système d'information à base des bonnes pratiques d'ITIL V3. JDTIC, Novembre 2012
11. Chakir, A., Medromi, H., Sayouti, A.: Une approche multi-agents pour la gouvernance d'un data warehouse à base des bonnes pratiques d'ITIL. JDSIRT (2013)
12. Chakir, A., Medromi, H., Sayouti, A.: Actions for data warehouse success. Journal - IJACSA 2013 **4**(8), August 2013
13. Sayouti, A., Medromi, H.: Book Chapter in the book, Multi-AgentSystems - Modeling, Control, Programming, Simulations and Applications. InTech, April 4, 2011. ISBN 978-953-307-174-9
14. Kimball, R.: Concevoir et déployer un data warehouse. Edition Eyrolles (2000)
15. Chakir, A., Medromi, H., Sayouti, A.: An approach multi-agent with the best practice of ITIL, to maintain the operability of a data warehouse. IJAIS **6**(7), February 2014
16. Elsawah, S., Guillaume, J.H.A., Filatova, T., Rook, J., Jakeman, A.J.: A methodology for eliciting, representing, and analysing stakeholder knowledge for decision making on complex socio-ecological systems: From cognitive maps to agent-based models
17. AChakir, A., Medromi, H., Sayouti, A.: Actions for data warehouse success. Journal - IJACSA 2013 **4**(8), August 2013
18. Martel, J.-M.: L'aide multicritère à la décision: méthodes et applications. In: CORS - SCRO 1999 Annual Conference, Windsor, Ontario, June 7–9, 1999
19. Evans, G.: An overview of techniques for solving multiobjective mathematical programs. Management Sciences **30**(11), 1268–1282 (1984)
20. Roy, B.: Méthodologie multicritère d'aide à la décision. Ed. Economica (1985)
21. Henriet, L.: Systèmes d'évaluation et de classification multicritères pour l'aide à la décision: Construction de modèles et procédures d'affectation. Computer Science, Université Paris Dauphine, Paris IX, French (2000)
22. Roy, B.: ELECTRE III : Un algorithme de classements fondé sur une représentation floue des préférences en présence de critères multiples. Cahiers du CERO **20**(1), 3–24 (1978)

An Abstract Framework for Introducing Computational Trust Models in JADE-Based Multi-Agent Systems

Youssef Mifrah, Abdeslam En-Nouaary and Mohamed Dahchour

Abstract Agents in open distributed systems rely on their peers to achieve their goals. In such situation, the trust management become a critical subject that must be handled. Formalizing trust as a computer concept will help agents improving the decision making process and decrease the risk of being unsatisfied. This research aims to develop a practical framework that introduce computational trust models in multi-agent systems applications. The framework presented in this paper is based on an abstract structure that handle different trust models, and provide a ready to use development tool for JADE applications.

Keywords Computational trust · Reputation · Multi-Agents System · JADE

1 Introduction

Multi-agent systems are represented as the next generation of distributed architectures. They are described as a distributed systems that consist of multiple interacting intelligent agents, used together to solve problems which are difficult to be solved by an individual agent. Agent systems are usually extensible, heterogeneous and unpredictable. In the last two decades, multi agents system based architectures has converted to a more openness structures with less restriction on internal behaviors of agents that interact inside the system. Many systems in use today by millions of users provides such features. Peer 2 peer network, online games and e-commerce platform are a kind of open dynamic network where users could enter and leave the system dynamically. These characteristics introduce new challenges while modeling and developing multi-agent systems. Trust management in agent systems is one of the most critical challenges that faces researchers in this area. In human societies, trust presents

Y. Mifrah(✉) · A. En-Nouaary · M. Dahchour
Institut National des Postes et Télécommunications, Rabat, Morocco
e-mail: {mifrah,abdeslam,dahchour}@inpt.ac.ma

E. Sabir et al. (eds.), *The International Symposium on Ubiquitous Networking*,
Lecture Notes in Electrical Engineering 366,
DOI: 10.1007/978-981-287-990-5_41

an ubiquitous concept involved implicitly in any interaction. Whether it's a human-human interaction or human-machine interaction, as human, we implicitly estimate a value of trustworthy of the partner. In the virtual world, Multi-agent systems are an analogy of human societies. Trust and reputation management has become a topic of research due to the fast growing that knows distributed architectures and multi agent system. Intelligent agents are intended to achieves specific goals with a minimum or without human intervention. Autonomy's feature that characterize agents present an advantage for task automation and coordination. However, when an agent initiator delegate a task to an agent processor, the result obtained depend on the capacities, intention and willingness of the agent processor. To address this concern, researchers have looked for new techniques and concepts that would improve agents interactions and help them achieve their goals in an efficient way, by proposing models that integrate trust and reputation management. With a trust model, agents can associate to their partners an evaluation value that estimate the level of trustworthiness of those partners. Each of the trust models, found in literature, presents its main components and the methodology to use for managing trust [1]. The multitude of trust models proposed and theirs various background theories make it difficult to choose the right model for a distributed system under development. On the one hand, the diversity of computational trust models make the user confused. Indeed, each model uses its specific terminology to describe elements that compose the model (trust, credibility, confidence..). Some models give a deep analysis about the trust management, and more elements that construct the trust than others. On the other hand, most of the trust models found in literature are not designed for developers of Agent Oriented Software Engineering. To address the previous limitations, we propose in this paper an abstract framework for modeling multi-agent systems trust, called GeFMAT, with an implementation in the JADE platform. This framework defines in an abstract level how to design a multi-agent system in an open environment. GeFMAT does not only give a structure of the system, but also how the workflow information is done, and how agents interact between each other and how to apply their trust metrics while managing trust and preparing for the phase of decision making.

2 Computational Trust Models

The process of formalizing trust and reputation can be done from different perspectives. Researchers propose models that handle agents concepts and introduce trust management from a social approach. Human societies is the most important source of inspiration for researchers in this field. But the way the trust assessments is managed differs from one trust model to another. There are many computational trust models in the literature, some of them are based on probability theory [2] [3] [4], others use some handcrafted formula and conditions to build trust evaluation [5]. Trust models can be classified following different classification dimensions. Sabater J. describe in [1], a survey of different classification. There is a classification based on the paradigm type, where models are classified as cognitive or probabilistic. Models that refers to the degree of belief and uncertainty and refers to a cognitive representation of trust are considered as cognitive model [6]. On the other side, models that use probabilistic algorithm without any aspect of cognitive representation are considered as probabilistic trust models. Another important classification of trust models related to this research and especially for the design of the framework is the trust representation, which was proposed by Patricia Vector et al. [7]. This classification focus on the representation of trust and whether it include the factor of distrust or not. Some models use a single value that represent the trust assessment [9], while others propose more than one value, such as Beta Reputation System [3] that formulate a trust assessment as a vector of tree value : belief, disbelief, and uncertainty, in addition to some constraints and formulas used while calculating those values. Guha et al. also propose a model that handle an assessment of trust and distrust, and put forward an algorithm for the propagation of those values inside the system. Those classifications shows how existing trust models differs from one another. The next section introduce an abstract framework that include trust management in multi-agent system. This framework helps users design and analyze multi-agent systems that integrate the concept of trust into the decision process.

3 The GeFMAT Framework

Many frameworks was proposed by researchers to design multi-agent systems by capturing common concepts used in such systems. Each of the proposed frameworks uses a specific domain model that design agents systems from different perspectives. Some of them are based on organizational and hierarchical perspectives, and include the notions of environment, hierarchy and role. Others focus on the interactional aspect of agent systems. There is also other works that combine between different perspectives. However no meta-model of the proposed framework handle explicitly to concept of trust. To introduce the concept of trust into multi-agent systems, we propose a meta-model for modeling multi-agent systems in open environment. This meta-model that captures the semantics of concepts of multi-agent systems involved in an open environment. It include also a set of components related to trust and reputation concepts. Figure 1 present the components of the meta-model and theirs relationships.

As presented in figure 1, The meta-model is in the form of a set of meta-classes with their relationships. Each meta-class represent a common concept used in existing multi-agent systems. The meta-class Agent presents an autonomous entity that communicates with others agents and provides one or more services. There is no constraint on the internally specification of the agent model. While designing a system, we specify features that will characterize agents in the system. An agent will have a list of features values. For example, in an e-commerce platform where some users could be represented by agents, each user has a profile composed of features (user name, country, registration time, experiences...etc). Those features constitute the agents profile, and they reflect its instance in the system. The Service

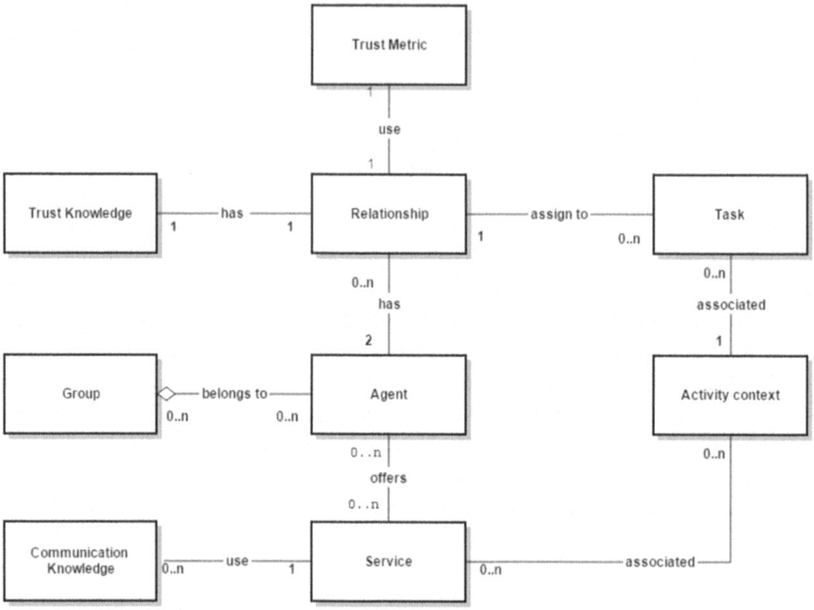

Fig. 1 The Meta-model of GeFMAT

meta-class, presents an expertise that an agent can propose to other agents to perform their tasks. An agent can provide one or more services. The service concept uses some communication knowledge, and it is associated to one or more activity contexts. Communication knowledge presents a policy of communication. This policy could be presented by an interaction protocol and knowledge used during the communication process. Its like how the deal between the agent and his partner will be confirmed, what requirements should be shared before task performing, and how the result will be received. Those concepts would be handled by this meta-class. The Activity context meta-class represent the boundary that group a set of common services. The Task meta-class presents the delegated task that agents exchange and delegate. It could present a request for service, or for information. A task is related to an activity context, which present the context of application of this task. The relationship meta-class present information that characterizes the relationship between two agents. It's a directional relation between an initiator agent and its partner. This relationship contains also passed experiences between those agents, the trust evaluation of the initiator about its partner, and the trust metric used. In general, if an initiator agent select a partner to process a task, then two new relationships will be created between them, one for the initiator which presents his relation to the partner, and another one for the partner which present his relation to the initiator. When the partner finishes task processing, the initiator will receive a result and use it with other parameters as an input to the trust metric to evaluate the trustworthiness of the partner, and then updates the associated trust knowledge. Trust knowledge is defined as the model that presents the trust assessments that an agent uses when evaluating his partner. Several works have been done to propose a formal presentation for trust. Some models present trust as a variable that takes one value from a finite enumeration space [11] while others present trust as a numeric variable that takes its value from a range [12]. There are also other models that present trust as a vector in a multidimensional space where we can set to each agents feature an evaluation value [20]. Others represent trust using subjective logic [3]. The approach of modeling trust using the subjective logic gives more control on the state of the unavailability of trust information (the case of a newcomer agent) because of the integration of the concept of uncertainty. The trust metric presents the logic applied to evaluate an agent. Each agent could associate a trust metric to one relationship between itself and his partner who has accepted a delegated task. For an autonomous agent interacting with other agents within the system, an agent could have a choice between many metrics. A group forms an aggregation of agents. Agents that are a parts of a group could share specific information, however, belonging to the same group is also an information in itself. A group could be specified by the designer statically while designing the agent architecture, or by agents at runtime. Those groups are dynamic, which means that members could change during the activities of the system, and at any time, the group should have an agent designated as the leader of the group. This form of aggregation would help agents to evaluate the trustworthiness of groups members more efficiently, especially when trust knowledge about the targeted agent is scarce or unavailable.

The presented meta-model reflect the structural part of the framework, the behavioral part is represented by a process model that define steps applied by agents to assess trust and improve the process of decision making. The framework adopts a workflow of information for the management of the trust. This workflow captures and presents in an abstract way how agents should manage information about the environment during the decision process, and how to use feedbacks after the decision process.

Figure 2 describe the framework workflow.The agent environment in the figure presents the scope that agent could reach. The communication between agents and the agent environment initiated by an agent called the initiator agent. The initiator starts by a call for proposals. Each agent in the environment could respond the initiator by a proposal as a feedback of its call of proposals or a refuse. In this step of information gathering, the initiator collects a set of propositions from its environment. Next, it uses the trust knowledge and the proposition of each of the proposer as an input to a trust metric in the analysis step. This trust metric is associated to the proposer. The initiator obtain an evaluation value using the evaluation function. In the rating and decision making step, the initiator classifies and sorts out the set of propositions collected using the trust knowledge of the agent. The best proposition that satisfies the delegated task with an accepted level of trustworthiness of the agent will be chosen, and the initiator will select its proposer to process the task. After making the decision and delegating the task, the initiator receives a feedback from the selected agent. This feedback will be used in an analysis process, where the agent compares

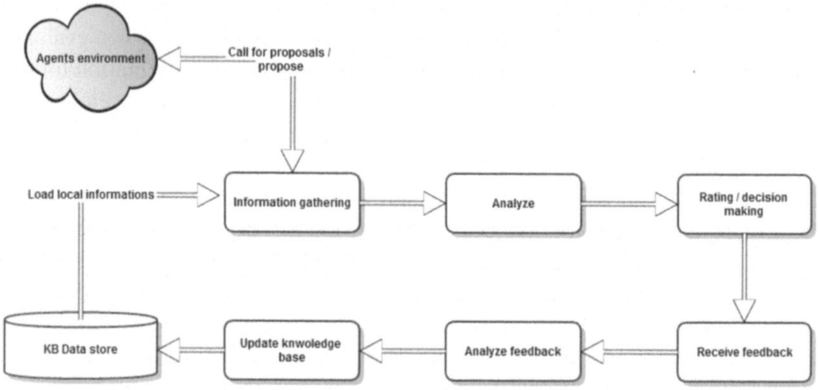

Fig. 2 GeFMAT Framework workflow

what was proposed to what is done. The analysis results in an opinion that will be used to update the agent trust knowledge. This sequence of action is based on the Contract Net Protocol [18].

4 Implementation and Experimentation

There are many development platforms dedicated to implement multi-agent systems. The purpose of those platforms is to simplify the development of agent applications by providing a set of tools that help users implementing, debugging and testing theirs systems. Those platforms are categorized into different classes .This classification is based on the kind of system that could be implemented using the platform. There are platform oriented middleware dedicated to implement interoperable agent systems such as ADK [19] and JADE [8]. There are social platforms that handle the organizational architecture and help expressing group behavior such as MadKit [10]. There are also reasoning platforms that focus on the internal processing of agents within the systems like JASON [15] and SOAR [17]. Lars Braubach et al. established a detailed classification of existing platforms [16]. Each platform is based on a set of standards and specifications such as MASIF [14], FIPA [18], JXTA [13], and web services. The advantage of the Java Agent DEvelopment Framework (JADE) over other platforms is that it complies with the FIPA specification for interoperable intelligent multi-agent systems and represents an agent middleware providing a set of graphical tool used during the development process. JADE use an agent abstraction to design agent in the system. It use also a task model to handle the behavioral part. A publish subscribe paradigm is supported to provide agent a tool for registering their service and searching available service. This paradigm is implemented using yellow pages and white pages service with the aim of helping agents to publish their

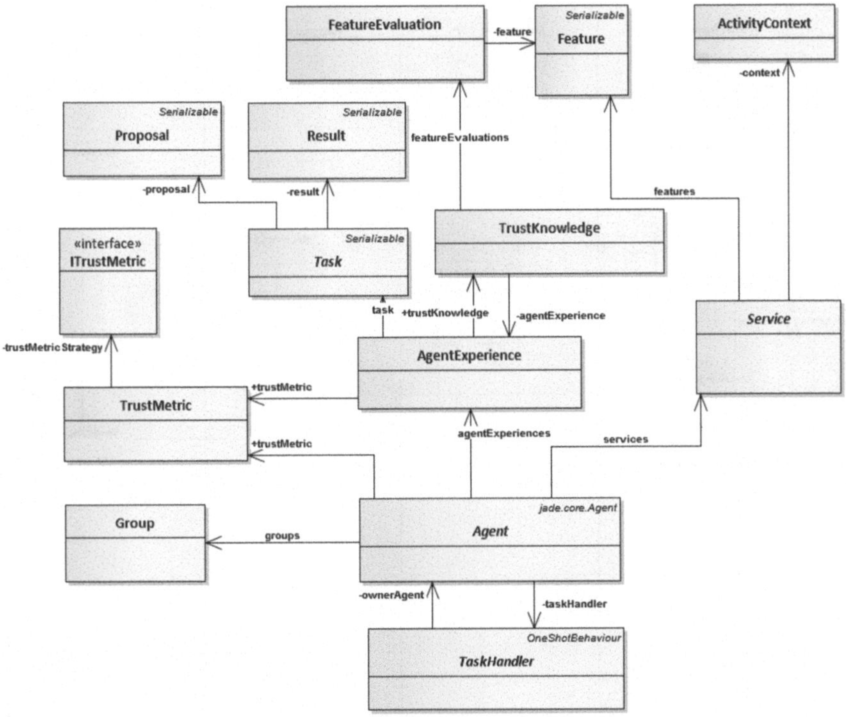

Fig. 3 GeFMAT meta-model implementation

services and to discover their peers. JADE Framework provide a set of classes that could be directly extended and used in our framework. Some meta-classes of the GeFMAT meta-model extends JADE classes, others are implemented from scratch. For example, jade.core.Agent class is extended to an Agent class that present an abstraction of the agent used in GeFMAT Framework. In addition to meta-classes presented in figure 1 new classes are introduced such as AgentExperience that present the acquired experience and TaskHandler that present a listener for task request. Trust model metrics are presented by the interface ITrustMetric. the diagram 3 shows the classes implementation of GeFMAT.

The workflow process is also implemented within methods used by agents while delegating tasks. Its composed by the set of steps described in previous section. Those steps are detailed in the activity diagram depicted in figure 4.

To verify the proposed implementation of GeFMAT Framework, we have built a testbed to compare behaviors of agents that use those trust models. In this testbed, we define a set of agents, and trust metrics to be used. Figure 5 shows a snapshot of a communication between agents.

As shown in Figure 5, an initiator agent called taskGeneratorStudent_0_B request for the list of registered agent in the Directory Facilitator, then it send call for proposals to all those agents. Each agent could respond with a proposal or a refuse.

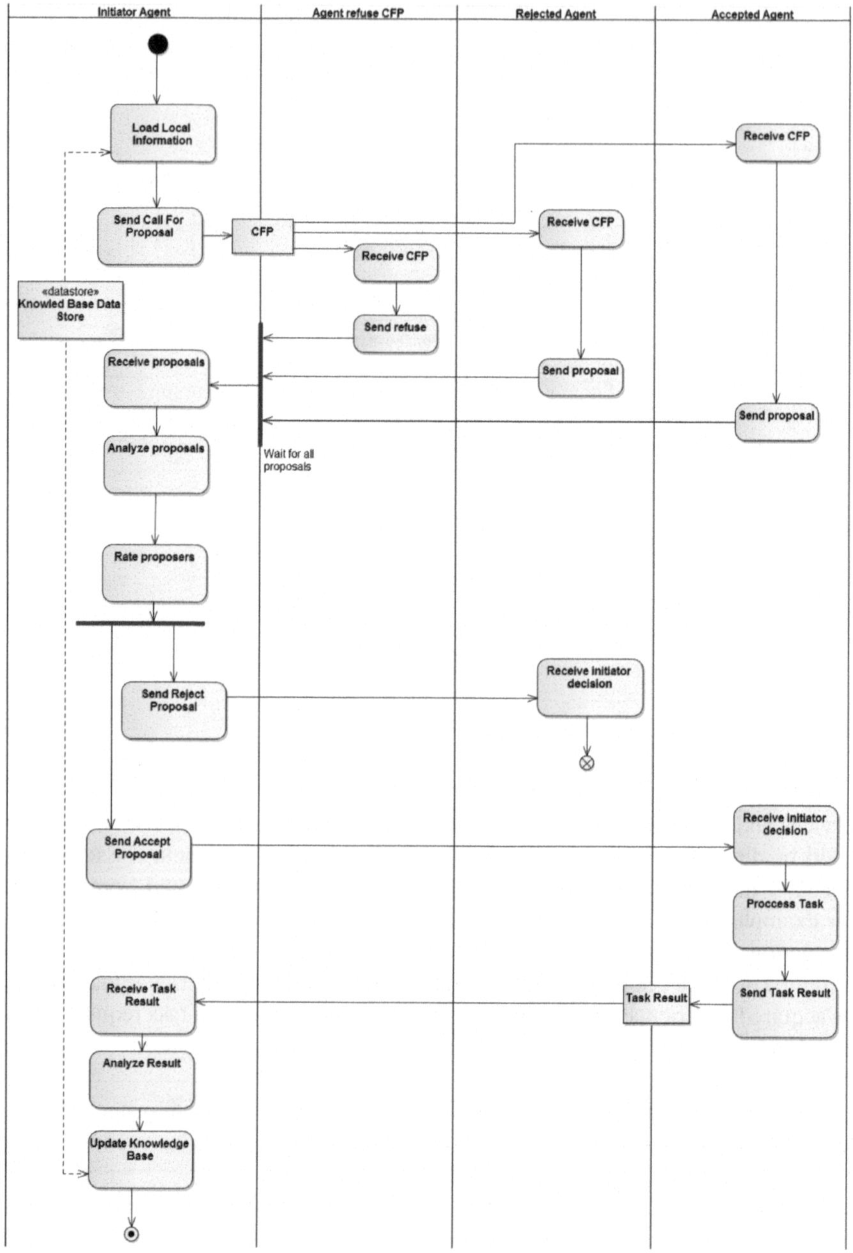

Fig. 4 GeFMAT workflow Implementation

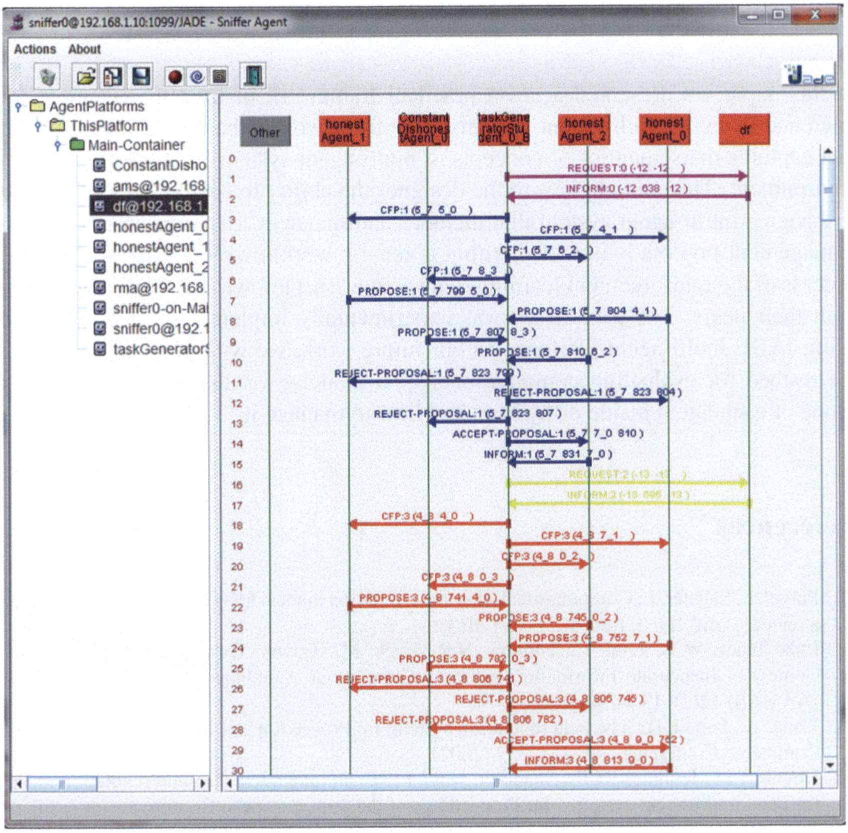

Fig. 5 GeFMAT Agents interaction in JADE Platform

The initiator agent handle those response and select the best proposer depending on his trust knowledge. The selected agent will receive an accept proposal with the task to be performed, the rest will receive a reject proposal response. When the selected agent perform the delegated task, it will send a result feedback to the initiator.The initiator will verify the validity of the result and update its trust knowledge about the selected agent.

Our proposed testbed provide some features that are not available in existing testbeds. It can evaluate more than one trust models at the same time in the same test case scenario. It can also evaluate those models against different attack strategies. Further details about our testbed are presented in a submitted article [21]. The source code of the GeFMAT Framework and the testbed is available online [1]. Classes related to the framework are grouped under core packages.

[1] https://github.com/mifmif/JADETrustTestbed

5 Conclusion

In this paper we presented a novel practical framework of trust management for open and dynamic multi-agent systems. This framework is based on a meta-model that captures the semantics of concepts of multi-agent systems involved in an open environment. This model gives to the designer the ability to customize it as needed to design a multi-agent system that includes and manages trust concepts. The trust management process is included within a generic workflow. It describes a generic process of the reinforcement learning of an agents trust knowledge while interacting with their peers. The framework was experimentally implemented and evaluated using JADE multi-agent platform. In our future work, we will work on improving the testbed for evaluating computational trust models. we also plan focus on the scope of each agent inside de system, and how to manage it.

References

1. Pinyol, I., Sabater, J.: Computational trust and reputation models for open multi-agent systems: a review. Artif. Intell. Rev. **40**, 1–25 (2013)
2. Luke Teacy, W.T., Patel, J., Jennings, N.R., Luck, M.: Travos: Trust and Reputation in the Context of Inaccurate Information Sources. Autonomous Agents and Multi-Agent Systems (AAMAS) **12**(2), 183–198 (2006)
3. Jsang, A., Ismail, R.: The beta reputation system. In: Proceedings of the 15th Bled Electronic Commerce Conference, pp. 324–337 (2002)
4. Huynh, T.D., Jennings, N.R., Shadbolt, N.R.: FIRE: an integrated trust and reputation model for open multi-agent systems. In: Proceedings of the 16th European Conference on Artificial Intelligence (2004)
5. Sierra, C., Debenham, J.: An information-based model for trust. In: Proceedings of 4th International Joint Conference on Autonomous Agents and MultiAgent Systems, pp. 497–504 (2005)
6. Sabater, J., Paolucci, M., Conte, R.: Repage: Reputation and image among limited autonomous partners. Journal of Artificial Societies and Social Simulation **9**(2) (2006)
7. Victor, P., Cornelis, C., De Cock, M.: Trust Networks for Recommender Systems, vol. 4 (2011)
8. The JAVA Agent DEvelopment Framework. http://jade.tilab.com/
9. Massa, P., Avesani, P.: Trust-aware collaborative filtering for recommender systems. In: Proc. of the Federated International Conference on the Move to Meaningful Internet, pp. 492–508 (2004)
10. Multiagent Development Kit. http://www.madkit.net
11. Jonker, C.M., Treur, J.A.: Dynamic perspective on an agent's mental states and interaction with its environment. In: Proceedings of the First International Joint Conference on Autonomous Agents and Multi-Agent Systems, AAMAS 2002, pp. 865–872. ACM Press (2002)
12. Yu, B. Singh, M.: An evidential model of distributed reputation management. In: Proceedings of the First International Joint Conference on Autonomous Agents and Multiagent Systems (AAMAS), Bologna, Italy, pp. 294–301 (2002)
13. The Juxtapose Project. https://jxta.kenai.com/
14. Milojicic, D., Breugst, M., Busse, I., Campbell, J., Covaci, S., Friedman, B., Kosaka, K., Lange, D., Ono, K., Oshima, M., Tham, C., Virdhagriswaran, S., White, J.: MASIF - the OMG mobile agent system interoperability facility. In: Rothermel, K., Hohl, F. (eds.) Proc. 2nd Int. Workshop Mobile Agents (MA 1998). Lecture Notes in Computer Science, vol. 1477, pp. 50–67. Springer, Stuttgart (1998)

15. http://jason.sourceforge.net
16. Kirn, S., Herzog, O., Lockemann, P., Spaniol, O.: Multiagent Engineering, Theory and Applications in Enterprises, pp. 503–530 (2006)
17. http://soar.eecs.umich.edu/
18. The Foundation for Intelligent Physical Agents. Fipa contract net interaction protocol (2002)
19. Tryllian's Agent Development Kit. http://www.tryllian.com/adk.html
20. Sabater, J., Sierra, C.: Social regret, a reputation model based on social relations. SIGecom Exchanges **3**(1), 44–56 (2002)
21. Youssef, M., Abdeslam, E.-N., Dahchour, M.: A JADE based testbed for evaluating computational trust models. In: 10th International Conference on Intelligent Systems: Theories and Applications (2015)

Including EAS-SGR IT Risk Framework in an IT GRC Global Framework

Hajar Iguer, Hicham Medromi, Adil Sayouti and Saadia Tallal

Abstract In the context of IT governance, different companies are using their expertise to come out with a new solution that helps their own governance. In fact, information systems managers struggle to comply with laws and regulations applied by their companies countries. With the variety of tools and frameworks, they need to know all processes in order to apply the only process that is going to be of help to their systems. In our case, we choose to discuss and study IT risk management which constitute an important component of the IT-GRC architecture. We have published several papers in this subject and we are still enhancing different aspects of the EAS-SGR architecture. In this paper, we will demonstrate the several interconnexions between IT-GRC components. These systems are based on multi-agent and expert systems knowing their integrated artificial intelligence. This expertise is one of our primal elements that were never used before in scientific research.

Keywords IT risks management · EAS-SGR architecture · Expert system · Multi-agent systems

1 Introduction

1.1 *IT Governance*

IT governance has evolved over many years, there are a number of third party IT governance framework and standards that an organization might deploy, many of

H. Iguer(✉)
EAS Team, LISER Laboratory, ENSEM, UIC, Casablanca, Morocco
e-mail: hajar.iguer@gmail.com

H. Medromi · A. Sayouti · S. Tallal
EAS Team, LISER Laboratory, ENSEM, Casablanca, Morocco
e-mail: hmedromi@yahoo.fr, s.tallal@gmail.com

E. Sabir et al. (eds.), *The International Symposium on Ubiquitous Networking*,
Lecture Notes in Electrical Engineering 366,
DOI: 10.1007/978-981-287-990-5_42

which have associated certification schemes. These frameworks fall, essentially, into two categories: proprietary frameworks and national/international standards [9].

This briefing provides a high level set of business arguments for IT Governance. It also explains how an IT Governance initiative can enable business and IT executives to:

- Be sure that that they are aware of all IT related risks likely having an impact on their organization;
- Know how to improve the management processes within IT to manage these risks;
- Ensure there are manageable relationships with suppliers, service providers and with the business (customers);
- Ensure there is a transparent and understandable communication of these IT activities and management processes to satisfy the Board and other interested stakeholders [7].

Most IT management frameworks and standards offer solutions and tools that can help with IT governance, but they are typically very detailed, and have narrow scopes. No single framework or standard provides a full set of IT governance tools and, collectively, they can provide a confusing picture that actually hinders the core purpose of IT governance. This purpose is to equip boards with information for controlling risk and the levers for directing, evaluating, and monitoring how well IT supports their core business [10].

In spite of these existing frameworks, enterprises felt the need to adjust them to everyday reality. That's why thought of combining multiple independent framework in global one.

IT Governance defines five important pillars [13]:

- Strategic Alignment
- Value Creation
- Risk Management
- Resource Management
- Performance measurement.

1.2 IT Risk

Risk management is a set of coordinated activities to direct and control organization towards the risk [1]. It is generally identifies three goals in the management of risks to SI:

- Improve the security of information systems.
- Justify the budget allocated to the security of the information system.
- Prove the credibility of the information system using the analysis.

Risk management methods and tools enable the organization to plan and implement programs to maximize their opportunities and to control the impact of potential threats [4].

Table 1 Risk Methods

Risk Method	
Au IT Security Handbook	IT Grundschutz
Cramm	Magerit
A&K Analysis	Marion
Ebios	Mehari
ISAMM	Migra
ISF Methods	Octave
SP800 30	Risk safe Assessment

From this list of Risk Methods shown in the table (Tab. 1), we have tested MEHARI and EBIOS since they are the majorly used. The rest are mostly commercialized tools, only available in free trial with a limit of time. These two methods are proved very detailed and elaborated.

Table 2 (Tab. 2) presents some of risk frameworks, standards and solutions used by companies all over the world. For Example CGE Risk Management is an industrial solution that has multiple modules and one of them is dedicated to risk management and which is not available for free. On the other hand ISO 27001 states a set of guidelines that allows ensuring the respect of the limits of risk exposure for your information system.

Table 2 Risk Framework, Standards and Solution

Risk Framework, Standards and Solutions	
ISO/IEC 27005 (Standard)	Aviva risk Management Solution
ISO/IEC 27001 (Standard)	Web2 Security Services
CGE Risk Management	

1.3 IT Compliance

The Sarbanes-Oxley Act of 2002 is a United States federal law passed in response to a number of major corporate and accounting scandals including those affecting Enron and WorldCom. Although the Act is focused on financial reporting, it has given IT issues increased importance [8]. IT controls such as such as security, incident management, disaster recovery and electronic records management can have a direct or indirect impact on the financial reporting process and they do affect the reliability and security of systems in which companies keep their financial records. There are several titles and sections in the Sarbanes Oxley Act that have a direct impact on internal controls, including IT control [3].

1.4 IT GRC

In literature, IT GRC means Information technology- Governance, Risk and Compliance which ensures the cohabitation of different major component for the organization. It is defined as global approach to the organization ensuring that it is consistent with aligning business objectives with its policies, regulations and risk management compliance [14].

IT GRC frameworks are considered as an application that combines IT governance, IT Risk and IT compliance in order to permit a reactive and executive management . In computer science literature there is a considerable offer in terms of frameworks that give an answer to all indecisive managers. These tools help companies to commit to decisions they won't regret [12].

2 Multi-Agent Systems

Our application would not have the intelligence and autonomy without our multi-agents systems. In fact, we are going to define an agent and a multi-agent system.

Agent: is a virtual or physical entity that cooperates with others agents in the same environment working towards the same objective. An agent had different characteristics; he can be cooperative, communicative and reactive [6].

Multi-Agent System (MAS): is an organized set of agents. It consists of one or more organizations which structure rules cohabitation and teamwork between agents. In the same system, an agent can belong to several organizations.

The inter-agent communication is fundamental to the realization of the agent paradigm, as is the development of human language was the key to the development of human intelligence and societies. To share information and knowledge, agents use ACL (Agent Communication Language).

The multi-agent system introduces a new approach to the implementation of several systems including independent and autonomous elements. This field of research is one of the most innovative; it shows many applications in several fields. Its wealth is a benefit derived by several companies or institutions which use multi-agent systems. However due to their efficiency, multi-agent systems are determined by several agent models and find with some complexity in their implementation. The use of these systems is often considered difficult.

3 Related Work

In the business world, there is a multitude of solution that gives guidance to information security managers in order to help to protect their systems.

In the introduction, we talked about the existing solutions in the IT Governance domain; we analyzed the different frameworks and methods that exist for helping IT managers to step up a methodology or procedures to follow in their risk issues.

Risk management methods and tools enable the organization to plan and implement programs to maximize their opportunities and to control the impact of potential threats. The previous section provides an overview of available security risk analysis frameworks, methods and tools. Each solution had its advantages and drawbacks. We tried in our research to discuss these tools and propose a customizable framework to be adapted to any context.

Put simply risk management is about looking at the risks that arise in the workplace and then putting in place sensible health and safety measures to control them. You could view a risk's workplace as 'an incident waiting to happen.

Effective risk management means understanding what these risks are and preventing these accidents from ever materializing. The workplace can be defined as both within your premises or any employees you may have on the road.

Whether your business is large or small, effective risk management shouldn't be ignored. Managing risk today means fewer surprises and unexpected consequences in the future.

Here is the list of non exhaustive solutions that are available to all enterprises [15]:

- ConrolCase IT GRC
- MetricStream IT GRC Management Software Solution
- RSA Archer GRC

IT GRC challenge includes different elements such as mapping policies and controls, security and audit and inefficiency. It is crucial to provide coordination between different department to ensure the standardization of policies and controls. The industry is seriously attempting to improve IT-GRC initiative to align with tough regulatory requirements so does scientific researchers. Our team researchers are trying to establish a new framework based on most IT GRC methods, frameworks or tools and multi-agent system.

4 Proposed Approach: EAS-IT Risk Architecture

The EAS-SGR architecture was created to bring a different answer to IT risk management while being focused on the real aspects of security.

This new version of our architecture is a part of a global architecture which covers all of the IT-GRC aspects. Fig. 1 is presented by five major parts.

- The updater is a system that looks for updates in the frameworks or methods that are used in IT risk management. As an example, we choose to work with EBIOS and ISO 27001.

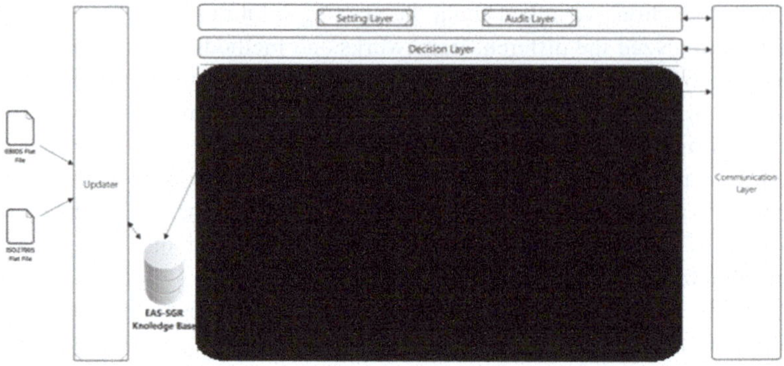

Fig. 1 EAS-IT Risk Architecture

- The setting layer reassembles all the general information that will be needed in this framework. The Audit Layer analyses the business objectives in order to understand the need of each organization.
- The decision layer decides on the right framework based on the environment and the need of the organization.
- The communication layer deals with the setup of the right type of communication
- Our EAS-SGR Architecture includes an expert system and three multi agent systems.

4.1 EAS-SGR Architecture

The EAS-SGR architecture is the black box as it is shown in figure 1.

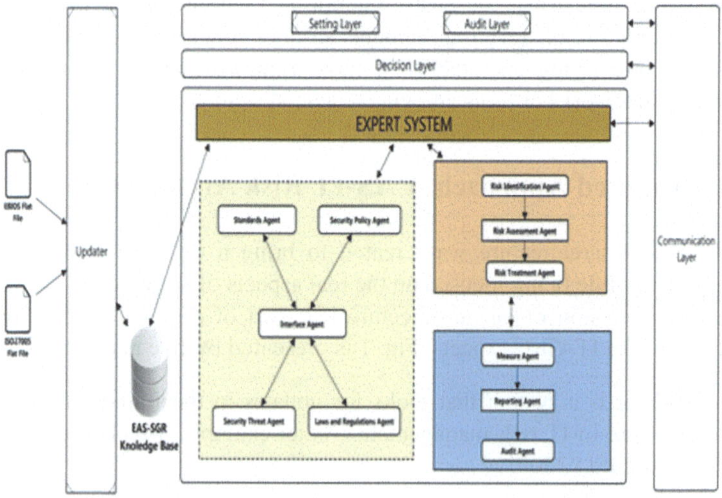

Fig. 2 EAS-SGR Architecture

Expert System

In a general way, an expert system can be conceived as a four-module system that acts as an information manager.

An expert system is a computer system that emulates the decision-making ability of a human expert. At the foundation of every expert system is a knowledge base representing ideas from a specific field of expertise. Because it's a collection of specialized knowledge, an expert system's knowledge base must be constructed by a user, an expert, or a knowledge engineer—a specialist who interviews and observes experts and painstakingly converts their words and actions into a knowledge base.

- The inference engine is an essential element for an expert system, since it works as the engine control that evaluates and applies rules. The inference engine uses forward chaining, a method which seeks to validate the assumptions in the rules and to complete the actions (consequences), not only as a logical conclusion.
- The rules library and the working memory form the so-called 'knowledge base', representing the knowledge captured from a human expert on the problem domain. The rules libraries are rules that manage knowledge. When an issue is submitted to the system evaluation, this rules library interacts with the user and the inference engine, allowing identification of the problem, possible solutions for it and the whole process that leads to conclusions. Rule-based systems are feasible for problems in which the solution process can be written in the form of 'IF-THEN' rules and for which the problem has no easy solution.
  - A set of facts to represent the memory of the initial work. This can be any relevant information related to the system's initial state;
  - A set of rules, a library built to deal with the set of facts. This should include any action that should be within the scope of the problem;
  - A condition stipulating that a solution was found or that no solution exists.

Controls Management MAS

Interface Agent: This agent offers a customizable assistance to four agents then collect and sensitize the data provides by them.

Laws and Regulation Agent: Its main role is to collect laws and regulations which are applied to the organization in order to prevent serious juridical problems.

Security Threats Agent: Collects security threats available in the frameworks knowledge base and adapt them to the environment.

Security Policy Agent: This agent obliges the organization to have a general security policy and a specific one if it is needed.

Standard Agent: It is the agent that is responsible of following the standards that have been chosen [1].

Risk Management MAS

This multi-agent system is constituted of three agents:

Risk Identification Agent: It is in charge of the identification of risk parameters.

Risk Assessment Agent: This agent evaluates threats and vulnerabilities in order to better understand and measure the risk impact.

Risk Treatment Agent: It chooses between different measures that were proposed at the main objective to alter risk and their impact on the organization [5].

Measure's Management MAS

Measure's Agent: It dictate a plan for the implementation of the chosen measure by the risk treatment agent.

Reporting Agent: realizes dashboard for the top management.

Audit Agent: It allows an audit of the information system to check the good application of the measure.

5 Conclusion and Future Work

In this paper, we introduced another version of our architecture EAS-SGR and illustrated the different aspects that were needed into its adaptation. It is flexible, customizable and updated to the original frameworks and methods that already exist in the market.

This particularity of our approach is to use multi agents systems that add the uniquely intelligence to our application. In addition, the combination of a method of risk management with ISO, internationally recognized, other frameworks and multi-agents systems provides the ability to secure and protect a system that represents the image of an organization. If the system is misused by the dramatic attempts of accessing restricted information by employees, it can harm its interests and the achievement of its business objectives.

This study is now limited to risk management in an organizational company. In our future work, we will start the audit of an information system. Then, we will detail the modeling, the realization of the application based on our framework. At the end, we will proceed for the test of the application on a case of study that includes an e-learning application. At last, we propose to adapt it to a global architecture which is EAS IT-GRC architecture.

In the context of creating a global platform, we plan to include other architecture produced by other researcher's who are working in the IT GRC domain.

References

1. Iguer, H., Medromi, H., Sayouti, A.: A new architecture multi-agents based combining EBIOS and ISO 27001 in IT risk management. In: Proc. ICEER 2013 (2013). paper 126
2. von Solmsa, B., von Solms, R.: From information security to business security? Computers & Security **24**, 271–273 (2005)
3. Kim, N.-Y., Robles, R.J., Cho, S.-E., Lee, Y.-S., Kim, T.-H.: Sox act and IT security governance. In: International Symposium on Ubiquitous Multimedia Computing (2008)
4. Shukla, N., Kumar, S.: A comparative study on information security risk analysis practices. In: On Issues and Challenges in Networking, Intelligence and Computing Technologies – ICNICT 2012, November 2012
5. Iguer, H., Faris, S., Medromi, H., Sayouti, A.: Conception d'une plateforme de gestion des risques basée sur les systèmes multi-agents et ISO 27005
6. Ferber, J.: Les systèmes multi-agents, vers une intelligence collective. InterEditions, pp. 63–144 (1995)
7. IT Governance: Developing a successful governance strategy: A best practice guide for decision makers in IT, ISACA
8. IT compliance and IT security— Part 1: Why is it necessary to comply with legal requirements? Privacy & Data Protection **7**(4)
9. Mangalaraj, G., Singh, A., Taneja, A.: IT governance frameworks and COBIT - a literature review. In: Twentieth Americas Conference on Information Systems, Savannah (2014)
10. Developing an IT governance framework: Alan Calder, National Computing Center
11. Racz, N., Panitz, J.C., Amberg, M., Weippl, E., Seufert, A.: Governance, Risk & Compliance (GRC) status quo and software use: results from a survey among large enterprises. In: ACIS 2010 Proceedings (2010). Paper 21, http://aisel.aisnet.org/acis2010/21 (retrieved December 13, 2010)
12. Kooper, M.N., Maes, R., Lindgreen, E.R.: On the governance of information: Introducing a new concept of governance to support the management of information. International Journal of Information Management: The Journal for Information Professionals **31**(3), 195–200 (2011)
13. Racz, N., Weippl, E., Seufert, A.: A process model for integrated IT governance, risk, and compliance management. In: Proceedings of the Ninth International Baltic Conference on Databases and Information Systems, Baltic DB&IS 2010, pp. 155–170. University of Latvia Press, Riga (2010)
14. Roiter, N.: CSO online article. IT GRC tools control your environment (2011)
15. Bonazzi, R., Hussami, L., Pigneur, Y.: Compliance management is becoming a major issue in IS design. In: D'atri, A., Saccà, D. (eds.) Information Systems: People, Organizations, Institutions, and Technologies (PDF), pp. 391–398. Springer (2009). doi:10.1007/978-3-7908-2148-2 (retrieved April 6, 2013)

Enterprise Architecture Complexity Component Based on Archimate Language

Jihane Lakhrouit and Karim Baïna

Abstract Enterprise architecture (EA) is a business and IT management tool that has grown in popularity during the last decade. It is based on models of business and IT with systematic frameworks. This paper provides an interesting overview on the agility measurement and dimensions. The proposed approach precises the components of each enterprise architecture layers and the complexity of the future architecture.

Keywords Agility · Enterprise architecture · EA · Evaluation of enterprise architecture · Archimate

1 Introduction

Enterprise architecture is the process of translating business vision and strategy into effective enterprise change by creating, communicating and improving the key principles and models that describe the enterprise's future state and enable its evolution. The scope of the enterprise architecture includes the people, processes, information and technology of the enterprise, and their relationships to one another and to the external environment.

The agility is the capability of surviving and prospering by reacting quickly and effectively to changing markets, driven by customer designed products and services [5]. The Federal Enterprise Architecture Framework (FEAF) defines EA as "a strategic information asset base, which defines the agency's mission and business activities supporting the mission, the information necessary for agency operations, the technologies necessary to support operations, and the transitional

J. Lakhrouit(✉) · K. Baïna
Alqualsadi Research Team on Enterprise Architecture, ENSIAS,
Mohammed V- Souissi University, Rabat, Morocco
e-mail: jihane.lakhrouit@gmail.com, baina@ensias.ma

E. Sabir et al. (eds.), *The International Symposium on Ubiquitous Networking*,
Lecture Notes in Electrical Engineering 366,
DOI: 10.1007/978-981-287-990-5_43

processes necessary for implementing new technologies in response to changing business needs". Combining EA and agility analysis gives the advantages of being able to analyze the IT systems in their enterprise wide context. In previous research, managing complexity has been identified as a major challenge in enterprise architecting [1], however IS literature has not mentioned a holistic, (i.e. applicable to all dimensions) and elements of an EA, approach to quantify complexity in EAs so far[2].

The goal of this paper is to precise the dimensions of agility and to identify the components of enterprise architecture based on archimate language. The paper is structured as follows: the second section presents the dimensions of agility, the third section describes the components of enterprise architecture, the fourth section presents the related works and the last section is dedicated to conclude our paper.

2 Dimensions of Agility

The Agility is defined by [11] [20] as a business-wide capability that embraces organizational structures, information systems, logistics processes and in particular, mindsets. Based on this definition we can remark that the agility affects organizational structures, information systems and processes. In others words the dimensions of agility includes the components of the enterprise as defined in the methods described below.

For Lui and Piccol [17] the agility is defined by: (1) technology, (2) process, (3) people, and (4) structure. Hence, they argue that the agility is not a simple summing of the agility of the four components, but it depends on their nonlinear relationship. (Tsourveloudis et al., 2002) proposed four aspects: (1) production infrastructure, (2) market infrastructure, (3) people infrastructure and (4) information infrastructure. Imache et al [10] proposed four dimensions: Process, Organization, Information, Resource and Environment. The weakness of these evaluation methods is that they don't include the dimensions of flexibility and capability of change, which must be considered because the agility is also defined as a rapid and proactive adaptation of enterprise elements to unexpected and unpredicted changes [12]. For these reasons several authors have introduced the flexibility in the dimensions of agility. [29] identified competitive foundations of agility as follows: (1) speed, (2) flexibility, (3) innovation, (4) proactivity, (5) quality, and (6) profitability. Sharifi et al. [23] identified four dimensions: responsiveness, competency, flexibility and quickness. Jackson and Johansson [9] divide agility capabilities into four main dimensions: (1) product-related change capabilities, (2) change competency within operations, (3) internal and external co-operation, and (4) people, knowledge, and creativity.

Some authors [6][3] noted that the complexity of organization in transition needs to be reduced in order to deal with the transition. The complexity of system hinders the ability of the enterprise to quickly react to change by re-configuration of products, processes, or organization structure. Considering that the less complex system is easier to change and is more agile, we consider the complexity of the system as the measure of agility.

All of the discussed dimensions are proposed by Sherehiy et al. [7] and which are 1) flexibility and adaptability, 2) responsiveness, 3) speed, 4) integration and low complexity, 5) mobilization of core competences, 6) high quality products and customized products, and 7) culture of change. In our research, we consider these seven dimensions to evaluate enterprise architecture agility; however, we concatenate the dimensions of mobilization of core competences and culture of change because both of them represent the ability to accept the change. In the remainder of this article we will evaluate the complexity dimension.

3 The Enterprise Architecture Components

In this approach, the enterprise architecture is composed of several components which are extracted of the archimate metamodels of the different layers of enterprise architecture. In this section we present the components of the business, the application and the technology layers.

3.1 The Layers of Enterprise Architecture

The layers of enterprise architecture are: The business architecture BA defines the business strategy, governance, organization, and key business processes. The application architecture AA provides a blueprint for the individual application systems to be deployed, their interactions, and their relationships to the core business processes of the organization. The technology architecture TA describes the logical software and hardware capabilities that are required to support the deployment of business, data, and application services.

3.2 Identify the Business Architecture Components Using the Meta-model of Business Architecture Layer

The Business architecture offers products and services to external customers, which are realized in the organization by business processes performed by business actors. This section presents firstly the meta-model of business architecture, secondly the definitions of each concepts in this meta-model and finally we present our business model and the different levels of granularity of this layer to calculate the complexity.

Business Architecture Meta-model

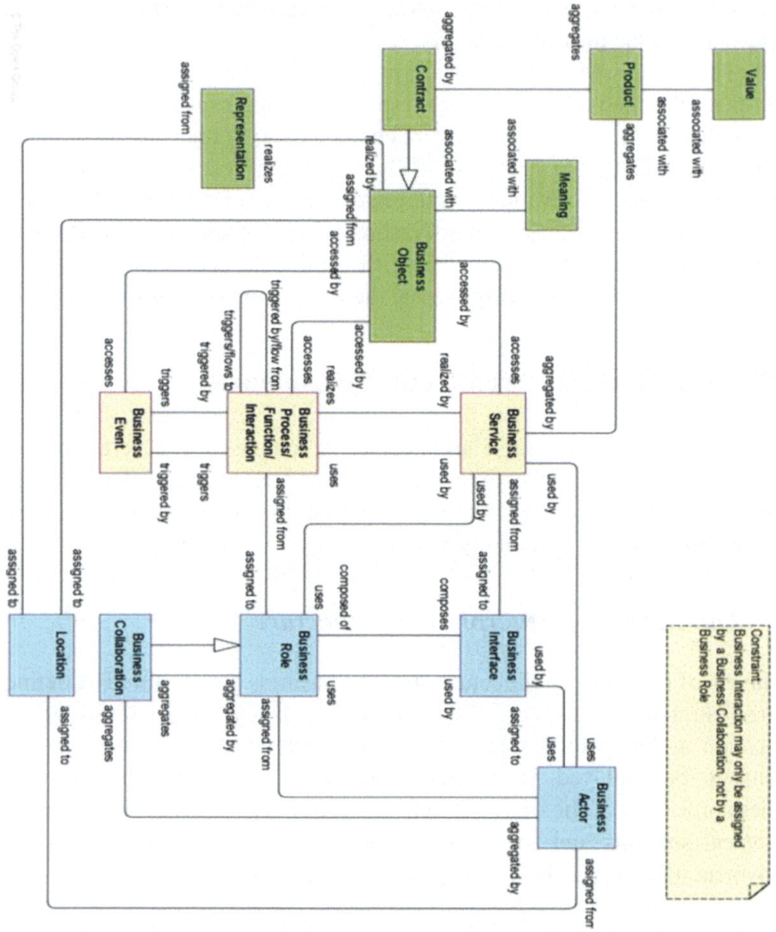

Fig. 1 Archimate business layer meta-model

Business Architecture Concepts. According to the archiMate language we define the most important classes that we will use in our approach. A business process is defined as a behavior element that groups behavior based on an ordering of activities. It is intended to produce a defined set of products or business services. A business function is defined as a behavior element that groups behavior based on a chosen set of criteria (typically required business resources and/or competences). A business service is defined as a service that fulfills a business need for a customer (internal or external to the organization). A product is defined as a coherent collection of services, accompanied by a contract/set of agreements, which is offered as a whole to (internal or external) customers.

Business Evaluation Model. In our business evaluation model we consider only the behavioral concepts which modeling the dynamic events of an enterprise (business service, business function and business process), and we consider also the product because is the collection of service which is offered as a whole to customers. We have defined the evaluation model, presented below, which contains four classes that represent the four levels of granularity that allow measuring the complexity of enterprise architecture.

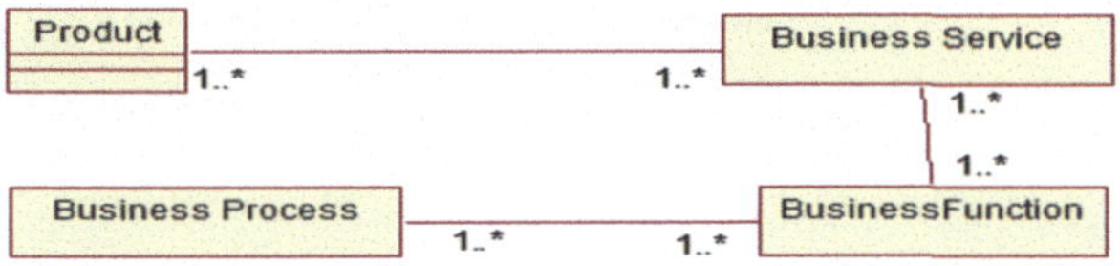

Fig. 2 Business evaluation model

Based on the model in the figure 2, we define four levels of granularity in order to define the complexity of business architecture:

Level 1: This is the first level where we measure the complexity of the product; however, this measure gives only global view of our business and it is not sufficient to specify the degree of complexity of our system. The next level is the complexity of the business service. Level 2: This is the second level where we measure the complexity of the business service. In this level we can get an important idea of the degree of the system complexity. Level 3: This is the third level where we measure the complexity of the business functionality. Level 4: This is the fourth level where we measure the complexity of the business process. This level is the most detailed level.

3.3 Identify the Application Architecture Components Using the Meta-model of Application Architecture Layer

The application architecture layer supports the business layer with application services which are realized by (software) applications. This section presents firstly the meta-model of application architecture, secondly the definitions of each concepts in this meta-model and finally we present our application model and the different levels of granularity of this layer to calculate the complexity.

Application Architecture Meta-model

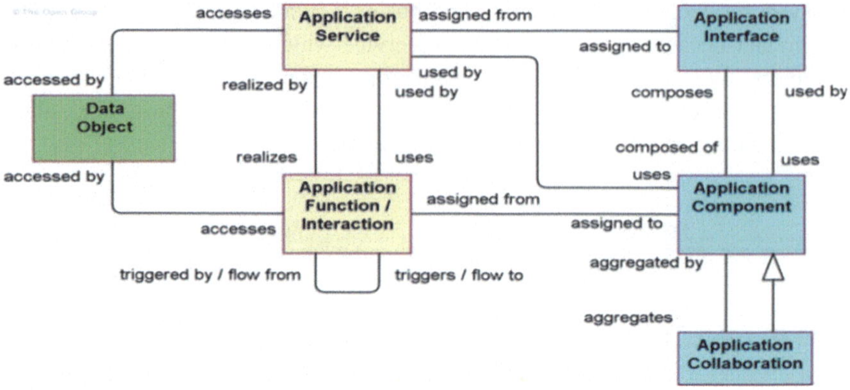

Fig. 3 Archimate application layer metamodel

Application Architecture Concepts. Based on the metamodel of archimate, the application layer is represented by: An Application Component as a modular, deployable, and replaceable part of a system that encapsulates its contents and exposes its functionality. A Component uses one or more services. An Application Service as an externally visible unit of functionality, provided by one or more components. A service uses one or more functions. An Application Function as a behavior element that groups automated behavior that can be realised by an application service. A data object is defined as a passive element suitable for automated processing. An Application collaboration (We name it in our model application) as an aggregate of two or more application components that work together to perform collective behavior.

Application Evaluation Model. In our approach we don't consider the class of data, because it is too detailed level and does not provide a profit on the assessment of agility, we don't consider either also the application interaction class because it describes only the behavior of an application collaboration. We have defined the evaluation model, presented below, which contains four classes that represent the four levels of granularity that allow measuring the complexity of enterprise architecture.

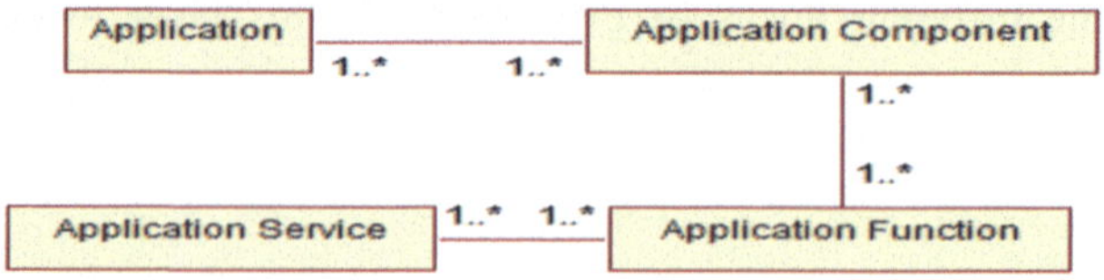

Fig. 4 Application evaluation model

Based on the model in the figure 4, we define four levels of granularity in order to define the complexity of application architecture:

Level 1: This is the first level where we measure the complexity of the applications; however, this measure gives only global view of the relationship which is not sufficient to specify the degree of complexity of our system. The next level is the complexity of the components. Level 2: This is the second level where we measure the complexity of the components. In this level we can get an important idea of the degree of the system complexity. Level 3: This is the third level where we measure the complexity of the functions. The granularity in this step is important because we can have a holistic view of the system complexity. Level 4: This is the fourth level where we measure the complexity of the services. This level is the most detailed level.

3.4 Identify the Technology Architecture Components Using the Meta-model of Technology Architecture Layer

The technology architecture offers infrastructure services (e.g., processing, storage, and communication services) needed to run applications, realized by computer and communication hardware and system software. This section presents firstly the meta-model of technology architecture, secondly the definitions of each concepts in this meta-model and the finally we present our technology model and the different level of granularity of this layers to calculate the complexity.

Technology Architecture Meta-model

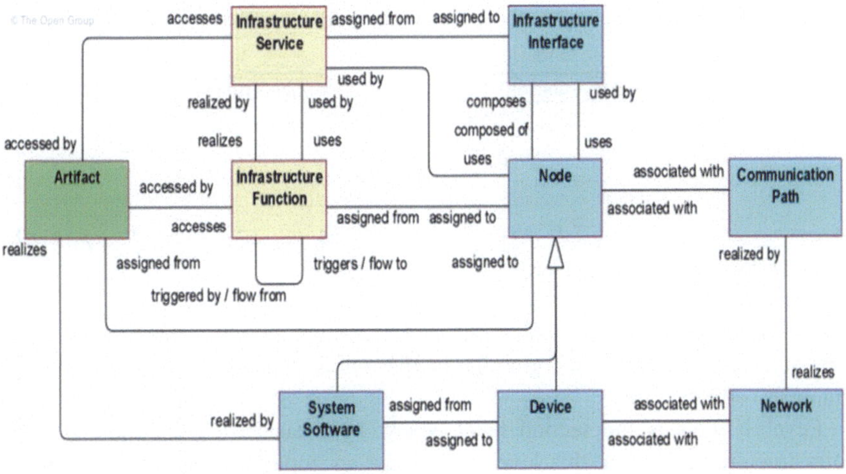

Fig. 5 Archimate technology layer metamodel

Technology Architecture Concepts. A device is defined as a hardware resource upon which artifacts may be stored or deployed for execution. A node is defined as a computational resource upon which artifacts may be stored or deployed for execution. System software represents a software environment for specific types of components and objects that are deployed on it in the form of artifacts. An infrastructure interface is defined as a point of access where infrastructure services offered by a node can be accessed by other nodes and application components. A network is defined as a communication medium between two or more devices. A communication path is defined as a link between two or more nodes, through which these nodes can exchange data. An infrastructure function is defined as a behavior element that groups infrastructural behavior that can be performed by a node. An infrastructure service is defined as an externally visible unit of functionality, provided by one or more nodes, exposed through well-defined interfaces, and meaningful to the environment. An artifact is defined as a physical piece of data that is used or produced in a software development process, or by deployment and operation of a system.

Technology Evaluation Model. In our approach we don't consider the class of devise and system Software, because it is too detailed level and does not provide a profit on the assessment of agility but we consider the Node class and it regroups the two concepts (Device and System Software), we don't consider either also the infrastructure interface class because it specifies only how the infrastructure services of a node can be accessed by other nodes. An artifact represents a concrete element in the physical world. It is typically used to model (software) products such as application components; in our technology model we consider only the artifact which represents the application components. The network represents the physical communication infrastructure; it represents the communication between the different elements in our technology model.

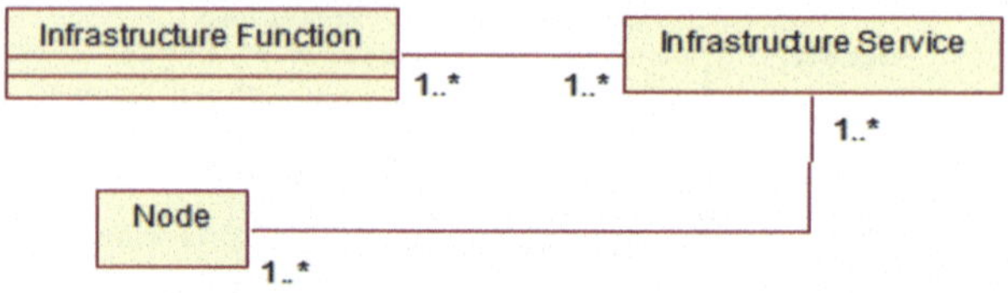

Fig. 6 Technology evaluation model

We have defined the evaluation model, which contains three classes that represent the three levels of granularity that allow measuring the complexity of enterprise architecture.

Level 1: This is the second level where we measure the complexity of the infrastructure service. In this level we can get an important idea of the degree of the system infrastructure. Level 3: This is the third level where we measure the complexity of the functions. The granularity in this step is important because we can have a holistic view of the infrastructure. Level 4: This is the fourth level where we measure the complexity of the infrastructure and especially the node. This level is the most detailed level.

3.5 Evaluation Model of Complexity Evaluation

In this section, we present the global model of components complexity evaluation.

When do you use the top-down approach? A Top-Down approach to change management implies imposed change as the initiative comes from the top layer. For example the change is from the business layer. The change of business layer will impact the application and business layers.

When do you use the bottom-up approach? A Bottom-Up approach to change management is triggered by the down layer. For example the change is started in the technology layer; for example, trends in technology demand rethinking in which markets we want to play and with what technology. The change of technology layer will impact the application and business layers.

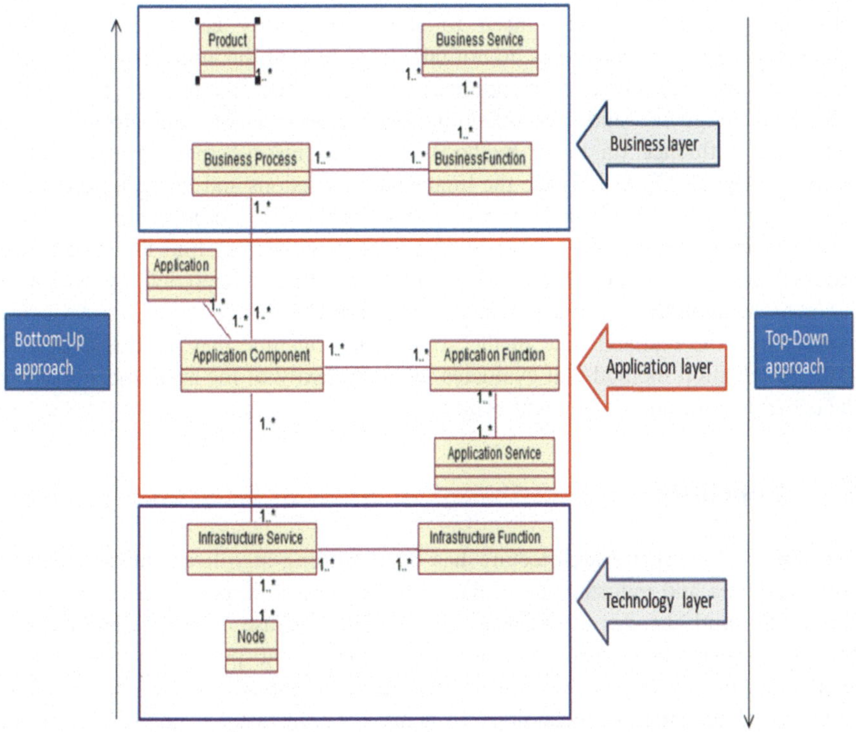

Fig. 7 Evaluation model

4 Related Works

In the following section, we present our literature review related to our research. In the literature, several different approaches to agility measurement can be found. Some authors [8][22][25][26][28] used an integrated agility index. The agility index was defined as combination of agile capabilities intensity levels. Also, the weighted index of agility, where the intensity of each agility capability was weighed by it importance was proposed [27].

Other methods of agility measurement [13] [21] were based on analytical hierarchical process (AHP) methodology. In this approach, a pairwise comparison technique was used to evaluate the agile capabilities. The obtained comparisons between all possible pairs of capabilities were synthesized to derive overall agility score. Liu and Zheng [13] combined AHP and BBN to evaluate the probabilities that an enterprise falls in high agility category in the respective decision factors. These methods are easy to implement; however, most agility measurements are described subjectively by linguistic terms, which are characterized by ambiguity and multi-possibility. Thus, the scoring of the above techniques can always be criticized, because the scale used to score the agility capabilities has two limitations: (1) such techniques do not take into account the ambiguity and multi-possibility associated with the mapping of one's judgment to a number, and (2) the subjective judgment and the selection and preference of evaluators have a significant influence on those methods. Thus, to resolve these problems The authors [14][15][22][24][25] use the linguistic expressions and fuzzy logic in their assessing models. A method of Arteta and Giachetti [3] used complexity as a surrogate measure of agility. Their model uses Petri Nets to find the state space probabilities needed for the complexity measure. In our approach we will also evaluate the complexity but using the network model.

Based on the realized state of the art we can conclude that there is a lack of agility evaluation method that evaluates the complexity of the system considering the context of the enterprise architecture.

5 Conclusion

The topic of enterprise architecture has been gaining significant attention from both academia and industry due to the inefficiencies of current IT architecture to cope with rapid changes in business environment because enterprise architecture can be a practical tool to increase the management power on information technology of any organization. It has the aim of creating harmony between the elements of information technology to achieve organizational goals. This paper has the aim of creating a model of the enterprise architecture components.

References

1. Eason, G., Noble, B., Sneddon, I.N.: On certain integrals of Lipschitz-Hankel type involving products of Bessel functions. Phil. Trans. Roy. Soc. London **A247**, 529–551 (1955)
2. Aitken, J., Christopher, M., Towill, D.R.: Understanding, implementing and exploiting agility and leanness. International Journal of Logistics: Research and Applications **5**(1), 59–74 (2002)
3. Arteta, B.M., Giachetti, R.E.: A measure of agility as the complexity of the enterprise system. Robotics and Computer-Integrated Manufacturing **20**(6), 495–503 (2004)
4. Bass, B.M.: Transformational leadership: Industry, military, and educational impact. Erlbaum, Mahwah (1998)
5. Cho, H., Jung, M., Kim, M.: Enabling technologies of agile manufacturing and its related activities in Korea. Computers & Industrial Engineering **30**(3), 323–334 (1996)
6. Dove, R.: Response Ability: the Language, Structure, and Culture of the Agile Enterprise. Wiley, New York (2001)
7. Gandomani, T.J., Zulzalil, H., Ghani, A., Sultan, B.M.: Important considerations for agile software development methods governance. Journal of Theoretical and Applied Information Technology **55**, 345–351 (2013)
8. Gandomani, T.J., Nafchi, M.Z.: Agility Assessment Model to Measure Agility Degree of Agile Software Companies. Indian Journal of Science and Technology **7**, 955–959 (2014)
9. Jackson, M., Johansson, C.: Agility analysis from a production system perspective. Intergarted Manufacturing Systems **14**(6), 482–488 (2003)
10. Imache, R., Izza, S., Nacer, M.: An Enterprise Information System Agility Assessment Model. Computer Science and Information Systems **9**(1), 107–133 (2012)
11. Katayama, H., Bennett, D.: Agility, adaptability and leanness: A comparison of concepts and a study of practice. International Journal of Production Economics, 43–51 (1999)
12. Kidd, P.T.: Agile Manufacturing: Forging New Frontiers. Addison-Wesley, Reading (1994)
13. Liu, Y., Zheng, H.: Measuring agility of enterprise using analytic hierarchy process and Bayesian belief networks. In: Proceedings of 2006 13th International Conference on Management Science and Engineering, ICMSE 2006, pp. 551–556 (2006)
14. Lin, C.T., Chiu, H., Chu, P.Y.: Agility index in the supply chain. International Journal of Production Economics **100**(2), 285–299 (2006)
15. Lin, C.T., Chiu, H., Tseung, Y.H.: Agility evaluation using fuzzy logic. International Journal of Production Economics **101**(2), 353–368 (2006)
16. Lapkin, A.: Gartner Defines the Term 'Enterprise Architecture.' Research Report, Gartner Group, July 2006
17. Lui, T.W., Piccoli, G.: Degrees of agility: Imlications from information systems desigh and firm strategy. In: Desouza, K.C. (ed.) Agile Information Systems: Conceptualization, Construction, and Management, Elsevier, Burlington, USA, pp 122–133 (2007). ISBN 10:0-7506-8235-3
18. Leenders, R.T.A., Van Engelen, J.M., Kratzer, J.: Virtuality, communication, and new product team creativity: a social network perspective. Journal of Engineering and Technology Management **20**(1), 69–92 (2003)

19. Overby, E., Bharadwaj, A., Sambamurthy, V.: Enterprise agility and the enabling role of information technology. European Journal of Information Systems, 120–131 (2006)
20. Power, D.J., Sohal, A., Rahman, S.U.: Critical success factors in agile supply chain management: An empirical study. International Journal of Physical Distribution & Logistics Management **31**(4), 247–265 (2001)
21. Ren, J., Yusuf, Y.Y, Burns, N.D.: A prototype of measurement system for agile enterprise. In: International Conference on Quality, Reliability, and Maintenance, Oxford, UK, pp. 247–252 (2000)
22. Rabah, I., Said, I., Mohamed, A.N.: An Enterprise Information System Agility Assessment Model. Computer Science and Information Systems **9**(1), 107–133 (2012)
23. Sharifi, H., Colquhoun, G., Barclay, I., Dann, Z.: Agile manufacturing a management and operational framework. Proceedings of the Institution of Mechanical Engineers Part B—Journal of Engineering Manufacture **215**(6), 857–869 (2001)
24. Shahraki, A., Yaghoobi, N., Fard, S.: Fuzzy evaluation to reach the required agility at manufacturing organizations. Journal of Basic Application Science **1**(9), 1112–1123 (2011)
25. Tsourveloudis, N.C., Valavanis, K.P.: On the measurement of enterprise agility. Journal of Intelligent and Robotic Systems **33**(3), 329–342 (2002)
26. Van Oyen, M.P., Gel, E.G.S., Hopp, W.J.: Performance opportunity for workforce agility in collaborative and noncollaborative work system. IIE Transactions **33**, 761–777 (2001)
27. Yang, S.L., Li, T.F.: Agility evaluation of mass customization product manufacturing. Journal of Materials Processing Technology **129**(1–3), 640–644 (2002)
28. Yusuf, Y., Sarhadi, M., Gunasekaran, A.: Agile manufacturing: the drivers, concepts and attributes. International Journal of Production Economics **62**(1–2), 33–43 (1999)
29. Yusuf, Y.Y., Adeleye, E.O.: A comparative study of lean and agile manufacturing with related survey of current practices in the UK. International Journal of Production Research **40**(17), 4545–4562 (2002)

A Multi-Agent Systems Contribution in Change of Information System

Nabil Benanbar, Laila Moussaid and Hicham Medromi

Abstract Over recent years, The progress made in best practices and the popularity of enterprise architecture systems have grown to give rise to new applications called Tools of Urbanization for Information Systems. The current information systems must adapt to these new environments by providing a transparent access to both information and services by using intelligent systems. The major challenge of these tools is to adapt to new market framework and to have strategies allowing them dynamically in the application.

In this paper, we will develop the mecanism of adaptation of framework in enterprise architecture and define the politic of change for the implementation of the new practices in Information System.

Our approach uses multi-agent systems to give the changes plan in order to implement the new IT governance framework.

Keywords IT governance · Information systems · Enterprise architecture · Multi-agent systems · IT framework

1 Introduction

The evolution of information systems since the 1960 has enabled the profound changes having evolved the needs of management tools in information systems. The appearance of the term Information System is indicative of the changes in attitudes : this stems resulted the emergence of new actors and new management entities of information systems. Appears in particular the challenge of the evolution on information system that must be directed and controlled.

N. Benanbar(✉) · L. Moussaid · H. Medromi
Team EAS, Laboratory of Computing, Systems and Renewable Energy,
National and High School of Electricity and Mechanic, Route d'El Jadida. BP 8118,
Oasis Casablanca, Morocco
e-mail: benanbar.nabil@gmail.com, moussaid@uh2c.ac.ma, hmedromi@yahoo.fr

E. Sabir et al. (eds.), *The International Symposium on Ubiquitous Networking*,
Lecture Notes in Electrical Engineering 366,
DOI: 10.1007/978-981-287-990-5_44

The complexity of information systems makes them less flexible in an economic environment where flexibility, adaptability and interoperability are the key words of the survival of businesses [1]. What are the components of Information Systems today? How did we go from the concept design of the computer system to the notion of the Urbanization of Information System? And finally, why this transformation have a need to adapt the evolution of the information system?

2 Urbanization of Information Systems

Start by defining the Urbanization independently of context of information systems. The urbanization [2] is "the science and technique of construction and planning of cities, towns and villages."

2.1 The Metaphor of the City

Usually, city planning is indeed grounded on geographical invariants. A city is divided into zones, quarters and blocks. City planning rules are then enacted and enforced for each area. Obviously, these divisions must be stable in the time, and the stakeholder of the city can therefore consider them as invariant [3].

The metaphor of the city and more specifically the vocabulary, rules and principles of the urbanism of towns has been widely used in the IT systems field because of the similarity of the initial issues: how to overhaul, modernize and judiciously profit from technological advances without erasing the existing IT, within the cost limits set, and do so while continuing live in the city while the work is carried out [4].

2.2 Urbanization for Information System (IS)

By analogy, we can define urbanism as "the science and technique of construction and development of IS". Urbanization enables:

- Federating the building blocks of an existing IT system around a whole architecture and following principles which will allow it to acquire the flexibility and reactivity necessary for it to be adapted to the constraints of the market or the environment.
- Managing the rapid and efficient taking into account by the thus "urbanized" IT system of critical development demands, using a rationalized approach.
- Concentrating development efforts on the new high added value functions and reusing, for the most, the existing system

3 State of Art

3.1 Strategic Allignement and Urbanization IS

Strategic alignment is an approach to align the strategy of information system on the business rules of company. This approach has purpose of enhancing the value in use of the information system and make it an asset to the company. To do this, a new system design was passed, it is a process of urbanization of information systems based on Modeling a Multi-Architecture environment.

3.2 Entreprise Architecture (EA)

Enterprise architecture is “a major view to organizational missions and functions, working processes, existing information, relational net, order of work performing which has the aim of making the data systems solid and efficient.” And “is a collection giving descriptive models about describing an organization in a way in which it can be in accordance with the established management requirements and be suitable and maintainable”. It is also “a base of strategy informational possessions which includes business mission, necessary data and technology needed to perform those missions and processes to serve modern technologies in replying to variable requirements of the mission” [5].

This concept “includes processes, mediums and necessary constructions to use informational technology solidly and harmoniously in the domain of organization to protect operations of the organization cycle in the present and in the future”. It should be ‘organized in a way that supports reasoning about the structure, properties and behavior of the system. Also the EA “defines the components that make up the overall system and provides a blueprint from which the system can be developed” [6]. EA face the same problem that deals with the urbanization of IS. It represents the global modeling of all enterprise resources [7] such as actors, processes, applications and technical architectures.

Enterprise architecture seeks to understand and identify the links between the elements of the business process (ie the links between processes that are running and the systems that support them).

Figure 1 presents the different layers of an enterprise architecture. It is composed of four levels:

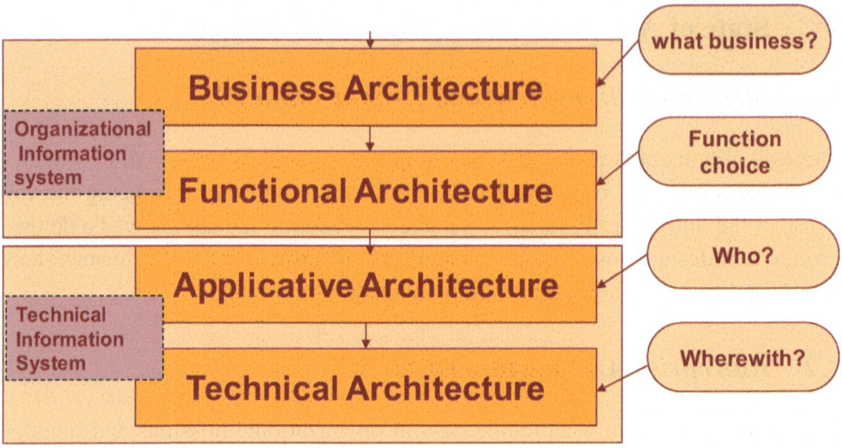

Fig. 1 Typical architecture for an information system

1. *Business Architecture :* The Business Architecture describes a structured way the missions, processes and activities. The business mapping is used to describe the business in a formalism understandable by a business actor.

2. *Functionnal Architecture :* The Functional Architecture describes the structure functions to carry out the activities of the Business Process View. The functional map used to describe the features the IT system and organization in urbanized architecture.This is a key element in a project of urbanization.

3. *Applicative Architecture :* The Application Architecture provides a description of the software applications used to achieve functional activities and services. The application mapping is used to describe the implementation of the services offered by the IS as applications or software components in an architecture components.

4. *Technical Architecture :* The technical architecture describes and organizes the various hardware, software and communication made to automate the activities. Mapping technique is used to describe infrastructure operating software components the IT system such as:

   - Different hardware and software (Servers, Workstations, ...)
   - Applications
   - Telecommunication tools (network).

3.3 TOGAF Framework

TOGAF is an architecture framework – The Open Group Architecture Framework. Put simply, TOGAF is a tool for assisting in the acceptance, production, use, and maintenance of architectures. It is based on an iterative process model supported by best practices and a reusable set of existing architectural assets [8].

TOGAF can be used for developing a broad range of different enterprise architectures. TOGAF complements, and can be used in conjunction with, other frameworks that are more focused on specific deliverables for particular vertical sectors such as Government, Telecommunications, Manufacturing, Defense, and Finance.

The Architecture Development Method is the core of TOGAF. It describes the TOGAF Architecture Development Method (ADM) – a step-by-step approach to developing an enterprise architecture [9].

The ADM describes how to derive an organization-specific enterprise architecture that addresses business requirements. The ADM is the major component of TOGAF and provides guidance for architects on a number of levels Fig. 2 :

1. Preliminary Phase: Prepare the organization for successful architecture projects. Undertake the preparation and initiation activities required to meet the business directive for a new enterprise architecture.
2. Phase A « Architecture Vision » : Set the scope, constraints, and expectations for a TOGAF project. Create the Architecture Vision. Define stakeholders. Validate the business context and create the Statement of Architecture Work.
3. Phase B « Business Architecture », Phase C« Information Systems Architectures», Phase D « Technology Architecture » : Develop architectures at three levels:

    - Business
    - Information Systems
    - Technology

   In each case, develop the Baseline and Target Architecture and analyze gaps.

4. Phase E « Opportunities and Solutions » : Perform initial implementation planning and the identification of delivery vehicles for the building blocks identified in the previous phases. Identify major implementation projects, and group them into Transition Architectures.

5. Phase F« Migration Planning » : Analyze cost benefits and risk. Develop detailed Implementation and Migration Plan.

6. Phase G: « Implementation Governance » Provide architectural oversight for the implementation. Prepare and issue Architecture Contracts (Implementation Governance Board). Ensure that the implementation project conforms to the architecture.

7. Phase H « Architecture Change Management »: Provide continual monitoring and a change management process to ensure that the architecture responds to the needs of the enterprise and maximizes the value of the architecture to the business.

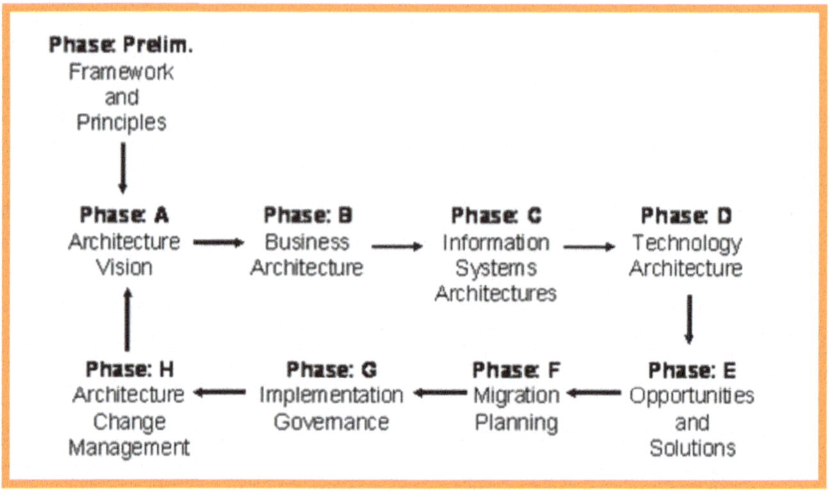

Fig. 2 Phases of the ADM approach

3.4 Critical Analysis of the Existing

Faced with a very versatile environment, change management is insufficiently addressed issue in the management of information systems. During our literature study on the urbanization of information systems, we found firstly that the communication and intelligence between the different layer of entreprise architecture is fully absent, secondly, the weak point of this architecture lies in the dynamic use of IT framework in different layers.

in the next section, we present our proposal to improve these architectures based on the characteristics related to intelligence and communication of a multi-agent system.

4 Improved Proposal

The best practices of the framework is changing over the years. This change facilitates access to new services via new processes and modify current practices of the company's stakeholders (Fig. 3).

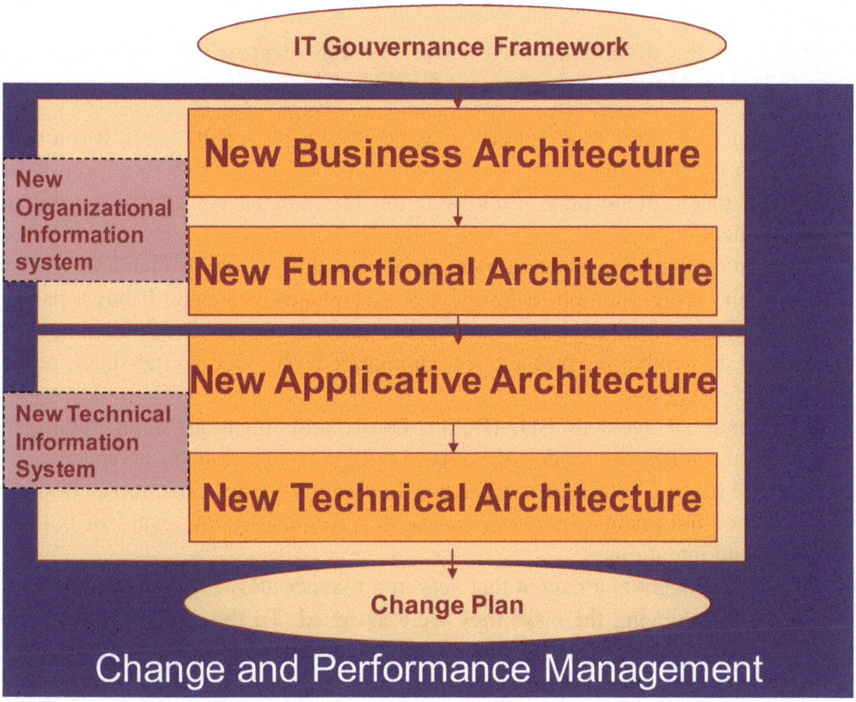

Fig. 3 Overall target architecture

This section aims to provide an improved approach to Entreprise architecture of information systems based on Multi-Agent Systems. This proposal defines strategies "smart" to control the changes made by IT governance framework.

we are interested in the first place in the global modeling of information systems, then analyze the differences based on decisions relating to the implementation of new framework in order to build a project plan to upgrade and lead the current information system of his state to the target state.

4.1 Multi-Agent Systems (MAS) Approach

Definitions

An agent is a computer system within an environment and with an autonomous behavior made for achieving the objectives that were set during its design [10].

A multi-agents system is a system that contains a set of agents that interact with communications protocols and are able to act on their environment. Different agents have different spheres of influence, in the sense that they have control (or at least can influence) on different parts of the environment. These spheres of influence may overlap in some cases; the fact that they coincide may cause dependencies reports between agents [11].

Types of Agent

Starting from the definitions cited above, we can identify the following agent types [12] (Fig. 4) :

- The reactive agent is often described as not being "clever" by itself. It is a very simple component that perceives the environment and is able to act on it. Its capacity meets mode only stimulus-action that can be considered a form of communication.
- The cognitive agent is an agent more or less intelligent, mainly characterized by a symbolic representation of knowledge and mental concepts. It has a partial representation of the environment, explicit goals, it is capable of planning their behavior, remember his past actions, communicate by sending messages, negotiate, etc..
- The intentional agent or BDI (Belief, Desire and Intention) is an intelligent agent that applies the model of human intelligence and human perspective on the world using mental concepts such as knowledge, beliefs, intentions, desires, choices, commitments. Its behavior can be provided by the award of beliefs, desires and intentions.
- The rational agent is an agent that acts in a manner allowing it to get the most success in achieving the tasks they were assigned. To this end, we must have measure of performance, if possible objective associated with a particular task that the agent should run.
- The adaptive agent is an agent that adapts to any changes that the environment can have. He is very intelligent as he is able to change its objectives and its knowledge base when they change.
- The communicative agent is an agent that is used to communicate information to all around him. This information can be made of his own perceptions as it may be transmitted by other agents.

4.2 Overview of the Proposed Solution

Our approach is based on on modeling the enterprise architecture in order to facilitate the transformation of IT Planning. Generally, This involves the development of:

- Description Guidelines architecture, showing the current status.
- Description of the target architecture, defining a future vision for IT.
- Gap analysis to identify issues and gaps and solutions

Our proposal is to use MAS to discover the changes made by the decision of implement the new process and activities.

This proposal proposes to list the process to implement and explore improvements to current practices in order to update the information system (Services, Roles, Knowledge, Skills, Resources).

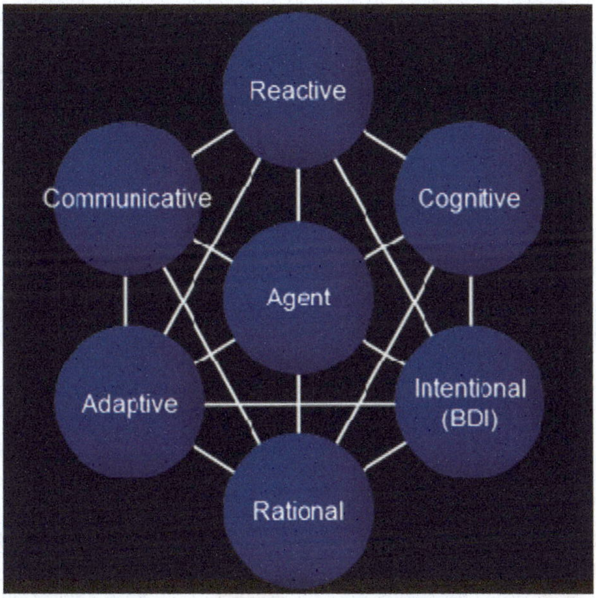

Fig. 4 Types of agents

Each process will be represented by an agent or MAS which will have the capacity to define its services according to the analysis of the current context and resources available.

So, we can use the multi-agent systems as an intelligent system to propose an implementation plan to transform the information system.

4.3 Proposed Architecture

In our proposal, we are interested in the study of the two main components of our target system that is the New directives and context.

New directives: This part represents the base of our system, it allows to define the new developments in the functioning of the information system.

The context: represents the environment of each architecture. We considered that the context of each architecture consists of a set of service provided to the user.

In the following, we will present our architecture based on MAS: the main objective of this system is to generate the new services of all the architectures so that each layer services meet the needs of the new politic of IT Gouvernance.

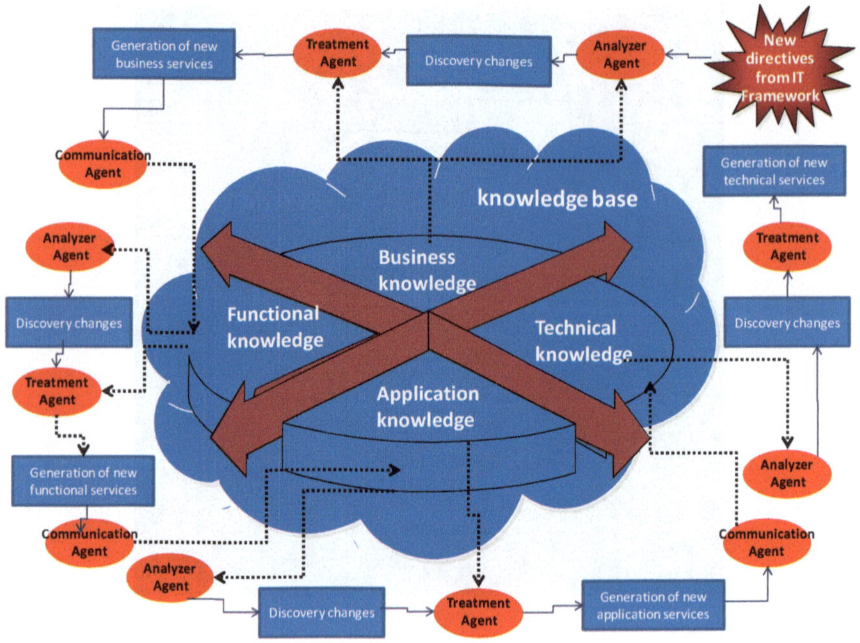

Fig. 5 Proposed Architecture

Agents involved in the running of the platform are :

- Analyzer agent: It is a type of cognitive agent, scans messages from other agents or from outside to discover the changes.
- Treatment agent: It is a cognitive or intelligent agent, it is the environment of each architecture and other agents, it takes into account the existing and analyzes the gap from the new changes. At this level, the agent generates the list of new services and updates the basic knowledge of architecture.
- Communication agent : It is a cognitive agent, enables filtering and distribution of information between architectures for change layer.

The system has a knowledge base powered by the company data which will allow to answer the following questions : What Activity? What organization? What functions? What Applicative? What infrastructure? It is a base which is the overall system of the company, it is consulted when needed by the agents to find the required information and to make changes plan, corrections and data consolidation.

The update service provided by the new directives involves a change in the global context. The interrogation of the context services and the gap analysis between existing services and new services offered by the decision will generate a list of new services for our information system. After the update, the change layer sends a notification to manger in order to generalize the IT transformation plan.

4.4 *Example of Change*

In our example, we are interested on IT support process of information system.

New directives : To implement all Support process.

The context : we consider that all process are implemented unless the Problem Management process and the implementation of IT service management is already in production.

The following table illustrates the role of each agent in the IT framework :

Agent	Mission	Result
Analyzer Agent 1	To Discover Support process to implement based on the knowledge base	Problem Management
Treatment Agent 1	To Generate of new business services	List of new business services
Communication Agent 1	To Notify the new services	Information communicated
Analyzer Agent 2	To discover changes for functions	Control problems and control errors
Treatment Agent 2	To generate new functionnal services of the process	• Registration Problem or Error • Classification Problem or Error • Investigation and diagnostic of Problem or Error • Solution
Communication Agent 2	To Notify the new services	Information communicated
Analyzer Agent 3	To discover the implemented functions	The IT service management application is already implemented including the functions of the problem management process (No change)
Treatment Agent 3	To Discover the activities to be implemented in the application architecture	No implementation
Communication Agent 3	To Notify the new services	Information communicated
Analyzer Agent 4	To discover the technical architecture requirements	Nothing to report
Treatment Agent 4	To Generate of new technical services	No change

5 Conclusion

The growing need for enterprise architecture tools in the world necessitates the establishment of a process of change for information systems.

In this paper, we proposed the use of a Multi-Agent System in a multi-system architectures to discover new practices brought by IT framework. The objective is to deploy new IT governance process in order to align the information system on the new changes.

The use of a knowledge base containing all the information about the company is required to explore the changes and thus be able to propose an action plan for the implementation of new practices.

Future work consists on extending our prototype by defining a detailed architecture of each agent and specifying communication between these agents in order to identify the operational methods of change.

References

1. Dubois, S.: Apports de l'urbanisme et des cartographies pour piloter l'évolution d'un système d'information, thèse professionnelle, Ecole des Mines de Paris, Octobre 2002
2. Longépé, C.: Le projet d'urbanisation du système d'information. Edition Dunod, Paris (2001)
3. Contini, I.: L'apport historique de l'urbanisme des villes pour l'urbanisme des systèmes d'information. Colloque Urbanisation des Systèmes d'Information, Paris (2002)
4. Longépé, C.: Le projet d'urbanisation du SI: Cas concret d'architecture d'entreprise. Dunod, Paris (2009)
5. Trabelsi, L., Abid, I.: Urbanization of Information Systems as a Trigger for enhancing Agility: A State in The Tunisian Firms. European Journal of Business and Management **5**(5) (2013)
6. Kaisler, S., Armour, F., Valivullah, M.: Enterprise architecting: critical problems. In: Proceedingsof the 38th Annual Hawaii International Conference on System Sciences. IEEE Computer Society Press, Hawaii (2005)
7. Club Urba-EA, Urbanisme des SI et Gouvernance. Edition DUNOD, Paris (2006). P8
8. TOGAF, The Open Group Architecture Framework. http://www.opengroup.org/togaf/
9. TOGAF™ Version 9 A Pocket Guide, Copyright © 2009, The Open Group
10. Wooldridge, M.: Intelligent Agents, Multi agent systems: A modern Approach to Distributed Artificial Intelligence, pp. 27–78. MIT Press, Cambridge (1995)
11. Wooldridge, M.: An Introduction to Multi-Agent Systems. Wiley & Sons (2000)
12. Maalal, S., Addou, M.: A practical application of a method of designing multi-agent systems based on the AUML language and the MDA approach. In: Proceedings of the Fourth Workshop on Information Technologies and Communication WOTIC 2011, Casablanca, Morocco, p. 104 (2011)

Designing Multi Agent System Architecture for Project Performance Based on PMBOK Standard

Houda Hammouch, Hicham Medromi and Adil Sayouti

Abstract PMBOK (project management body of knowledge) is the recognized and the most accepted standard for project management, however many projects still fail by using or misusing PMBOK guidelines, due to poor communications, lack of monitoring and non performance measurement. In this paper we use the advantages of agent technology to make implementation and use of PMBOK processes more efficient and to have insights on the progress of projects and to anticipate possible problems that can lead the project failure.

Keywords Multi Agent System · PMBOK · Project performance · Project management · PRINCE2 · IPMA competence baseline · ISO 21500

1 Introduction

A project is a process, consisting of a set of coordinated and controlled activities with start and finish dates, undertaken to achieve an objective conforming to specific requirements including constraints of time, cost and resources [1].

The origin of the idea of project governance is related to solve problem about project failures, uncontrolled cost inflation, outages on critical systems. A poor estimation is often the cause of errors that can cause harm to the organization, projects with excessive costs from the expected initial budget, major delays in relation to delivery dates originally scheduled, resource requirements misconfigured, both in terms of numbers and qualifications, many studies have demonstrated, that most projects do not meet time and budget goals, or fail to

---

H. Hammouch(✉) · H. Medromi · A. Sayouti
Systems Architecture Team (EAS), Laboratory Informatics,
Systems and Renewable Energy (LISER), University Hassan II Ain Chock,
ENSEM, Oasis BP 8118, Casablanca, Morocco
e-mail: houda.hammouch@gmail.com, sayouti@gmail.com, hmedromi@yahoo.fr

E. Sabir et al. (eds.), *The International Symposium on Ubiquitous Networking*,
Lecture Notes in Electrical Engineering 366,
DOI: 10.1007/978-981-287-990-5_45

satisfy customer and company expectations[2][3][4]. The report of Stndish Group, Chaos Manifesto 2012, shows that only 37% of projects are succeeding, while the big percentage concern challenged project 43% of projects are late, over budget, and/or with less than the required features and functions[5].

For several years, project management has become a must in organizations because it promotes the success of projects and the achievement of objectives. The time spent on a project is considered a gain for the other projects. Management of project consist in [6]:

- Identify requirements
- Addressed during the planning and execution of the project, the various needs, concerns and expectations of stakeholders
- Balance the competing project constraints: the content, quality, schedules filed the budget, resources and risk

Unfortunately, the management are often ill-equipped although there are many project management software that make easy project management work, the scope of the software is limited to the automation of backup tasks and time management, such as version control systems or configuration management systems, an important part of project management software take care of project planning and task scheduling, other software project management rather attach to the monitoring results, to identify in advance the risk of slippage. But does not participate in the implementation of best practice and measure the deviation from accepted practices to finally be able to audit the existing projects and identify significant weaknesses in these controls and advise the management on measures to implement.

Agent technology offers a talented solution in order to address project management problems in a distributed environment[7], where software agents can support the project manager and team members to monitor and coordinate tasks, apply control measures quality, validation and verification.

The proposed model is based on the system's ability to learn, from a database of past projects to better manage projects and evaluate the implementation of standards for determining process that requires regular monitoring, the idea is to capitalize on new projects and consider the time spent on a project a gain for the new one, and to increase business performance through project management. The objective of this paper is to study the practices and tools used in project management field, existing standards and norms, identify the weaknesses in order to propose a new model of decision support allow management to have insights on the progress of projects and to anticipate possible problems that can lead the project failure.

2 Project Management: Existing Standards

Although there is no single framework for the complete project management, there is a number of frames available that can serve as starting points [6][8][9], the use of IT repositories have several advantages:

- Capitalize on best practices
- Save time
- Having proven and mature practice
- Facilitate dialogue between actors
- Avoid project failures
- Make the most successful projects, more efficient
- Being competitive
- Go quickly and reduce risks
- Speaking a common language
- Do not reinvent everything every time

There are several recognized standards for training project managers:

- PMBOK : PMI, Project Management Institute [6]
- PRINCE2 : OGC, Organisation of Government Commerce [10]
- ICB: IPMA , International Project Management Association [11]
- ISO 21500 Best practice [12]

2.1 The ICB Model

ICB (IPMA Competence Baseline) model is based on a repository of jurisdiction by IPMA: International Project Management Association. 46 skills that a professional project management must be identified in three categories [11]:

- Technical Skills for Project Management (20 items)
- Behavioural Competencies for project management (15 items)
- Contextual skills for project management (11 elements).

These competences are seen through the eye of a Project Management, known as Eye of Competence.

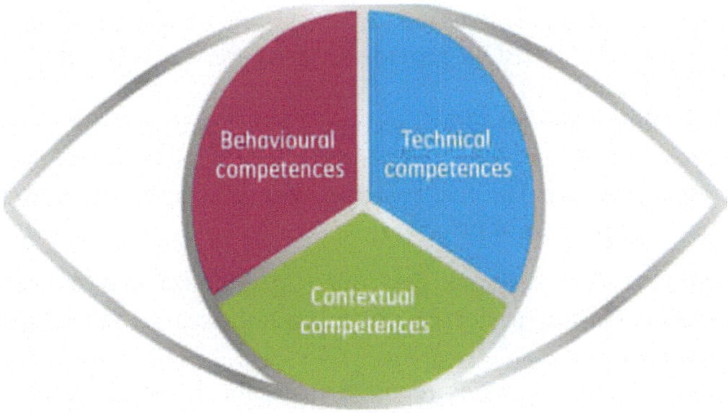

Fig. 1 IPMA Model [11]

2.2 PRINCE2 Methodology

PRINCE2 (PRojects IN Controlled Environments) is a structured approach to project management [10], it is a process-oriented guide and themes that are of management disciplines. Also an annex containing 26 pre formatted documents for the proper control of the project. There are also seven principles that determine the great values in terms of project management, if we respect these principles, then we are in a project managed by PRINCE2.

2.3 ISO 21500

The ISO 21500: 2012 - the Project Management Guidelines provides project management guidelines [13] can be used by any type of organization, whether public, private or as an association, as well as any type of project, regardless of its complexity, size or duration. It gives a macroscopic level description of the concepts and processes that are considered as good practice in project management. However, it does not give detailed recommendations for the management of programs or project portfolios.

2.4 Project Management Body of Knowledge

PMBOK is a Guide to the Project Management Body of Knowledge or project management body of knowledge. This guide is published by the Project Management Institute (PMI), he had at least 5 editions (PMI, 1996) (PMI, 2000) (PMI, 2004) (PMI, 2008). The Project Management Body of Knowledge (PMBOK) is a recognized standard for the project management profession, a formal document that describes established norms, methods, processes, and practices.

The PMBOK Guide provides guidelines for managing individual projects. It defines project management and related concepts and describes the project management life cycle and the related processes. This full body of knowledge includes knowledge of proven, traditional practices, which are widely applied, as well as knowledge of innovative and advanced practices, which may have seen more limited use. The PMBOK provides a common lexicon within the Project Management profession. [6]

In the last edition the PMBOK recognizes 5 basic process groups and 10 knowledge areas typical of almost all projects. The basic concepts are applicable to projects, and programs. The five basic process groups are, the following figure illustrates the group process of project:

- Initiating: Those processes performed to define a new project or a new phase of an existing project by obtaining authorization to start the project or phase.
- Planning: Those processes required to establish the scope of the project, refine the objectives, and define the course of action required to attain the objectives that the project was undertaken to achieve.

- Executing : Those processes performed to complete the work defined in the project management plan to satisfy the project specifications.
- Monitoring and Controlling: Those processes required to track, review, and regulate the progress and performance of the project; identify any areas in which changes to the plan are required; and initiate the corresponding changes.
- Closing: Those processes performed to finalize all activities across all Process Groups to formally close the project or phase.

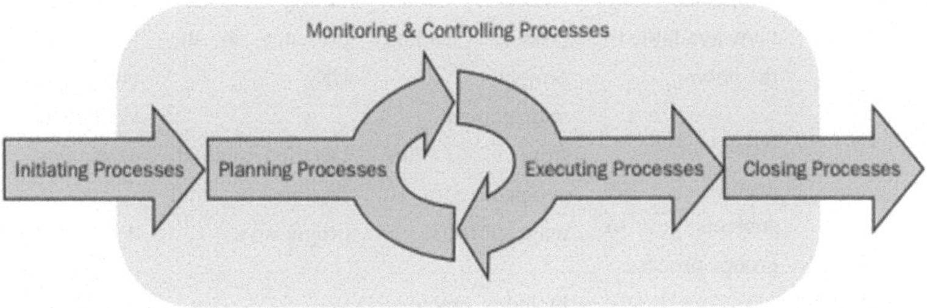

Fig. 2 Project Phases according to PMBOK [6]

The Ten knowledge Areas are

- Project Integration Management
- Project Scope Management
- Project Time Management
- Project Cost Management
- Project Quality Management
- Project Human Resource Management
- Project Communications Management
- Project Risk Management
- Project Procurement Management
- Project Stakeholder Management

2.5 Analysis and Justification of Selection

We present in the next section a comparison of different project management standards: ISO 21500 standard, IPMA certification, the method PMBOK, Prince 2. The aim is to highlight the values of PMBOK and better situate the referential compared to those already existing.

Table 1 Comparison of project management standards [14]

	ISO 21500	PMBOK	Prince2	IPMA
Application	Any type of organization and actor.	Any type of organization and actor.	Any type of organization and actor.	Certification of a person on these project management skills.
Strengths	Simplified view available to the public Affordable groups by subjects or groups process easy to understand certification possible	Details the process leading project management from the perspective of 10 main subjects. Includes practical approved by thousands of project management professionals in the Model Basic certification of work "Project Management Professional" (PMP)	Strong presence in the UK Detailed method, flexible and progressive Does not require a great experience certification possible	Exam behavioral and contextual skills, made by examiners 4 levels of certification recognized in Europe
Weaknesses	Adapting to project dimensions Lack of guides for plugs Lack of clarification tools	great work 400 PAGES	Weak presence outside the UK Few general knowledge / theory in for project management	The exams are different in each country Long for certification expensive Poorly recognized in the US Not tied to a specific method

The repositories are based on the same basic concepts of management, planning, budget estimates, monitoring. The particularity of the PMBOK is that it goes into detail on the processes and tools that project managers must know used, it is the most accepted standard. However many projects still fail by using or misusing PMBOK guidelines. Poor communications, lack of monitoring, non performance measurement are the primary reasons of project failure. To answer this problematic, we will associate agent technology to the PMI standard in order to propose a model of supervision and project management audit to assess the maturity of project management, an intelligent model based on humans reasoning, explicitly: autonomy, intelligence, responsiveness and learning, qualities that we want to implement using the Multi agents systems, to validate the choice of MAS, we present in the following paragraph agents and MAS.

3 Presentation of MAS

The Agent: physical or software, autonomous entity that is pro-active, reactive, social, able to take part to an organized activity, in order to achieve its goals, by interacting with other agents and users[15].

The notion of multi agent system immediately releases the idea of a system composed of several agents. An agent is an autonomous real or abstract entity, which is able to act on itself and its environment, in a multi-agent can communicate with other agents, whose behavior is the result of its observations, knowledge and interactions with other agents.[16] [17]. The use of Multi-Agent Systems (MAS) has a number of advantages [18]:

- Dynamic System, SMA can model a set of agents that interact.
- Large number of agents
- A distributed problem solving
- The SMA can "respond" to the individual failure of one element without degrading the system as a whole.
- Robustness and reliability

4 Proposed Approach

Learning and intelligence provided by the multi-agent systems is a promising approach for the development of a formal project management. It has the following advantages:

- It provides a valuable aid to good project management
- It automates the process of managing
- The model provides opportunities for evaluators to improve the accuracy of the data provided.

The application of agents in the management of business processes has been studied in many research [18][19] [20]. The model affords the project manager, and the management an interim view of what has gone well and what needs to be improved with the project to successfully complete it. If this diagnostic is done at the close of a project, a project can be used to develop success criteria for future projects by providing a forensic review. This review will provide an opportunity to learn what elements of the project were successfully managed and which ones presented some challenges. This will help the organization identify what it needs to do so that mistakes are not repeated on future projects.

Our approach is based on the theory of multi-agent systems and PMI standards, the idea is to propose a model of supervision and project management audit to assess the maturity of project management, an intelligent model based on humans reasoning, autonomy, intelligence, responsiveness and learning, qualities that we want to implement using the MAS. The figure below shows the EAS PRGOV Architecture based on the principle of MAS which is: everyone must cooperate to achieve the same goal and that implements processes within the PMBOK standard.

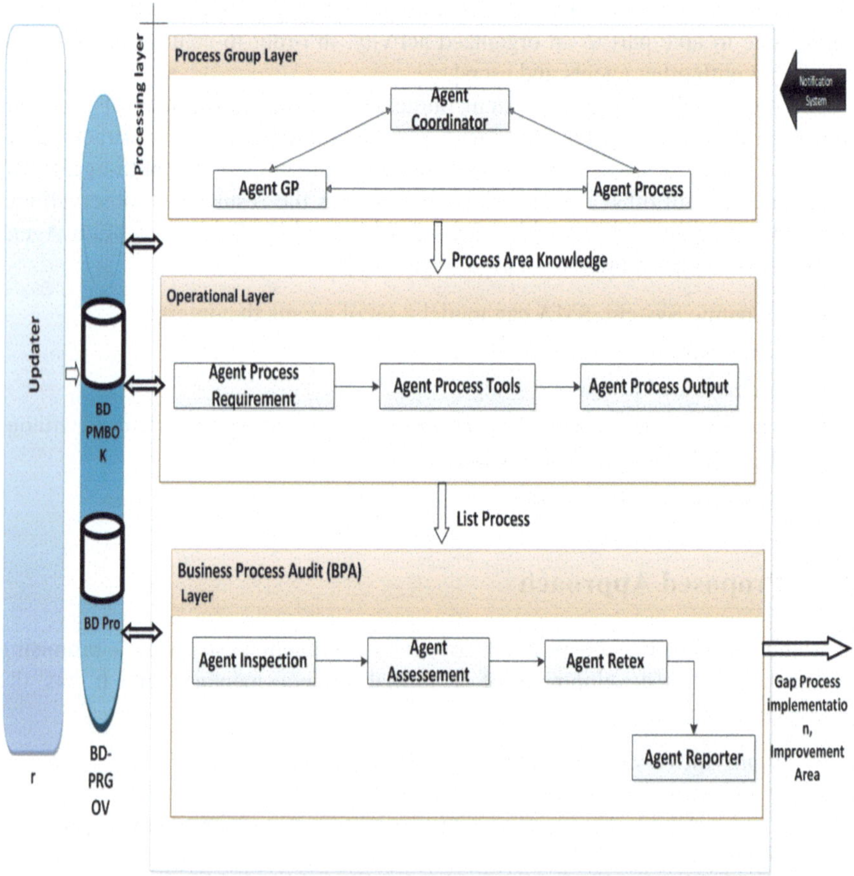

Fig. 3 The global architecture

EAS-PRGOV is a platform for the audit of IT projects for collecting objective data to determine whether the projects are properly managed, the review process is carried out to measure the implementation of good project management practices by the PMBOK standard. The objective is to determine for each organization the degree of implementation of the standard, the most important process and to assess the weight for each process and criticality.

EAS-PRGOV is made of tree essential MAS; Process Layer, Operational Layer and Audit Layer, each part is a multi-agent system. The system has a knowledge base enriched by the project data, consulted by agents to highlight the desired information and to make a change and correct data consolidation.

❖ *Process Group Layer*

This layer is interested in project management standard depending on the phase of the project, this layer is composed by 3 agents: the Coordinator Agent, GP Agent and the Process Agent.

- The Coordinator Agent starts the PMBOK processes and coordinate between the GP Agent and the Process Agent, he receives the system notification, identifies areas of knowledge related to the notification and send the request to the GP agent.
- GP Agent: this agent is responsible of the areas of knowledge, according to the standard there are five phasing; initiating, planning, executing, monitoring and closing, according to the area of knowledge the GP agent creates its instance and sends the request to the agent process.
- Process Agent active processes related to the knowledge area and process group. The table 2 present the identification process related to the case management of integration.

Table 2 Identification process, case of management of integration

Number	Knowledge area	Process group	Processes
04	Management of the Project Integration	Initiating	Develop the Project Charter
04	Management of the Project Integration	Planning	Develop project management plan
04	Management of the Project Integration	Executing	Direct and control the execution of the project
04	Management of the Project Integration	Monitoring	Monitor and master the project work
04	Management of the Project Integration	Closing	Close the project or phase

❖ *Operational Layer*

In this layer, plug process must be completed each process has input and output, a process is a set of action and activity in relation to each other, the processes are interacting, the diagram below show this interaction process groups in a project

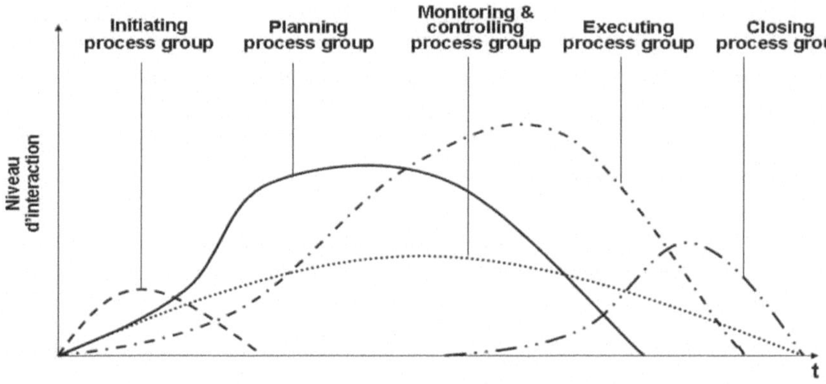

Fig. 4 Interaction process groups in a project [6]

The requirement Agent recovered the inputs of the process, the Tools Agent determines the technical tools and manipulated and the Output Agent determines the output of the process and the repository, the location where we store and manage the business processes

❖ *Audit Layer*

This layer allows the evaluation of the implementation of the framework; this work is highlighted by scenario inspection-Assessement -Retex and Report in order to provide control and constant process improvement:

- Inspection Agent verifies the implementation of processes and scripts
- Assessment Agent Identify the weight and importance of this process compared to the current phase and the project
- Retex Agent for filtering information and updating data. Feeds the database following the progress of spots
- Reporter Agent provides control of the company by a set of, Key Performance Indicators (KPIs) These KPIs are used to compare the progress of activities based on the process with the expected results during this phase analysis, we identify the problems we want to solve. We also offer vis-à-vis improvements in process efficiency.

5 Conclusion and Perspectives

In this paper, we studied the current practices in project management, referential and software, we also identified the cause of misusing guidelines that is related to poor communications, lack of monitoring and non performance measurement, as a solution we proposed a model allowing both to visualize the implementation of good practices and then to measure performance.

As perspective we plan to detail the architecture based on MAS and describe in details the job and scenarios that involve the participation of every agents and multi-agent system, future work consists on extending our prototype by defining a detailed architecture of each agent and specifying communication between these agents. Our objective is to validate the architecture that we propose in this paper by developing a distributed platform that will allow companies to establish effective governance of projects.

References

1. ISO, International Organization for Standardization (ISO). ISO 10006:2003, Quality management systems–Guidelines for quality management in projects, 2nd edn. ISO (2003)
2. Stepanek, G.: Software Programming Secrets: Why Projects Fail, p. 184. APress (2005)
3. Liew, Z.: Project Management: Why Projects Fail (2008)
4. Hartman, F., Ashrafi, R.: Development of the SMARTTM project planning framework. Int. J. Proj. Manag. **22**(6), 499–510 (2004)
5. The Standish Group International.: CHAOS MANIFESTO 2012 The Year of the Executive Sponsor (2012)
6. PMI, A Guide to the Project Management Body of Knowledge: PMBOK Guide, 5th edn. (2013)
7. Nienaber, R.C., Barnard, A.: A Generic Agent Framework to Support the Various Software Project Management Processes. Interdiscip. J. Information, Knowledge, Manag. **2** (2007)
8. ISO 10006: International Organization for Standardization, Quality Management. Guidelines to Quality in Project Management (1997)
9. Turner, J.R.: The global body of knowledge, and its coverage by the referees and members of the international editorial board of this journal. Int. J. Proj. Manag. **18**, 1–6 (2000)
10. OGC: PRINCE 2–Managing successful Projects with PRINCE 2. TSO (The Stationnery Office), p. 327 (2009)
11. IPMA, IPMA–International Project Management Association, ICB–IPMA Competence Baseline, Version 3 (2006)
12. Zandhuis, A., Stellingwerf, R.: ISO 21500 guidance on project management, pp. 1–51 (2013)
13. Afnor: NF ISO 21500, lignes directrices sur le management de projet
14. Phan, R., Domingues, S., Liu, S., Liu, X., Carpentier, C.: Processus et autodiagnostic sur le management de projet selon L'ISO 21500 (2013)

15. Ferber, J.: Les Systèmes Multi Agents: vers une intelligence collective. InterEditi (1995)
16. Medromi, H., Sayouti, A.: Les Systèmes Multi-Agents: Application au Contrôle sur Internet, Éditions Univ. Eur (2012)
17. Alkhateeb, F., Al Maghayreh, E., Abu Doush, I.: Multi-AgentSystems - Modeling, Control, Programming, Simulations and Applications, InTech (2011)
18. Chiu, D., Li, Q., Karlapalem, K.: Cooperative Exception Handling in ADOME Workflow Management System. Information Systems: an International Journal (special issue on Web Information Systems Engineering) **26**(2), 93–120 (2001)
19. Wang, H., Mylopoulos, J., Liao, S.: Intelligent Agents and Financial Risk Monitoring Systems. Communications of the ACM **45**(3), 83–88 (2002)
20. O'Brien, P.D., Wiegand, W.E.: Agent based process management: applying intelligent agents to workflow. The Knowledge Engineering Review **13**(2), 1–14 (1998)

Author Index